Transition Metal Oxides for Electrochemical Energy Storage

Transition Metal Oxides for Electrochemical Energy Storage

Edited by Jagjit Nanda and Veronica Augustyn

WILEY-VCH

Editors

Dr. Jagjit Nanda
Oak Ridge National Laboratory
Chemical Sciences Division
1 Bethel Valley Road
37831 Oak Ridge TN
United States

Prof. Veronica Augustyn
North Carolina State University
Department of Materials Science & Engineering
911 Partners Way
27606 Raleigh NC
United States

Cover Images: © Antoine2K/Shutterstock; Smile Fight/Shutterstock

Library of Congress Card No.: applied for

British Library Cataloguing-in-Publication Data
A catalogue record for this book is available from the British Library.

Bibliographic information published by the Deutsche Nationalbibliothek
The Deutsche Nationalbibliothek lists this publication in the Deutsche Nationalbibliografie; detailed bibliographic data are available on the Internet at <http://dnb.d-nb.de>.

Print ISBN: 978-3-527-34493-2
ePDF ISBN: 978-3-527-81722-1
ePub ISBN: 978-3-527-81724-5
oBook ISBN: 978-3-527-81725-2

Typesetting Straive, Chennai, India
Printed and bound by CPI Group (UK) Ltd, Croydon, CR0 4YY

C086066_160322

Contents

Foreword

Since their introduction into portable electronic devices more than 30 years ago, lithium-ion batteries have grown to dominate the energy storage sector and are now the technology of choice for powering portable devices, electric vehicles, and increasingly, smart grids. At the heart of lithium-ion battery technology is the mechanism of lithium-ion intercalation into solid-state host materials for both the cathode and the anode. Intercalation reactions into transition metal compounds were of great interest in the 1970s because of the possibility of controlled structural and electronic modification. Our team at Exxon made seminal discoveries into the mechanism of electrochemical Li^+ intercalation forming the basis of Li-ion batteries. These studies developed the important distinction between intercalation vs. conversion or decomposition reactions, and showed the importance of intercalation reactions for extended high-capacity cycling. This work led to the first generation of lithium-ion batteries (Exxon and Moli-Energy), and identified the challenges of working with lithium metal anodes.

Following the seminal studies of lithium intercalation into transition metal chalcogenides, attention turned to transition metal oxides due to their higher intercalation potentials in a number of laboratories. John Goodenough's group identified the layered oxides such as $LiCoO_2$ as outstanding replacements for the layered sulfides. Today, transition metal oxides are the preferred materials for lithium-ion battery cathodes because of their high redox potentials and good lithium-ion intercalation capacity. Besides their use in existing Li-ion battery technology, transition metal oxides also form the basis of emerging beyond Li-ion chemistries such as Na-ion, where they have much higher capacities than the sulfides.

This book provides a timely reference into the solid-state chemistry, characterization, and modeling of transition metal oxides for electrochemical energy storage written by international experts. Chapters are devoted to mechanisms of energy storage (such as coupled cation/electron transfer reactions; Na^+ intercalation; conversion reactions; and anion redox), characterization (such as nuclear magnetic resonance and X-ray techniques), and modeling. Emerging hot topics, such as the role of artificial intelligence and machine learning methods in predicting properties of energy storage materials and cobalt-free transition metal oxides cathodes for Li-ion, are also covered. Overall, the book provides a great reference for novices and experts to gain a deeper understanding into the solid-state chemistry of transition metal oxides of potential use for energy storage.

Vestal, NY
8 September 2021

M. Stanley Whittingham
2019 Nobel Laureate in Chemistry

1

An Overview of Transition Metal Oxides for Electrochemical Energy Storage

Ethan C. Self[1], *Devendrasinh Darbar*[1], *Veronica Augustyn*[2], *and Jagjit Nanda*[1]

[1] *Oak Ridge National Laboratory, Chemical Sciences Division, Oak Ridge, TN 37831, USA*
[2] *North Carolina State University, Department of Materials Science and Engineering, Raleigh, NC 27695, USA*

Transition metal oxides (TMOs) are used in many commercial and research applications, including catalysis, electrochemical energy storage/conversion, electronics, and thermoelectrics. This book focuses on TMOs for electrochemical energy storage devices with particular emphasis on intercalation-based secondary (rechargeable) batteries. This introductory chapter provides a broad overview of such applications, and detailed treatments of specific subjects are given in Chapters 2–16.

1.1 Fundamentals of Electrochemical Cells

An electrochemical cell consists of two electrodes (denoted as cathode/anode or positive/negative) separated by an ionically conductive, electronically insulating electrolyte. Batteries convert chemical energy into electrical energy through Faradaic charge transfer processes where: (i) oxidation/reduction reactions occur within anode/cathode active materials and (ii) electrons are transported through an external circuit to maintain charge neutrality at each electrode. These reactions are irreversible in primary batteries (e.g. Zn–MnO_2 and Li–MnO_2) designed for single-use applications. On the other hand, secondary batteries (e.g. lead-acid, nickel-metal hydride, and Li-ion) leverage reversible redox processes and can be repeatedly charged/discharged, a requirement for many end-use applications (e.g. electric vehicles).

Electrochemical capacitors are another form of energy storage devices which provide specific energy and power between that of dielectric capacitors and rechargeable batteries. Supercapacitors store/deliver energy through non-Faradaic processes where ions are stored in the electrochemical double layer near the electrode surfaces. On the other hand, pseudocapacitive materials store energy through charge transfer reactions which may include: (i) oxidation/reduction of the electrode surface and/or (ii) intercalation of ions into a host active material. Hybrid configurations utilizing pseudocapacitive materials approach the specific energy of rechargeable batteries.

Transition Metal Oxides for Electrochemical Energy Storage, First Edition.
Edited by Jagjit Nanda and Veronica Augustyn.

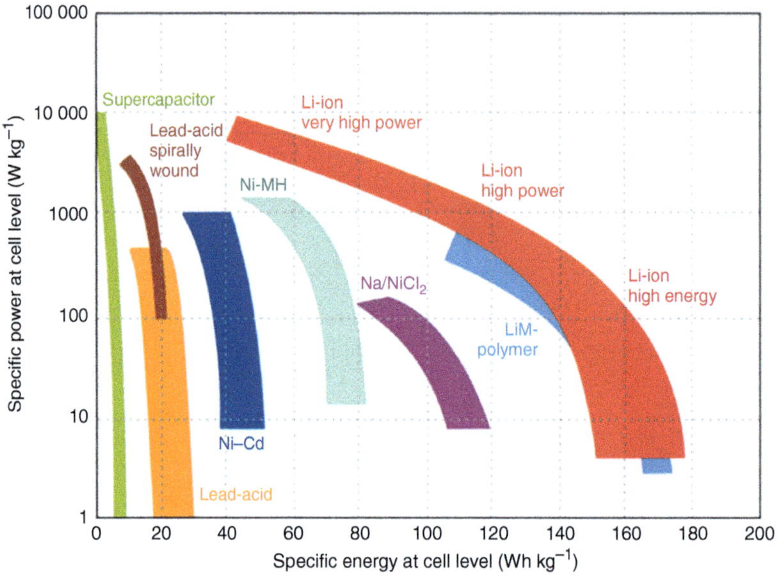

Figure 1.1 Ragone plot showing the specific energy and power of several electrochemical energy storage systems. Source: Image reproduced from Hayner et al. [1], International Energy Agency, Technology Roadmaps: Electric and Plug-in Hybrid Electric Vehicles, 2009, p. 12. (Original source: Johnson Control – SAFT 2005 and 2007.)

Two important performance metrics of energy storage devices are the specific energy ($Wh\,kg^{-1}$) and specific power ($W\,kg^{-1}$), which describe *how much* and *how quickly* energy can be stored/delivered, respectively. Analogous quantities normalized to system volume (i.e. energy/power densities with units of $Wh\,L^{-1}$ and $W\,L^{-1}$) are also commonly used. Ragone plots (Figure 1.1) summarize these energy/power relationships and are useful to assess the viability of different energy storage platforms for a given application. Figure 1.1 shows a fundamental tradeoff between a system's specific energy and power. For example, supercapacitors exhibit: (i) high specific power due to rapid ion adsorption/desorption near electrode surfaces but (ii) low specific energy since charge storage only occurs within the electrochemical double layer. On the other hand, batteries store energy within the bulk structure of active materials, enabling high specific energy. The rate of energy storage/delivery in batteries is generally limited by solid-state diffusion or phase nucleation kinetics in the active material, resulting in lower specific power than supercapacitors. With these trends in mind, it should be emphasized that the energy/power characteristics of an electrochemical device are also highly dependent on design factors such as material selection, cell format, and electrode architecture.

Sections 1.2–1.5 provide basic overviews of electrochemical energy storage devices where TMOs play critical roles in device operation. The importance of advanced characterization and computing resources on guiding material development, understanding degradation mechanisms, and optimizing system performance is also highlighted.

1.2 Li-Ion Batteries: Basic Principles and TMO Electrodes

Over the last four decades, Li-ion batteries have successfully transitioned from research and development to commercial applications, including portable electronics, electric vehicles, and grid storage. The foundation of this technology is based on cation intercalation reactions wherein Li^+ is stored in TMO cathodes and graphite anodes [2]. These intercalation reactions are highly reversible, and today's Li-ion batteries can undergo hundreds or thousands of cycles with minimal chemical and/or structural changes to the active material (see Chapters 9 and 10 for detailed discussion on degradation mechanisms of Li-ion batteries).

The working principles of Li-ion batteries are illustrated in Figure 1.2. During charge, Li^+ deintercalate from the TMO cathode (e.g. $LiCoO_2$), transport through the electrolyte, and intercalate into the anode active material (e.g. graphite). To maintain charge neutrality, electrons are simultaneously extracted from the cathode (typically via transition metal oxidation), transported through an external circuit, and inserted in the anode (electrochemical reduction of graphite). During discharge, these processes are reversed, and Li^+ ions and electrons are transported back to the cathode. Figure 1.2b shows qualitative cathode/anode voltage profiles as a function of capacity. Commercial Li-ion batteries typically have cell voltages ~3.6 V and specific energies ~200 Wh kg^{-1}, although these values depend on active material selection and cell design.

Commercial Li-ion electrodes are prepared by casting a slurry containing active material (the host material which reversibly stores Li^+), electronically conductive carbon additives, and polymer binder onto a current collector (typically Cu for the anode and Al for the cathode). Key electrochemical properties, including operating voltage, reversible capacity, and cyclability, are strongly dependent on the active material's crystallographic structure and transition metal selection. Conventional cathode active materials include: (i) lithium TMOs with layered or spinel crystallographic structures (e.g. $LiCoO_2$ and $LiMn_2O_4$, respectively) and (ii) olivine structures containing polyanionic groups (e.g. $LiFePO_4$). A wide range of related compositions

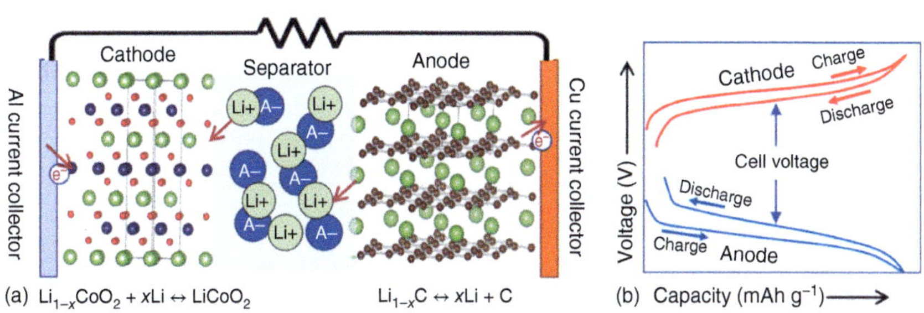

Figure 1.2 (a) Schematic of the working principle of a lithium-ion battery containing a $LiCoO_2$ cathode and graphite anode. (b) Qualitative voltage profiles during charge and discharge. Source: Belharouak et al. [3].

containing transition metal substitutions (e.g. $LiNi_xMn_yCo_{1-x-y}O_2$) have been developed to maximize the energy density and cycling stability of TMO cathodes. While most commercial Li-ion batteries contain graphite anodes, some systems also utilize TMO anodes such as $Li_4Ti_5O_{12}$. The vast compositional landscape of TMOs explored as Li-ion active materials is detailed in Chapters 2, 3, 5, 7, and 16.

1.3 Brief History of Lithium-Ion Batteries

In the 1960s, intercalation chemistry was a prominent method used to alter materials' electronic and optical properties [4, 5]. For instance, the electronic conductivity of WO_3 can be varied several orders of magnitude by intercalating monovalent cations into the structure [6]. Rechargeable batteries which utilize ion intercalation reactions were first demonstrated in the 1970s by Whittingham [7] at Exxon Corporation. These prototype cells contained a layered TiS_2 cathode, lithium metal anode, and liquid electrolyte (e.g. $LiClO_4$ dissolved in a mixture of dimethoxyethane and tetrahydrofuran) [7]. One limitation of this system was the Li metal anode which forms dendrites that can penetrate the separator and internally short-circuit the cell. As an alternative anode, Yazami demonstrated reversible Li^+ intercalation in graphitic carbons using a polymer electrolyte in the 1980s [8]. However, in liquid electrolytes these intercalation anodes were hindered by solvent co-intercalation which resulted in graphite exfoliation and electrode degradation during cycling. In 1985, a group led by Yoshino at Asahi Kasei Corporation (Japan) identified petroleum coke anodes [9] which were stable during Li^+ insertion/removal, and these anodes were incorporated into Li-ion full cells containing liquid electrolyte. These discoveries ultimately led to the first commercial Li-ion batteries introduced by Sony in 1991. At around the same time, suitable electrolyte solvents (e.g. ethylene carbonate) which do not co-intercalate in graphite were also identified. Compared to disordered carbons derived from petroleum coke, graphitic anodes operate at more negative potentials which enables higher cell voltages and energy densities.

In addition to carbonaceous anodes, the 1980s also witnessed advancements in Li-ion intercalation cathodes. Notable examples include $LiCoO_2$ (Goodenough and coworkers) [10], $LiMn_2O_4$ (Thackeray and coworkers) [11], and polyanionic cathodes such as $Li_yFe_2(XO_4)_3$, X = S, Mo, and W (Manthiram and coworkers) [12]. These scientific breakthroughs enabled Li-ion batteries to dominate the rechargeable battery market for decades, and these systems are now ubiquitous in commercial devices ranging from portable electronics to electric vehicles. The 2019 Nobel Prize in Chemistry celebrated this achievement with an award to Profs. John Goodenough, Stanley Whittingham, and Akira Yoshino.

1.4 The Role of Advanced Characterization and Computing Resources

The rich history of Li-ion batteries demonstrates that translating scientific discoveries into viable energy storage systems requires decades of research and development.

Central to this development process are fundamental scientific discoveries (e.g. identifying structure/property correlations and understanding degradation mechanisms) needed to optimize material selection and system performance. Such efforts often require combining routine material characterization (e.g. electroanalytical methods, vibrational spectroscopy, and lab-source X-ray diffraction) with world-class instrumentation (e.g. X-ray/neutron beamlines). Several chapters in this book describe advanced characterization methods that enabled critical insights on TMOs for electrochemical energy storage, and examples include: (i) solid-state nuclear magnetic resonance spectroscopy (NMR) and electron paramagnetic resonance (EPR) to probe local bonding structure (Chapter 11), (ii) *in situ* and *operando* neutron diffraction to identify phase transformations during material synthesis and battery operation (Chapter 12), and (iii) X-ray spectroscopy and imaging to probe transition metal valence states (Chapter 13). In addition to these experimental efforts, advanced computing resources and machine learning (Chapters 14 and 15) have accelerated material discovery and provided mechanistic insight into processes occurring at length/time scales which are difficult to probe experimentally.

1.5 Beyond Lithium-Ion Batteries

The performance of Li-ion batteries is largely dictated by the active materials, and replacing the graphite anode with Li metal can enable batteries with specific energies >400 Wh kg^{-1} (compared to ~200 Wh kg^{-1} for today's commercial Li-ion batteries). As previously mentioned, formation of Li dendrites during charging presents major safety and reliability concerns, and to date, liquid electrolytes have been unable to effectively suppress Li dendrite formation. To combat this issue and stabilize the Li metal anode, there is growing interest in solid-state batteries (SSBs) containing solid electrolyte separators. Representative classes of solid-state Li^+ conductors include garnet oxides (e.g. $Li_7La_3Zr_2O_{12}$), sulfides (e.g. β-Li_3PS_4 and Li_6PS_5Cl), polymers (e.g. poly(ethylene oxide) containing a Li^+ salt), and inorganic glasses (e.g. Lipon). The opportunities for Li metal are tremendous, but enabling practical SSBs requires overcoming major challenges related to material synthesis, processing, and interfacial stabilization as discussed in Chapter 8.

Energy storage costs are a primary factor in determining the viability of battery technologies for grid storage applications. While the cost of Li-ion batteries has decreased substantially in the past decade (e.g. from ~US $1000/kWh in 2008 to ~US $130/kWh in 2020), further cost reductions are needed to enable widespread adoption for large-scale stationary storage. Recent developments on low-cost rechargeable batteries discussed in this book include: (i) aqueous electrolytes for Li-ion systems (Chapter 6) and (ii) Na-based chemistries which utilize active material hosts that undergo intercalation, conversion, or alloying reactions (Chapters 3 and 4).

Development of next-generation energy storage systems is critical to combat climate change in the coming decades. Global energy consumption rates [13] are expected to increase to 40.8 TW by 2050, and electrochemical energy storage systems represent one of the most promising platforms to bridge the supply/demand gap for intermittent renewable resources (e.g. solar and wind). In this regard, TMOs are expected to continue playing critical roles as components for low-cost, reliable energy storage to enable a sustainable future.

Acknowledgments

The authors acknowledge support from Energy Efficiency and Renewable Energy (EERE), Vehicle Technologies Office (VTO), Department of Energy.

This chapter has been authored in part by UT-Battelle, LLC, under contract DE-AC05-00OR22725 with the US Department of Energy (DOE). The US government retains and the publisher, by accepting the article for publication, acknowledges that the US government retains a nonexclusive, paid-up, irrevocable, worldwide license to publish or reproduce the published form of this manuscript, or allow others to do so, for US government purposes. DOE will provide public access to these results of federally sponsored research in accordance with the DOE Public Access Plan (http://energy.gov/downloads/doe-public-access-plan).

References

1 Hayner, C.M., Zhao, X., and Kung, H.H. (2012). Materials for rechargeable lithium-ion batteries. *Annual Review of Chemical and Biomolecular Engineering* 3: 445–471.
2 Dunn, B., Kamath, H., and Tarascon, J.M. (2011). Electrical energy storage for the grid: a battery of choices. *Science* 334 (6058): 928–935.
3 Belharouak, I., Nanda, J., Self, E.C., et al. (2020). Operation, manufacturing, and supply chain of lithium-ion batteries for electric vehicles. https://doi.org/10.2172/1737479.
4 Gamble, F.R., Osiecki, J.H., Cais, M. et al. (1971). Intercalation complexes of Lewis bases and layered sulfides: a large class of new superconductors. *Science* 174: 493–497.
5 Aronson, S., Salzano, F.J., and Bellafiore, D. (1968). Thermodynamic properties of the potassium-graphite lamellar compounds from solid-state emf measurements. *Journal of Chemical Physics* 49 (1): 434–439.
6 Whittingham, M.S., Michael, S., and Jacobson, A.J. (1982). *Intercalation Chemistry*. New York: Academic Press.
7 Whittingham, M.S. (1976). Electrical energy storage and intercalation chemistry. *Science* 192: 1126–1127.
8 Yazami, R. and Touzain, P. (1983). A reversible graphite-lithium negative electrode for electrochemical generators. *Journal of Power Sources* 9: 365–371.

9 Yoshino, A., Sanechika, K., Nakajima, T. (1987). Secondary battery. US Patent 4,668,595, filed 9 May 1986 and issued 26 May 1987, https://patents.google.com/patent/US4668595A/en.

10 Mizushima, K., Jones, P.C., Wiseman, P.J., and Goodenough, J.B. (1980). Li_xCoO_2 ($0<x\leq1$): a new cathode material for batteries of high energy density. *Materials Research Bulletin* 15: 783–789.

11 Thackeray, M.M., David, W.I.F., Bruce, P.G., and Goodenough, J.B. (1983). Lithium insertion into manganese spinels. *Materials Research Bulletin* 18: 461–472.

12 Manthiram, A. (2020). A reflection on lithium-ion battery cathode chemistry. *Nature Communications* 11 (1): 1550.

13 Lewis, N.S. and Nocera, D.G. (2006). Powering the planet: chemical challenges in solar energy utilization. *Proceedings of the National Academy of Sciences of the United States of America* 103 (43): 15729–15735.

2

Metal–Ion-Coupled Electron Transfer Kinetics in Intercalation-Based Transition Metal Oxides*

Victoria A. Nikitina and Keith J. Stevenson

Skolkovo Institute of Science and Technology, Center for Energy Science and Technology, Nobel Straße 3, 121205 Moscow, Russia

2.1 Introduction

Intensive research in the field of lithium ion intercalating systems over the last several decades resulted in the design of hundreds of active material and electrolyte systems for practical battery applications [1–3]. Given the high priority of achieving maximum capacity, energy density, and rate capability characteristics of the Li-ion batteries, the focus of the majority of studies on the coulombic efficiency, cyclability, and charge–discharge (*C*-rate) characteristics of the active materials is not surprising. Still, a detailed mechanistic understanding on the metal–ion-coupled electron transfer (ET) kinetic and thermodynamic properties is needed to realize improved performance for next-generation energy storage such as for transportation and grid-scale applications [4–6].

Most of the modern electrochemical characterization methods in the assessment of metal–ion battery kinetics and thermodynamic processes focus on the analysis of charge/discharge curves performed under galvanostatic conditions or the differential capacity (derivative of the charge Q with respect to voltage V) dQ/dV vs. V curves, which results from a simple differentiation of the experimental charge/discharge curves [7]. Yet, classical electrochemical kinetic models, which offer a variety of techniques to treat the electrochemical data for redox processes in solution, for adsorbed reactants, and as well as for very complex cases of multistep processes [8, 9], are rarely applied for a quantitative analysis of the intercalation processes, which occur at the solid electrode/electrolyte interface. In this chapter, we review the processes, which involve redox reactions in transition metal oxide electrodes coupled with metal–ion intercalation steps, with emphasis on the specifics of metal–ion-coupled electron transfer kinetics in multiparticle composite systems, which are indeed the key electrode architectures in metal–ion battery research. We aim to broaden the well-known classical electrochemical concepts concerning the

*This chapter was originally published in Advanced Energy Materials, 2021, Volume 10, Issue 22; DOI: 10.1002/aenm.201903933. Reproduced with permission of WILEY-VCH GmbH.

Transition Metal Oxides for Electrochemical Energy Storage, First Edition.
Edited by Jagjit Nanda and Veronica Augustyn.

reaction-limiting steps to encompass ion intercalation processes as well as discuss the conditions required for a given slow step to manifest itself. The point of interest in this discussion is the criteria, which should allow for facile and reliable diagnostics of the intercalation rate control regime for new materials or for unconventional solvent/metal–ion salt systems, as the knowledge of the rate-determining factors is essential for choosing the optimal strategy to enhance the rate capability and capacity characteristics of a metal–ion battery (e.g. by decreasing the particle size, tuning the surface layers composition and structure, and the solvating properties of the medium). This discussion also focuses specifically on "battery-type" intercalation metal oxide systems and not "pseudocapactive-type" metal oxide materials. Several recent reviews that provide formal definition and distinction between these two types of energy storage mechanisms can be found here [10–12].

At its most basic level, an intercalation process can be regarded as comprising two steps: electron transfer (ET) from the current collector to the redox-active material (faradaic), which results in the change of structure due to the transition metal oxidation/reduction, and ion transfer (IT) into the metal oxide crystal lattice from the bulk electrolyte. The second process of IT also involves several stages: (i) ion diffusion in the solvent bulk; (ii) transition through surface films (generally regarded as solid electrolyte interphase [SEI] layers in conventional carbonate-based electrolytes); (iii) charge transfer at the solution/electrode interface or SEI/electrode interface; (iv) ion diffusion in the electrode material. Apart from these simplest steps, the step of ionic desolvation in the vicinity of the electrode/solution or electrode/SEI interface should generally occur, as in most cases the ion enters the crystal lattice without its solvation shell. More complex cases also involve the step of nucleating a new phase, which is typical for materials with a wide miscibility gap (such as $LiFePO_4$), where the rate of initiating the phase transition turns out to be slower than the charge-transfer kinetics or ionic diffusion [13]. As the complex nucleation dynamics should be considered in this case together with the charge-transfer and diffusion equations, in this contribution we limit ourselves to the analysis of the metal oxide electrode materials with wider single-phase regions, where the events of nucleation do not determine the intercalation kinetics.

We primarily address the voltammetric and electrochemical impedance responses of the oxide electrode materials due to the relative simplicity of the data analysis and highly informative shape of the electrochemical responses, which allows for both the initial qualitative-level rate-limiting step diagnostics as well as for the advanced quantitative estimation of the key kinetic parameters, which characterize both the transport and the kinetic limitations in an intercalation material/electrolyte system. However, we note that even for a qualitative-level analysis the basic requirements for the experimental setup are to be fulfilled to ensure the reproducibility and the reliability of the electrochemical data obtained.

(1) The electrochemical measurements should be carried out in a three-electrode cell with a reliable reference electrode with a high exchange current density to ensure its reversibility. While a Li metal reference electrode generally demonstrates rather fast ET/IT kinetics [14], Na and K metal electrodes due to their higher reactivity with the electrolyte cannot be used as reference electrodes

in a three-electrode setup in organic carbonate electrolytes [15]. Alternative redox electrodes are recommended for accurate electrochemical kinetic studies of Na^+ and K^+ intercalation (e.g. Ag^+/Ag electrodes or reference electrodes based on the partially charged phase transforming materials with a wide miscibility gap [16]).

(2) The responses of inhomogeneous thick porous electrodes cannot be analyzed quantitatively within the framework of simple models [17]. Hence, the mass loading of the electrode material powder should be minimized to 1–0.5 mg cm^{-2} of the current collector.

(3) Minimization of the electrolyte/material ratio (i.e. using small amounts of electrolyte) is essential in case of reductive/oxidative electrolyte decomposition with the SEI layers formation to stabilize the electrochemical interfaces, as low volume of the electrolyte minimizes the dissolution of the formed protective surface layers.

In this discussion, we rely on the phenomenological electrochemical kinetics theory of intercalation materials, as formulated in earlier works of Levi and Aurbach [18, 19], as this approach provides a transparent way to estimate the key kinetic parameters of an intercalation reaction, such as diffusion coefficients, apparent rate constants, and ion–host interaction parameters. A more advanced treatment by M. Bazant is based on the phase-field theory of electrochemical kinetics, which generalizes the classical Butler–Volmer and Marcus equations [20, 21]. The latter approach allows to describe the kinetic and thermodynamic events in the course of intercalation within the framework of highly accurate and physically adequate formulism, which possesses a high explanatory power. However, the application of this approach to treat experimental kinetic data for intercalation systems requires knowledge of a large number of parameters, such as reorganization energies, activity coefficients, strain energies, for reactants, products, and intermediates, which can hardly be estimated quantitatively. The phenomenological treatment does not grant information on the physical origin of charge transfer activation barriers, but allows for the elucidation of the nature of the intercalation rate-limiting step and comparative analysis of the kinetic parameters for different ion–host systems, which are essential for the rational control of intercalation kinetics in energy storage and conversion devices.

2.2 Thermodynamic Control

The simplest case in the metal–ion-coupled electron transfer kinetics is when both the rate of diffusion and the rate of ion transfer are very fast and do not affect the shape of the cyclic voltammograms (CVs), charge/discharge curves, or current transients. In this case, the current (under potentiostatic conditions) or the potential (under galvanostatic conditions) are determined by the shape of the intercalation isotherm, i.e. by the dependence of the host lattice free energy on the amount of the intercalated metal ions, which is characterized by the intercalation level θ (the ratio of the number of the ions in the host lattice to the number of the available intercalation sites). The most direct analogy between the intercalation and the adsorption processes relies on the Nernst equation for the intercalation of an ion inside an empty site in the host lattice: $M^+ + e^- + [\ ] = [M]$, which can be modified to account for the repulsive or attractive interactions by introducing a phenomenological parameter "g" (negative for the attraction between ions in the host and positive for repulsion). The resulting equation represents a Frumkin intercalation isotherm, as introduced by Levi and Aurbach [18]:

$$E(\theta) = E_{eq} + \frac{RT}{F} \ln\left(\frac{1-\theta}{\theta}\right) + \frac{RT}{F} g(0.5 - \theta) \tag{2.1}$$

where E_{eq} is the standard redox potential (V) (the equilibrium potential of a real or a hypothetical phase having a composition of $\theta = 0.5$); g – dimensionless constant, which corresponds to the interaction parameter in the Frumkin isotherm formulism.

The variation in the value of the g parameter induces changes in the slope of the galvanostatic charge/discharge profiles (Figure 2.1a). The limiting case of g = −4 results in a typical plateau, which marks the loss of system stability in a single-phase intercalation process. Within the framework of the phenomenological approaches, the attractive interactions in the system become strong enough to result into separation of the system into cation-rich and cation-poor phases. Indeed, further increase in the negative g value results in an S-shaped isotherm, which clearly implies the coexistence of two phases in some interval of θ values. The corresponding cyclic voltammetry curves (Figure 2.1b) calculated for different g values show broad peaks for the repulsive interactions (g = 0, 2, 4) and sharper peaks for the attractive interactions, which is so far completely analogous to the adsorption processes, and can be treated using Laviron analysis for surface confined redox-active species [22] (the equations for the current/potential curves calculation can be found in [18]). An experimental example of a system without any significant charge transfer or diffusional limitations (this case is designated as "thermodynamic control") is the slightly overlithiated lithium-manganese oxide $Li_{1+x}Mn_2O_4$ [14, 23]. At low scan rates (<0.1 mV s^{-1}), the experimental CVs show two pairs of peaks, centered at 4.0 and 4.13 V (vs. Li^+/Li) corresponding to the consecutive de/insertion of 0.5 Li per f.u. (Figure 2.1c). The symmetry of the cathodic and anodic peaks and negligible peak-to-peak separation implies the thermodynamic control of Li-ion intercalation. Indeed, the basic features of experimental curves can be reproduced using g values of 1.1 and −0.88 for the broader and the sharper peaks, respectively.

However, the experimental isotherms are very rarely as simple as imposed by Eq. (2.1). Figure 2.1d shows the experimental isotherm for $LiCoO_2$ in a conventional 1 M $LiPF_6$ ethylene carbonate/dimethyl carbonate (EC/DMC) electrolyte [24]. It is established that a phase transition from hexagonal to monoclinic phase occurs upon Li-ion deinsertion from the $Li_\theta CoO_2$ structure at $\theta \sim 0.78$ [25]. Thus, two isotherms for the lithiated and the delithiated phases are required (solid lines in Figure 2.1d). While the initial part of the isotherm in the Li-rich phase can be approximated using a single g value, this is impossible for the exponential rise of the curve at $0.5 < \theta < 0.75$, which simply reflects the very strong dependence of the crystal lattice energy on the amount of the intercalated ions. This can be overcome

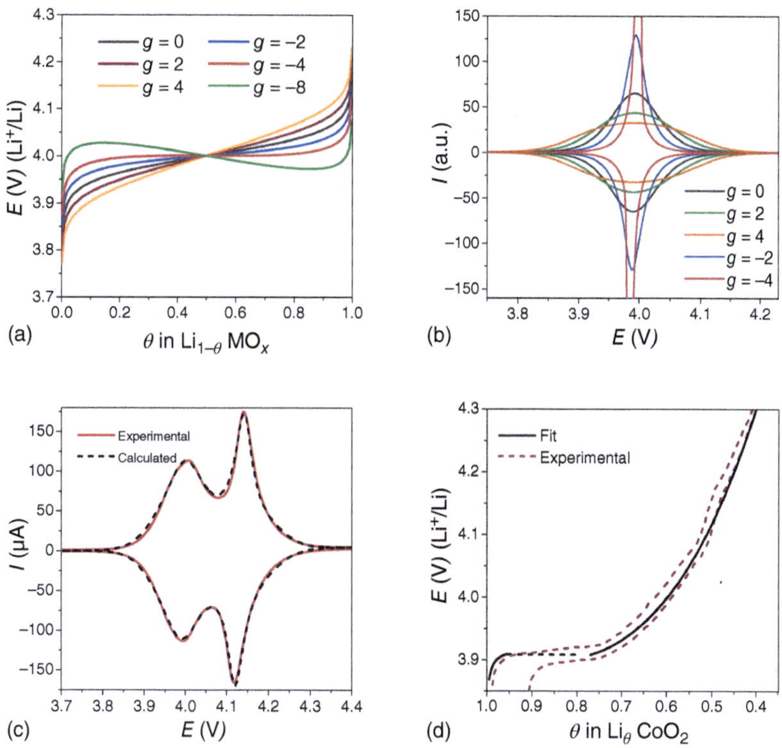

Figure 2.1 Charging curves (deintercalation isotherms) (a) and CVs (b) calculated for different values of the interaction parameter g. CV at 0.1 mV s^{-1} and the fit for $LiMn_2O_4$ electrode in 1 M $LiPF_6$ in EC/DMC electrolyte (c). Experimental (de)intercalation isotherm for $LiCoO_2$ electrode (dashed lines) and the fits for Li-rich and Li-poor phases (solid lines, d). Source: Vassiliev et al. [14]/with permission of Elsevier.

by introducing a series of $g_1 \ldots g_n$ values, which modifies Eq. (2.1):

$$E(\theta) = E_{eq} + \frac{RT}{F}\ln\left(\frac{1-\theta}{\theta}\right) + \frac{RT}{F}[g_1(0.5-\theta) + g_2(0.5-\theta)^2 + \ldots + g_n(0.5-\theta)^n] \tag{2.2}$$

The fitting parameters g_n introduce very little uncertainty into the experimental data modeling, as these are evaluated directly from the experimental charge/discharge curves and are used only to construct an analytical approximation of the experimental intercalation isotherm. For instance, the data for $LiMn_2O_4$ electrodes can be accurately reproduced by approximating the experimentally derived $E(\theta)$ using three g values for each redox process (process at 4.0 V: $g_1 = 1.13, g_2 = -0.45, g_3 = 2.34$; process at 4.13 V: $g_1 = -0.81, g_2 = 1.57, g_3 = 9.58$) [14]. The thermodynamic control is also typical for sodium- and potassium-ion intercalation into Prussian blue compounds [26] and Na^+ intercalation into $AVPO_4F$ materials [15, 27]. As the diagnostic criteria for the thermodynamic regime, one can use the symmetry of the anodic and cathodic curves and the proportionality of the current peak to the scan rate value, while the conclusions on the type of interactions between the ions in the host matrix can be formulated based on the peaks' shape.

2.3 Diffusional Control

The cation mobility inside a solid structure is often the key parameter to be addressed upon the introduction of a new metal–ion battery material. The apparent diffusion coefficient D is used to characterize the ionic mobility inside a solid structure:

$$D = \frac{l^2}{\tau} \tag{2.3}$$

where l is the diffusion length (cm) (half of the particle diameter d or the thin-film thickness) and τ is the diffusion time constant (s). Certain specifics of ion intercalating materials are determined by the need to consider the dimensionality of the diffusion geometry: planar (or 1D, as in channels in the $LiFePO_4$ structure), cylindrical (2D, diffusion in layers of $LiCoO_2$ and other layered materials), spherical (3D, diffusion in $LiMn_2O_4$), and the finite nature of the diffusion inside a material particle. An excellent treatise on the determination of the diffusion coefficient from chronoamperometry data with an account of the typical hindering experimental factors (such as slow charge transfer kinetics, uncompensated ohmic drop, and wide particle size distribution) is available in [28–30].

The general treatment of the diffusion problem is based on a standard set of Fick's equations for a given diffusion geometry (e.g. planar) with boundary conditions at the particle surface and at its center [31]:

$$\frac{\partial \theta}{\partial t} = D\frac{\partial^2 \theta}{\partial x^2};\quad \left(\frac{\partial \theta}{\partial x}\right)_{x=1} = 0;\quad \theta_{x=0} = \theta(E);\quad \left(\frac{\partial \theta}{\partial x}\right)_{x=0} = \frac{I(E)}{S \cdot F \cdot D \cdot 10^6 \cdot \frac{\rho \cdot n_M}{M_r}} \tag{2.4}$$

where x is the relative coordinate ($0 \leq x = 2l/d \leq 1$), E is the electrode potential (V), I is the current (μA), F is the Faraday constant, $\theta(E)$ or $E(\theta)$ is the intercalation isotherm, S is the working area of the particle surface (cm^2) (only planes contributing to lithium diffusion are relevant), which is calculated according to the equation:

$$S = \frac{n \cdot 10^4 \cdot m}{\rho \cdot d} \tag{2.5}$$

where $n = 2$, 4, and 6 for planar, cylindrical, and spherical diffusion geometries, m is the mass of the intercalating material in the electrode (g), ρ is the phase density (g cm^{-3}).

As most intercalation materials undergo phase transformations in the course of charge/discharge, a more complex diffusion problem should be solved for the phase boundary movement, which is discussed in detail in Refs. [14, 32].

Diffusional control in intercalation materials is generally unavoidable at high scan or charge/discharge rates due to the finite size of the material particles and the generally low ionic mobilities in intercalation materials [33]. To illustrate the effect of the diffusion coefficient value on the shape of the CVs, a set of simulated voltammograms for a simple model intercalation process is shown in Figure 2.2a.

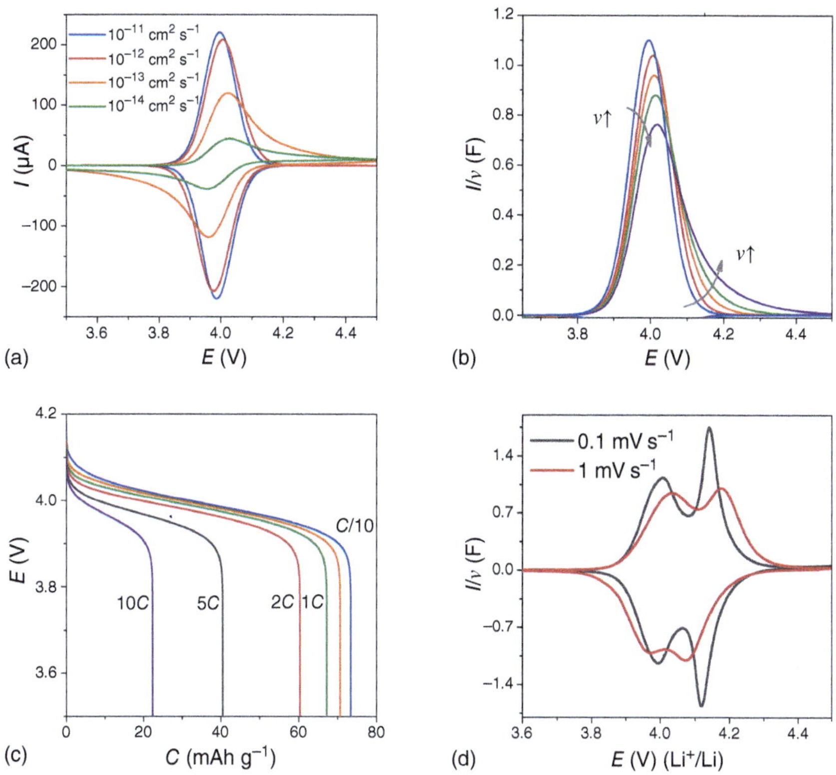

Figure 2.2 (a) Simulated CVs for a simple intercalation process (E_{eq} = 4.0 V, g_1 = 1, m = 0.5 mg, v = 0.2 mV s^{-1}, d = 1 μm) for different D values (indicated in the plot). (b) Normalized per scan rate CVs for scan rate values: 0.05, 0.2, 0.4, 0.6, 1 mV s^{-1} (D = 10^{-12} cm^2 s^{-1}, g = 1, E_{eq} = 4.0 V, m = 0.5 mg). (c) Charge/discharge curves calculated for different C rates: C/10, C/5, 1C, 2C, 5C, 10C. (d) Experimental CVs of a $LiMn_2O_4$ electrode in 1 M $LiPF_6$ EC/DMC electrolyte at 0.1 and 1 mV s^{-1}. Source: Vassiliev et al. [14]/with permission of Elsevier.

The model redox process occurs in a material with the kinetic parameters similar to those for the (de)intercalation step of $LiMn_2O_4$ (E_{eq} = 4.0 V, g_1 = 1, m = 0.5 mg, scan rate v = 0.2 mV s^{-1}, d = 1 μm). Even for the large diffusion length of a micron-sized particle, completely reversible behavior (i.e. thermodynamic control) is observed when the diffusion coefficient reaches 10^{-11} to 10^{-12} cm^2 s^{-1}. Diffusional limitations appear when the D value drops to 10^{-13} cm^2 s^{-1} (Figure 2.2a). The peak-to-peak separations in the CV reach the typical value for diffusion-controlled processes (59 mV), while the symmetry of the peaks is decreased and the characteristic current "tails" (slow decays of current after the maxima) appear after the peak potentials. Note that when the diffusion coefficient is further decreased, the peaks' height and the total intercalation charge decrease, which is a consequence of incomplete discharge (lithiation) of a particle during the time of the potential sweep. In experiments with polydisperse powder electrodes, the long-time current decays would also reflect the contributions of larger particles with higher diffusion lengths [14].

Naturally, the increase in the scan rate also results in the appearance of asymmetric current tails and the decrease of the peak's height (Figure 2.2b). Under these conditions, the reversible capacity on charge or on the discharge of the electrodes decreases with the increase in the *C* rate value, e.g. for a micron-sized particle with $D = 10^{-12}\ \mathrm{cm^2\,s^{-1}}$ at a 5*C* charge/discharge rate, almost half of the reversible capacity will be lost (Figure 2.2c). The experimental CVs of a $LiMn_2O_4$ electrode show the transition from thermodynamic control to diffusional control upon increase of the scan rate from 0.1 to $1\ \mathrm{mV\,s^{-1}}$ (Figure 2.2d).

Basic features of diffusional control, which can be used for the kinetic regime diagnostics are: (i) the proportionality of the current maxima to $v^{1/2}$, (ii) 59-mV peak-to-peak separation between the anodic and the cathodic peaks, and (iii) decrease in the intercalation capacity with increase in the charge/discharge rate.

2.4 Kinetic Control

Despite being mentioned quite often, diffusional control is by far not the only way to impose rate limitations on an intercalation material charge/discharge rate. The intercalation process involves the interfacial transfer of both the electron (from the current collector to the material particle) and the ion (from the solution side to the site inside the crystal lattice). Both processes involve the transfer of charge across the boundary between two phases, which can be phenomenologically described within the framework of Butler–Volmer kinetics. It is generally not yet established whether it is the transfer of an ion or of an electron that determines the rate of the process. However, at least for a number of well-studied materials ($LiCoO_2$ and $LiMn_2O_4$), changing the solvent results in quite pronounced changes in the intercalation rate [15, 26, 28, 29, 34, 35]. It is thus more natural to suggest that the intercalation rate (at least for the studied systems) is limited by the interfacial transfer of a metal ion, as the solvent effect can hardly be expected for a solid-state electron transfer reaction. Regardless of the type of species which traverses the interface, the Butler–Volmer equation can be used to describe the dependence of the (de)intercalation rate on the overvoltage $\eta = E - E_0(\theta)$ [14]:

$$I(E) = SF \cdot 10^6 \cdot \frac{\rho \cdot n_M}{M_r} \cdot k_s \cdot \\ \times \left\{ \theta \cdot \exp\left[\frac{(1-\alpha)F(E - E_0(\theta))}{RT}\right] - (1-\theta) \cdot \exp\left[-\frac{\alpha F(E - E_0(\theta))}{RT}\right] \right\} \tag{2.6}$$

where α is the transfer coefficient, M_r is the molecular mass of the material ($\mathrm{g\,mol^{-1}}$), n_M is the number of intercalating M^+ ions per f.u., k_s is the apparent heterogeneous rate constant ($\mathrm{cm\,s^{-1}}$), which includes the intercalating metal ion concentration in solution c_M, i.e. $k_s = k_s^0 c_M^\alpha$ to keep the standard dimensionality.

For a material with a relatively high charge transfer rate constant ($10^{-8}\ \mathrm{cm\,s^{-1}}$) and fast diffusional transport ($10^{-9}\ \mathrm{cm^2\,s^{-1}}$), the kinetic limitations would not impact the shape of the CV response (Figure 2.3a), while decreasing the rate constant value will result in an increase in the peak-to-peak separations and a moderate decrease in the

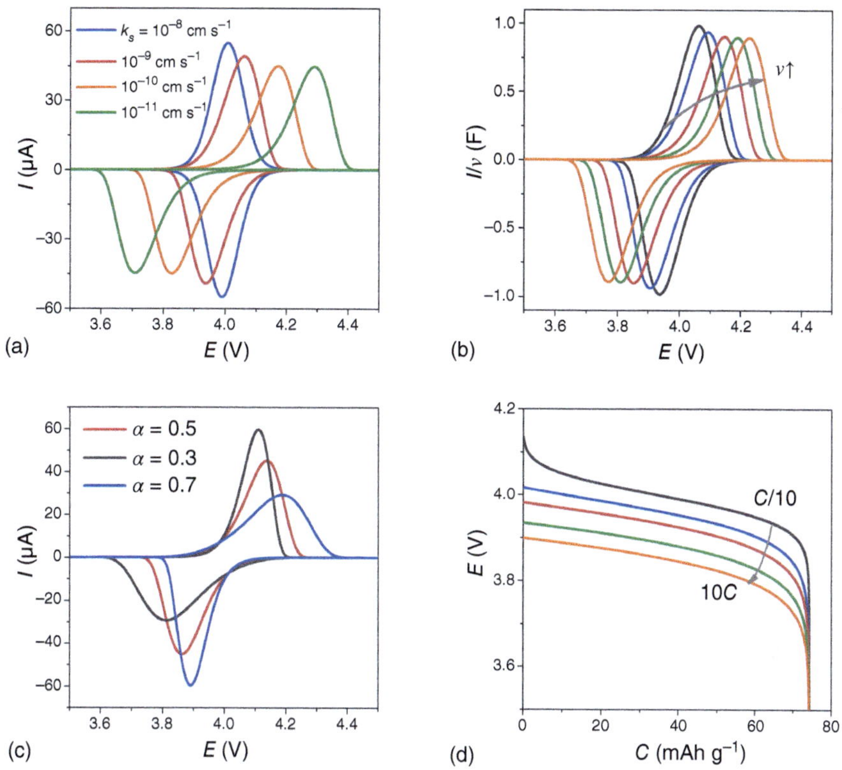

Figure 2.3 (a) Simulated CVs ($E_{eq} = 4.0$ V, $g_1 = 1$, $m = 0.5$ mg, $v = 0.1$ mV s^{-1}, $D = 10^{-9}$ cm^2 s^{-1}, $\alpha = 0.5$, $d = 1$ μm) for different values of k_s. (b) The effect of the scan rate (0.05, 0.1, 0.3, 0.7, 1.5 mV s^{-1}) on the CV shape (current is normalized to the scan rate). (c) The effect of the transfer coefficient value on the CV peaks symmetry ($D = 10^{-9}$ cm^2 s^{-1}, $k_s = 10^{-9}$ cm s^{-1}). (d) Charge/discharge curves for the case of rate control by slow charge transfer ($D = 10^{-9}$ cm^2 s^{-1}, $k_s = 10^{-9}$ cm s^{-1}) calculated for different C rates: C/10, 1C, 2C, 5C, 10C.

peaks' heights (at low enough scan rates). A set of normalized CV curves shows the changes in the differential intercalation capacity (current divided be the scan rate v) vs. potential curves upon increasing the scan rate for a rate constant value of 10^{-9} cm s^{-1}, which is close to the typical value for Li^+ (de)intercalation into $LiCoO_2$ materials (Figure 2.3b). The increase in the potential difference between the anodic and cathodic current maxima results in a visible hysteresis on both the CV curves and charge/discharge profiles. The transfer coefficient value α is not a transparent parameter in case of ion transfer reactions, as in contrast to simple outer-sphere redox processes α should not necessarily be close to 0.5 at low over potentials [36]. The α value determines the symmetry of the anodic and cathodic CV branches, which is illustrated in Figure 2.3c.

Kinetic limitations (given the absence of any diffusional limitations) do not result in a pronounced reversible capacity loss at high charge/discharge rates (Figure 2.3d). However, the hysteresis between the charge and discharge curves decreases the average voltage (for our model case, from 4.05 to 3.85 V), which decreases the energy density of the material. Correspondingly, if both the diffusion and kinetic limitations are operative, both the capacity and the average working potential would decrease to some extent, depending on the *C*-rate and D and k_s values.

"Mixed" intercalation rate control regime refers to the situation when both diffusion and charge transfer kinetics are sufficiently slow to contribute to the intercalation rate. A typical voltammogram of a model material under mixed kinetic control is shown in Figure 2.4a. The characteristic features of both the kinetic and diffusional cases of control manifest themselves in the increase of peak-to-peak separations and current decay times with the increase in scan rate. An experimental example of mixed control by charge transfer kinetics and solid-state diffusion is Li^+ intercalation into $LiCoO_2$ (Figure 2.4b), which is complicated by the phase transformation proceeding at $E = 3.91$ V (vs. Li^+/Li). The treatment of phase boundary movement is

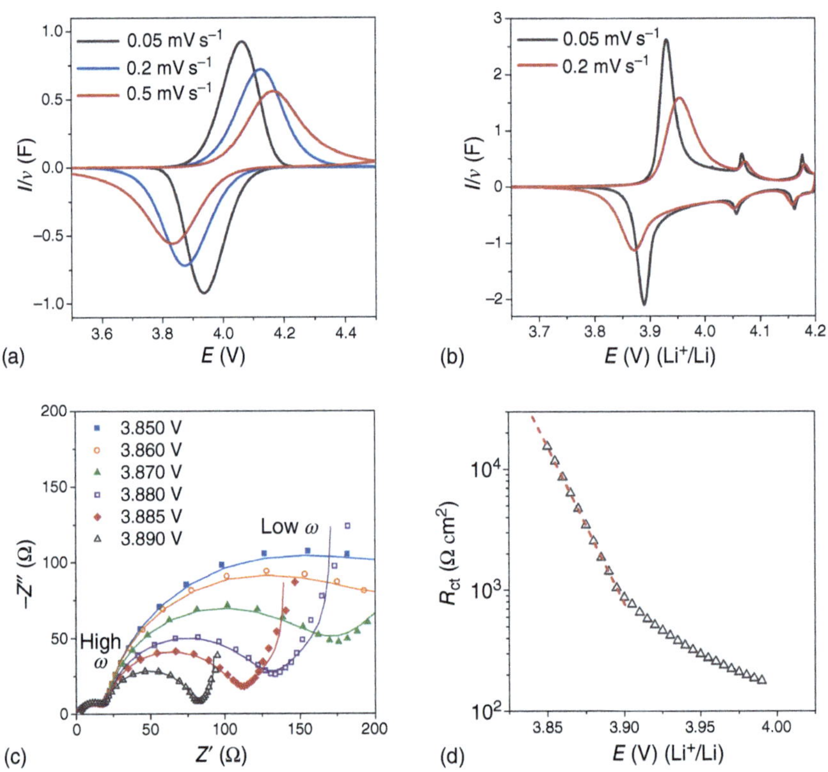

Figure 2.4 (a) Calculated CVs for the case of "mixed" reaction rate control ($k_s = 5 \times 10^{-9}$ cm s^{-1}, $D = 10^{-12}$ cm^2 s^{-1}). (b) Experimental CVs of $LiCoO_2$ electrode in 1 M $LiPF_6$ EC/DMC electrolyte at scan rates 0.05 and 0.2 mV s^{-1}. Source: Reproduced with permission from Ref. Levin et al. [34], Copyright (2017), Elsevier. (c) Experimental impedance spectra of $Li_{1-\theta}CoO_2$ electrodes in 1 M $LiClO_4$ propylene carbonate solution at a set of potentials, corresponding to $0.003 < \theta < 0.015$. (d) R_{ct} vs. E dependence for the case of rate control by slow interfacial charge transfer. Source: Vassiliev et al. [37]/with permission of Elsevier.

somewhat more complicated, but the general features of the control by both kinetics and diffusion can be observed, with the anodic and cathodic peaks asymmetry being attributed to the different kinetic parameters (D, k_s, and α) in the two phases [14].

In the mixed diffusion and charge-transfer rate control regimes, electrochemical impedance spectra (EIS) analysis is an efficient tool for characterizing the rate-limiting step. In a typical experiment, the charge transfer resistance R_{ct} is evaluated from the medium-frequency semicircle diameter based on the appropriate equivalent circuit analysis [38], if such a semicircle is present in the experimental Nyquist plots. Generally, a higher-frequency semicircle is also present in the impedance spectra of intercalation electrodes in nonaqueous solvents, which is most often attributed to the resistance of the surface layers (R_{SL}), formed from the products of reductive (SEI) or oxidative (cathode/electrolyte interphase, CEI) electrolyte decomposition. One can use the dependence of the semicircle diameter on the potential value to differentiate between the R_{SL} and R_{ct} semicircles. The R_{SL} generally does not depend on the potential (provided that the layer is stable and does not dissolve/redeposit during the measurement time), while R_{ct} vs. E dependence should show large variation in the resistance values in the beginning of the (de)intercalation, as the exchange current density i_0 depends strongly on θ:

$$R_{ct}(E) = \frac{RT}{n_M F i_0(E)} \tag{2.7}$$

$$i_0(E) = nFk_s S c_O(E)^{\alpha} c_R(E)^{1-\alpha} \tag{2.8}$$

$$c_O(E) = \frac{\rho n_M \theta(E)}{M_r}, \quad c_R(E) = \frac{\rho n_M (1-\theta(E))}{M_r} \tag{2.9}$$

c_O, c_R are the concentrations of the oxidized and reduced forms (vacancy [ ] and metal ion [M] in the host lattice). Thus, R_{ct} is expected to vary by orders of magnitude with the potential (i.e. with θ), if the slow step is associated with the ion (or electron) transfer across the boundary between the two phases. Figure 2.4c shows the variation in the diameters of the R_{ct} semicircles (at medium to low frequencies) at a set of potentials, which correspond to low θ values ($0.003 < \theta < 0.015$) in the $Li_{1-\theta}CoO_2$ material in an EC/DEC(diethyl carbonate)-based electrolyte [37]. A constant value of the high-frequency semicircle diameter indicates the lack of potential dependence, while the diameter of the medium-frequency semicircle shows a linear variation on a logarithmic scale in the region of very low θ values (Figure 2.4d), as expected based on Eqs. (2.7)–(2.9) for a kinetically limited intercalation process.

One of the major complications in the electrochemical data analysis for multiparticle electrodes concerns the width of the particle size distribution. Even if the distribution is not particularly wide, each group of particles with a certain diffusion length would contribute to the electrochemical signal in whichever type of electrochemical experiment – CV [14], chronoamperometry [30, 34], or EIS [30]. An example of a CV for a model material with an unphysical though illustrative particle size distribution, 50% (wt) of 2 μm particles and 50% of 5 μm particles, is shown in Figure 2.5a. The partial voltammograms of the two fractions differ greatly in their

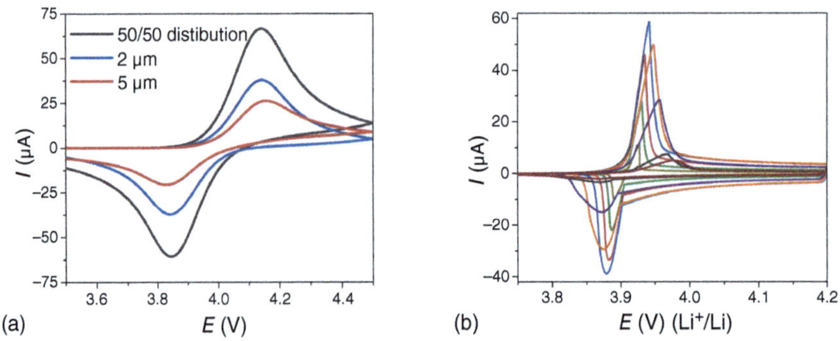

Figure 2.5 (a) Partial voltammograms for a bidisperse sample (50% by weight of 2-µm particles, 50% of 5-µm particles) and the complete curve for the sample with this particle size distribution ($k_s = 5 \times 10^{-9}$ cm s^{-1}, $D = 10^{-12}$ cm^2 s^{-1}). (b) Reconstructed partial voltammograms for $LiCoO_2$ electrodes at 0.075 mV s^{-1}. Source: Vassiliev et al. [14]/with permission of Elsevier.

shape and peak potentials, while the resulting voltammogram would be a sum of partial traces, which complicates the data analysis. On the positive side, a careful numerical analysis makes it possible to reconstruct the unknown particle size distribution for a polydisperse material from a set of CVs, which is advantageous, as real particle size distribution cannot be easily evaluated from microscopy, given the ambiguity of the diffusion length determination in particle agglomerates. As an example, the reconstructed CV responses of major fractions in a $LiCoO_2$ powder material produced via a solid-state route are shown in Figure 2.5b [14].

2.5 Effect of Surface Layers on Ion Transfer Kinetics

The SEI, a layer with distinct chemical composition, which forms at the anode side of a lithium-ion battery, enables the high-voltage operation of lithium-ion batteries, as surface layers prevent continuous reductive electrolyte decomposition by blocking the electronic transport from the electrode to the electrolyte components [39]. At the cathode/electrolyte interface, much thinner layers are formed at high (oxidizing) potentials, which also adds to the stabilization of the electrode/solution interface, although not as efficiently as in the case of the anodic SEI [40]. As it became clear from the recent experimental work on the electrochemistry of the cathode/electrolyte interface, any type of SEI/CEI layer at the electrode surface blocks not only the electron transfer to/from the solvent and salt in the electrolyte, but also creates kinetic hindrances for the transfer of metal ions across the material/solvent interface [15, 27, 34, 35].

For instance, the (de)intercalation kinetics can be compared for the same electrode material ($LiCoO_2$) in a standard commercial electrolyte (1 M $LiPF_6$ in EC/DMC), where CEI layers are formed and in an alkaline aqueous solution, where no dense surface layers are formed (except for surface hydroxides monolayers). In aqueous media, the CV shows sharp peaks (Figure 2.6a), which are broadened significantly upon the transition to the carbonate-based electrolyte, which is indicative

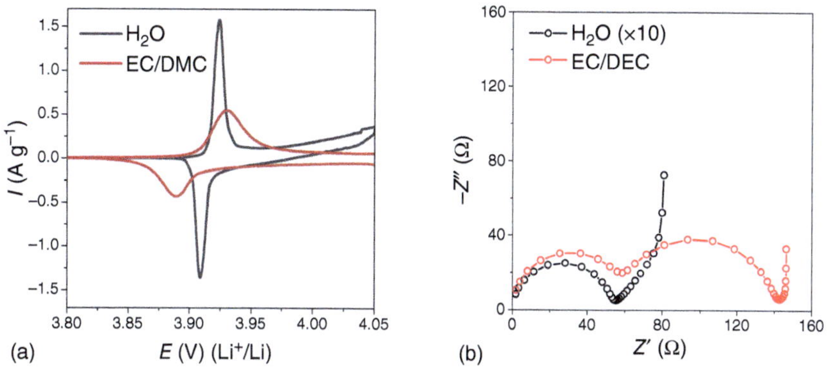

Figure 2.6 CVs at 0.05 mV s^{-1} (a) and impedance spectra at 3.94 V (Li^+/Li) of $LiCoO_2$ electrodes (b) in aqueous and EC/DMC electrolytes. The resistances in H_2O are multiplied by 10 for better visibility. Source: Levin et al. [34]/with permission of Elsevier.

of much faster charge transfer kinetics in aqueous media [34]. The impedance spectrum in EC/DMC exhibits high-frequency semicircles, which can be attributed to the resistance of the CEI layers, while such semicircles are absent in aqueous media (Figure 2.6b). Correspondingly, the intercalation rates (characterized by the diameter of the medium-frequency semicircle R_{ct}) are 1 order of magnitude higher in aqueous media, which implies that barrier layers in the EC/DEC electrolyte impede ion transfer across the electrode/electrolyte interface.

Another observation relates to the correlation between R_{CEI} and R_{ct} resistances. The resistivity of the CEI layers on the surface of $LiMn_2O_4$ electrodes in EC/DEC electrolyte can be varied by increasing the concentration of lithium bis(trifluoromethane)sulfonimide (LiTFSI) salt [35]. If the CVs for the different ratios (11 : 1, 4 : 1, 2 : 1) of EC(DEC):LiTFSI are compared, a pronounced distortion of the CV shape for the lowest solvent-to-salt molar ratio is observed (Figure 2.7a). Drastic changes in both R_{CEI} and R_{ct} can be detected by electrochemical impedance spectroscopy, with higher resistances being observed for higher salt-to-solvent ratios (Figure 2.7b). If both resistances are evaluated from the impedance spectra by fitting to an adequate equivalent circuit, it becomes evident that R_{ct} and R_{CEI} values change codirectionally (Figure 2.7c). This observation can be explained by assuming that in the case of CEI layers formation, the limiting step is associated with the transfer of the ion across the *electrode material/CEI interface*, which changes dramatically with the change in the surface layer structure (the charge transfer rates drop by 40 times when the R_{CEI} increases by a factor of 30) [35]. We note that no chemical, spectroscopic, or structural analysis can provide direct information on the SEI layers resistivity and its impact on the charge transfer kinetics, thus making electrochemical methods the most reliable tool to characterize CEI/material interfaces.

The effects of the structure of the CEI layers can also be illustrated by the dependence of the nature of the alkali cation on the intercalation rates. It was shown that the $AVPO_4F$ (A = Li^+, Na^+, K^+, Rb^+) structure can reversibly accommodate up to 0.5 alkali ions per f.u. in a continuous system of spacious tunnels and cavities, similar

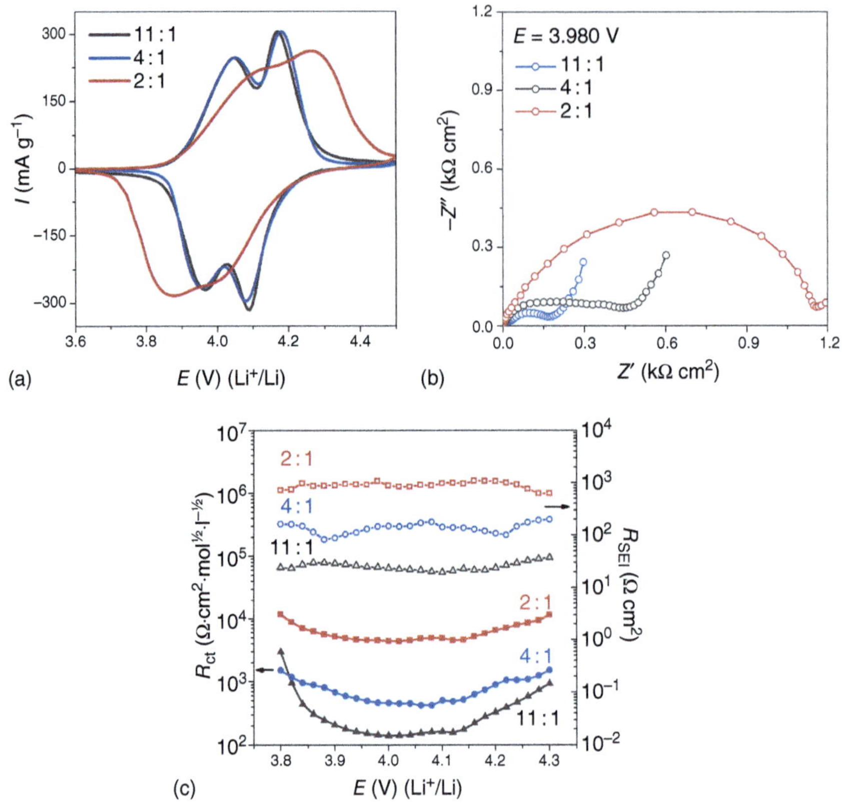

Figure 2.7 (a) CVs of $LiMn_2O_4$ electrodes in 11 : 1, 4 : 1, and 2 : 1 EC/DEC:LiTFSI solutions at 0.1 mV s^{-1}. (b) Nyquist plots at 3.980 V. (c) R_{ct} (filled symbols) and R_{CEI} (open symbols) potential dependencies. Source: Nikitina et al. [35]/with permission of American Chemical Society.

to the Prussian blue systems [27, 41–43]. $AVPO_4F$ can be regarded as a model system for the analysis of the effect of the cation's radius on the charge transfer rates in carbonate-based electrolytes. CVs of $AVPO_4F$ electrodes in EC/DEC solutions are depicted in Figure 2.8a showing reversible behavior for all the cations with only Li^+ (de)intercalation demonstrating typical features of a phase transformation (cathodic and anodic asymmetry of the curves) [27]. If charge transfer resistances in the Li^+, Na^+, and K^+ electrolytes are analyzed, the apparent trend shows significant kinetic limitations for the Li^+ ion transfer, while the resistances for K^+ and Na^+ transfer are 1.5 orders of magnitude lower (Figure 2.8b). The much more pronounced kinetic limitations for Li^+ (de)intercalation might be partially attributed to the more resistant cathode/CEI layer interface, as LiF and other lithium CEI-forming salts would have a minimum solubility in nonaqueous solvents as compared to the K^+ and Na^+ salts [44]. Another factor is thought to refer to the higher activation barrier for the transfer of a small Li^+ ion inside a spacious cavity, due to the ion "pinning" effect. Indeed, this suggestion is justified by the analysis of the alkali ions diffusion coefficients (Figure 2.8c), with the lowest D values being observed for the transport of the Li^+ ion: $D_{Li^+} < D_{Na^+} < D_{Rb^+} < D_{K^+}$.

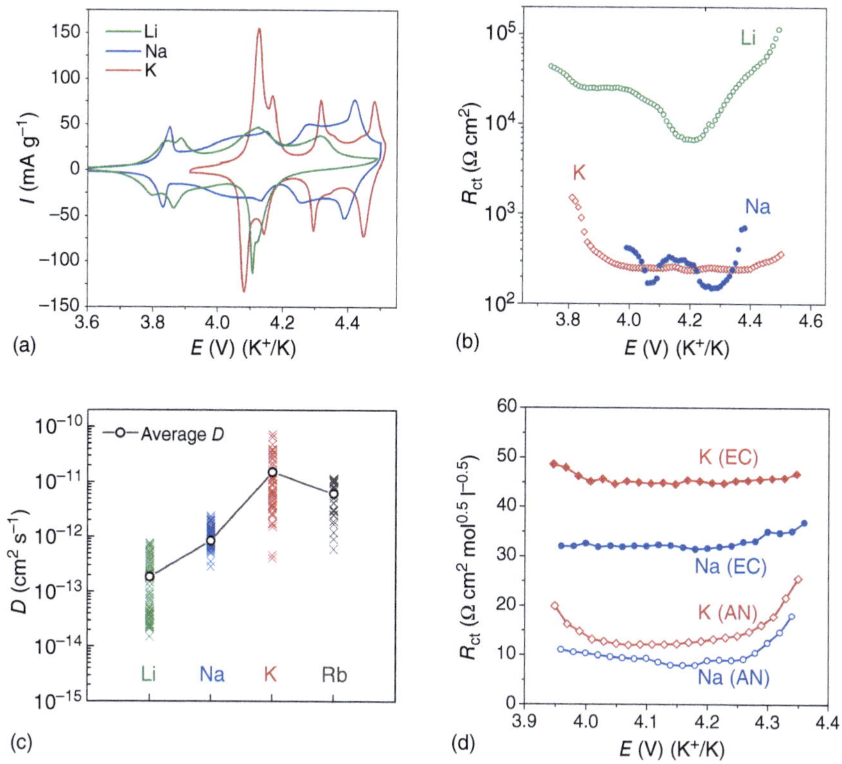

Figure 2.8 (a) CVs of $AVPO_4F$ electrodes in the Li^+, Na^+, and K^+ electrolytes at $0.05\ mV\,s^{-1}$. (b) Dependence of the charge transfer resistance values on the potential in the Li^+, Na^+, and K^+ electrolytes, normalized to the surface area of the material particles. [27]. (c) Alkali ions diffusion coefficients in $AVPO_4F$ structure. Source: Fedotov et al. [41]/with permission of Royal Society of Chemistry. (d) Normalized per surface area and alkali ion solution concentration charge transfer resistances of Na^+ and K^+ intercalation into $AVPO_4F$ electrodes from the electrolytes based on acetonitrile (AN) and EC/DEC. Source: Nikitina et al. [15]/with permission of Elsevier.

No pronounced rate limitations by slow charge transfer can be detected for the intercalation of K^+, Na^+, or Rb^+ into the $AVPO_4F$ host [15, 27, 41]. However, important information on the effect of the CEI/material interface structure on the (de)intercalation kinetics can be obtained by contrasting Na^+ and K^+ (de)intercalation kinetics in CEI-forming electrolytes and in solvents, which are stable enough to not decompose significantly on the material's surface at oxidative potentials [15]. Acetonitrile (AN) is an example of such a solvent, where impedance spectra do not show the characteristic R_{CEI} semicircles, which, of course, does not rule out the existence of some surface layers with a substantially lower resistivity. In Figure 2.8d, the R_{ct} values for Na^+ and K^+ transfer into $AVPO_4F$ material are compared in AN and EC/DEC solvents. Since different salt concentrations were used in [15] (0.1 M in AN and 0.5 M in EC/DEC), an additional normalization per ionic concentration is required to compare the data in the two solvents (according to Eqs. (2.6) and (2.8)). Two important observations can be deduced: (i) the R_{ct}

values for Na^+ and K^+ ions in AN are in a very close agreement, with $R_{ct(Na^+)}$ being only $5\,\Omega\,cm^2\,mol^{0.5}\,l^{-0.5}$ lower than $R_{ct(K^+)}$; and (ii) in EC/DEC, the R_{ct} values for K^+ are significantly higher (by $20\,\Omega\,cm^2\,mol^{0.5}\,l^{-0.5}$) than R_{ct} for Na^+ transfer. These findings point to the existence of material/CEI interface structure effects on the intercalation rates for large ions even when there are no other kinetic hindrances for the ion incorporation into the host structure. Apparently, the CEI/material interface structure determines the activation barrier height for the larger ions. Alternatively, as less resistive to electron transfer CEI layers are formed in the K^+ and Na^+ electrolytes, it possibly accelerates electrolyte decomposition reactions at high potentials for Na^+ and K^+ cathodes, which makes high-voltage K-ion batteries far less feasible than, for instance, Li-ion batteries [15].

2.6 Slow Desolvation as a Limiting Intercalation Step

The accumulation of a large set of experimental data on the charge transfer rates and their potential dependencies not only in ethylene or propylene carbonate-based solvents, but also in other solvents (water, AN, glymes, dimethyl sulfoxide) suggests that the Butler–Volmer type potential dependence of the charge transfer rates is not a characteristic parameter for the intercalation processes in solvents where surface layers formation is not observed, e.g. in aqueous solutions. For instance, in the case of $LiCoO_2$ and $LiMn_2O_4$ electrodes in aqueous media the charge-transfer rates show only minor potential dependence [34, 37] with the apparent activation energies for the slow process being significantly lower [45] (c. $25\,kJ\,mol^{-1}$), than those for Li^+ intercalation from organic carbonate solvents ($50–60\,kJ\,mol^{-1}$) [46]. As low activation energies were also reported for solvated Li^+ intercalation into graphite [47] (compared to the much higher activation energies for the intercalation of desolvated ion), the idea of the desolvation step being rate-determining would be compatible with a large set of experimental results on the activation energy solvent dependence.

Figure 2.9a shows the EIS of $LiMn_2O_4$ electrodes in an aqueous electrolyte at sufficiently low deintercalation levels ($0.001 < 1 - \theta < 0.02$), which, according to Eqs. (2.7)–(2.9), would correspond to the variation of the charge transfer resistance by at least 1 order of magnitude [37]. This is clearly not the case. The R_{ct} values were evaluated from the diameters of the semicircles in the Nyquist plots and the potential dependence of R_{ct} does not exhibit an exponential rise, as predicted by Butler–Volmer kinetics (the expected trend is shown by dashed lines in Figure 2.9b). The same situation is observed for $LiMn_2O_4$ electrodes in AN (Figure 2.9b), while the behavior in propylene carbonate–based electrolyte is much closer to the classical Butler–Volmer kinetic control, as that observed for $LiCoO_2$ in commercial electrolytes (Figure 2.4c,d). The lack of the characteristic charge transfer rate dependence on the applied potential implies that the quantity, which is estimated from the medium-frequency region of the impedance spectra, is not in fact the *charge transfer resistance*, but rather it refers to the kinetic resistance (R_{kin}) of some other step (chemical, and not electrochemical) in the complex intercalation mechanism. If this chemical step relates indeed to the desolvation of the ion in the

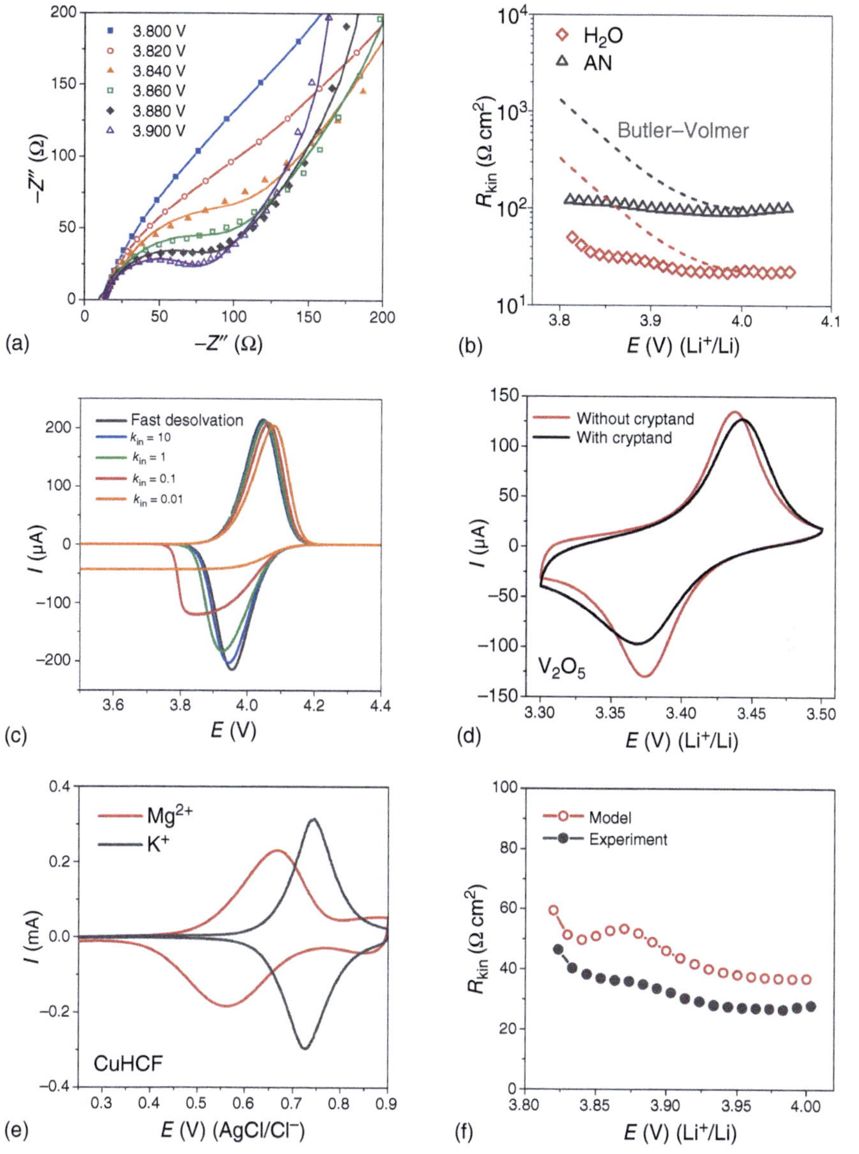

Figure 2.9 (a) Impedance spectra of $LiMn_2O_4$ electrode at a set of potentials (indicated by numbers), corresponding to low deintercalation levels ($0.001 < 1 - \theta < 0.025$) in aqueous 1 M $LiClO_4$. (b) Experimental kinetic resistance values (R_{kin}) vs. E for $LiMnO_4$ in aqueous and AN-based electrolytes (1 M $LiClO_4$) and the R_{ct} values calculated under the assumption of Butler–Volmer type of reaction rate control (dashed lines, $\alpha = 0.5$). (c) Model CVs calculated for different rates of desolvation k_{in} ($K_p = 1$) at 0.1 mV s^{-1}. (d) CVs of V_2O_5 electrode in 0.2 M $LiClO_4$ in PC and 0.2 M $LiClO_4$ + 0.2 M cryptand 2.2.2 in PC at 0.5 mV s^{-1}. (e) CVs of copper hexacyanoferrate electrode in 1 M $MgSO_4$ and 0.5 M K_2SO_4 aqueous solutions at 0.5 mV s^{-1} scan rate. (f) Calculated and experimental potential dependencies of R_{kin} for the case of rate-determining (de)solvation step ($K_p/k_{in} = 1/10$) for $LiMn_2O_4$ electrode. Source: Vassiliev et al. [37]/with permission of Elsevier.

vicinity of the electrode/electrolyte interface, then the intercalation process can be modeled as a charge transfer step with a preceding desolvation chemical reaction (CE mechanism):

$$LiS_n^+ \underset{k_{out}}{\overset{k_{in}}{\rightleftharpoons}} Li^+ + nS \tag{2.10}$$

$$Li^+ + [\ \ ]_{solid} + e^- \underset{k_s^{out}}{\overset{k_s^{in}}{\rightleftharpoons}} [Li]_{solid} \tag{2.11}$$

where S is the solvent molecule and n is the number of solvent molecules in solvation shell.

Under these conditions, the rate of the electrochemical step (2.11), which is described in the framework of Butler–Volmer kinetics, depends on the concentration of the desolvated Li^+ in the vicinity of the interface, and not on the Li^+ concentration in the solution bulk:

$$\begin{aligned} I(E) = SF \cdot 10^6 \cdot \frac{\rho n_{Li}}{M_r} \Bigg\{ & k_s^{out}\theta \exp\left[\frac{(1-\alpha)F(E-E_0(\theta))}{RT}\right] \\ & - k_s^{in}[Li](1-\theta)\exp\left[-\frac{\alpha F(E-E_0(\theta))}{RT}\right]\Bigg\} \end{aligned} \tag{2.12}$$

where k_s^{in}, k_s^{out} are the apparent heterogeneous rate constants for the forward and reverse electrochemical steps, and [Li] is the surface concentration of the desolvated Li^+.

For a system in equilibrium, the rates of the forward and the reverse electrochemical reactions are equal, i.e. $k_s^{out} = k_s^{in}[Li]_{eq}$. Then, Eq. (2.12) can be written in a simpler form:

$$\begin{aligned} I(E) = SF \cdot 10^6 \cdot \frac{\rho n_{Li}}{M_r} \Bigg\{ & k_s^{out} \cdot \theta \cdot \exp\left[\frac{(1-\alpha)F(E-E_0(\theta))}{RT}\right] \\ & - k_s^{out}\theta_{Li}(1-\theta)\exp\left[-\frac{\alpha F(E-E_0(\theta))}{RT}\right]\Bigg\} \end{aligned} \tag{2.13}$$

where $\theta_{Li} = \frac{[Li]}{[Li]_{eq}}$. In this case, the dimensionality of the apparent heterogeneous rate constant k_s^{out} remains conventional: (cm s^{-1}).

The rate of the chemical reaction (2.10) within the formal kinetics approach can be described by the relations:

$$v_{out} = k_{out}[Li][S]^n; \quad v_{in} = k_{in}[LiS_n^+]; \quad K_p = \frac{k_{in}}{k_{out}} = \frac{[Li][S]^n}{[LiS_n^+]} \tag{2.14}$$

where k_{in}, k_{out} are rate constants for the forward (desolvation) and reverse (solvation) reactions, respectively; [Li], [S], $[LiS_n^+]$ are concentrations of desolvated Li^+ ion, complexing agent (solvent), and solvated Li^+ ion; and K_p is the equilibrium constant of the (de)solvation reaction. Further analysis of the problem is analogous to the treatments of preceding homogeneous chemical reactions [9].

Qualitative analysis of the sequential reactions (2.10) and (2.11) allows to conclude that for the forward intercalation reaction, when the desolvated Li^+ is transferred from the interfacial solution layer into the bulk of the material and is slowly supplied by reaction (2.10), its surface concentration becomes diminished, which results in the electrochemical reaction rate decrease. In the limiting case of exceptionally slow desolvation, one could expect the appearance of the cathodic limiting current, which would be determined by the kinetics of reaction (2.10). In the reverse deintercalation process, the surface concentration of desolvated lithium would increase. However, the bulk concentration of Li^+ in solution in most cases is significantly lower than the solubility limit, which means that the number of "vacancies" for the Li^+ ion in solution exceeds the number of desolvated ions by orders of magnitude. Under these conditions, no strong influence of the slow (de)solvation process on the deintercalation reaction rate can be predicted. Numerical modeling of the CVs for different k_{in}/K_p ratios (Figure 2.9c) shows that the decrease in the desolvation rate (k_{in}) results in pronounced distortions of the cathodic branch of the CV and even in the appearance of the limiting cathodic current (e.g. for $K_p = 1$, $k_{in} = 0.01$). The deviations in the anodic part of the curve are far less pronounced.

To test whether slow desolvation can indeed account for the unusual type of R_{kin} vs. E dependencies, one can design a model system, where solvating interactions are indeed strong enough to be manifested in both the shape of the CV and impedance spectra. As Li^+ is strongly complexed by cryptand-type ligands, we employed high concentration of cryptand 2.2.2 in propylene carbonate (PC) as a model solvating medium. The high stability constant of the [Li^+Cr] complex ($\log\beta = 6.9$) results in negligible concentrations of "free" Li^+ in 0.2 M $LiClO_4$ + 0.2 M cryptand 2.2.2 solutions. Figure 2.9d shows the CV of V_2O_5 electrodes at 0.5 mV s^{-1} scan rates in the model electrolyte with and without the complexing agent. The pronounced asymmetry of the cathodic and anodic branches appears in the cryptand solution, as expected for a slow desolvation step. Moreover, similar distortions can be frequently observed in the processes of ion intercalation into Prussian blue analogues (Figure 2.9e). The observed distortions are very similar to those in the model cryptand-containing electrolyte. Based on these findings, one may conclude that the slow desolvation rate-limiting step is indeed rather probable in intercalation reactions.

Finally, if the ideas discussed are applied to treat the data for $LiMn_2O_4$ electrodes in aqueous media, it becomes clear that no diagnostic CV shape distortions occur, which indicates that the desolvation rates, though determining the intercalation kinetics, should be quite fast (e.g. the case of $k_{in} = 10$ in Figure 2.9c). If the kinetic resistances are calculated from the computed impedance spectra within the framework of the slow desolvation model, the similarity between the experimental and calculated R_{kin} vs. E dependencies is obvious (Figure 2.9f). Additionally, the impedance spectra for the systems, which demonstrate a non-Butler–Volmer type of R_{kin} vs. E dependency, cannot be approximated to an equivalent circuit with pure capacitive elements – the introduction of constant phase elements is required [48]. The calculated spectra show the characteristic distortions in the high-frequency limit, which arise from chemical reaction kinetics that closely resemble the deviations in the experimental spectra [37].

The observed surprisingly good agreement between the predictions of the slow desolvation model and the experimental spectra implies that in cases when no hindering layers form at the surface, the charge transfer step is very fast, which allows for the manifestations of other (chemical) steps of the complex intercalation process.

2.7 Concluding Remarks

Based on the presented instructive examples of different kinetic regime manifestations in the shape of voltammograms and experimental impedance spectra, it is possible to both inquire into the nature of the rate-limiting step on a qualitative level and to estimate the key kinetic parameters (the rate constant and the diffusion coefficient), which govern the rate of metal ion incorporation into various transition metal oxide materials. The basic limitations of the intercalation reactions can be reliably deduced from relatively simple voltammetric tests, which provide for the precise control of battery performance. Basic understanding of the nature of the slow step allows to precisely alter the factors, which are responsible for a given intercalation mechanism. For instance, diffusional control would require reducing the particle size, while kinetic control, which originates from the CEI layers at the oxide surface, would require minimizing the surface area of the particles and tuning the chemistry (composition and thickness of the surface layers) between the electrode material and the solvent/electrolyte salt. In more complex cases with the interference of chemical reactions, the solvent/ion interactions need to be analyzed. We believe that this type of analysis is preferential for the optimization of the performance of battery materials as compared to more empirical research routes.

References

1 Li, M., Lu, J., Chen, Z., and Amine, K. (2018). 30 Years of lithium-ion batteries. *Advanced Materials* 30 (33): 1800561.

2 Winter, M., Barnett, B., and Xu, K. (2018). Before Li ion batteries. *Chemical Reviews* 118 (23): 11433–11456.

3 Nitta, N., Wu, F., Lee, J.T., and Yushin, G. (2015). Li-ion battery materials: present and future. *Materials Today* 18 (5): 252–264.

4 Dunn, B., Kamath, H., and Tarascon, J.M. (2011). Electrical energy storage for the grid: a battery of choices. *Science* 334 (6058): 928–935.

5 Notter, D.A., Gauch, M., Widmer, R. et al. (2010). Contribution of Li-ion batteries to the environmental impact of electric vehicles. *Environmental Science and Technology* 44 (17): 6550–6556.

6 Cano, Z.P., Banham, D., Ye, S. et al. (2018). Batteries and fuel cells for emerging electric vehicle markets. *Nature Energy* 3 (4): 279–289.

7 Zhang, Z. and Zhang, S.S. (2015). *Rechargeable Batteries. Materials, Technologies and New Trends*. Switzerland: Springer International Publishing.

8 Compton, R.G. and Banks, C.E. (2010). *Understanding Voltammetry*, 2e. London: Imperial College Press.

9 Bard, A.J. and Faulkner, L.R. (2001). *Electrochemical Methods: Fundamentals and Applications*, 2e. New York: Wiley.
10 Brousse, T., Bélanger, D., and Long, J.W. (2015). To be or not to be pseudocapacitive? *Journal of the Electrochemical Society* 162 (5): A5185–A5189.
11 Augustyn, V., Simon, P., and Dunn, B. (2014). Pseudocapacitive oxide materials for high-rate electrochemical energy storage. *Energy & Environmental Science* 7 (5): 1597–1614.
12 Xie, J., Yang, P., Wang, Y. et al. (2018). Puzzles and confusions in supercapacitor and battery: theory and solutions. *Journal of Power Sources* 401: 213–223.
13 Chueh, W.C., El Gabaly, F., Sugar, J.D. et al. (2013). Intercalation pathway in many-particle $LiFePO_4$ electrode revealed by nanoscale state-of-charge mapping. *Nano Letters* 13 (3): 866–872.
14 Vassiliev, S.Y., Levin, E.E., and Nikitina, V.A. (2016). Kinetic analysis of lithium intercalating systems: cyclic voltammetry. *Electrochimica Acta* 190: 1087–1099.
15 Nikitina, V.A., Kuzovchikov, S.M., Fedotov, S.S. et al. (2017). Effect of the electrode/electrolyte interface structure on the potassium-ion diffusional and charge transfer rates: towards a high voltage potassium-ion battery. *Electrochimica Acta* 258: 814–824.
16 La Mantia, F., Wessells, C.D., Deshazer, H.D., and Cui, Y. (2013). Reliable reference electrodes for lithium-ion batteries. *Electrochemistry Communications* 31: 141–144.
17 Levi, M.D. and Aurbach, D. (2004). Impedance of a single intercalation particle and of non-homogeneous, multilayered porous composite electrodes for Li-ion batteries. *Journal of Physical Chemistry B* 108 (31): 11693–11703.
18 Levi, M.D. and Aurbach, D. (1999). Frumkin intercalation isotherm – a tool for the description of lithium insertion into host materials: a review. *Electrochimica Acta* 45 (1–2): 167–185.
19 Levi, M.D., Salitra, G., Markovsky, B. et al. (1999). Solid-state electrochemical kinetics of Li-ion intercalation into $Li_{1-x}CoO_2$: simultaneous application of electroanalytical techniques SSCV, PITT, and EIS. *Journal of the Electrochemical Society* 146 (4): 1279–1289.
20 Bai, P. and Bazant, M.Z. (2014). Charge transfer kinetics at the solid–solid interface in porous electrodes. *Nature Communications* 5: 3585.
21 Bazant, M.Z. (2013). Theory of chemical kinetics and charge transfer based on nonequilibrium thermodynamics. *Accounts of Chemical Research* 46 (5): 1144–1160.
22 Laviron, E. (1979). General expression of the linear potential sweep voltammogram in the case of diffusionless electrochemical systems. *Journal of Electroanalytical Chemistry* 101 (1): 19–28.
23 Xia, Y. (1996). An investigation of lithium ion insertion into spinel structure Li–Mn–O compounds. *Journal of the Electrochemical Society* 143 (3): 825–833.
24 Shao-Horn, Y., Levasseur, S., Weill, F., and Delmas, C. (2003). Probing lithium and vacancy ordering in O_3 layered Li_xCoO_2 ($x \approx 0.5$). *Journal of the Electrochemical Society* 150 (3): A366–A373.

25 Reimers, J.N. and Dahn, J.R. (1992). Electrochemical and in situ X-ray diffraction studies of lithium intercalation in Li_xCoO_2. *Journal of the Electrochemical Society* 139 (8): 2091–2097.
26 Chen, W., Tang, J., and Xia, X.-H. (2009). Composition and shape control in the construction of functional nickel hexacyanoferrate nanointerfaces. *Journal of Physical Chemistry C* 113 (52): 21577–21581.
27 Nikitina, V.A., Fedotov, S.S., Vassiliev, Y.S. et al. (2017). Transport and kinetic aspects of alkali metal ions intercalation into $AVPO_4F$ framework. *Journal of the Electrochemical Society* 164 (1): A6373–A6380.
28 Montella, C. (1999). Discussion on permeation transients in terms of insertion reaction mechanism and kinetics. *Journal of Electroanalytical Chemistry* 465 (1): 37–50.
29 Montella, C. (2006). Apparent diffusion coefficient of intercalated species measured with PITT. *Electrochimica Acta* 51 (15): 3102–3111.
30 Diard, J.P., Le Gorrec, B., and Montella, C. (2001). Influence of particle size distribution on insertion processes in composite electrodes. Potential step and EIS theory. *Journal of Electroanalytical Chemistry* 499 (1): 67–77.
31 Britz, D. (2005). *Digital Simulation in Electrochemistry*. Berlin Heidelberg New York: Springer.
32 Khandelwal, A., Hariharan, K.S., Senthil Kumar, V. et al. (2014). Generalized moving boundary model for charge–discharge of $LiFePO_4$/C cells. *Journal of Power Sources* 248: 101–114.
33 Park, M., Zhang, X., Chung, M. et al. (2010). A review of conduction phenomena in Li-ion batteries. *Journal of Power Sources* 195 (24): 7904–7929.
34 Levin, E.E., Vassiliev, S.Y., and Nikitina, V.A. (2017). Solvent effect on the kinetics of lithium ion intercalation into $LiCoO_2$. *Electrochimica Acta* 228: 114–124.
35 Nikitina, V.A., Zakharkin, M.V., Vassiliev, S.Y. et al. (2017). Lithium ion coupled electron-transfer rates in superconcentrated electrolytes: exploring the bottlenecks for fast charge-transfer rates with $LiMn_2O_4$ cathode materials. *Langmuir* 33 (37): 9378–9389.
36 Schmickler, W. (1995). A unified model for electrochemical electron and ion transfer reactions. *Chemical Physics Letters* 237 (1–2): 152–160.
37 Vassiliev, S.Y., Sentyurin, V.V., Levin, E.E., and Nikitina, V.A. (2019). Diagnostics of lithium-ion intercalation rate-determining step: distinguishing between slow desolvation and slow charge transfer. *Electrochimica Acta* 302: 316–326.
38 Thomas, M.G.S.R., Bruce, P.G., and Goodenough, J.B. (1985). AC impedance analysis of polycrystalline insertion electrodes: application to $Li_{1-x}CoO_2$. *Journal of the Electrochemical Society* 132 (7): 1521–1528.
39 Peled, E. and Menkin, S. (2017). Review – SEI: past, present and future. *Journal of the Electrochemical Society* 164 (7): A1703–A1719.
40 Gauthier, M., Carney, T.J., Grimaud, A. et al. (2015). Electrode-electrolyte interface in Li-ion batteries: current understanding and new insights. *Journal of Physical Chemistry Letters* 6 (22): 4653–4672.

41 Fedotov, S.S., Samarin, A.S., Nikitina, V.A. et al. (2018). Reversible facile Rb^+ and K^+ ions de/insertion in a $KTiOPO_4$-type $RbVPO_4F$ cathode material. *Journal of Materials Chemistry A* 6 (29): 14420–14430.

42 Fedotov, S.S., Khasanova, N.R., Samarin, A.S. et al. (2016). $AVPO_4F$ (A = Li, K): a 4 V cathode material for high-power rechargeable batteries. *Chemistry of Materials* 28 (2): 411–415.

43 Chihara, K., Katogi, A., Kubota, K., and Komaba, S. (2017). $KVPO_4F$ and $KVOPO_4$ toward 4 volt-class potassium-ion batteries. *Chemical Communications* 53 (37): 5208–5211.

44 Wynn, D., Roth, M.M., and Pollard, B.D. (1984). The solubility of alkali-metal fluorides in non-aqueous solvents with and without crown ethers, as determined by flame emission spectrometry. *Talanta* 31 (11): 1036–1040.

45 Nakayama, N., Nozawa, T., Iriyama, Y. et al. (2007). Interfacial lithium-ion transfer at the $LiMn_2O_4$ thin film electrode/aqueous solution interface. *Journal of Power Sources* 174 (2): 695–700.

46 Yamada, I., Miyazaki, K., Fukutsuka, T. et al. (2015). Lithium-ion transfer at the interfaces between $LiCoO_2$ and $LiMn_2O_4$ thin film electrodes and organic electrolytes. *Journal of Power Sources* 294: 460–464.

47 Abe, T., Fukuda, H., Iriyama, Y., and Ogumi, Z. (2004). Solvated Li-ion transfer at interface between graphite and electrolyte. *Journal of the Electrochemical Society* 151 (8): A1120–A1123.

48 Lasia, A. (2014). *Electrochemical Impedance Spectroscopy and Its Applications.* Switzerland: Springer.

3

Next-Generation Cobalt-Free Cathodes – A Prospective Solution to the Battery Industry's Cobalt Problem*

Nitin Muralidharan[1], *Ethan C. Self*[2], *Jagjit Nanda*[2], *and Ilias Belharouak*[1]

[1] *Oak Ridge National Laboratory, Electrification and Energy Infrastructures Division, Oak Ridge, TN 37831, USA*
[2] *Oak Ridge National Laboratory, Chemical Sciences Division, Oak Ridge, TN 37831, USA*

This manuscript has been authored by UT-Battelle, LLC, under Contract No. DE-AC0500OR22725 with the U.S. Department of Energy. The United States Government retains and the publisher, by accepting the article for publication, acknowledges that the United States Government retains a non-exclusive, paid-up, irrevocable, world-wide license to publish or reproduce the published form of this manuscript, or allow others to do so, for the United States Government purposes. The Department of Energy will provide public access to these results of federally sponsored research in accordance with the DOE Public Access Plan (http://energy.gov/downloads/doe-public-access-plan).

3.1 Introduction

Since their commercial debut in 1991, lithium-ion batteries (LIBs) have enabled revolutionary advances in a range of applications, including portable electronics, grid storage, and electric vehicles (EVs) [1, 2]. Academic and industrial R&D on LIBs largely focuses on next-generation systems with higher energy density and lower cost. Recently, the global mobility revolution has accelerated EV demands with projections as high as 200 million EVs on the road by 2050 [3, 4]. Considering that profitability of battery industries is hedged on economies of scale, the companies that can effectively secure battery material supply chains are poised to dominate this market space for decades to come. In modern LIBs, the cathode accounts for about 30% of the overall battery cost [2], and mainstream LIB cathodes (e.g. $LiNi_xMn_yCo_{1-x-y}O_2$ [NMC] and $LiNi_{0.8}Co_{0.15}Al_{0.05}O_2$ [NCA]) contain substantial amounts of cobalt. The cost of cobalt has nearly tripled in recent years

* Nitin Muralidharan and Ethan C. Self are equally contributed.

*This chapter was originally published in Advanced Energy Materials, 2022, Volume 12, Issue 9; DOI: 10.1002/aenm.202103050. Reproduced with permission of WILEY-VCH GmbH.

Transition Metal Oxides for Electrochemical Energy Storage, First Edition.
Edited by Jagjit Nanda and Veronica Augustyn.

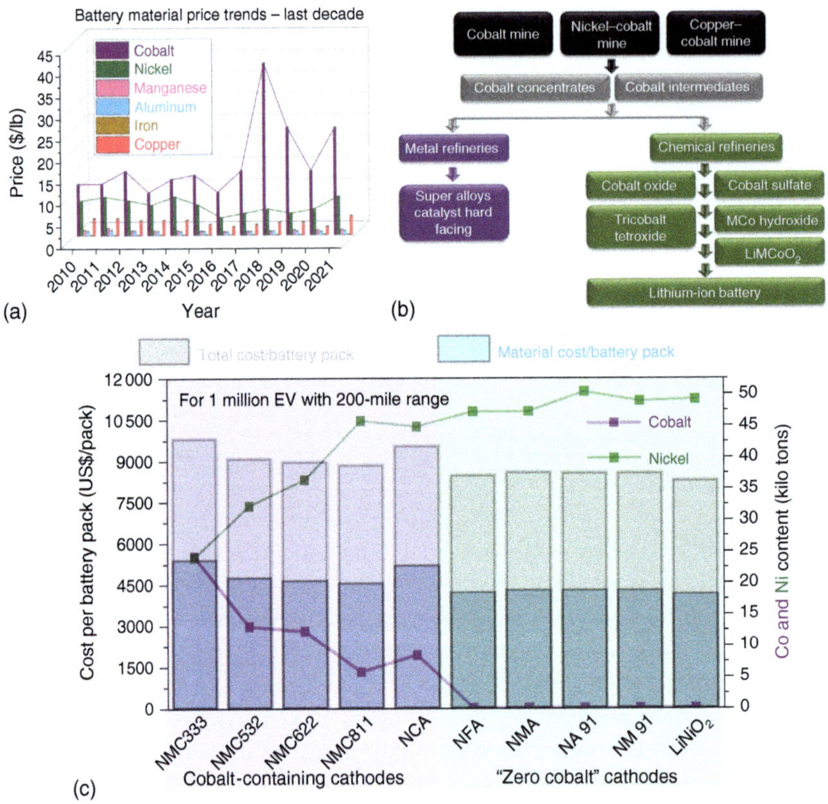

Figure 3.1 (a) Trends in the prices of battery materials over 2010–2021, (b) Cobalt market flow chart, and (c) BatPaC model highlighting the potential of cobalt-free cathode materials for low-cost LIBs.

(see Figure 3.1a) due to increased demand and supply chain constraints, which creates a major problem for the LIB industry. Notably, cobalt prices are nearly 60% higher than nickel, the second most widely used cathode constituent. Recently, the Cobalt Development Institute (CDI) reported that ~58% of global cobalt production is used in diverse applications, including super alloys, catalysts, magnets, and pigments (see Figure 3.1b) [5]. For battery manufacturing, Co availability is crucial to meet ever-growing demands. Moreover, the US Department of Energy's goal for EVs includes reducing LIB costs to <US\$80 kWh^{-1}, a target that can only be met if Co loading is lowered to <50 mg Wh^{-1} at the cell level. Even if new cathode compositions that reduce or eliminate Co are developed, global battery-manufacturing infrastructures are not set up to handle such adoptions without major facility overhauls. Therefore, next-generation low Co/Co-free cathodes should be designed to remain compatible with existing manufacturing lines.

This chapter highlights several classes of low Co/Co-free cathodes, including layered, spinel, olivine, and disordered rocksalt (DRX) structures. Challenges related to large-scale manufacturing of these materials are also discussed. Overall, the goal of this work is to provide readers with a comprehensive understanding of challenges and opportunities for next-generation LIBs based on low Co/Co-free cathodes.

3.2 Potential of Cobalt-Free Cathode Materials

It is unrealistic to expect LIBs with cobalt-based cathodes to meet EV market projections in the coming decades. To illustrate this point, consider the following scenario: to produce 1 million EVs each with a 200-mile range, ~10 kiloton cobalt is required if all EVs housed LIBs containing NCA cathodes with 15% Co (see Figure 3.1c). Furthermore, if NMC111 cathodes are utilized, then the amount of cobalt required increases to more than 20 kilotons. With known global cobalt reserves estimated at ~7 million tons, a projection of 200 million EV on the road by 2050 results in exhaustion of global battery specific cobalt resources if NMC111 cathodes are used. It is thus unrealistic to rely on such a rapidly depleting raw material to meet future EV demands. Moreover, it is also cost-prohibitive to the battery-manufacturing industry to acquire this amount of cobalt, especially at a time of rapid fluctuations in cobalt prices. Transitioning to cobalt-free cathodes also offers significant cost savings for the battery-manufacturing industry compared to mainstream cobalt-containing cathode materials (NCA and NMC). In the same proposed scenario, battery packs with the 15% cobalt containing NCA cathodes would be ~US$10 000 (material costs ~US$5000), whereas packs containing a novel cobalt-free alternative would be ~US$8500 (material costs ~US$3500), providing ~15% cost savings. Thus, it is evident that under these scenarios, LIBs containing cobalt-free cathodes would be able to solve the battery industry's cobalt problem and power millions of future EVs. Figure 3.2a,b depicts the crystal structures and representative voltage profiles of several prominent LIB cathodes. Sections 3.3–3.5 describe the performance and technical challenges of several classes of Co-free cathodes, including layered, spinel, olivine, and DRX systems [2, 8, 9].

3.3 Layered Cathodes

3.3.1 Conventional Layered Cathodes

Layered cathode materials are arguably the most commercially feasible due to their ease of manufacturing and handling. Lithium cobalt oxide ($LiCoO_2$ or LCO) was the first layered transition metal oxide to be commercialized and is still used in many applications (theoretical capacity, 274 mAh g^{-1}). As discussed previously, the high cost of cobalt led to the pursuit of lithium nickel oxide ($LiNiO_2$ or LNO). However, this material is difficult to synthesize and suffers from structural changes that lead to poor cyclability. Moreover, due to the similar sizes of Li^+ and Ni^{2+} ions, intermixing of these cations can also occur during phase formation resulting in ion diffusion bottlenecks that lower the material's practical capacity. Over the years, various metal substitutions, including aluminum, cobalt, and/or manganese, were

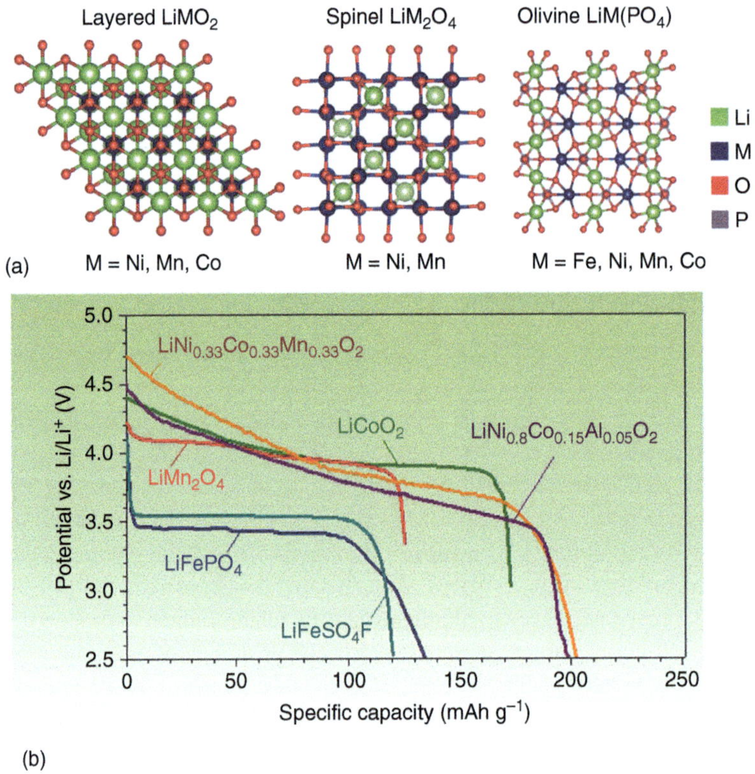

Figure 3.2 (a) Crystallographic structures of conventional Li-ion cathodes, including layered (space group *R*-3*m*), spinel (*Fd*-3*m*), and olivine (space group *Pnma*) materials, as viewed along the *a*-axis. (b) Voltage profiles for conventional Li-ion cathodes. Crystallographic structures were prepared using VESTA software [7]. Source: (a) Based on Momma and Izumi [7]; (b) Nitta et al. [6]/Reprinted with permission of Elsevier.

found to significantly improve the stability and electrochemical performance of LNO [10, 11]. Although aluminum is electrochemically inactive, adding it to the material along with cobalt reduces the amount of cationic mixing in the Li-slab and stabilizes the layered structure, preventing phase transitions and O_2 release that can cause thermal runaway. Sections 3.3.2 and 3.3.3 detail the structure and electrochemical properties of various layered cathodes containing one or more transition metal substitutions.

3.3.2 Binary Layered Ni-Rich Cathode Materials

When considering binary systems for layered cobalt-free cathode materials, manganese and aluminum substitution are an obvious choice. Manganese substitution in LNO systems has been widely reported in the literature. It is also widely reported that Mn doping enhances the thermal stability of the cathode, especially in the highly delithiated state. The substituted Mn atoms in their tetravalent Mn^{4+} are electrochemically inactive and thus mitigate the Jahn–Teller distortions that plague the

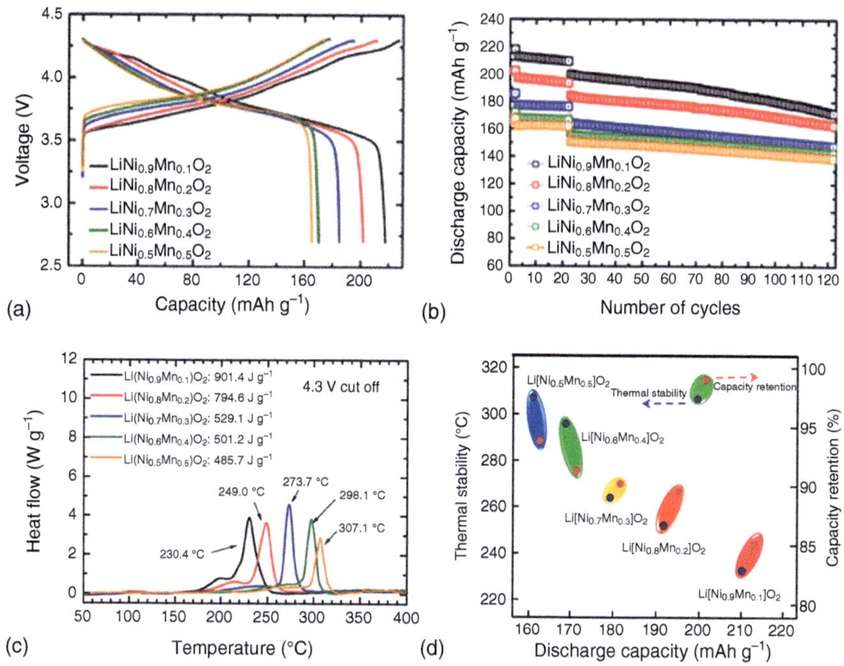

Figure 3.3 (a) Galvanostatic charge/discharge profiles of the Mn-doped LNO system, (b) cycling performance of Mn-doped LNO system, (c) heat flow profiles of Mn-doped LNO, and (d) thermal stability variation with discharge capacity for the Mn-LNO compositions investigated. Source: Sun et al. [12]/with permission of American Chemical Society.

LNO structure during electrochemical cycling. Despite stabilizing the structure, Mn substitutions have been known to enhance the cation mixing in the LNO system thus leading to decreased capacity. Several groups have reported enhanced cycling and thermal stabilities with the incorporation of Mn atoms in a wide range of compositions – $LiNi_{0.5}Mn_{0.5}O_2$, $LiNi_{2/3}Mn_{1/3}O_2$, and $LiNi_{0.9}Mn_{0.1}O_2$. Y.K. Sun et al. [12] systematically investigated Mn doping of LNO system in a range of compositions ($LiNi_yMn_{1-y}O_2$, between $0.5 \leq y \leq 0.9$) as shown in Figure 3.3.

Galvanostatic charge/discharge profiles of Co-free Mn-substituted LNO are shown in Figure 3.3. Notably, as the Mn content increases, the cycling stability improves at the expense of lower discharge capacities (e.g. 212 mAh g^{-1} [10% Mn] vs. 164 mAh g^{-1} [50% Mn]) since Mn^{4+} is not electrochemically active. Y.K. Sun et al. also reported the thermal stabilities of the Mn-LNO system in a delithiated state (Figure 3.3c,d). Cathodes with low Mn content generated less heat during thermal decomposition (901.4 J g^{-1} for 10% Mn vs. 485.7 J g^{-1} for 50% Mn). The authors speculated that the 10% Mn sample's delayed onset of exothermic reaction was a result of spinel phase formation that improved the material's structural integrity. Despite these promising results, further systematic studies are required prior to commercial deployment of these cathode materials.

Mg and Al substitutions have also been widely reported to improve the performance of LNO cathodes [13]. It has been shown that Mg substitutions mitigate

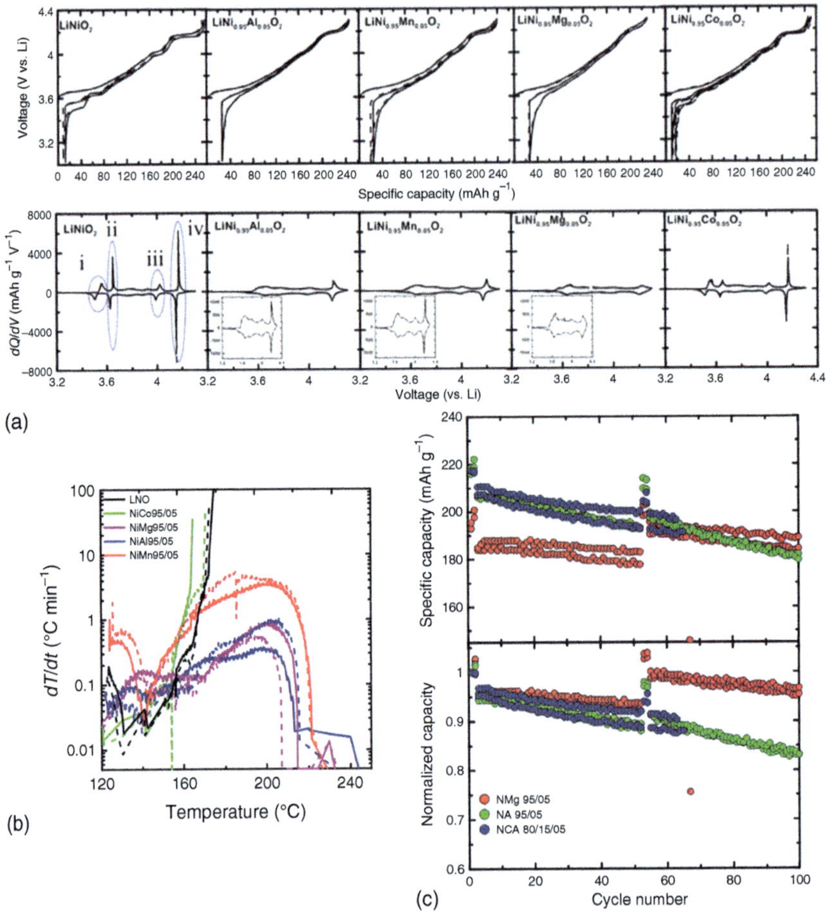

Figure 3.4 (a) Charge/discharge profiles (i) and *dQ/dV* plots (ii) showing the electrochemical behavior of doped LNO systems, (b) accelerated rate calorimetry tests showing the potential thermal stability benefits of dopants for the LNO system, (c) charge/discharge cycling performance assessment of the doped LNO compared with commercial NCA. Source: Li et al. [14].

particle cracking during cycling, thereby enhancing cycling stability. H. Li et al. [14] systematically investigated Al ($LiNi_{0.9}Al_{0.1}O_2$)-, Mg ($LiNi_{0.9}Mg_{0.1}O_2$)-, and Mn ($LiNi_{0.9}Mn_{0.1}O_2$)-doped LNO systems (Figure 3.4a–c). They observed that the addition of Al, Mn, and Mg substituents suppressed detrimental phase transitions during cycling as evidenced through the *dQ/dV* plots shown in Figure 3.4a.

Through accelerated rate calorimetry tests (Figure 3.4b), it was determined that Mn, Mg, and Al substitution also improve the thermal stability of LNO. It was observed that the self-heating rates of delithiated LNO and 5% Co-doped LNO rapidly increased beyond 160 °C, triggering subsequent thermal runaway. On the other hand, delithiated LNO doped with 5% Al, Mg, or Mn did not reach the self-heating rate threshold throughout the investigated temperatures. These results clearly show that appropriate dopants can substantially lower the reactivity of layered nickel-rich cathodes.

Charge/discharge cycling tests of the Al-, Mn-, and Mg-doped LNO systems (Figure 3.4c) when compared to the commercial 15% Co-doped NCA material revealed that comparatively better capacity retention was achieved with the Al-doped LNO system when the upper cut-off voltage was limited to 4.3 V. Despite the promise of binary cobalt-free nickel-rich layered systems, practical challenges related to scalability, processability, and safety warrant further investigations. As such, research efforts are increasingly aimed at developing these binary Co-free cathodes for commercial applications.

3.3.3 Ternary Layered Ni-Rich Cathode Materials

In recent years, Co-free analogues to NCM and NCA have also been developed. The motivation for developing such materials stems from the expectation that multiple metal substitutions may resolve multiple issues (e.g. capacity fade, thermal stability, and cation mixing) simultaneously. In this context, our team is pursuing several R&D thrusts on rapid development and commercial deployment of $LiNi_xFe_yAl_zO_2$ (where $x \geq 0.8, x+y+z=1$, denoted as NFA) [15, 16]. Our approach was motivated based on the premise that although high nickel content generally yields high capacity, the electrochemical performance can be significantly improved by replacing some of the nickel with small amounts of trivalent aluminum and iron, which have similar ionic radii to Ni^{3+} (0.54 Å for Al^{3+} and 0.55 Å for Fe^{3+} compared to 0.56 Å for Ni^{3+}) and give rise to better structural stability, less cationic disorder, enhanced safety, and improved cycle life. A schematic showing the benefits of iron and aluminum incorporation and the resulting crystal structure (rhombohedral, with space group $R\text{-}3m$, corresponding to an α-$NaFeO_2$-like crystal) of several NFA variants is shown in Figure 3.5a,b, respectively. It is widely reported that nickel-rich cathode systems suffer from cation-mixing issues owing to the similarities in the ionic radii of Li^+ and Ni^{2+}. Through Rietveld refinement analysis of the neutron diffraction patterns, it was determined that the extent of cation mixing or antisite defect formation was ~4% in the synthesized NFA variants.

For upscaling the NFA cathode materials, we employed the coprecipitation process as it is usually the method of choice from an industrial standpoint. Figure 3.5c shows a schematic representation of the coprecipitation process employed using continuous stirred-tank reactors (CSTR) for upscaling NFA cathode material, which are initially synthesized in the precursor form (hydroxides or carbonates based on the choice of reagents). A typical coprecipitation process involves addition of transition metal reagents (usually metal sulfates, carbonates, or chlorides) and a chelating agent (e.g. NH_4OH) to the CSTR while controlling stirring speed (500–1000 rpm) and temperature (>50 °C). NaOH is generally pumped into the reactor to maintain the pH in the coprecipitation zone (usually >10 for Ni-rich cathode precursors). Figure 3.5d shows the precipitation behavior (variation in pH profiles) of the individual transition metal reagents along with the coprecipitation behavior of the NFA precursor as a function of NaOH content. These findings suggest that coprecipitation

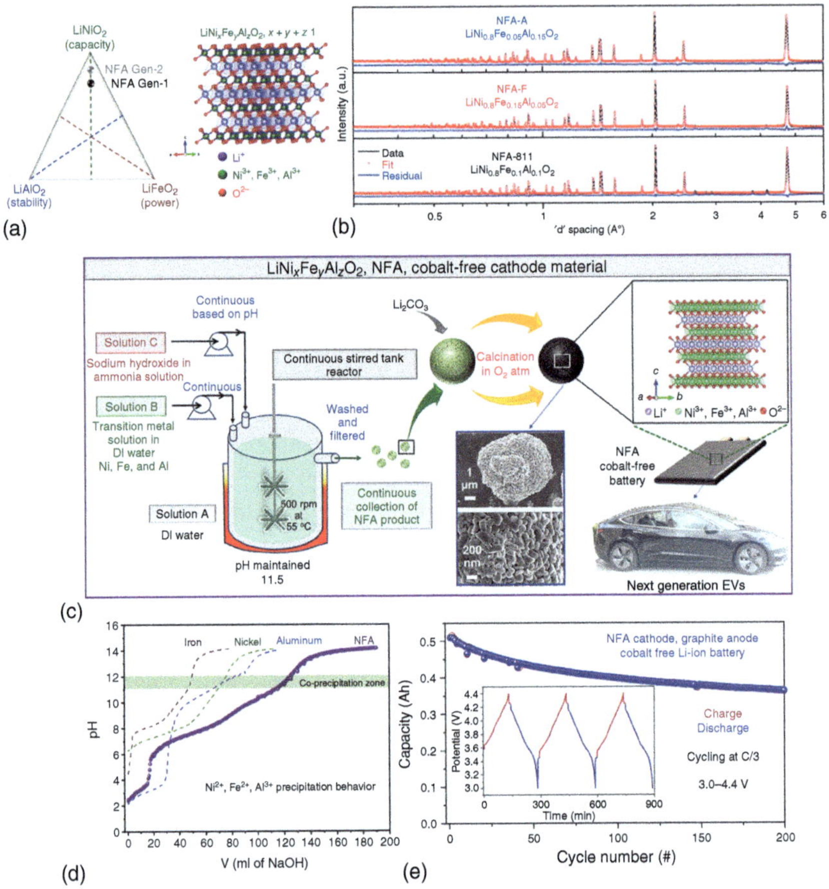

Figure 3.5 (a) Schematic showing the benefits of the NFA class with Al and Fe incorporation, (b) neutron powder diffraction patterns showing the *R*-3*m* crystal structure of the NFA compositional variants, (c) schematic representation of the upscaling process for the NFA cathode system, (d) precipitation behavior of iron, nickel, aluminum, and the NFA system, and (e) cycling performance of a cobalt-free LIB enabled by the NFA class of cobalt-free cathode. Source: (a, b) Muralidharan et al. [16]/with permission of John Wiley & Sons; (c–e) Muralidharan et al. [15]/with permission of Elsevier.

of Ni^{2+}, Fe^{2+}, and Al^{3+} is challenging due to differences in the solubility product K_{sp} of the respective hydroxides. More specifically, aluminum hydroxides start to precipitate at lower pH values compared to nickel and iron hydroxides, which usually precipitate at pH values >6, thus narrowing the co-precipitation window. Based on these inferences, we fixed a pH window between 10 and 12 as the coprecipitation window for the successful synthesis of NFA hydroxide precursors. The synthesized precursor outflow from the reactor is filtered, washed, and dried following which is then mixed with a suitable lithium source (LiOH, Li_2CO_3, $LiNO_3$, etc.) prior to calcination in oxygen-rich atmospheres at elevated temperatures (>600 °C). Following calcination, NFA electrodes were fabricated using the slot-die coating process, and cobalt-free LIB pouch cells were assembled and tested. The pouch cells were cycled between 3 and 4.4 V at $C/3$ following initial formation cycle protocols. The assembled cobalt-free 0.5 Ah LIB pouch cell enabled by the NFA cathode exhibited reasonable capacity retention, ~72% after 200 continuous charge/discharge cycles ($C/3$) at a broader voltage window of 3–4.4 V (see Figure 3.5e). The obtained results broadly highlight the feasibility and the potential of the NFA layered system for use in next-generation cobalt-free LIBs.

Several other types of such cobalt-free ternary systems have been reported in literature during the past few years. Successful cathode classes include $LiNi_xMn_yMg_zO_2$ (NMM) [17], $LiNi_xMn_yTi_zO_2$ (NMT) [17], and $LiNi_xMn_yAl_zO_2$ (NMA) [18] (in all cases, $x+y+z=1$ and $x>0.60$), which show high capacities >200 mAh g^{-1} when operating with an upper cut-off voltage of 4.5 V (Figure 3.6a,b). Cycling performance assessment (Figure 3.6c) of these novel cathode classes reveals that these cobalt-free systems demonstrate high-capacity retentions >80% after 100 cycles under these conditions. Owing to the recent discovery of these ternary Co-free cathode, there is insufficient understanding of the thermal stabilities of these systems. W. Li et al. [18] recently reported the thermal characteristics of delithiated nickel-rich NMA cathode (NMA-89) using differential scanning calorimetry measurements. It was observed that the NMA class outperforms other cobalt-containing cathode variants owing to the incorporation of Mn and Al. Overall, a wide range of Co-free binary and ternary layered cathodes have been developed, which show great promise for next-generation LIBs with improved energy density and lower cost.

3.4 Spinel and Olivine Cathodes

Co-free lithium metal oxides with spinel structures (LiM_2O_4, M = Mn and/or Ni) have been extensively studied as alternatives to layered compositions (e.g. LCO, NMC, and NCA) [9, 19, 20]. The spinel structure (see Figure 3.1a) is related to that of layered α-$NaFeO_2$, with both classes containing cubic close-packed oxygen anions but different cation arrangements among the octahedral and tetrahedral sites [21]. First reported by Thackeray, $LiMn_2O_4$ cathodes have reversible capacities ~140 mAh g^{-1} (corresponding to $0<x<0.8$ in $Li_{1-x}Mn_2O_4$) and specific energies ~450 Wh kg^{-1} (see Figure 3.2b) [22–24]. Partial substitution of Mn with Ni (i.e. $LiNi_{1/2}Mn_{3/2}O_4$, LNMO) enables one to leverage $Ni^{2+/3+/4+}$ redox centers, resulting in a high-voltage cathode with a nominal capacity ~130 mAh g^{-1} and average potential ~4.7 V vs. Li/Li$^+$ [19, 21]. These properties make LNMO very attractive for high-energy battery applications, but a primary limitation of such systems is the lack of compatible electrolytes (e.g. oxidative decomposition of carbonate-based electrolytes at potentials >~4.5 V vs. Li/Li$^+$), which leads to formation of an unstable cathode/electrolyte interface at high states of charge. Other issues for spinel cathodes include Mn^{2+} dissolution in the electrolyte and structural degradation caused by Jahn–Teller distortions at high degrees of discharge.

Beyond layered and spinel structures, phospho-olivine cathodes with the general formula $LiTMPO_4$ (TM = Fe, Mn, Ni, and/or Co) represent the third class of

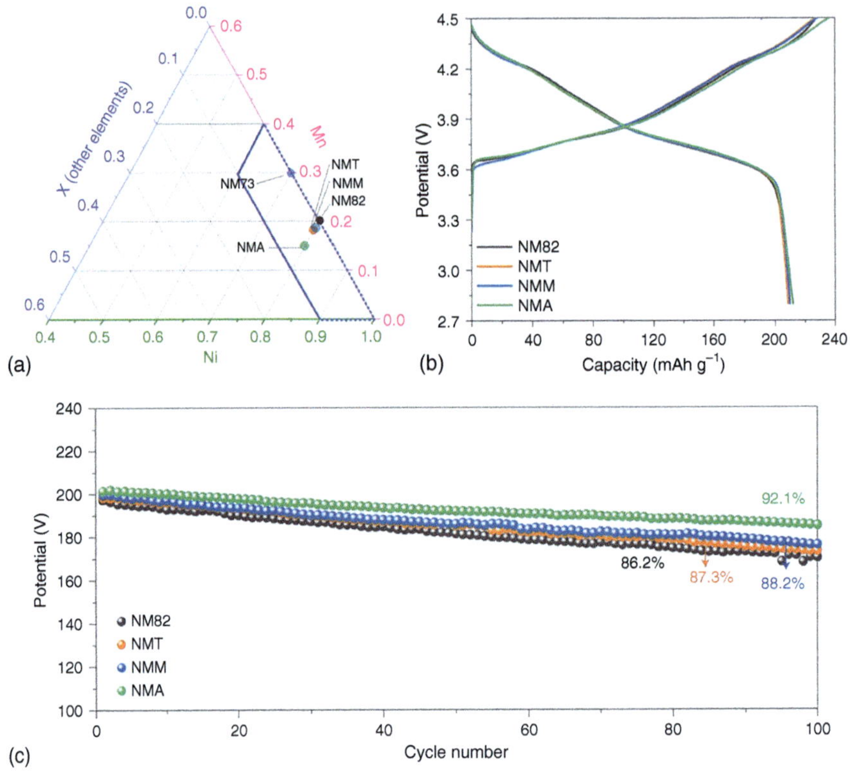

Figure 3.6 (a) Phase diagram representing potential cobalt-free cathode materials. (b) Charge/discharge profiles of common cobalt-free cathode materials reported in literature NM82, NMT, NMM, and NMA in the voltage range of 2.8–4.5 V (0.1 C). (c) Cycling performance of the cathode materials at C/3. Source: Liu et al. [17]. With permission of Springer Nature.

traditional Li-ion cathodes. The PO_4^{3-} polyanion lowers the energy of the TM redox couple, resulting in higher voltages compared to oxide cathodes based on the same TM. The most prominent olivine cathode, $LiFePO_4$ (LFP), was first reported by Goodenough, Padhi, and coworker in 1997 [25]. Providing charge compensation through an Fe^{2+}/Fe^{3+} redox couple, LFP has a practical reversible capacity of 170 mAh g^{-1}, albeit at a modest operating voltage (3.4 V vs. Li/Li$^+$), which results in a lower specific energy compared to layered and spinel cathodes (see Figure 3.1b). Due to the low electronic conductivity of olivine materials, carbon coatings are required to obtain high-power LFP cathodes [26, 27]. This finding was critical to enable commercialization of LFP cathodes in several applications, including electric bus battery packs, power tools, and grid-scale storage. In addition to LFP, studies have demonstrated reversible cycling of high-voltage $LiCoPO_4$, $LiMnPO_4$, and $LiNiPO_4$ cathodes with operating potentials up to 4.8 V vs. Li/Li$^+$ [28–30]. However, similar to voltage LNMO spinel cathodes, widespread adoption of these materials has largely been hindered by unstable cathode/electrolyte interfaces.

3.5 Disordered Rocksalt (DRX) Cathodes

Overcharging traditional Li-ion cathodes results in capacity and voltage fade due to: (i) irreversible O_2 evolution and (ii) TM migration to Li^+ vacancies, which impedes Li^+ transport [31]. As such, a widely held belief in the community has been that high degrees of Li/TM ordering are required to produce cathodes with high specific energy and cycle life [24].

Recent reports on cation DRX cathodes have challenged the conventional wisdom that cation ordering is a prerequisite for high-performance Li-ion cathodes. DRX and layered $LiTMO_2$ cathodes are described by two different forms of the rocksalt structure based on their cation ordering. While layered compounds contain alternating TM and Li layers in the (111) direction (α-$NaFeO_2$ structure, see Figure 3.2a), DRX cathodes contain a random Li/TM arrangement (α-$LiFeO_2$ structure, see Figure 3.7a). DRX systems therefore have a broader distribution of local bonding environments, which results in unique Li^+ diffusion pathways not accessible in layered systems (Figure 3.7b). Computational studies have shown that Li-excess DRX cathodes (i.e. $Li_{1+x}TM_{1-x}O_2$) contain percolating 0-TM channels, which enable long-range Li^+ diffusion through the active material [32]. These predictions have been supported by promising electrochemical performance for a wide range of DRX compositions (see Figures 3.7c and 3.8). While DRX phases typically have moderate operating potentials (c. 2.5–3.5 V vs. Li/Li^+), their high capacities (up to 400 mAh g^{-1}) result in specific energies, which meet/exceed that of traditional Li-ion cathodes.

While most oxide cathodes rely on Mn, Ni, and/or Co redox centers, DRX systems are compatible with a vast combination of TMs. The formation of ordered vs. disordered $LiTMO_2$ structures is largely driven by electronic configuration of the TMs. Cations with no d-orbital electrons (e.g. V^{5+}, Ti^{4+}, Zr^{4+}, Nb^{5+}, and Mo^{6+}) have low-energy penalties for MO_6 octahedral distortions, and as such, DRX phases generally contain a mixture of d^0 cations (to promote cation disorder) and redox-active TMs (e.g. Mn, Ni, and V) for charge compensation [31]. Some notable DRX cathodes without a d^0 cation (e.g. ball-milled $LiFeO_2$ and $LiTiO_2$) [37] have also been reported. This broad compositional landscape makes DRX systems excellent candidates for Co-free cathodes, which utilize low-cost TMs that are incompatible with conventional layered, spinel, and olivine structures. Furthermore, the disordered nature of DRX cathodes results in small, isotropic volume changes during Li extraction/reinsertion, which may reduce strain and cracking during extended cycling [31].

DRX cathodes are much less mature than conventional Li-ion cathodes, and several challenges must be overcome before DRX cathodes are viable for commercial adoption. DRX phases are typically synthesized using energy-intensive mechanochemical and solid-state reactions, which are not readily scalable and provide little control over particle morphology. Therefore, developing alternate synthesis pathways (e.g. coprecipitation) to produce high-quality DRX cathodes is critical. Additional research is also needed to better understand the redox mechanisms of these materials. For example, several DRX cathodes have demonstrated

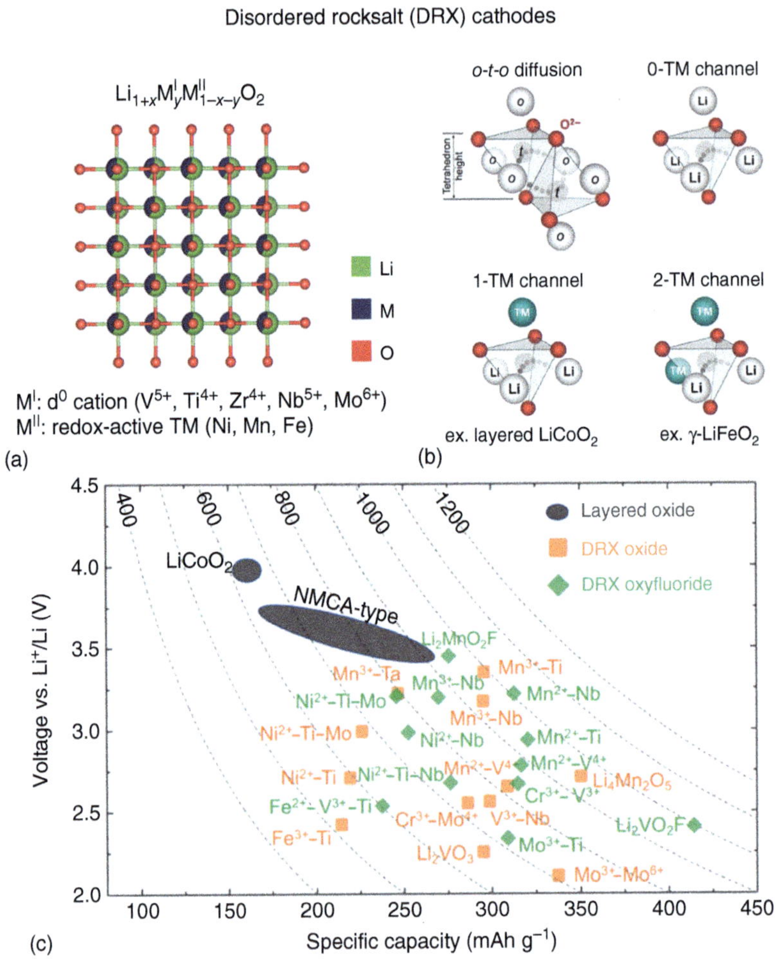

Figure 3.7 (a) Crystallographic structure of DRX cathodes in which Li^+ and TM cations are randomly distributed (generated using VESTA software [7]). The majority of reported DRX cathodes contain one or more d^0 cations (M^{I}) to promote disorder and a redox-active TM cation (M^{II}) for charge compensation. (b) Various local structures in DRX cathodes wherein Li^+ can diffuse between adjacent MO_6 octahedra through an intermediate tetrahedral site. (c) Summary of electrochemical performance of several DRX cathodes and their comparison with conventional layered systems. Dashed lines show the active material's specific energy. Source: (a) Based on Momma and Izumi [7]; (b) Lee et al. [32]/with permission of AAAS; (c) Clément et al. [31]/Republished with permission of Royal Society of Chemistry.

high capacity by leveraging anionic charge compensation, but oxygen evolution typically compromises the cathode's reversibility. New design strategies (e.g. partial fluorine substitution [31, 38]) may provide a means to mitigate performance degradation. Other limitations of DRX cathodes include: (i) the low electronic conductivity of DRX phases, which requires excessive carbon in composite electrodes (e.g. ~20 wt% compared to ~5 wt% carbon in commercial Li-ion cathodes), (ii) the materials' sloping voltage profiles, which are related to the degree of cation disorder [39], and (iii) the need for new electrolyte formulations to stabilize the cathode/electrolyte interface at high states of charge. Despite these challenges, DRX

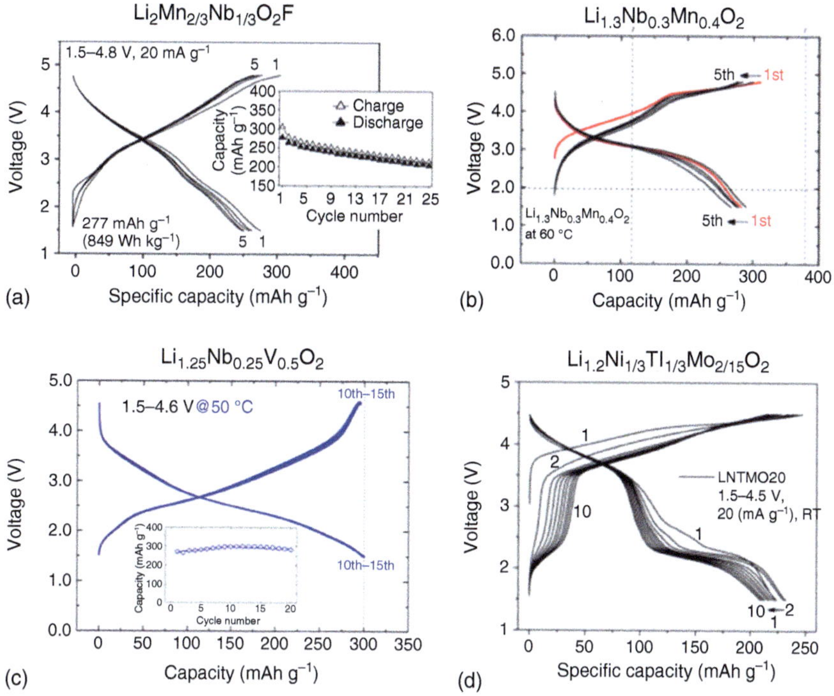

Figure 3.8 Voltage profiles for several DRX cathodes including (a) $Li_2Mn_{2/3}Nb_{1/3}O_2F$. Source: Lee et al. [33]/Reprinted with permission of Springer Nature. (b) $Li_{1.3}Nb_{0.3}Mn_{0.4}O_2$. Source: Yabuuchi et al. [34]/Reprinted with permission of Proceedings of the National Academy of Sciences of the United States of America. (c) $Li_{1.25}Nb_{0.25}V_{0.5}O_2$. Source: Nakajima and Yabuuchi [35]/Reprinted with permission of American Chemical Society. (d) $Li_{1.2}Ni_{1/3}Ti_{1/3}Mo_{2/15}O_2$. Source: Lee et al. [36]/Reprinted with permission of Royal Society of Chemistry.

cathodes have outstanding potential for high-energy LIBs due to their compositional flexibility, low-cost, and high capacity.

3.6 Challenges in Commercial Adoption of New Cobalt-Free Chemistries

Despite the multitude of cobalt-free cathode options available, upscaling and commercial deployment require facile integration into existing manufacturing processes [40]. LIB production involves two key manufacturing industries with unique challenges: (i) cathode-manufacturing industry and (ii) battery-assembling industry. Figure 3.9a shows a schematic representation of a typical industrial-scale cathode-manufacturing process. The coprecipitation process using CSTR is a widely adopted technique for the large-scale manufacturing of cathode material precursors, which are then converted to their final forms through further processing steps such as milling and calcination. Figure 3.9b depicts electrode fabrication and battery assembly approaches widely employed in battery-manufacturing industries. Large-scale manufacturing of cobalt-free cathode materials that can

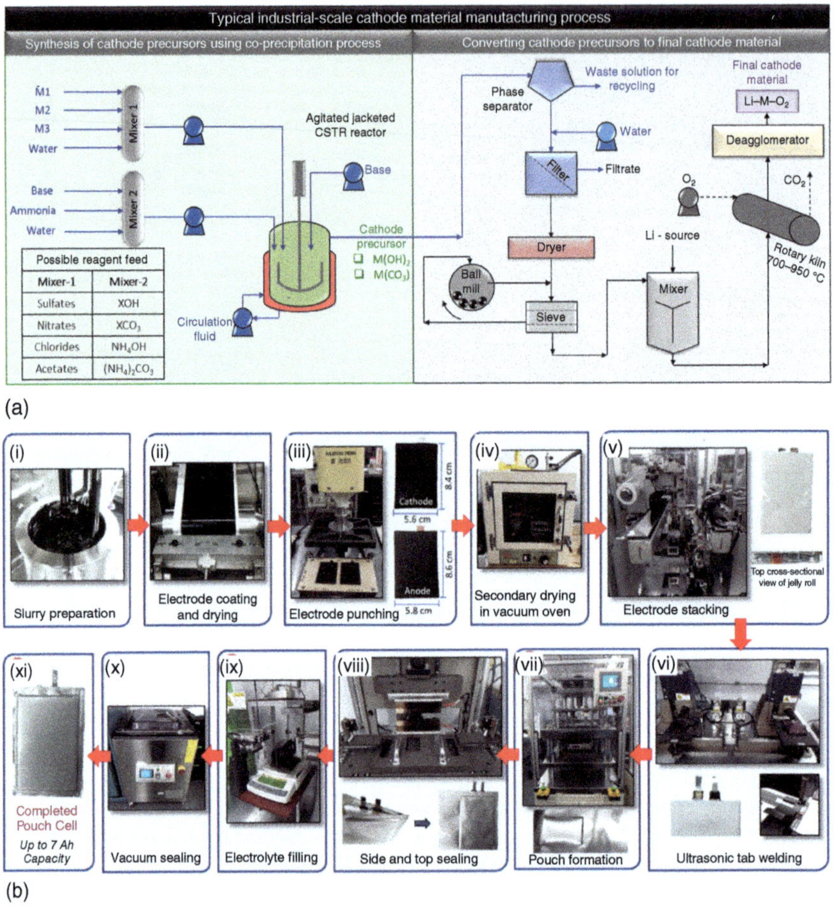

Figure 3.9 (a) Schematic representation of a typical industrial-scale cathode material-manufacturing process and (b) electrode coating ((i) to (iv)) and cell fabrication techniques and protocols ((v) to (xi)) employed in commercial battery-manufacturing industries. Source: Wood et al. [41] / with permission of Elsevier.

be integrated into next-generation LIBs requires optimization of synthesis and processing parameters as outlined in Sections 3.6.1–3.6.4.

3.6.1 Synthesis of Cathode Precursors

Cathode materials are usually synthesized using coprecipitation synthesis routes to produce hydroxide or carbonate precursors. Several synthesis variables must be optimized to produce high-quality precursors including: (i) coprecipitation pH range, (ii) flow rates of metal reagents (sulfates, nitrates, acetates, chlorides, carbonates), chelating agent (ammonium hydroxides and carbonates), and base (NaOH, KOH), (iii) process temperatures, and (iv) cover gas (e.g. N_2, Ar, and air). Additionally, today's commercial cathodes are usually synthesized in spherical morphologies to enable effective lithium diffusion during calcination while ensuring that the final product has high tap densities. All these parameters are critical for the assembled batteries to deliver high-energy densities.

3.6.2 Synthesis of Final Cathode Powders

Effective lithium incorporation in the synthesized cathode precursors is vital to achieve the desired electrochemical performance. Some of the calcination step variables, which warrant optimizations, include: (i) choice of Li source (LiOH, Li_2CO_3, and various Li salts), (ii) temperature profiles (e.g. heating/cooling rates and hold times), (iii) nature of material packing (e.g. packed bed and fluidized bed), and (iv) choice of cover/reactive gas and the corresponding flow rates. Intimate mixing of the lithium source and the cathode precursor is necessary to achieve the desired final cathode compositions. Typically, 2–5 wt% excess lithium is added to the precursor prior to the calcination process to compensate for the lithium loss at the elevated temperatures (>650 °C). Notably, cobalt-free nickel-rich formulations (LNO, NFA, NMA, NM, NF, NA, etc.) typically require an oxygen-rich environment during calcination to achieve good compositional homogeneities while minimizing cation mixing. Moreover, today's mainstream nickel-rich Co-containing cathodes usually contain surface coatings such as Al_2O_3, ZrO_2, TiO_2, SiO_2 to mitigate parasitic side reactions with the electrolyte [42, 43]. As such, surface engineering will likely play a critical role in optimizing the performance of next-generation Co-free cathodes.

3.6.3 Electrode Fabrication

Li-ion cathodes are prepared by casting slurries containing active material, conductive carbon, and polymer binder onto an Al foil current collector. During electrode fabrication, several processing variables must be carefully controlled including: (i) slurry formulation and viscosity, (ii) choice of solvent (organic or aqueous), (iii) coating thickness and line speeds, (iv) electrode drying times and temperatures, and (v) calendering process. Optimizing these processes is critical to develop homogenous electrodes with good adhesion to the current collector. For example, nickel-rich layered cathodes may sometimes have lithium carbonate impurities on the surface, which could cause a rise in pH of the slurry, causing undesirable gelation [41]. Thus, certain slurry stabilizers that can modulate the pH, viscosity, and other relevant rheological parameters may be necessary for novel cathode chemistries. Tuning the drying and solidification protocols is also necessary to achieve cathode architectures that deliver high-energy and power densities. Additionally, optimized coating line speeds are crucial to achieve electrodes with uniform morphologies and minimal coating defects (pinholes, streaks, and agglomerates) [41].

3.6.4 Battery Assembly

LIBs can be prepared in several form factors (e.g. pouch, prismatic, or cylindrical cells) depending upon the specific end use. To produce high-performance LIBs containing Co-free cathodes, several key variables need optimization including:

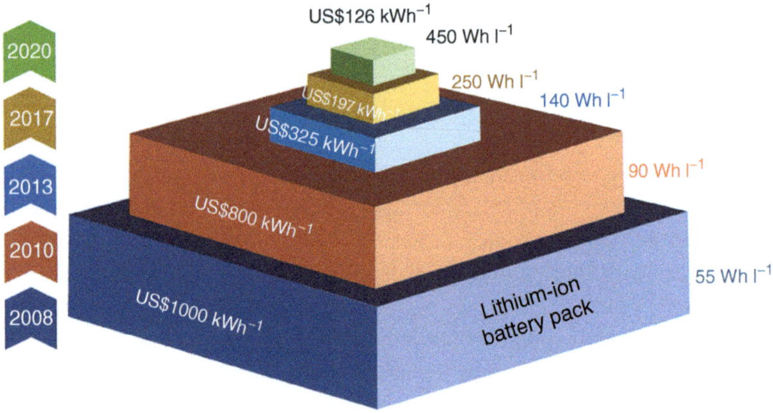

Figure 3.10 Cost and energy density of Li-ion battery packs from 2008 to 2020.

(i) choice of electrolyte solvent and Li salt, (ii) electrolyte additives, and (iii) formation of cycling protocols [44]. For next-generation layered cathodes, protocols established for the large-scale manufacturing of mainstream NMC and NCA cathodes can likely be implemented with minimal changes. On the other hand, manufacturing LIBs containing emerging cathode chemistries (e.g. DRX systems) will likely require extensive optimization due to a lack of established manufacturing protocols. Thus, systematic R&D efforts directed toward bridging this knowledge gap are essential for successful commercial deployment of such novel cathode systems.

3.7 Summary and Perspective

Since their commercial debut in 1991, the specific energy of LIBs has increased from 150 to 250 Wh kg^{-1}. Likewise, cost has decreased from US$1000 kWh^{-1} in 2008 to US$126 kWh^{-1} in 2020 (see Figure 3.10), which has opened new opportunities in EV propulsion and grid storage applications. Despite these exciting trends, most commercial LIBs rely on Co-based cathodes (e.g. LCO, NMC, and NCA), which is undesirable due to the high cost and low abundance of Co. As such, next-generation Li-ion cathodes, which utilize inexpensive, earth-abundant materials, are needed to meet projected energy demands.

This review highlights low-Co/Co-free cathodes for high-energy LIBs. $LiFePO_4$ (olivine) and $LiMn_2O_4$ (spinel) have been commercialized for some applications, but these systems have inferior energy density compared to $LiCoO_2$ (layered). On the other hand, Ni-rich layered oxides have shown impressive performance (e.g. 200 mAh g^{-1} at ~3.9 V vs. Li/Li$^+$ for NMC811). It should be noted that overreliance on Ni-based cathodes is likely unsustainable as energy storage demands continue to grow. Beyond conventional Li-ion cathodes containing high degrees of Li/TM ordering, DRX cathodes are a promising class of next-generation cathodes due to their compatibility with a wide range of transition metals. Several DRX systems with specific energies >800 Wh kg^{-1} have been developed, but challenges related to their synthesis and electrode fabrication need to be resolved before these systems are commercially viable. To realize the full potential of next-generation low-Co/Co-free cathodes, the influence of particle morphology and microstructure (e.g. single crystal [45] vs. polycrystalline powders [46]) on several performance criteria (e.g. specific energy, cycling stability, safety, and calendar life) requires further investigation. Likewise, developing new electrolyte additives and formulations is a promising approach to stabilize cathode/electrolyte interfaces, especially for high-voltage systems. Importantly, as demand for LIBs continues to grow, recycling technologies must be developed to reduce waste and recover valuable resources. Combining effective recycling programs with next-generation cathode development can provide a two-pronged strategy toward reducing reliance on Co [47–49].

Beyond traditional LIBs prepared with liquid electrolytes, there is growing interest in solid-state batteries (SSBs) containing a Li metal anode, high-energy cathode, and Li^+-conducting solid electrolyte. SSBs can potentially enable specific energies >400 Wh kg^{-1}, although achieving this goal requires overcoming several materials and processing challenges as recently highlighted [42, 50–53]. While the performance of solid-state cathodes lags far behind advanced Li-ion systems, developing low-Co/Co-free cathodes for SSBs offers tremendous opportunities for safe, low-cost, and high-energy batteries. Overall, developing high-performance cathodes with minimal/no Co will be critical to meet ever-growing energy demands and work toward a sustainable energy landscape in the coming decades.

Acknowledgments

This research at Oak Ridge National Laboratory, managed by UT-Battelle, LLC, under contract DE-AC05-00OR22725 with the US Department of Energy (DOE), was sponsored by the Energy Efficiency and Renewable Energy (EERE), Vehicle Technologies Office (VTO), (Program Manager: Peter Faguy, and Office Director: David Howell). The authors also acknowledge Dr. Rachid Essehli, Dr. Zhijia Du, Dr. Ruhul Amin, and the members of the Emerging and Solid-State Batteries Group, in the Electrification and Energy Infrastructures Division, and the Energy Storage and Conversion Group, in the Chemical Sciences Division, at Oak Ridge National Laboratory for the useful discussions and feedback.

Conflict of Interest

The authors report no conflict of interest.

References

1 Asl, H.Y. and Manthiram, A. (2021). Toward sustainable batteries. *Nature Sustainability* 4: 379–380.

2 Li, W., Erickson, E.M., and Manthiram, A. (2020). High-nickel layered oxide cathodes for lithium-based automotive batteries. *Nature Energy* 5 (1): 26–34.

3 U.S. DRIVE Grid Integration Technical Team (GITT) and Integrated Systems Analysis Technical Team (ISATT) (2019). Summary Report on EVs at Scale and the U.S. Electric Power System.
4 Yu, X. and Manthiram, A. (2021). Sustainable battery materials for next-generation electrical energy storage. *Advanced Energy and Sustainability Research* 2 (5): 2000102.
5 Gandon, S. (2017). No cobalt, no Tesla?. https://techcrunch.com/2017/01/01/no-cobalt-no-tesla/ (accessed 11 October 2021).
6 Nitta, N., Wu, F., Lee, J.T., and Yushin, G. (2015). Li-ion battery materials: present and future. *Materials Today* 18 (5): 252–264.
7 Momma, K. and Izumi, F. (2011). VESTA 3 for three-dimensional visualization of crystal, volumetric and morphology data. *Journal of Applied Crystallography* 44 (6): 1272–1276.
8 Kim, Y., Seong, W.M., and Manthiram, A. (2020). Cobalt-free, high-nickel layered oxide cathodes for lithium-ion batteries: progress, challenges, and perspectives. *Energy Storage Materials* 34: 250–259.
9 Amin, R., Muralidharan, N., Petla, R.K. et al. (2020). Research advances on cobalt-free cathodes for Li-ion batteries: the high voltage $LiMn_{1.5}Ni_{0.5}O_4$ as an example. *Journal of Power Sources* 467: 228318.
10 Seong, W.M. and Manthiram, A. (2020). Complementary effects of Mg and Cu incorporation in stabilizing the cobalt-free $LiNiO_2$ cathode for lithium-ion batteries. *ACS Applied Materials & Interfaces* 12 (39): 43653–43664.
11 Xie, Q. and Manthiram, A. (2020). Long-life, ultrahigh-nickel cathodes with excellent air storage stability for high-energy density lithium-based batteries. *Chemistry of Materials* 32 (17): 7413–7424.
12 Sun, Y.-K., Lee, D.-J., Lee, Y.J. et al. (2013). Cobalt-free nickel rich layered oxide cathodes for lithium-ion batteries. *ACS Applied Materials & Interfaces* 5 (21): 11434–11440.
13 Zhou, K., Xie, Q., Li, B., and Manthiram, A. (2021). An in-depth understanding of the effect of aluminum doping in high-nickel cathodes for lithium-ion batteries. *Energy Storage Materials* 34: 229–240.
14 Li, H., Cormier, M., Zhang, N. et al. (2019). Is cobalt needed in Ni-rich positive electrode materials for lithium ion batteries? *Journal of the Electrochemical Society* 166 (4): A429.
15 Muralidharan, N., Essehli, R., Hermann, R.P. et al. (2020). $LiNi_xFe_yAl_zO_2$, a new cobalt-free layered cathode material for advanced Li-ion batteries. *Journal of Power Sources* 471: 228389.
16 Muralidharan, N., Essehli, R., Hermann, R.P. et al. (2020). Lithium iron aluminum nickelate, $LiNi_xFe_yAl_zO_2$ – new sustainable cathodes for next-generation cobalt-free Li-ion batteries. *Advanced Materials* 32 (34): 2002960.
17 Liu, T., Yu, L., Liu, J. et al. (2021). Understanding Co roles towards developing Co-free Ni-rich cathodes for rechargeable batteries. *Nature Energy* 6 (3): 277–286.
18 Li, W., Lee, S., and Manthiram, A. (2020). High-nickel NMA: a cobalt-free alternative to NMC and NCA cathodes for lithium-ion batteries. *Advanced Materials* 32 (33): 2002718.

19 Santhanam, R. and Rambabu, B. (2010). Research progress in high voltage spinel $LiNi_{0.5}Mn_{1.5}O_4$ material. *Journal of Power Sources* 195 (17): 5442–5451.

20 Yu, X., Yu, W.A., and Manthiram, A. (2021). Advances and prospects of high-voltage spinel cathodes for lithium-based batteries. *Small Methods* 5 (5): 2001196.

21 Zhang, Z. and Zhang, S.S. (2015). *Rechargeable Batteries*. Springer International Publishing.

22 Thackeray, M.M., David, W.I.F., Bruce, P.G., and Goodenough, J.B. (1983). Lithium insertion into manganese spinels. *Materials Research Bulletin* 18: 461–472.

23 Thackeray, M.M., Johnson, P.J., Picciotto, L.A. et al. (1984). Electrochemical extraction of lithium from $LiMn_2O_4$. *Materials Research Bulletin* 19: 179–187.

24 Manthiram, A. (2020). A reflection on lithium-ion battery cathode chemistry. *Nature Communications* 11 (1): 1550.

25 Padhi, A.K., Nanjundaswamy, K.S., and Goodenough, J.B. (1997). Phospho-olivines as positive-electrode materials for rechargeable lithium batteries. *Journal of the Electrochemical Society* 144: 1188–1194.

26 Ravet, N., Gauthier, M., Zaghib, K. et al. (2007). Mechanism of the Fe^{3+} reduction at low temperature for $LiFePO_4$ synthesis from a polymeric additive. *Chemistry of Materials* 19: 2595–2602.

27 Reddy, M.V., Mauger, A., Julien, C.M. et al. (2020). Brief history of early lithium-battery development. *Materials (Basel)* 13 (8): 1884.

28 Zhou, F., Cococcioni, M., Kang, K., and Ceder, G. (2004). The Li intercalation potential of $LiMPO_4$ and $LiMSiO_4$ olivines with M = Fe, Mn, Co, Ni. *Electrochemistry Communications* 6 (11): 1144–1148.

29 Amine, K., Yasuda, H., and Yamachi, M. (2000). Olivine $LiCoPO_4$ as 4.8 V electrode material for Li batteries. *Electrochemical and Solid-State Letters* 3 (4): 178–179.

30 Okada, S., Sawa, S., Egashira, M. et al. (2001). Cathode properties of phospho-olivine $LiMPO_4$ for Li secondary batteries. *Journal of Power Sources* 97–98: 430–432.

31 Clément, R.J., Lun, Z., and Ceder, G. (2020). Cation-disordered rocksalt transition metal oxides and oxyfluorides for high energy lithium-ion cathodes. *Energy & Environmental Science* 13 (2): 345–373.

32 Lee, J., Urban, A., Li, X. et al. (2014). Unlocking the potential of cation-disordered oxides for rechargeable lithium batteries. *Science* 343 (6170): 519–522.

33 Lee, J., Kitchaev, D.A., Kwon, D.H. et al. (2018). Reversible Mn^{2+}/Mn^{4+} double redox in lithium-excess cathode materials. *Nature* 556 (7700): 185–190.

34 Yabuuchi, N., Takeuchi, M., Nakayama, M. et al. (2015). High-capacity electrode materials for rechargeable lithium batteries: Li_3NbO_4-based system with cation-disordered rocksalt structure. *Proceedings of the National Academy of Sciences of the United States of America* 112 (25): 7650–7655.

35 Nakajima, M. and Yabuuchi, N. (2017). Lithium-excess cation-disordered rocksalt-type oxide with nanoscale phase segregation: $Li_{1.25}Nb_{0.25}V_{0.5}O_2$. *Chemistry of Materials* 29 (16): 6927–6935.

36 Lee, J., Seo, D.-H., Balasubramanian, M. et al. (2015). A new class of high capacity cation-disordered oxides for rechargeable lithium batteries: Li–Ni–Ti–Mo oxides. *Energy & Environmental Science* 8 (11): 3255–3265.

37 Obrovac, M.N., Mao, O., and Dahn, J.R. (1998). Structure and electrochemistry of $LiMO_2$ (M = Ti, Mn, Fe, Co, Ni) prepared by mechanochemical synthesis. *Solid State Ionics* 112: 9–19.

38 Lee, J., Papp, J.K., Clement, R.J. et al. (2017). Mitigating oxygen loss to improve the cycling performance of high capacity cation-disordered cathode materials. *Nature Communications* 8 (1): 981.

39 Abdellahi, A., Urban, A., Dacek, S., and Ceder, G. (2016). Understanding the effect of cation disorder on the voltage profile of lithium transition-metal oxides. *Chemistry of Materials* 28 (15): 5373–5383.

40 Darga, J., Lamb, J., and Manthiram, A. (2020). Industrialization of layered oxide cathodes for lithium-ion and sodium-ion batteries: a comparative perspective. *Energy Technology* 8 (12): 2000723.

41 Wood, D.L. III, Wood, M., Li, J. et al. (2020). Perspectives on the relationship between materials chemistry and roll-to-roll electrode manufacturing for high-energy lithium-ion batteries. *Energy Storage Materials* 29: 254–265.

42 Nisar, U., Muralidharan, N., Essehli, R. et al. (2021). Valuation of surface coatings in high-energy density lithium-ion battery cathode materials. *Energy Storage Materials* 38: 309–328.

43 Darbar, D., Self, E.C., Li, L. et al. (2020). New synthesis strategies to improve Co-free $LiNi_{0.5}Mn_{0.5}O_2$ cathodes: early transition metal d^0 dopants and manganese pyrophosphate coating. *Journal of Power Sources* 479: 228591.

44 Zhang, X., Zou, L., Cui, Z. et al. (2021). Stabilizing ultrahigh-nickel layered oxide cathodes for high-voltage lithium metal batteries. *Materials Today* 44: 15–24.

45 Langdon, J. and Manthiram, A. (2021). A perspective on single-crystal layered oxide cathodes for lithium-ion batteries. *Energy Storage Materials* 37: 143–160.

46 Wang, Y., Wang, E., Zhang, X., and Yu, H. (2021). High-voltage "single-crystal" cathode materials for lithium-ion batteries. *Energy and Fuels* 35 (3): 1918–1932.

47 Bai, Y., Muralidharan, N., Sun, Y.-K. et al. (2020). Energy and environmental aspects in recycling lithium-ion batteries: concept of battery identity global passport. *Materials Today* 41: 304–315.

48 Bai, Y., Muralidharan, N., Li, J. et al. (2020). Sustainable direct recycling of lithium-ion batteries via solvent recovery of electrode materials. *ChemSusChem* 13 (21): 5664–5670.

49 Bai, Y., Hawley, W.B., Jafta, C.J. et al. (2020). Sustainable recycling of cathode scraps via Cyrene-based separation. *Sustainable Materials and Technologies* 25: e00202.

50 Albertus, P., Anandan, V., Ban, C. et al. (2021). Challenges for and pathways toward Li-metal-based all-solid-state batteries. *ACS Energy Letters* 6: 1399–1404.

51 Dudney, N.J., West, W.C., and Nanda, J. (2016). *Handbook of Solid State Batteries*, 2e. Singapore: World Scientific Publishing Co. Pte. Ltd.

52 Dixit, M.B., Parejiya, A., Muralidharan, N. et al. (2021). Understanding implications of cathode architecture on energy density of solid-state batteries. *Energy Storage Materials* 40: 239–249.

53 Parejiya, A., Essehli, R., Amin, R. et al. (2021). $Na_{1+x}Mn_{x/2}Zr_{2-x/2}(PO_4)_3$ as a Li^+ and Na^+ super ion conductor for solid-state batteries. *ACS Energy Letters* 6 (2): 429–436.

4

Transition Metal Oxide Anodes for Electrochemical Energy Storage in Lithium- and Sodium-Ion Batteries*

Shan Fang[1], *Dominic Bresser*[2], *and Stefano Passerini*[1,2]

[1] *Helmholtz Institute Ulm (HIU), Helmholtzstrasse 11, 89081 Ulm, Germany*
[2] *Karlsruhe Institute of Technology (KIT), P.O. Box 3640, 76021 Karlsruhe, Germany*

4.1 Introduction

Rechargeable LIBs have become the common power source for portable electronics since their first commercialization by Sony in 1991 and are, as a consequence, also considered the most promising candidate for large-scale applications like (hybrid) electric vehicles and short- to mid-term stationary energy storage [1–4]. Due to the resulting great interest in this technology, extensive efforts have been made to achieve further improved performance. As a result, the energy density of LIBs has continuously increased at a rate of 7–8 Wh kg^{-1} per year, already passing 250 Wh kg^{-1} at cell level (for 18650-type cells). Simultaneously, the overall cost decreased substantially from initially around €1000 per kWh to less than €200 per kWh [5], while a further reduction to less than €150 per kWh is anticipated within the next 5–10 years [6] – or, in fact, might have been realized already following some recent newspaper articles. Nonetheless, further improvement is required for realizing a fully electrified transportation sector and eventually succeeding in transitioning to renewable energy sources. For this reason, there is a great quest for alternative inactive and active materials, including *inter alia* the anode – not least because graphite, the state-of-the-art for LIBs, which is intrinsically limiting the fast charging of the full-cell [7, 8]. Another important concern is related to the availability of the required elements, including *inter alia* lithium [9, 10], which has led to a rapidly increasing interest in alternative charge carriers – in particular, sodium [11–13]. In fact, the two technologies share several similarities and, hence, room-temperature SIBs are considered a "drop-in technology," as many of the achievements obtained for LIBs can be readily implemented for SIBs. This has resulted in rapid progress for the development of SIBs within only a few years. However, there are some fundamental differences between the two systems due to

*This chapter was originally published in Advanced Energy Materials, 2021, Volume 10, Issue 1; DOI: 10.1002/aenm.201902485. Reproduced with permission of WILEY-VCH GmbH.

Transition Metal Oxides for Electrochemical Energy Storage, First Edition.
Edited by Jagjit Nanda and Veronica Augustyn.

Table 4.1 Comparison of lithium and sodium regarding selected physicochemical properties and cost.

	Lithium	**Sodium**
Cation radius (Å)	0.76	1.02
Relative atomic mass	6.94	22.98
$E°$(vs. SHE) (V)	−3.04	−2.71
Cost, carbonates	US$5000/ton	US$150/ton
Theoretical capacity of metal electrodes (mAh g^{-1})	3829	1165
Coordination preference	Octahedral and tetrahedral	Octahedral and prismatic
Desolvation energy in PC (kJ mol^{-1})	218.0	157.3

SHE, standard hydrogen electrode.

the different charge carriers such as, for instance, the size of the cation, the standard redox potential, or simply the different cost for the corresponding precursors – as summarized briefly in Table 4.1. These affect the diffusion and transport properties, the maximum achievable energy density, or the price of the cell, respectively. Furthermore, the different reactivity of the two systems eventually also has an effect on the (decomposition) reactions at the interface between the electrode and the electrolyte, including the charge transfer and cation desolvation, before entering the host structure. With respect to the potential host structure for the negative electrode – or in other words, the electrode-active material – the material classes of choice are frequently carbons (e.g. graphite or hard carbons) or metal oxides. For the latter, there are essentially three different alkali cation storage mechanisms: (i) insertion (including intercalation in case of layered structures), (ii) alloying, and (iii) conversion. In case of (i) insertion-type materials, the Li^+ and Na^+ cations can be reversibly stored in the host material without any severe (irreversible) deterioration of the initial crystal structure and very low volume variation, which commonly allows for excellent cyclability and good capacity retention; also due to the commonly rather low number of Li^+/Na^+ inserted by unit weight and volume. The most classic insertion-type metal oxide anodes are based on titanium (e.g. TiO_2 [14] and $Li_4Ti_5O_{12}$ [15–17] or $Na_2Ti_3O_7$ [18–21]) as redox-active center, which is reduced to Ti^{3+} upon lithiation/sodiation and reoxidized to Ti^{4+} when the alkali metal (AM) cations are subsequently deinserted again. Considering the rather high mass of these compounds in combination with the relatively limited lithium/sodium uptake, however, the use of these materials is essentially limited to high-power rather than high-energy applications [8] or applications for which cycling stability is more important than energy density [22, 23]. Differently, (ii) alloying-type materials (like Si, Sn, Ge, or Zn) provide high to very high – and frequently fast – lithium and sodium storage capacities, commonly exceeding one lithium or sodium per atom of the alloying element (e.g. $Li_{15}Si_4$ or $Li_{4.4}Sn$).

Nevertheless, the accompanying extensive volume variation limits the cycle life of such electrodes – not least as a result of the continuous exposure of fresh surfaces to the electrolyte, resulting in an ongoing electrolyte decomposition and solid electrolyte interphase (SEI) formation and, as a consequence, the drying out of the cell [4, 24, 25]. In an attempt to (partially) overcome this issue, the use of the corresponding metal oxides had been proposed (e.g. SiO_x, SnO_2, GeO_2, ZnO) [26–29]. The initially formed Li_2O was supposed to buffer the volume variation occurring upon the subsequent alloying of the elemental metal/metalloid and provide an *in situ*-generated ion-conducting matrix. However, such matrix does not prevent another issue of alloying-type materials, i.e. the continuous aggregation of the elemental particles upon de-/lithiation/sodiation, which eventually still leads to rapid capacity degradation [27, 30]. This issue is avoided for transition metal oxides (TMOs), for which the elemental transition metal (TM) does not alloy with lithium/sodium. Instead, the thus-created metallic nanonetwork allows for a good electron conduction within the original primary particle and, by this, allows for the reversible cycling of the initially formed Li_2O [4, 31, 32]. Advantageously, this comes with significantly less volume changes upon de-/lithiation/sodiation while also allowing for capacities approaching 1000 mAh g^{-1} (e.g. Fe_2O_3) thanks to the multielectron transfer reaction per TM occurring upon the conversion of TMOs to TM^0 and lithium/sodium oxide [4, 33–35]. These unique features in combination with the richness of materials synthesis and design for such TMOs has triggered a remarkable attention in the past almost 20 years since the first report on this mechanism by Tarascon and coworkers [31]. Accordingly, there has been a series of review articles in the past years on this subject [4, 33, 36–38]. Cabana et al. [33], for instance, provided a very comprehensive overview on the conversion mechanism for lithium-ion batteries until 2010, which has been updated with a clear focus on new mechanistic insights by Bresser et al. [4] in 2016. An overview organized by following the periodic table has been presented by Nitta and Yushin [34], for example. Similarly, very comprehensive review articles on (conversion-type) anode materials for sodium-ion batteries have been published in the past – partially more generally covering all potential negative material candidates [39, 40], partially with a particular focus on (nanostructured) conversion and alloying materials [32]. A comprehensive comparison of TMO negative electrodes with a clear focus on the conversion-type lithium and sodium storage, however, has been missing so far. Accordingly, this review article is organized as follows: First, we provide the reader with a general introduction into the reaction mechanism as such, including the major advantages and remaining challenges. Subsequently, we provide an overview on the most relevant TMOs for use as negative electrodes in lithium-ion and sodium-ion batteries with particular attention toward recent publications that have not been covered by the abovementioned review articles so far. Finally, we conclude this review with a brief summary and an opinionated perspective toward the potential application of TMO negative electrodes in commercial lithium-ion and sodium-ion battery cells.

4.2 Potential Advantages and Challenges of the Conversion Mechanism

In 2000, Tarascon and coworkers [31] reported a new Li^+ storage mechanism involving the reversible electrochemical lithiation of TMOs (TM = Co, Ni, Fe, Cu). According to the main reaction, the process is generally referred to as a "conversion reaction," formally as follows:

$$TM_aX_b + (b \cdot n)AM \leftrightarrows aTM + bAM_nX \tag{4.1}$$

with TM being a transition metal (e.g. Fe, Co, Mn, Cu, Ni), AM an alkali metal (e.g. Li, Na), X an anion (such as H, N, P, O, S, F), and n as the formal oxidation state of X. Since the TM is generally reduced to the metallic state, involving the transfer of several electrons per unit formula of the starting compound, conversion reactions commonly show much higher capacities than intercalation reactions. In case of Co_3O_4, for example, the complete reduction of Co^{2+}/Co^{3+} in a Li^+-containing electrolyte leads to the formation of Co^0 and Li_2O, which corresponds to an overall specific capacity of 890.4 mAh g^{-1} (based on Co_3O_4). This value is about three times higher than the capacity based on a classical intercalation reaction as in case of $LiCoO_2$ (273.8 mAh g^{-1}) for which Co^{3+} is reversibly oxidized to Co^{4+}. In practice, in fact, even less capacity is achieved, as only about 0.5 lithium can be reversibly extracted without severely deteriorating the crystalline structure of the host lattice. Adelhelm and coworkers [35] investigated the impact of the alkali metal on the conversion reaction. The standard cell potential E° can be calculated using:

$$E^\circ = -\frac{\Delta rG}{z \cdot F}. \tag{4.2}$$

Accordingly, they conducted a direct comparison of the cell voltages for lithium and sodium. It has been recognized that the change in cell voltage is constant for oxides, fluorides, hydrides, etc. when replacing lithium with sodium. For example, the change in cell voltage for oxides is 0.96 V, as shown in (4.3):

$$2Li + Na_2O \leftrightarrows Li_2O + 2\,Na \quad \Delta E(Li - Na) : 0.96\,V \tag{4.3}$$

The same calculation applies generally to other conversion reactions with X = H, O, S, F, Cl, Br, and I (Figure 4.1a). When lithium is substituted by sodium in conversion reactions, the cell voltage shifts to lower values for hydrides, oxides, sulfides, and fluorides. For chlorides, the cell voltage is nearly the same, while for iodides and bromides the sodium conversion reaction shows even higher cell voltages compared to the corresponding reaction with lithium. This is remarkable, as sodium-based cells are intuitively considered to provide a lower cell voltage compared to their lithium analogues. This behavior can be reasonably explained in consideration of the associated Born–Haber cycles [35]. The lithium compounds have larger lattice energies than the corresponding sodium versions, resulting in greater negative Gibbs reaction energies, which then translates to higher cell voltages. Nevertheless, this difference might be compensated or even overcompensated in case of the chlorides and bromides/iodides, respectively, as a result of the lower ionization and cohesive energy for the sodium analogues.

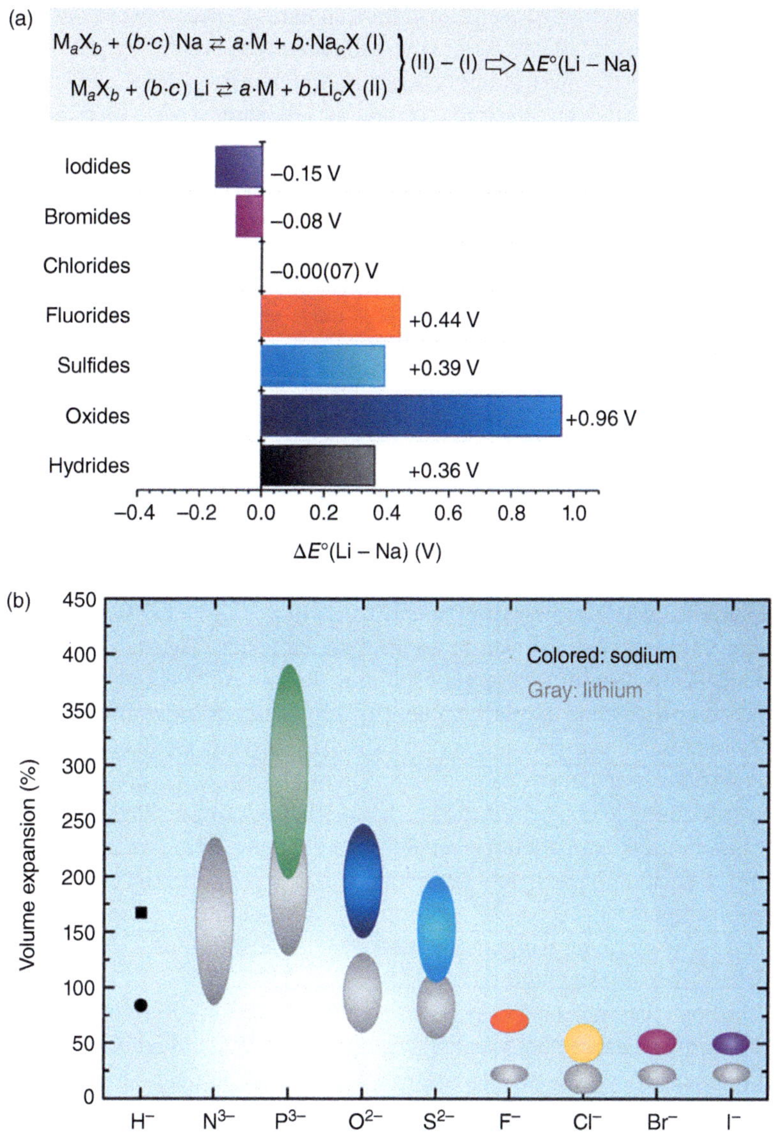

Figure 4.1 (a) Effect of the replacement of Li by Na in conversion reactions on the corresponding redox potential for various compounds: For fluorides, sulfides, oxides, and hydrides, the reaction with lithium shows higher cell voltages in a theoretical LIB configuration. For chlorides, the difference is essentially zero and for the even heavier bromides and iodides, a hypothetical SIB is anticipated to deliver a higher cell voltage. (b) Calculated volume expansions for Li- and Na-based conversion reactions. Source: Klein et al. [35]/with permission of Royal Society of Chemistry.

With respect to their employment in commercial battery cells, however, several fundamental obstacles still remain before these materials can become a viable alternative, despite the great promise of this concept in general. Among these challenges, one of the most relevant is the low initial coulombic efficiency (CE), which is commonly below 75% and associated *inter alia* with the strong structural reorganization. The initial lithiation results in the formation of the corresponding metallic nanoparticles being distributed in the simultaneously formed amorphous AM_nX matrix [31]. During the first lithiation, also the SEI layer occurs on the surface of the electrode, which is composed of decomposition products of the utilized electrolyte. Depending on the nature and structure of the electrode-active material as well as the current applied, the lithiation reaction may include one or more intermediate phases. Another main obstacle, and even more important limitation, is the poor energy efficiency of the AM storage process due to large polarization effects and/or different reaction pathways [4]. Overpotentials appear to be intrinsic to conversion reactions and their magnitude increases with the bond polarity. For fluorides, for example, the combined overpotentials can exceed 1 V [5], which results in a rather low round-trip energy efficiency. In addition, the conversion reaction is accompanied by considerable volume changes. The effect on halides is minimal (generally less than 50%), while for sulfides or oxides, the volume expansion may exceed 100%, potentially causing significant mechanical deterioration of the electrode (Figure 4.1b). These volume changes are even more severe for sodium, i.e. approximately twice as high. Thus, it is anticipated that the development of suitable conversion materials for SIBs will be even more demanding than for LIBs. At this point, we may note again that the conversion mechanism, as generalized in (4.1), frequently involves an initial Li^+ insertion [41], leading characteristic nanostructures. Accordingly, diffusion distances stay short, enhancing the reversibility of the conversion reaction [42]. In contrast, the formation of relatively large TM^0 particles may negatively affect the rechargeability. Replacing Li^+ by Na^+ will certainly have an important influence on this initial insertion mechanism and the subsequent reaction steps – even though this requires further investigation.

In fact, the hope that replacing Li by Na might help to overcome the challenges for Li-involving conversion reactions has not become true so far, and only a few studies have specifically focused on such a direct comparison between Li and Na. In general, the capacities obtained for the Na analogues are lower than those for the corresponding LIB configuration. For Fe_2O_3 nanoparticles, for instance, a high capacity of 1000 mAh g^{-1} was reported in the Li-configuration, which is close to the theoretical specific capacity of 1006 mAh g^{-1}, but only around 350 mAh g^{-1} were obtained for the Na-configuration. Nonetheless, this value is still much higher than common intercalation-related capacities – especially in case of sodium [43]. Remarkably, a comparison of the reactivity with Li and Na when studying the conversion reaction for the rather little explored carbodiimide anion exhibits the reversed behavior, i.e. the capacities observed for Na exceed those for Li [44], though the definite reason remains to be elucidated. Thus, despite the remaining challenges, conversion-type reactions are still considered a promising field of research, particularly with regard to the potential use of cost-efficient and abundant materials like iron oxide or sulfides.

4.3 Transition Metal Oxides as Anode Materials

TMOs are, by far, the class of conversion materials that attracted most attention – presumably as a result of their ease of handling and their high capacities. Also, with respect to sodium storage, TMOs have received considerable attention recently, including Fe_2O_3, Co_3O_4, MnO, CuO, and NiO. Nevertheless, the reversible capacities of oxides as sodium-ion active materials are much lower than their theoretical capacities and, as mentioned earlier, when investigated as reversible Li^+ hosts. The reaction potential, however, is commonly lower for the reversible sodium storage [45], rendering them more suitable as anode materials compared to their lithium analogues, which is consistent with theoretical calculations based on thermodynamic data. According to the reported literature, the reversible capacities of oxides are obtained at voltages lower than 2 V. Beside such "intrinsic" values, in both cases, modifying the electrode structure or incorporating additional conductive materials is an effective strategy to improve the reaction kinetics.

In the following, the most promising oxide materials as conversion anodes for LIBs and SIBs are reviewed and, as far as possible, we will provide a direct comparison of their performance. Nonetheless, due to the lack of information available in literature, such comparative analysis remains rather difficult – also as some studies are focusing less on "basic electrochemical information" and more on factors essential for the final performance, such as the areal-active material loading and capacity (in $mg\,cm^{-2}$ or $mAh\,cm^{-2}$) or the weight fraction of the nonactive components in nanostructured composites.

4.3.1 Iron Oxide (Fe_3O_4, Fe_2O_3)

Iron oxides, such as α-Fe_2O_3 and Fe_3O_4, are considered to be attractive candidates for the next generation of anode materials due to their abundance, nontoxicity, and low cost. Fe_2O_3, for instance, following the general reaction mechanism $Fe_2O_3 + 6Li^+ + 6e^- \leftrightarrows 2Fe + 3Li_2O$ exhibits a high theoretical specific capacity of $1006\,mAh\,g^{-1}$, as reported firstly by Tarascon and coworkers [31]. They reported that each formula unit allows for the uptake of 0.5 mol of Li being reversibly inserted into nano-α-Fe_2O_3 (20 nm) in the voltage range between 1.5 and 4.0 V. When extending the voltage range to 0.005–3.0 V, 8.5 mol of Li per mol of Fe_2O_3 react, resulting in the degradation of the crystal structure and the formation of metallic iron nanoparticles and Li_2O based on the common conversion reaction. In addition, a polymeric layer on the particles was also observed, which was assigned to the electrolyte (solvents and salt) decomposition. Nanostructured metal oxides have been used to enhance the rate performance and reversible capacity, taking advantage of the shortened Li^+ transport pathways and the decreased volume changes due to the electrochemical reaction [46]. In a subsequent study, Chen et al. [47] reported the synthesis of α-Fe_2O_3 nanotubes using alumina membranes as the template. The as-prepared nanotubes with uniform size and shape and high specific surface area, displayed excellent electrochemical activity, including a very high discharge capacity ($1415\,mAh\,g^{-1}$ at $100\,mA\,g^{-1}$). Chowdari and coworkers

[48] synthesized α-Fe_2O_3 nanoflakes on copper foil by a hydrothermal method. The electrochemical tests revealed that such Fe_2O_3 nanoflakes show a reversible capacity of 680 ± 20 mAh g^{-1}, equivalent to 4.05 ± 0.05 mol of Li per mole of Fe_2O_3 with negligible capacity decay up to 80 cycles for a voltage range of 0.005–3.0 V and at a specific current of 65 mA g^{-1}. The average lithiation and delithiation potential, though, were 1.2 and 2.1 V, respectively. Cho's group [49] employed an iron-based metal organic framework (MOF) as template and prepared spindle-like porous α-Fe_2O_3. This material showed greatly improved electrochemical performance. After 50 cycles at 0.2*C*, a capacity of 911 mAh g^{-1} was maintained and, even when increasing the current to 10*C*, a capacity of 424 mAh g^{-1} was delivered by this electrode material. Balducci and coworkers [50] investigated the electrochemical charge/discharge mechanism of hematite and found that at the end of the delithiation process, lithium iron oxide (α-$LiFeO_2$) was formed, thus providing an additional source of irreversible capacity loss (Figure 4.2a–c). Backert et al. [53] reported a graphene-wrapped Ni@Fe_2O_3 composites, employing sulfonated reduced graphene oxide (rGO). The developed complex material showed highly improved electrochemical performance, offering a capacity of 1051 mAh g^{-1} after 40 cycles at 50 mA g^{-1}. Lou and coworkers [54] synthesized Fe_2O_3 hollow structures, including hollow microboxes and spheres. When evaluated as anode material for LIBs, these well-defined hollow structures exhibited remarkable cycling performance with high specific capacity. Such hierarchical hollow architectures allow for shortening the diffusion pathways for the Li cations and may, moreover, accommodate the volume variation occurring upon cycling. Various carbon additives have also been mixed with the metal oxide particles to enhance their electronic conductivity [55]. Wang et al. [56] reported an advanced carbon-coated CNT@Fe_2O_3 hierarchical nanostructure. This material was constructed through the bottom-up assembly of β-FeOOH nanospindles on the carbon nanotube (CNT) framework. After the thermal treatment, carbon-coated hollow α-Fe_2O_3 nanohorns were obtained. Thanks to the greatly improved kinetics and electrode stability, this hybrid structure exhibited a stable capacity retention of 800 mAh g^{-1} after 100 cycles at a specific current of 500 mA g^{-1} and excellent rate performance. Following these promising studies on Fe_2O_3 as a potential anode candidate, a few studies analyzed the performance in lithium-ion full cells. For example, Aravindan et al. [57] fabricated LIBs using one-dimensional (1D) Fe_2O_3 as the anode-active material and $Li_{1.33}Ni_{0.5}Mn_{1.5}O_4$ for the cathode. Before assembling the full cells, the authors overlithiated the cathode with a certain amount of Li (0.33 mol) to compensate the irreversible capacity loss occurring at the anode. Such full cells delivered a specific energy of 193 Wh kg^{-1} and an average cell voltage of 3.27 V and showed a capacity retention of 88% after 60 cycles. Verrelli et al. [58] combined an Fe_2O_3–mesocarbon microbead (MCMB) composite with $Li_{1.35}Ni_{0.48}Fe_{0.1}Mn_{1.72}O_4$ as cathode, resulting in a Li-ion full cell with an operating voltage of c. 3 V, high CE, and a stable capacity of about 100 mAh g^{-1}, which translates into a theoretical gravimetric energy density of 300 Wh kg^{-1}.

Besides Fe_2O_3, also Fe_3O_4 has been intensively studied as anode material in lab-scale LIBs, since it has a high theoretical capacity as well (925 mAh g^{-1}). Just

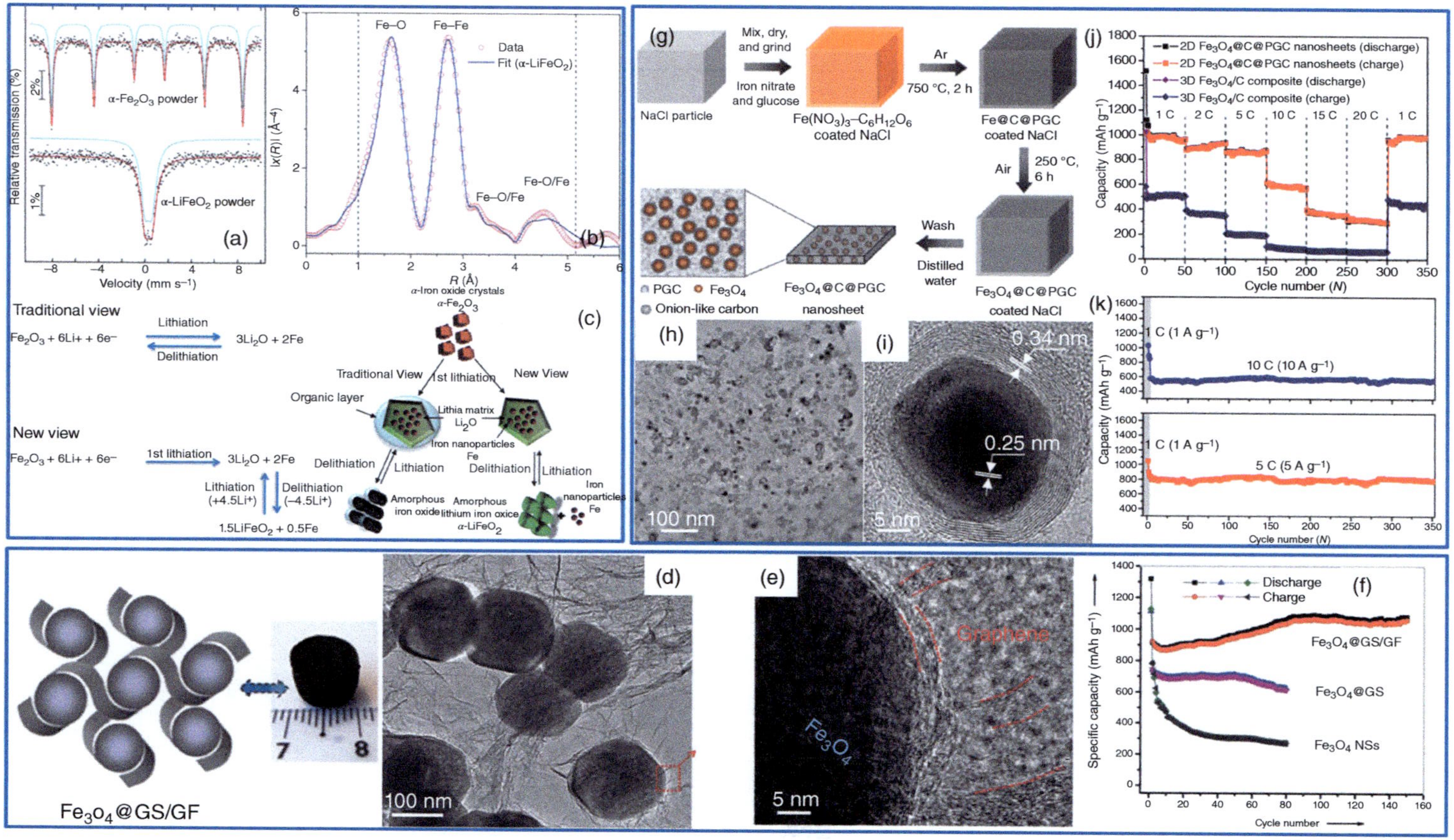

Figure 4.2 (a) Experimental and simulated ^{57}Fe Mossbauer spectra of pristine α-Fe_2O_3 and the cycled (delithiated) electrode at 298 K in comparison with powdered α-$LiFeO_2$, indicating the formation of α-$LiFeO_2$ upon cycling, as also confirmed by (b) EXAFS analysis and fitting. (c) Summary of the de-/lithiation reaction mechanism for carbon-coated Fe_2O_3. Source: Brandt et al. [50]/with permission of Royal Society of Chemistry. (d) Schematic illustration, photograph, TEM, and (e) high-resolution transmission electron microscopy (HRTEM) images of Fe_3O_4@GS/GF and (f) its cycling performance. Source: Wei et al. [51]/with permission of John Wiley & Sons. (g) The fabrication procedure of 2D Fe_3O_4@C@PGC. (h) TEM and (i) HRTEM images of the resulting composite. (j) Rate capability and (k) constant current cycling performance of Fe_3O_4@C@PGC electrodes. Source: He et al. [52]/with permission of Royal Society of Chemistry.

like Fe_2O_3, the realization of nanostructured particle architectures has been considered as an efficient way to enhance its electrochemical properties. For instance, Liu et al. [59] developed uniform pomegranate-like nanoclusters composed of ultrafine Fe_3O_4@nitrogen-doped carbon (Fe_3O_4@N–C) subunits with a diameter of around 4 nm prepared by a facile one-pot method. Compared with the reference core–shell nanoparticles, this unique structure provided even shorter Li^+ and electron diffusion pathways, further enhanced structural stability during cycling, and high electronic conductivity. As a result, this pomegranate-like Fe_3O_4@N–C possessed good rate performance and capacity retention upon long-term charge–discharge tests at specific currents of 1 A g^{-1} (1063.0 mAh g^{-1} and 98.4% capacity retention after 1000 cycles), 10 A g^{-1} (606.0 mAh g^{-1} with 92.0% capacity retention after 1000 cycles), and 20 A g^{-1} (417.1 mAh g^{-1} with 91.7% capacity retention after 1000 cycles). Three-dimensional (3D) graphene foams (GF) have also been widely used in energy storage materials, since the highly porous structure can alleviate the pulverization of the metal oxides induced by volumetric changes and contribute to the electronic conductivity [60]. Feng and coworkers [51] described a novel approach to synthesize 3D GF cross-linked with Fe_3O_4 nanospheres and encapsulated in graphene sheets (Fe_3O_4@GS; Figure 4.2d,e). In such hierarchical Fe_3O_4@graphene-sheets/graphene-foam (Fe_3O_4@GS/GF) hybrids, the double protection helps to alleviate the volume changes occurring during the electrochemical processes. The graphene shells suppress the aggregation of Fe_3O_4 NSs and buffer the volume changes, while the interconnected 3D graphene network reinforces the core–shell structure of Fe_3O_4@GS and improves the overall electronic conductivity of the electrode. Therefore, such a composite achieves a high reversible capacity of 1059 mAh g^{-1} over 150 cycles with excellent rate capability (Figure 4.2f). Besides these complex hierarchical structures, simple carbon coating is a widely used approach to modify the surface of active materials, as it can significantly improve the electronic conductivity of the electrode materials and reduce, if not avoid, side reactions at the particles surface [61, 62]. Wan and coworkers [63] synthesized carbon-coated Fe_3O_4 nanospindles via the partial reduction of carbon-coated hematite nanospindles. The resulting Fe_3O_4–C nanospindles displayed a stable capacity of ~600 mAh g^{-1} at $C/2$, which was fully retained after 80 cycles. He et al. [52] fabricated carbon-encapsulated Fe_3O_4 nanoparticles embedded in 2D highly conducting porous graphitic carbon nanosheets (Fe_3O_4@C@PGC nanosheets) using NaCl as template (Figure 4.2g–i). In such an architecture, the thin carbon layers prevent the direct contact between the Fe_3O_4-active material and the electrolyte, thus, preserving the structural and interfacial characteristics of the Fe_3O_4 nanoparticles. At the same time, the electronically conductive and flexible PGC nanosheets can alleviate the volume changes of the Fe_3O_4@C nanoparticles and prevent their deleterious aggregation through maintaining the electrical and overall structural integrity of the composite electrode upon cycling. Electrodes, based on this composite, exhibit very high cycling stability and rate capability (Figure 4.2j,k). A full cell of porous carbon–Fe_3O_4/$Li[Ni_{0.59}Co_{0.16}Mn_{0.25}]O_2$ was developed by Ming et al. [64]. The cell showed a high specific capacity of ~150 mAh g^{-1} and high energy density of 483 Wh kg^{-1} (working voltage: ~3.2 V). An anode consisting of Fe_3O_4 nanoparticles coated with a layer of F-doped carbon (Fe_3O_4@CF_x) and coupled with a $LiNi_{0.5}Mn_{1.5}O_4$ cathode has also been investigated [65]. Such full cell provides a gravimetric energy density of 371 Wh kg^{-1} accompanied by a capacity retention of 66.8% after 300 cycles at 100 mA g^{-1}.

Following these promising results, iron oxides have also been studied as anode materials in SIBs. Chen and coworkers [66], for instance, reported the synthesis of a porous 3D γ-Fe_2O_3@C nanocomposite, using an aerosol-assisted method, and demonstrated its promising performance for SIBs. The resulting electrodes showed a reversible capacity of 740 mAh g^{-1} after 200 cycles at 200 mA g^{-1}. Even after 1400 cycles at high currents (i.e. 2 A g^{-1}), the electrodes maintained a reversible capacity of 358 mAh g^{-1} (Figure 4.3a–d). Jiang et al. [69] investigated a series of thin-film TMOs, such as Fe_2O_3, NiO, Co_3O_4, and Mn_3O_4, all showing high electrochemical activity as anodes in SIBs. Especially Fe_2O_3 delivered a high capacity of 386 mAh g^{-1} at 100 mA g^{-1} after 200 cycles. A reversible capacity of 233 mAh g^{-1} was achieved even at a very high specific current of 5 A g^{-1}. Park and Myung [70] examined carbon-coated Fe_3O_4 dispersed on CNTs for both lithium and sodium storage. Used in SIBs, this material displayed a good rate capability at high current (196 mAh g^{-1} at 2.4 A g^{-1}). Zhou et al. [71] developed Fe_3O_4 and Co_3O_4 nanocrystals embedded in carbon spheres prepared by a biochemistry method using recombinant elastin-like polypeptides (containing hexahistidine) and further carbonization. The active material particles with a diameter of c. 5 nm were homogeneously distributed in the carbon matrix. This unique structure exhibited encouraging sodium storage capacity, i.e. 309 mAh g^{-1} after 100 cycles at 0.5 A g^{-1}. Liu et al. [67] reported the synthesis of a 3D–0D graphene-Fe_3O_4 quantum dot composite, constituted by Fe_3O_4 quantum dots (~4.9 nm) anchored on the surface of 3D structured graphene nanosheets. The composite exhibited ultrahigh sodium storage capacity and outstanding cycling performance (Figure 4.3e–j). Qin et al. [68] successfully prepared chain-like Fe_3O_4/C/red-P by employing a magnetic field. This treatment helped to tune the orientation of the chain-like arrays on the current collector to optimize the electrochemical performance, providing superior rate performance (692 mAh g^{-1} at 2 A g^{-1}) and good cycle-life performance (Figure 4.3k). Qi et al. [72] reported the synthesis of carbon-coated Fe_3O_4 nanoparticles anchored on rGO (Fe_3O_4@C/rGO). This composite delivered a capacity of 356 mAh g^{-1} as SIB anode material after 300 cycles at 0.1 A g^{-1}. Slightly superior performance was obtained for Fe_2O_3 nanoparticles encapsulated in an N-doped carbon matrix, which was synthesized using an iron-containing MOF as precursor and provided a specific capacity of 474 mAh g^{-1} at 0.1 A g^{-1} after 100 cycles [73]. Chen et al. [74] designed 1D nanocomposites with γ-Fe_2O_3 comprised in porous carbon fibers (γ-Fe_2O_3/PCF) by electrospinning. They compared the performance as Li^+ and Na^+ host and found an inferior performance in case of sodium; the capacity retention was only c. 78% after 50 cycles at 0.1*C*. Better results for a similar particle morphology were obtained by Ni et al. [75], who synthesized Fe_2O_3 nanotubes, which were subsequently treated by surface sulfurization, i.e. the formation of FeS_2 (S-Fe_2O_3). The realization of such heterostructure led to a built-in electric field, which enhances the charge transfer and reduces the activation energy. When tested as anode materials

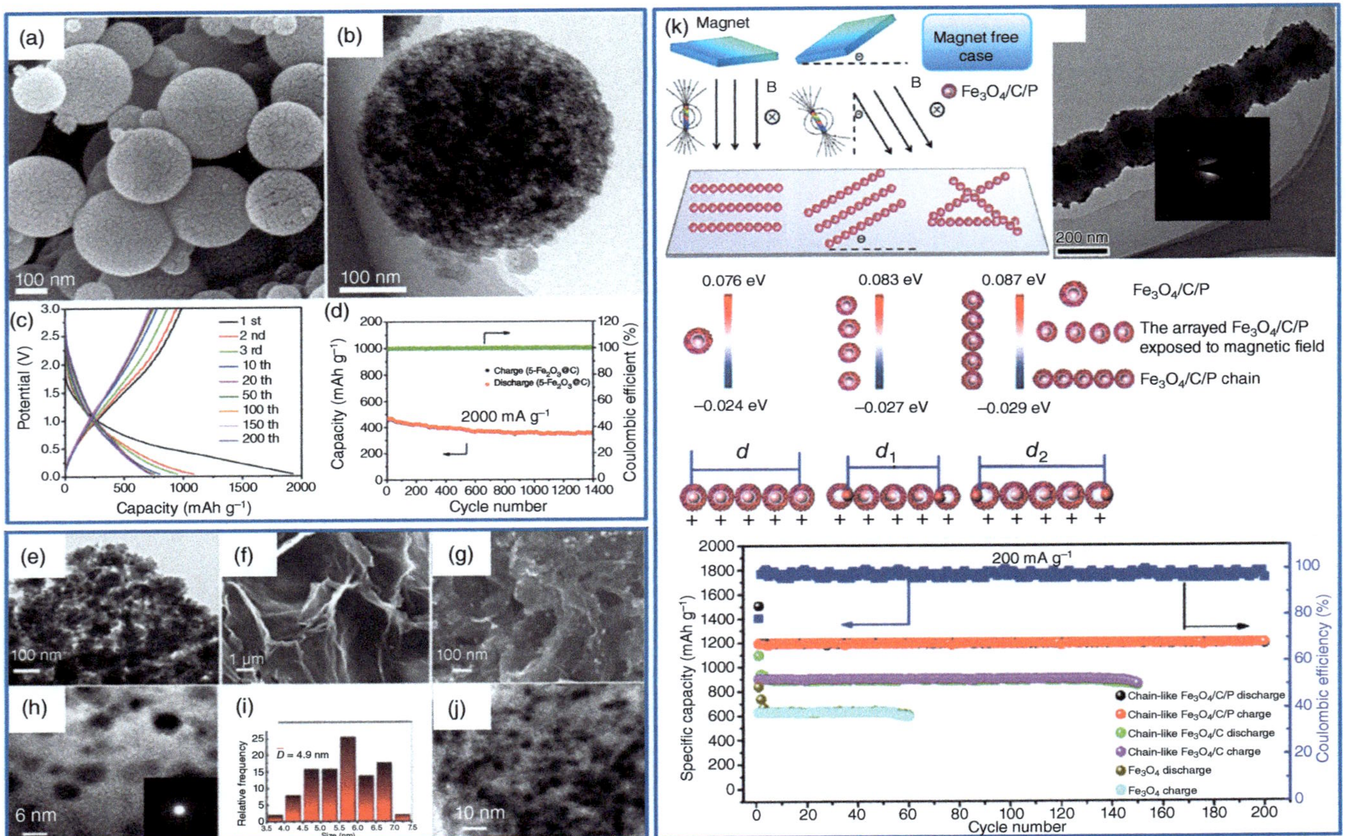

Figure 4.3 (a, b) Scanning electron microscopy (SEM) and TEM images of Fe_2O_3@C and its electrochemical performance as a sodium-ion anode: (c) The dis-/charge profiles of Fe_2O_3@C between 0.04 and 3.0 V and (d) the long-term constant current cycling performance at a voltage range of 0.04–3.0 V. Source: Zhang et al. [66]/with permission of John Wiley & Sons. (e–j) SEM and TEM images of graphene–Fe_3O_4 hybrids and (i) the particle size distribution of the embedded Fe_3O_4 quantum dots (QDs). Source: Liu et al. [67]/with permission of American Chemical Society. (k) Chain-like Fe_3O_4/C/red-P assembled in the presence of a magnetic field and its long-term cycling performance. Source: Qin et al. [68]/with permission of American Chemical Society.

for SIBs, the composite showed an excellent electrochemical performance of 91% capacity retention after 200 cycles at 5 A g^{-1}. The final evaluation in sodium-ion full cells with a $Na_{0.67}(Mn_{0.67}Ni_{0.23}Mg_{0.1})O_2$ cathode revealed a rather promising gravimetric energy density of 142 Wh kg^{-1}. The findings of the studies reviewed herein are comparatively summarized in Table 4.2. The high abundance, potentially low cost, environmental friendliness, facile synthesis, and richness in chemistry, including several different oxidation states available, render iron oxides a promising energy storage material. Nonetheless, the poor electronic conductivity (specifically in the de-/lithiated/sodiated state) and the continuous cleavage and reformation of ionic bonds remain as major hurdles for long-term stable cycling, despite the great progress in designing nanostructured material composites, including electron conducting secondary phases. In fact, as shown in Table 4.2, the specific capacity of iron oxides ranges between 500 and 900 mAh g^{-1} for lithium storage and ~300 mAh g^{-1} for sodium storage. However, these high specific capacities are commonly achieved for low mass-loading electrodes and relatively limited cycle numbers. Commercial cells have to provide stable capacity for >1000 cycles ideally and capacity loadings of ~4 mAh cm^{-2} – challenges, which remain to be addressed for this material family and which require proper solutions to ensure good adhesion of the coating layer to the current collector, high mechanical stability, fast and efficient electrolyte wetting, as well as high ionic and electronic conductivities. Nevertheless, prior to addressing these "practical" issues, the main hurdles remain: (i) the investigation of the electrochemical reaction mechanism to overcome the large voltage hysteresis and, by this, the improvement of the relatively low-energy storage efficiency; (ii) the wide operational voltage range to obtain high-capacity values, limiting the overall energy density of the resulting full cells, which would eventually provide significantly varying output voltages.

4.3.2 Cobalt Oxide (CoO, Co_3O_4)

Also, the cobalt oxides CoO and Co_3O_4 have attracted great attention as potential electrode-active materials due to their advanced redox properties. As an anode material in LIBs, CoO can deliver a high theoretical specific capacity (716 mAh g^{-1}) according to the reaction: $CoO + 2Li^+ + 2e^- \leftrightarrows Co + Li_2O$. As an example, Cao and coworkers [76] fabricated CoO nanowire clusters (NWCs) consisting of ultra-small nanoparticles (c. 10 nm) following a hydrothermal synthesis method. These nanowires (NWs) were directly grown on the copper current collector. At a dis-/charge rate of 1*C* (i.e. 716 mA g^{-1}), the resulting electrode delivered a high capacity of 1516 mAh g^{-1}, which was almost maintained (c. 1331 mAh g^{-1}) even at 5*C*. Passerini and coworkers [41] fabricated a CoO–Co–C nanocomposite via the *in situ* carbothermal reduction of Co_3O_4. When compared with the original Co_3O_4 nanoparticles, the CoO–Co–C composite showed enhanced long-term cycling stability and CE. Based on an *in situ* X-ray diffraction (XRD) analysis, they found that the metallic cobalt remains electrochemically inactive within the composite, while the obtained specific capacity is mainly derived from the comprised CoO.

Table 4.2 Overview of reported electrochemical performances for selected iron oxide–based active materials and composites.

Materials	LIBs or SIBs	Active material mass loading ($mg\ cm^{-2}$)	Composite material loading ($mg\ cm^{-2}$)	Specific capacity ($mAh\ g^{-1}$)	Specific current ($mA\ g^{-1}$)	Capacity retention (based on charge capacity)/number of cycles	References
α-Fe_2O_3 nanotube	LIBs	NA	NA	510	100	36%/100	[47]
α-Fe_2O_3 nanoflakes	LIBs	0.175	0.175	680	65	83%/80	[48]
Spindle-like porous α-Fe_2O_3	LIBs	0.8	0.8	911	200	72%/50	[49]
srGO/Ni@Fe_2O_3	LIBs	2.2–2.6	NA	1051	50	76%/40	[53]
Hierarchical Fe_2O_3 microboxes	LIBs	NA	NA	945	200	80%/30	[54]
Carbon-coated α-Fe_2O_3	LIBs	1.35	1.5	820	130	99%/50	[55]
Carbon-coated CNT@Fe_2O_3	LIBs	NA	NA	~800	500	75%/100	[56]
Fe_3O_4@N–C	LIBs	NA	NA	1063	1000	98.4%/1000	[59]
Fe_3O_4@GS/GF	LIBs	NA	NA	1059	93	80%/150	[51]
Fe_3O_4–C nanospindles	LIBs	NA	NA	530	462	70%/80	[63]
Fe_3O_4@C@PGC	LIBs	NA	NA	998	1000	97%/100	[52]
3D porous γ-Fe_2O_3@C	SIBs	0.6	1.0	740	200	38%/200	[66]
Fe_2O_3	SIBs	NA	NA	386	100	41%/200	[69]
Fe_3O_4@C	SIBs	0.75	1.0	309	500	36%/100	[71]
Fe_3O_4@C/rGO	SIBs	NA	NA	306	100	44%/100	[72]
Fe_2O_3@NC	SIBs	~1.5	~1.875	473.7	100	>100%/100	[73]
γ-Fe_2O_3/PCF	LIBs SIBs	1.5	NA	980 291	100 100	97%/100 78%/50	[74]
S-Fe_2O_3	SIBs	0.4	NA	367	5000	91%/200	[75]

srGO, sulfated and reduced graphene oxide.

The authors reported that the metallic cobalt enhanced the electronic conductivity of the composite electrode, but, unfortunately, had a negative effect on the long-term cycling performance by catalytically inducing the electrolyte continuous decomposition (Figure 4.4a–f).

Despite these promising studies for CoO, Co_3O_4 has been investigated far more extensively, since it possesses a markedly higher theoretical capacity (890 mAh g^{-1}) and is easier to prepare. Almost any cobalt salt, such as hydroxides, carbonates, nitrates, oxalates, acetates, or sulfates, generates Co_3O_4 above 300–400 °C heat treatment in air. Li et al. [79], e.g. reported Co_3O_4 nanotubes synthesized using an anodic aluminum oxide (AAO) template, which exhibited very high cycling stability when compared with Co_3O_4 nanorods and nanoparticles. Chen and coworkers [80] reported a facile method to fabricate a composite of Co_3O_4 nanoparticles (NPs) (30–40 nm) embedded in electrically conductive graphene. When tested as anode-active material for LIBs, the Co_3O_4/graphene electrode exhibited a reversible capacity of 935 mAh g^{-1} after 30 cycles, remarkable cycling stability with high CE (above 98%), and good rate performance. Lou et al. [81] reported a facile self-supported topotactic transformation method for the synthesis of needle-like Co_3O_4 nanotubes. The as-prepared Co_3O_4 nanotubes showed ultrahigh capacity of 918 mAh g^{-1} at 50 mA g^{-1} after 30 cycles. Similarly, Li et al. [82] reported mesoporous Co_3O_4 nanowire (NW) arrays, exhibiting high capacity and good rate performance, thus, being capable of retaining 50% of the initial capacity when increasing the dis-/charge rate to 50*C*. Yan et al. [83] further modified such 1D morphologies and developed C-doped Co_3O_4 hollow nanotubes (HNTs). Such heterogeneous doping was reported to generate a local built-in electric field due to an unbalanced charge distribution, which greatly facilitates the charge transfer. As a consequence, these Co_3O_4 HNTs, assembled out of less than 10-nm nanocrystals benefitting from a "surface-locking" effect and *in situ* topotactic conversion, demonstrated remarkable lithium uptake reversibility and a capacity of 950 mAh g^{-1} after 300 cycles at 500 mA g^{-1}.

The reversible capacity of cobalt oxides in Na configuration is much lower than their theoretical capacity and the capacities obtained as Li-ion hosts. Rahman et al. [77], by using *ex situ* XRD and cyclic voltammetry experiments, proposed that the mechanism for the reversible conversion reaction of Co_3O_4 with Na would be: $Co_3O_4 + 8Na^+ + 8e^- \rightarrow 4Na_2O + 3Co$. According to the XRD analysis, after the 1st and 20th cycles, the conversion reaction is not completed in the first lithiation to 0.01 V, but progresses upon cycling. Nonetheless, the capacity maintained at 447 mAh g^{-1} after 50 cycles with about 86% capacity retention (Figure 4.4g,h). In another study, monodispersed hierarchical Co_3O_4 spheres intertwined with CNTs, thus, forming a Co_3O_4@CNTs hybrid, showed good cycling performance and rate capability with a capacity of up to 230 and 184 mAh g^{-1} at 1600 and 3200 mA g^{-1}, respectively [84]. A binder-free hierarchical CoO-active material grown on carbon nanofibers (CNFs) has been synthesized by Jiang et al. [85]. They achieved an improved electrochemical performance by controlling the amount of CoO in the CNFs. However, the capacity remained rather low with 193 mAh g^{-1} at 50 mA g^{-1} after 50 cycles. In fact, the CoO particles appeared rather large with a size of about

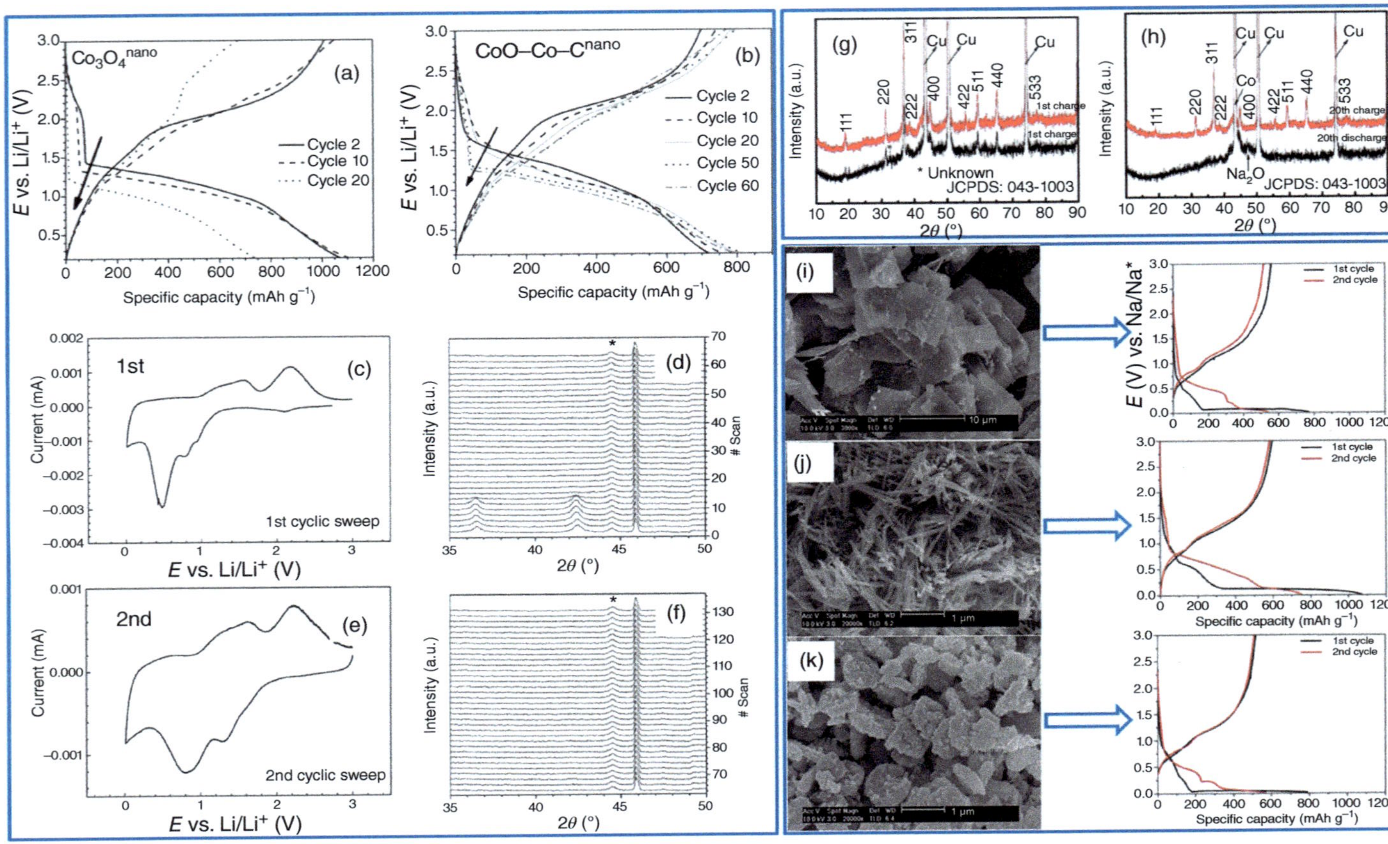

Figure 4.4 (a, b) Selected dis-/charge profiles for nanoparticulate Co_3O_4 and CoO–Co–C, revealing a better cycling stability for the latter, and (c–f) an *in situ* XRD analysis of such CoO–Co–C nanocomposite: The corresponding 1st and 2nd cyclic voltammograms and the simultaneous evolution of the XRD pattern. Source: Bresser et al. [41]/with permission of John Wiley & Sons. (g, h) *Ex situ* XRD patterns of Co_3O_4 electrodes in the fully discharged and charged states during the 1st and 20th cycles. Source: Rahman et al. [77]/Royal Society of Chemistry/CC BY 3.0. (i–k) SEM images of Co_3O_4 particles with different morphologies and the corresponding dis-/charge profiles. Source: Longoni et al. [78]/with permission of Elsevier.

5 μm and highly aggregated. Some of them lost the contact with the CNFs, which presumably led to further performance deterioration. Such issue was prevented by Li et al. [86] when fabricating a Ti-doped CoO@C composite by annealing a Co–Ti MOF precursor. The Ti-doping and the porous carbon matrix enabled high electronic conductivity, thus, contributing to a fast electron transfer and mass transport. The composite electrode delivered a capacity of 285 mAh g^{-1} at 100 mA g^{-1} after 100 cycles. Similarly, Ding and coworkers reported MOF-derived 2D networks of CoO and N-doped CNT (CoO-NCNT) as anode material for LIBs and SIBs. This composite showed an excellent cycling performance with a capacity of about 580 and 450 mAh g^{-1} at a specific current of 0.5 A g^{-1} after 2000 and 300 cycles, respectively. Comparable sodium storage capacity was reported by Xu and coworkers [87] for Co_3O_4 nanoparticles confined in rambutan-like hollow carbon spheres. The material revealed a high capacity of 409 mAh g^{-1} after 500 cycles at 0.5 A g^{-1}, corresponding to a capacity retention of 74.5%. When paired with a $Na_{0.5}MnO_2$ cathode, the resulting sodium-ion full cell delivered a capacity of 514 mAh g^{-1} at 0.1 A g^{-1} based on the anode mass loading. Generally, these studies demonstrate the advantageous effects of designing tailored (hierarchical) composites to achieve improved cycling stability and rapid electrode kinetics. Investigating the impact of the active material particle shape itself, Longoni et al. [78] synthesized Co_3O_4 with different morphologies and found that the electrochemical properties strongly are affected by the particle morphology. According to their results, a needle-like morphology allows for the realization of high specific capacities, exceeding 500 mAh g^{-1} after 50 cycles. In addition to their findings of this morphology-performance correlation, the authors unveiled that the reconversion proceeds only to the CoO phase, instead of the fully oxidized one, i.e. Co_3O_4. Further, they demonstrated that the CoO phase is more stable than Co_3O_4 during the charge/discharge process (Figure 4.4i–k) and presented a presodiation-treated electrode for conversion materials, sensibly decreasing the first cycle irreversibility and improving the cyclability. The findings of the studies reviewed herein are comparatively summarized in Table 4.3. Generally, cobalt oxides are probably the most advanced conversion-type electrode materials thanks to the high reversibility of the conversion reaction, accompanied by high specific capacities. The use of large amounts of cobalt, however, essentially excludes any reasonable application in commercial cells for cost, abundance, and toxicity issues. Besides, as for iron oxides, the main hurdle still remains: the large voltage hysteresis and the wide operational potential range. Nonetheless, the extensive work on cobalt oxides published already may render this material a good choice for a suitable model compound to better understand the underlying mechanisms.

4.3.3 Manganese Oxide (MnO, Mn_3O_4, MnO_2)

Compared with other TMOs, manganese oxide is an attractive anode material due to its abundancy in the earth's crust, low redox potential, relatively low polarization, high theoretical capacity, and environmental friendliness. However, manganese oxides are commonly affected by poor cyclability as a result of the large volume

Table 4.3 Overview of reported electrochemical performances for selected cobalt oxide-based active materials and composites.

Materials	LIBs or SIBs	Active material mass loading (mg cm^{-2})	Composite material loading (mg cm^{-2})	Specific capacity (mAh g^{-1})	Specific current (mA g^{-1})	Capacity retention (based on charge capacity)/number of cycles	References
CoO–Co–C	LIBs	2.2–2.4	NA	800	44.5	87.5%/80	[41]
Co_3O_4 nanotubes	LIBs	NA	NA	500	50	58.8%/100	[76]
Co_3O_4/graphene	LIBs	NA	NA	~935	50	>100%/30	[79]
Needlelike Co_3O_4 nanotubes	LIBs	NA	NA	918	50	97%/30	[80]
Co_3O_4 NW arrays	LIBs	NA	NA	700	111	NA	[81]
C-doped Co_3O_4 HNTs	LIBs	1	1.25	950	500	NA	[82]
Nanostructured Co_3O_4	SIBs	NA	NA	447	25	85%/50	[83]
Co_3O_4@CNTs	SIBs	0.8	1	487	NA	47%/20	[77]
CNF/CoO-4	LIBs SIBs	NA	NA	530 193	200 50	NA	[84]
ZnO-Co_3O_4@C	LIBs SIBs	1.5–1.7	NA	1785 684	200 200	NA	[85]
Ti-doped-CoO@C	LIBs SIBs	1–1.5	1.42–2.14	1108 285	200 100	>100%/150 ~100%/100	[86]
CoO-NCNTs	LIBs SIBs	1.05	1.5	583 450	500	87.4%/2000 86.8%/2000	[87]
Co_3O_4	SIBs	0.5–1	0.67–1.33	500	89	NA	[78]

NA, not available.

changes during the conversion reaction and the low electronic conductivity, both degrading the electrochemical performance. Therefore, improving the mechanical strength and electronic conductivity of manganese oxide-based electrodes is essential. An N-doped, cauliflower-like porous MnO@C/N composite prepared through a one-pot hydrothermal reaction, for example, showed good electrochemical performance as anode material for both LIBs and SIBs [88]. When employed as Li host, the material retained a capacity exceeding 830 mAh g^{-1} after 300 cycles at 500 mA g^{-1}. For SIBs, the reversible capacity was much lower. In fact, only 123 mAh g^{-1} were achieved after 200 cycles at 0.1 A g^{-1}. In a different approach, Huang and coworkers [89] synthesized MnO nanoparticles encapsulated in 3D mesoporous interconnected carbon networks (3D MnO-MICN) via a microwave-polyol process. When studied as anode-active material for LIBs, the 3D MnO-MICN nanocomposite exhibited a highly reversible capacity above 1220 mAh g^{-1} and an average CE of ~99% for more than 200 cycles at a specific current of 0.2 A g^{-1}. Chu et al. [90] reported MnO@Mn_3O_4 nanoparticles embedded in an N-doped porous carbon framework (NPCF) derived from Mn-organic clusters (MnO@Mn_3O_4/NPCF). Such composite displayed an excellent lithium-storage performance (1500 mAh g^{-1} at 0.2 A g^{-1} after 270 cycles), though a detailed analysis indicated that this high capacity essentially originated from capacitive Li^+ accumulation. Additionally, first-principle calculations revealed that a strong polarization and electronic interaction existed at the interfaces in the Mn_3O_4/NPCF composite, which effectively strengthens the interaction between the MnO@Mn_3O_4 nanoparticles and the NPCF. Meanwhile, the presence of defects in NPCF decreased the diffusion barrier, thus, enhancing the pseudo-capacitive Li^+ storage, leading to higher reversible capacities and an enhanced long-term cycling stability (Figure 4.5a–e). Diao et al. [92] developed bicomponent $(FeO)_x(MnO)_{1-x}$ encapsulated in an amorphous carbon matrix as negative active material for LIBs. The authors controlled the calcination temperature of the precursor to tune the bicomponent compound with varying stoichiometry. The thus-optimized $(FeO)_{0.198}(MnO)_{0.802}$, obtained at 800 °C, displayed the best performance, which they assigned to optimized electronic interactions between the single TMOs FeO and MnO as well as the carbon matrix. Good electrochemical performance has also been reported for coaxial manganese oxide/carbon nanotube arrays (MnO_2/CNT), which had been synthesized by using porous alumina as the template [93]. The highly conductive CNT core offers enhanced electron transport to the MnO_2 shell and effectively buffers the occurring volume changes. Employing rGO instead of CNTs, Dai and coworkers [91] reported a two-step solution-phase method for fabricating Mn_3O_4/rGO. This composite showed an unprecedented high capacity (~900 mAh g^{-1}), based on the mass of Mn_3O_4, as well as good rate capability and cycling stability (Figure 4.5f–j). Similarly, N-doped carbon-coated MnO embedded in graphene ribbons (IGR–MnO–C) showed good cycling stability and high reversible lithium storage capacity [94]. Xiao et al. [95] synthesized yolk-shell MnO@C nanodiscs via a carbothermal reduction process. The *in situ*-formed void space allows for buffering the occurring volume changes of MnO and the uniform carbon shell improved the electronic conductivity. As a result, such yolk-shell MnO@C displayed excellent cycling stability and high reversible capacity with

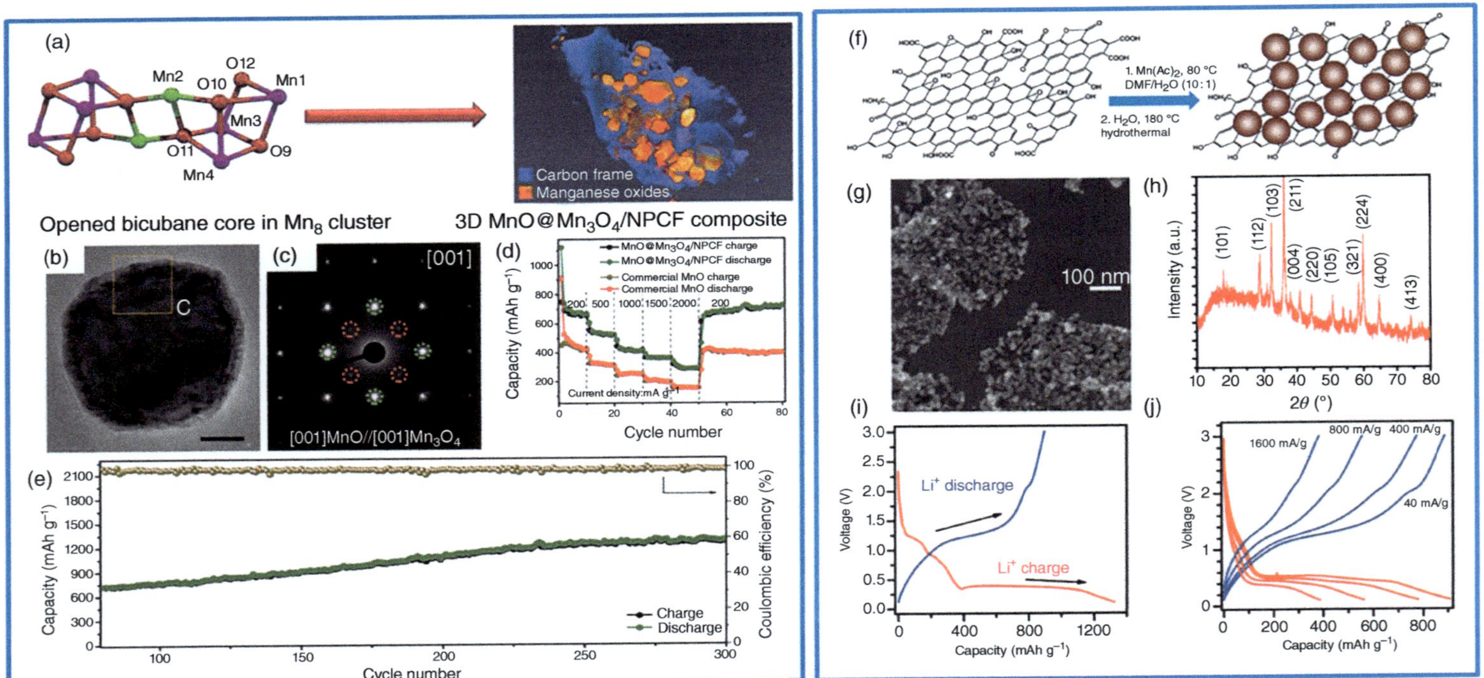

Figure 4.5 (a) Schematic illustration of the synthesis process of the MnO@Mn_3O_4/NPCF composite; (b) TEM image of a single MnO@Mn_3O_4 and (c) the corresponding selected area electron diffraction (SAED) pattern; (d) Rate capability and (e) long-term constant current cycling performance. Source: Chu et al. [90]/with permission of John Wiley & Sons. (f) Schematic presentation of the two-step synthesis of Mn_3O_4/rGO: (g) The corresponding SEM image and (h) XRD pattern of such Mn_3O_4/rGO hybrid; The dis-/charge profiles for (i) the first cycle and (j) at varying specific currents. Source: Wang et al. [91]/with permission of American Chemical Society.

605 mAh g^{-1} after 600 cycles at 1 A g^{-1}. Focusing on the structure of the manganese oxide itself rather than the incorporation of secondary electronically conductive phases, Abruña and coworkers [96] synthesized sponge-like nanosized Mn_3O_4 as Li-ion host. The initial reversible capacity was 869 mAh g^{-1} with a first cycle CE of 65%. Even after 40 cycles, a high reversible capacity of ~800 mAh g^{-1} was achieved.

Similarly, Weng et al. [97] developed a novel approach to prepare void-confined ultrasmall MnO_2 using porous SiO_2 as sacrificial template. The resulting material was characterized by a flake-like morphology with a flake width of several hundred nanometers. Electrodes based on these MnO_2 flakes exhibited a high reversible capacity of 570 mAh g^{-1} as sodium host and a capacity retention of about 70% after 500 cycles when applying a specific current of 150 mA g^{-1}. Based on an *in situ* X-ray absorption near-edge spectroscopy (XANES) study, the authors showed that the electrochemical reaction occurs via a two-phase conversion reaction, i.e. (Mn(III)-$O_{1.5}$ + Na^+ + e^- $\leftrightarrows$ 1/2Na_2O + Mn(II)-O) and non-Mn-centered redox reactions, such as the formation of a polymer-like film and surface space-charge layer (Figure 4.6a–f).

Porous MnO@C nanorods were synthesized via carbonization of a Mn-based MOF precursor. This MnO@C hybrid exhibited a reversible capacity of 260 mAh g^{-1} after 100 cycles at a specific current of 50 mA g^{-1} [99]. Li et al. [100] reported feather-like MnO_2 grown on carbon paper by a hydrothermal method and applied this composite as negative active material for SIBs, providing a rather high reversible capacity of 300 mAh g^{-1} after 400 cycles at 0.1 A g^{-1}. Peng and coworkers [98] synthesized ultrafine MnO nanoparticles, with a particle size of 4 nm, anchored on nitrogen-doped CNTs (NDCT@MnO) as anode-active material for SIBs. They found that the structure and the Na-storage behavior of the resulting nanocomposites was highly influenced by the carbonization temperature. The composite provides high rate capability and long-term cycling performance. In particular, the composite showed a reversible specific capacity (709 mAh g^{-1} at a specific current of 0.1 A g^{-1}) and high capacity retention of 536 mAh g^{-1} after 250 cycles at 0.2 A g^{-1}. Even at higher specific currents (5 A g^{-1}), a capacity of 273 mAh g^{-1} was retained after 3000 cycles (Figure 4.6g–k). Table 4.4 comparatively summarizes the findings of the herein reviewed studies on manganese oxides. In fact, given that the general issues of conversion-type materials (voltage hysteresis, limited cycle life, etc.) and the relatively low electronic conductivity of such materials might be overcome, manganese oxides would be an ideal candidate for commercial applications thanks to the high abundance, low cost, environmental friendliness, and biocompatibility of manganese. Moreover, manganese oxide shows the lowest de-/lithiation/sodiation potentials within a rather narrow voltage range, favoring energy density and stable output voltages on the full-cell level. Especially for sodium storage, though, the capacity retention is still too low and the cycle life is rather limited. Consequently, the detailed reaction mechanism and the underlying thermodynamics and kinetics remain to be elucidated particularly for sodium battery applications.

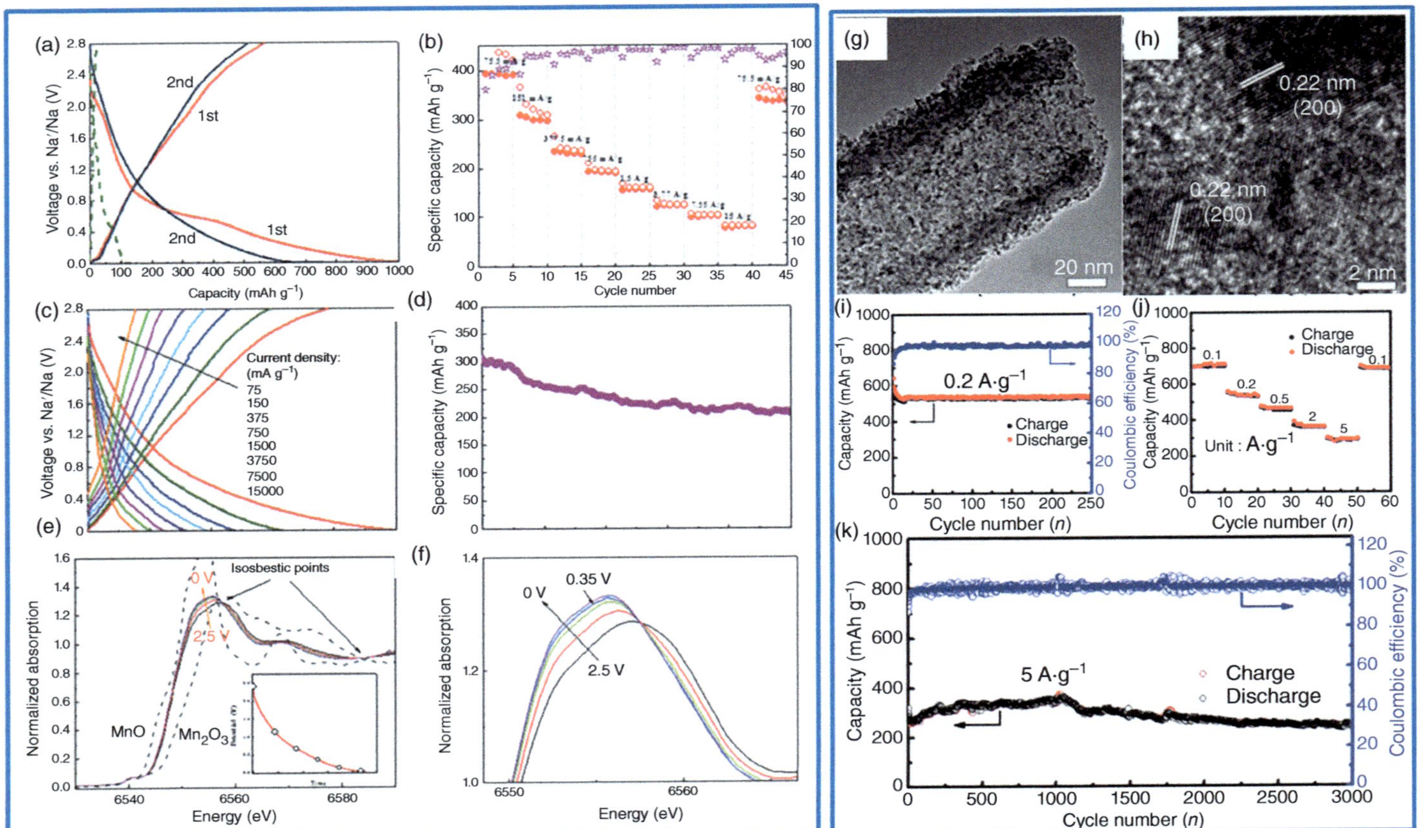

Figure 4.6 (a–d) Electrochemical performance of ultrafine MnO_2 flakes as active material for SIBs, including the dis-/charge profiles, rate capability, and long-term constant current cycling. (e, f) Operando synchrotron XANES spectra acquired during the second discharge (inset) of the ultrafine-MnO_2 electrode. Source: Weng et al. [97]/with permission of Royal Society of Chemistry. (g, h) TEM and HRTEM images of NDCT@MnO. (i–k) Electrochemical performance of NDCT@MnO. Source: He et al. [98]/with permission of American Chemical Society.

Table 4.4 Overview of reported electrochemical performances for selected manganese oxide–based active materials and composites.

Materials	LIBs or SIBs	Active material mass loading (mg cm^{-2})	Composite material loading (mg cm^{-2})	Specific capacity (mAh g^{-1})	Specific current (mA g^{-1})	Capacity retention (based on charge capacity)/ number of cycles	References
MnO@C/N	LIBs SIBs	NA	NA	837 123	500 100	>100%/300 91.1%/200	[88]
MnO-MICN	LIBs	NA	NA	1224	200	NA	[89]
MnO@Mn_3O_4/NPCF	LIBs	1 ± 0.2	NA	1500	200	>100%/300	[90]
$(FeO)_{0.198}$–$(MnO)_{0.802}$	LIBs	~0.6	~0.75	1523	200	119%/300	[92]
MnO_2/CNT	LIBs	NA	NA	500	50	NA	[93]
Mn_3O_4/rGO	LIBs	NA	NA	~900	40	NA	[91]
IGR-MnO-C	LIBs	~1	~1.25	904	500	>100%/500	[94]
MnO@C-YS	LIBs	~1	~1.42	605	1000	111%/600	[95]
Mn_3O_4	LIBs	NA	NA	800	30	92%/40	[96]
UF-MnO_2	SIBs	NA	NA	567	150	>70%/500	[97]
MnO@C	SIBs	NA	NA	260	50	46.7%/100	[99]
MnO_2/CP	SIBs	~2	NA	~300	100	63%/400	[100]
NDCT@MnO-7	SIBs	0.8	NA	536	200	~100/250	[98]

UF, ultrafine; CP, carbon paper.

4.3.4 Copper Oxide (Cu_2O, CuO)

Copper oxides, Cu_2O and CuO, are promising conversion-type anode materials due to their environmentally benignity and reversible theoretical capacities of 375 and 674 mAh g^{-1}, respectively. Studies on Cu_xO in various morphologies have been reported. Spherical and phase-pure CuO nanoparticles were synthesized by the controlled pyrolysis of Cu-based MOFs, as reported by Banerjee et al. [101]. Similarly, Chen and coworkers [102] fabricated CuO/Cu_2O composite hollow polyhedrons from Cu-based MOFs. CuO hollow nanocubes have been synthesized by a controlled oxidation reaction of Cu_2O nanocubes [103]. Unique Cu_2O–CuO–TiO_2 hollow nanocages were synthesized via a self-templated hydrothermal method [104]. The large void space in this material could solve the critical volume expansion issue during the charge/discharge cycles, although it reduces the volumetric capacity. As anode materials for LIBs, the fabricated Cu_2O–CuO–TiO_2 hollow nanocages exhibited superior reversible capacity (700 mAh g^{-1} at 50 mA g^{-1}) for over 80 cycles. In a slightly different approach, combining two different conversion materials, 3D hierarchical Co_3O_4/CuO nanowire (NW) heterostructure arrays have been successfully fabricated based on CuO NW arrays directly grown on Ni foam (Figure 4.7a–e), which are further evaluated as carbon- and binder-free electrodes for high-performance LIBs. When compared with the single Co_3O_4 nanosheets and CuO NWs, these composites exhibited a synergistic effect with enhanced electrochemical performance, maintaining a specific capacity of 1191 mAh g^{-1} after 200 cycles at a specific current of 200 mA g^{-1}, corresponding to a capacity retention of 90.9%, which is higher than that of the single Co_3O_4 nanosheets, providing only 664 mAh g^{-1} after 100 cycles – just like the single CuO NWs with 554 mAh g^{-1} after 100 cycles [105].

Following a comparably simple approach, nanocomposites with CNTs or graphene have been investigated to attain a stable cycling performance [107, 108]. For instance, a ternary Cu_2O/CuO/rGO composite with a 3D hierarchical nanostructure was fabricated by a hydrothermal method. The electrochemical reaction mechanism of this composite was studied by *in situ* Raman spectroscopy. The characteristic Raman peak of CuO becomes less intense upon reduction, indicating that CuO transforms to Cu_2O and Cu. Upon reoxidation, the characteristic peak gradually increases, which supports the (partial) reversibility of the conversion process (Figure 4.7f–k) [106].

The employment of copper oxide–based anodes in lithium-ion full cells has been investigated *inter alia* by Verrelli et al. [109, 110]. In a CuO–MCMB/$LiNi_{0.5}Mn_{1.5}O_4$ full cell, ~110 mAh g^{-1} was achieved after 100 cycles at a specific current of 148 mA g^{-1} and the gravimetric energy density was 420 Wh kg^{-1}. When using $Li_{0.85}Ni_{0.46}Cu_{0.1}Mn_{1.49}O_4$ as the cathode, the battery operates for over 50 cycles at a specific current of 48.7 mA g^{-1} with an average voltage of 3.4 V and a stable capacity of 100 mAh g^{-1} [110]. These full cells showed good cycling stability and remarkable rate capability. When combining such anodes with an ionic liquid-based electrolyte, the CuO-MCMB electrode delivered a high capacity of 580 mAh g^{-1} with a CE exceeding 98%. For the full cell, comprising a $LiNi_{0.5}Mn_{1.5}O_4$ cathode,

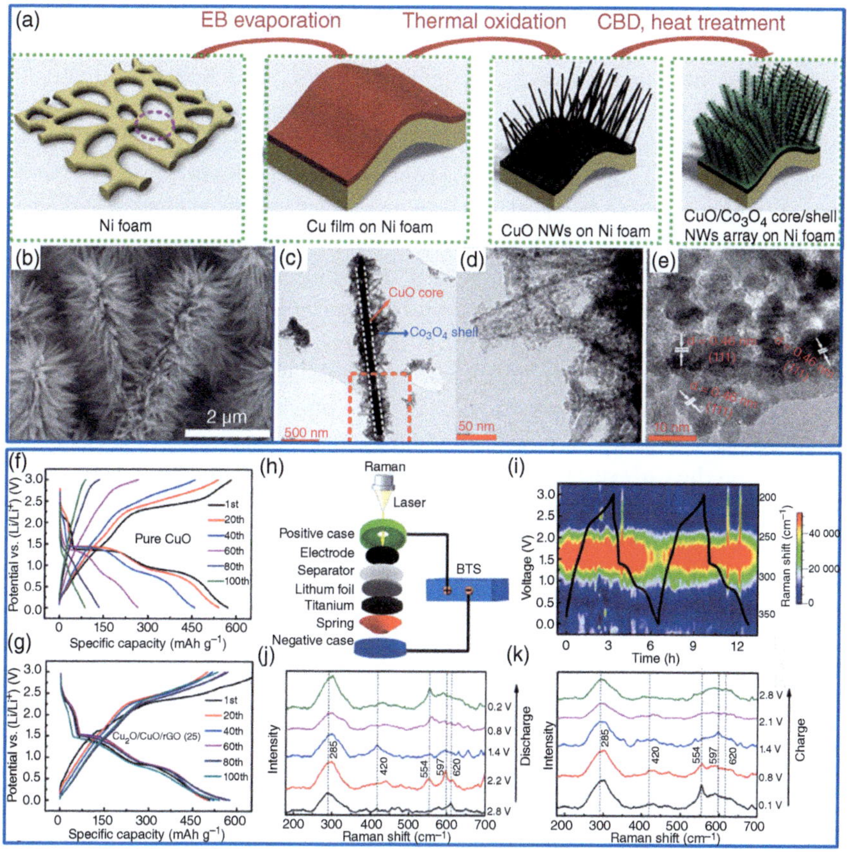

Figure 4.7 (a) Schematic illustration of the fabrication procedure of CuO/Co_3O_4 core/shell heterostructure nanowire arrays. (b) SEM and (c–e) TEM images of such CuO/Co_3O_4 core/shell nanowires. Source: Wang et al. [105]/with permission of Elsevier. (f, g) Discharge and charge profiles of electrodes based on (f) CuO and (g) $Cu_2O/CuO/rGO$ for selected cycles. (h) Setup for the *in situ* Raman analysis of the ternary $Cu_2O/CuO/rGO$ composite with (i) the resulting 2D Raman data including the dis-/charge profiles and the transformed 1D spectra for the (j) discharge and (k) charge process. Source: Wu et al. [106]/with permission of John Wiley & Sons.

the use of the ionic liquid electrolyte resulted in a full cell with an average operating voltage of 3 V and specific capacity about 120 mAh g^{-1} [111]. However, the cycling performance and voltage-profile retention remained limited. To address these issues, the authors further modified the conversion-type electrode material by combining CuO and Fe_2O_3 in the composition. Such a mixed conversion anode coupled with a spinel $Li_{1.35}Ni_{0.48}Fe_{0.1}Mn_{1.72}O_4$ cathode eventually allowed for a substantially enhanced full-cell performance [112]. The working voltage of the cell was 3.6 V and a capacity of 110 mAh g^{-1} was achieved after 100 cycles at specific current of 148 mA g^{-1} with an average CE of more than 99%. In another study on lithium-ion full cells, Zhang et al. [113] combined a CuO nanorod array (CNA) anode and a high-voltage spinel $LiNi_{0.5}Mn_{1.5}O_4$ cathode. These full cells exhibited good cycling performance (84% capacity retention after 100 cycles at 0.5*C*) and good rate performance (~240 mAh g^{-1} at 10*C*; 1*C* = 674 mA g^{-1}).

Following these promising results, copper oxides were investigated also for sodium-ion applications. Flexible and porous CNAs, for example, synthesized by simply etching commercial Cu foils, were used as binder-free anodes for SIBs, delivering a specific capacity of 291 mAh g^{-1} after 450 cycles at 200 mA g^{-1} and a capacity retention of 45.2% [114]. Similarly, N-doped carbon-coated CuO nanorod arrays (NC-CuO) were grown on Cu net to realize freestanding electrodes for both LIBs and SIBs [115]. For SIBs, the NC-CuO-based electrodes provided a reversible capacity of 214.97 mAh g^{-1} after 100 cycles at a specific current of 0.5 A g^{-1}, benefitting from the extensive space between the single CuO nanorods and the carbon coating surface layer, which enabled high structural stability and electronic conductivity. In a very comparable study, Ni et al. [116] reported CuO nanoarrays grown on 3D Cu foam. Instead of applying a carbon coating, however, they deposited TiO_2 thin films on the CuO surface by atomic layer deposition (R-CuO), with the goal to decrease the volume variation upon de-/sodiation. As a matter of fact, the resulting R-CuO electrodes showed good cycling performance and excellent rate capability (306 mAh g^{-1} after 200 cycles at 1.2 A g^{-1} and 155 mAh g^{-1} after 1000 cycles at 3.0 A g^{-1}). When coupled with $Na_{0.67}(Ni_{0.23}Mg_{0.1}Mn_{0.67})O_2$ as cathode-active material, the resulting full-cell delivered a specific energy of 142 Wh kg^{-1}. Liu et al. [117] found that the surface coating with carbon and Au allows for controlling the elongation rate of the nanowires along the <110> growth direction and for increasing the electrochemical reaction kinetics, which then led to an enhanced sodium storage performance. These results highlight the significant impact of surface coatings on the electrochemical performance. Additionally, according to an *in situ* TEM characterization, the authors observed that the electrochemical conversion of CuO is partially irreversible: Part of the Cu that is formed during the sodiation remains upon desodiation, while both oxide phases, i.e. CuO and Cu_2O, were detected, which explains at least part of the large initial capacity loss. Wang et al. [118] reported porous CuO nanowires as anode for SIBs with a first-cycle sodiation capacity of 640 mAh g^{-1} (at 50 mA g^{-1}) and maintaining and 303 mAh g^{-1} after 50 cycles. Based on an *ex situ* characterization, the authors proposed that, upon sodiation, Na^+ ions are inserted into CuO to form Cu_2O and Na_2O. In a second step, Cu_2O decomposes into Cu nanoparticles embedded in the Na_2O matrix. During the desodiation process, Cu nanoparticles are first oxidized to form Cu_2O and then converted back to CuO. However, Liu et al. [119] proposed a rather different kinetic evolution of the morphology and phase conversion in CuO nanowires during sodiation (Figure 4.8). According to the *in situ* TEM investigation, the authors proposed that the eventual sodiation products of CuO were $Na_6Cu_2O_6$, Na_2O, and Cu according to the following reactions:

$$2CuO + 2Na^+ + 2e^- \rightarrow Cu_2O + Na_2O \tag{4.4}$$

$$Cu_2O + Na_2O \rightarrow 2NaCuO \tag{4.5}$$

$$7NaCuO + Na^+ + e^- \rightarrow Na_6Cu_2O_6 + Na_2O + 5Cu \tag{4.6}$$

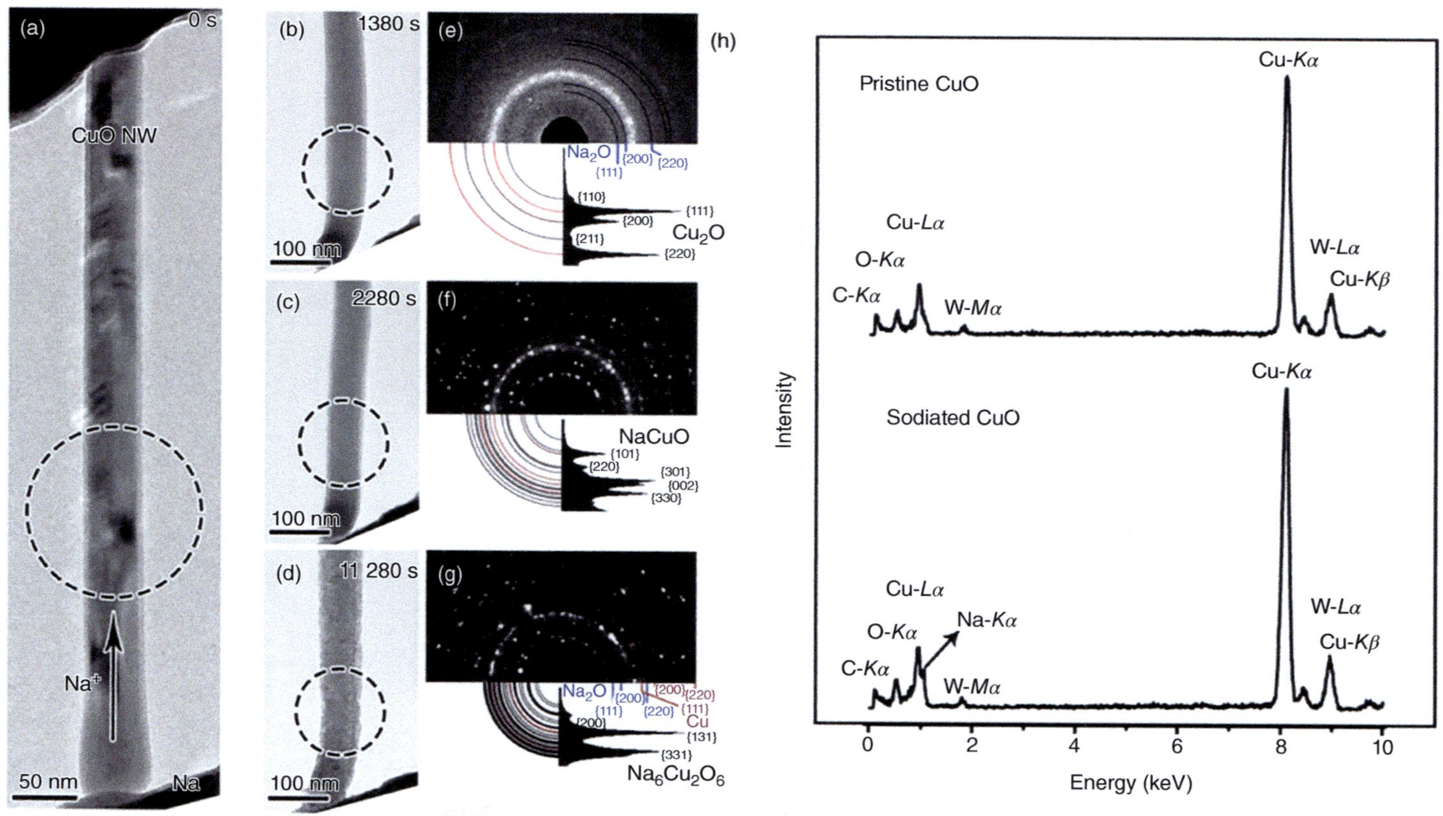

Figure 4.8 (a) TEM image of a pristine CuO NW. (b–d) TEM images of the sodiated CuO NW at different degrees of sodiation and the corresponding electron diffraction patterns of the circled area. (e–h) EDX spectra of pristine and sodiated CuO NWs. Source: Liu et al. [119]/with permission of Royal Society of Chemistry.

This discrepancy may originate from the different electrolytes used. In the latter work, in fact, a solid electrolyte was utilized instead of a liquid electrolyte. Further work is required to clarify whether different types of electrolytes may influence the sodiation process or if other parameters are affecting the reaction mechanism. Targeting to increase the electronic conductivity of the active material, Chen and coworkers [120] fabricated submicron CuO/C sphere-shaped particles by means of an aerosol spray pyrolysis. The CuO nanoparticles (~10 nm) were uniformly embedded in the carbon matrix, which significantly enhanced the electronic conductivity of the composite and helped to buffer the volume variations of CuO during discharge. Based on the unique structure, the composite delivered a capacity of 402 mAh g^{-1} after 600 cycles at 200 mA g^{-1}. Table 4.5 provides a comparative summary of the findings reviewed herein on copper oxides. Just like manganese or iron oxides, they are characterized by nontoxicity and facile synthesis – though of much higher cost – and beside the high electronic conductivity of metallic copper, the corresponding oxides are essentially insulators. To address this issue, researchers have introduced CNTs or graphene as secondary phases and developed well-designed nanostructures like nanorods, nanowires, or hollow structures (see Table 4.5) to improve the performance. However, hollow structures generally result in low volumetric energy densities and the required (template-assisted) synthesis procedures are commonly complicated and, thus, costly. Moreover, the cycle life and capacity retention of copper oxide anodes remains improvable – also for lithium-ion anode applications. The development of effective strategies to improve the rate capability and cycling performance is needed.

4.3.5 Nickel Oxide (NiO)

Also NiO, a p-type wide band gap semiconductor, has been considered as an attractive anode candidate for LIBs. The theoretical specific capacity of NiO is 718 mAh g^{-1} based on the reversible reduction of the oxide to metallic nickel and lithium oxide, i.e. $NiO + 2Li^+ + 2e^- \leftrightarrows Ni + Li_2O$ [121]. Nevertheless, just as for other conversion-type anodes, the cycling stability and rate capability of NiO remain to be improved. To overcome these issues, for instance, 2D sheet-like nanoparticles were synthesized, offering promising performance as anode material in LIBs. Vertically aligned NiO nanowalls [122] showed a reversible capacity of ~638 mAh g^{-1} after 85 cycles at a specific current of 895 mA g^{-1} in the voltage range of 0.005–3.0 V vs. Li. NiO–graphene sheet-on-sheet [123] nanostructures have been prepared and showed good cycling performance. After 40 cycles, at a specific current of 71.8 mA g^{-1}, a high reversible capacity of 1031 mAh g^{-1} was achieved, corresponding to a capacity retention of 97.6% with respect to the initial capacity. In consideration of the excellent electronic conductivity and mechanical flexibility of graphene, further studies, e.g. by Chen and coworkers [124] found that NiO nanosheets (NSs) strongly interact with graphene via a C–O–Ni bridge. According to their findings, this was the main reason for achieving higher reversible capacities and remarkable rate capability for such composites. Based on first-principle calculations, the oxygen bridges mainly derived from the interaction of epoxy/hydroxyl

Table 4.5 Overview of reported electrochemical performances for selected copper oxide–based active materials and composites.

Materials	LIBs or SIBs	Active material mass loading (mg cm^{-2})	Composite material loading (mg cm^{-2})	Specific capacity (mAh g^{-1})	Specific current (mA g^{-1})	Capacity retention (based on charge capacity)/ number of cycles	References
Pure CuO nanoparticles	LIBs	5	6.5	210	2000	~90%/40	[101]
CuO/Cu_2O hollow polyhedrons	LIBs	NA	NA	480	100	66%/248	[102]
CuO nanostructures	LIBs	NA	NA	560	150	>100%/50	[103]
Cu_2O–CuO–TiO_2	LIBs	NA	NA	700	50	58%/85	[104]
Co_3O_4/CuO	LIBs	NA	NA	1191	200	90.9%/200	[105]
CuO-CNT nanomicrospheres	LIBs	~2.5	~3.125	500	67	NA	[107]
CuO/CNT	LIBs	NA	NA	650	NA (0.1*C*)	100%/100	[108]
Cu_2O/CuO/rGO	LIBs	NA	NA	550	500	NA	[106]
CuO–MCMB	LIBs	~3.5	~4.375	500–400	52	NA	[109]
CuO–MCMB	LIBs	0.8	1	400	50	77%/50	[111]
CuO–Fe_2O_3–MCMB	LIBs	1.6–2.8	2.0–3.5	500	120	NA	[112]
CuO nanorod array	LIBs	0.8	1	666	500	91%/100	[113]
CuO nanorod arrays	SIBs	~2.0	NA	290.6	200	45.2%/450	[114]
NC-CuO array	LIBs SIBs	1.8	NA	551.66 214.97	500 500	>100%/200 NA	[115]
R-CuO nanoarray	SIBs	6	NA	155	3000	82%/1000	[116]
Porous CuO nanowires	SIBs	NA	NA	303	50	47.3%/50	[118]
10-CuO/C	SIBs	NA	NA	402	200	NA	[120]

groups, present on graphene, and the Ni atoms of the NiO NSs. This intensive and strong interaction of Ni with oxygenated graphene was found to promote the rapid transition of electrons from graphene to NiO, thus, allowing for the enhanced reversibility of de-/lithiation reaction. As a result, the NiO NS/graphene electrode showed high reversible capacity (883 mAh g^{-1} after 50 cycles at a specific current of 50 mA g^{-1}) and good rate performance. Introducing a second conversion-type TMO, Guo et al. [125] synthesized CuO@NiO ball-in-ball microspheres starting from Cu–Ni bimetallic organic frameworks. The resulting product displayed an elemental gradient (Cu-rich surface and Ni-rich core), matching the lithium reactivity sequence of the two metal oxides (CuO and NiO). The composite showed a reversible, larger-than-theoretical capacity of 1061 mAh g^{-1}, and the capacity was fully maintained after 200 cycles. Using metallic nickel as precursor, Wang et al. [126] employed a low-temperature thermal oxidation of Ni foam to fabricate nanostructured NiO electrodes with excellent rate capability for high-power LIBs. The electrode achieved a capacity of more than 375 mAh g^{-1} at a dis-/charge rate as high as 10*C*. Moreover, the extensive void spaces in the electrode helped to buffer the huge volume changes during cycling.

Comparing the lithiation and sodiation mechanism and kinetics with a set of complementary experimental and theoretical techniques, He et al. [127] found that only the "shrinking-core mode" was observed during sodiation process. A passivation layer of Na_2O, which formed at an early stage, starting from the outer particle surface, blocked the further sodiation and, thus, resulted in sluggish kinetics. Molecular dynamic simulation revealed that the sodiation pathway, leading to this shrinking-core mode, derives from a layer-by-layer reaction that occurs during sodiation. During the discharge process, however, the Li antisite defects will significantly distort the local NiO lattices, which facilitated the further insertion of and conversion with Li (Figure 4.9). In another direct comparison, Zou et al. [128] reported the preparation of a hierarchical NiO/Ni/graphene composite with a hollow ball-in-ball nanostructure, which revealed outstanding electrochemical performance as anode-active material for both LIBs and SIBs (962 mAh g^{-1} after 1000 cycles at 2 A g^{-1} for LIBs and ~200 mAh g^{-1} after 200 cycles at 200 mA g^{-1} for SIBs). The advantageous performance was assigned to the well-designed hierarchical hollow structure, which mitigates the volume changes of NiO during cycling while the graphene matrix provides percolating electron-conductive pathways to facilitate charge transfer and enable the formation of a stable SEI. Another hierarchical hybrid structure has been prepared by Yang et al. [129], constructing hollow Ni–NiO nanoparticles embedded in porous carbon nanosheets (Ni-NiO/PCNs). When testing this composite as anode material for SIBs, they recorded a specific capacity of 235 mAh g^{-1} after 5000 cycles at 1 A g^{-1}. The good electrochemical performance was attributed to the beneficial hollow core–shell structure and the high conductivity of the metallic Ni particles and carbon sheets, both providing continuous electron transfer pathways, which facilitates the electrochemical reaction kinetics. Table 4.6 provides a summary of the findings reviewed herein. Generally, NiO has a kind of "intermediate" position in the list of conversion-type anode materials. The theoretical specific capacity of 714 mAh g^{-1} is higher than for

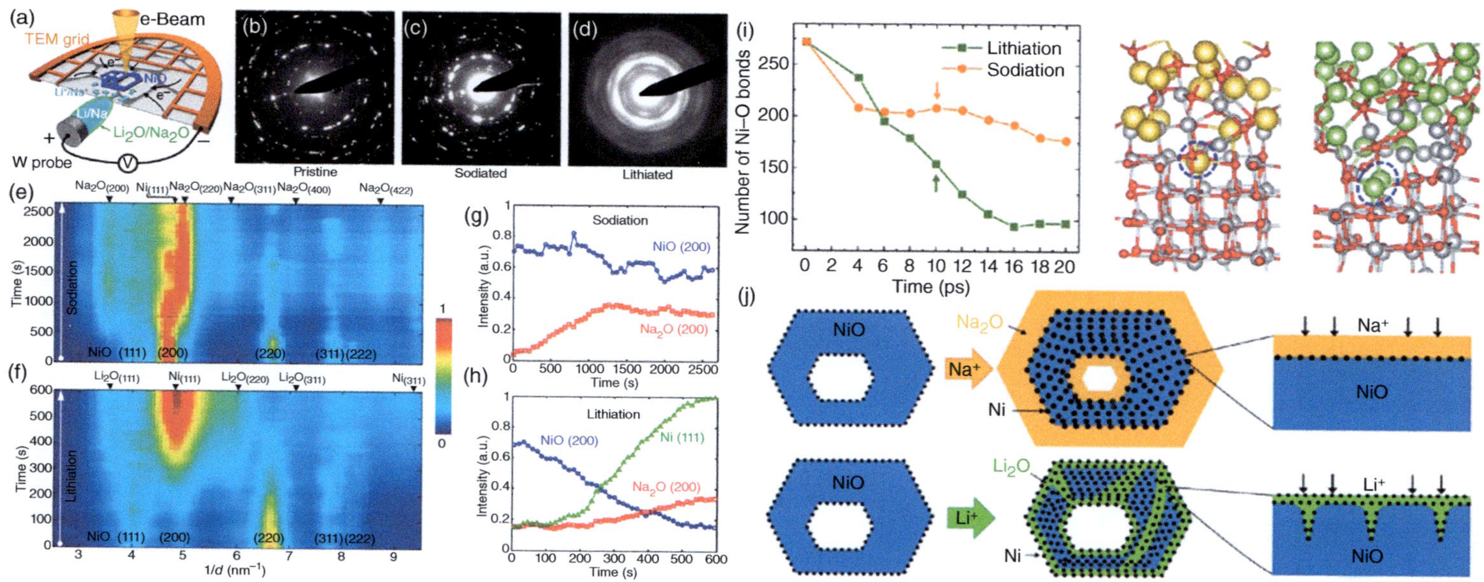

Figure 4.9 (a) Schematic illustration of the *in situ* TEM setup: (b–d) Electron diffraction patterns for the (b) pristine, (c) sodiated, and (d) lithiated phase and (e, f) colored intensity profiles as a function of reaction time for the (e) sodiation and (f) lithiation process and (g, h) the corresponding radial integration of the electron diffraction data as a function of time for the (g) sodiation and (h) lithiation with the Bragg peaks labeling. (i) Ab initio molecular dynamics (MD) simulations for the sodiation (in yellow) and lithiation (in green) of NiO. (j) Schematic illustrations of the different reaction modes for the electrochemical sodiation and lithiation. Source: He et al. [127]/with permission of American Chemical Society.

copper oxides, but lower than for iron oxides, for instance. Nickel is not as costly as cobalt or copper, but more expensive than iron or manganese. Moreover, nickel is not as toxic as cobalt, but less biocompatible than iron or copper. It might provide high volumetric capacities thanks to its high density of 6.67 $g\,cm^{-3}$, but suffers rather slow reaction kinetics; and while a variety of appealing strategies have been used to mitigate these issues, including the synthesis of very thin nanosheets or the incorporation of secondary carbon or metal oxide-based materials (see Table 4.6), the performance of NiO-based electrodes remains so far below expectation in terms of specific capacities and cycle life.

4.3.6 Ruthenium Oxide (RuO_2)

RuO_2 has a high theoretical capacity of 806 $mAh\,g^{-1}$ and metallic-like conductivity. Nonetheless, its high price rules out any application in large-scale commercialized batteries. As one of the pioneers for the investigation of this material as active material in LIBs, Maier and coworkers [130, 131] investigated the electrochemical reactivity with lithium over a wide voltage range from 0.05 to 4.3 V. They showed that this material can homogeneously and heterogeneously host lithium, offering a high capacity of 1130 $mAh\,g^{-1}$ and extremely high first cycle CE of 98%. Based on *ex situ* Raman and XRD studies, they revealed that the Li storage in RuO_2 involves three reaction steps: (i) Ru/Li_2O nanocomposite formation, (ii) Li-containing surface film formation, and (iii) the interfacial deposition of Li within the Ru/Li_2O matrix. The cycling performance, however, remained stable only for the first three cycles in this voltage range, before it faded abruptly because of the large volume variation. In a later study, they found that amorphous nanosized RuO_2 offers enhanced performance compared with bulk crystalline RuO_2. As a case study, the open-circuit voltage of amorphous RuO_2 has an excess potential of 0.6 V in comparison with the potentials of crystalline RuO_2 with 60-nm and 10-μm sizes [132]. Nevertheless, to find out how far and in which way the quasi-amorphicity is kinetically linked with the electrochemical performance still requires further clarification. Gregorczyk et al. [133] used *in situ* TEM to investigate the reaction mechanism during the lithiation process of single-crystalline RuO_2 nanowires. They found that a large volume expansion about 95% occurred upon lithiation, and 26% remained irreversible, i.e. was not recovered after delithiation. Furthermore, they observed a noticeable surface roughening and lithium embrittlement, while the initial reaction from crystalline RuO_2 via an intermediate phase of Li_xRuO_2 to the fully lithiated mixed phase of Ru/Li_2O appeared only partially reversible. In line with this finding, Kim et al. [134] investigated the reaction mechanism of RuO_2 and found that, at the beginning of the discharge process, an intermediate phase-assisted transformation of RuO_2 to $LiRuO_2$ takes place and that the latter subsequently decomposes to nanosized Ru and Li_2O. In addition to its theoretical capacity, an excess reversible capacity was observed for lithiation voltages lower than 0.5 V, although the comprised Ru did not show any redox activity. Following the earlier work by Maier and coworkers, they assigned this to the Li storage at the grain boundaries between nanosized Ru^0 and Li_2O.

Table 4.6 Overview of reported electrochemical performances for selected nickel oxide-based active materials and composites.

Materials	LIBs or SIBs	Active material mass loading ($mg\,cm^{-2}$)	Composite material loading ($mg\,cm^{-2}$)	Specific capacity ($mAh\,g^{-1}$)	Specific current ($mA\,g^{-1}$)	Capacity retention (based on charge capacity)/number of cycles	References
NiO nanowall	LIBs	0.1675	NA	~638	895	~98%/85	[122]
NiO-GNS	LIBs	NA	NA	1031	71.8	97.6%/40	[123]
NiONS/graphene composite	LIBs	1–2	1.25–2.25	1000	50	90%/50	[124]
CuO@NiO spheres	LIBs	~1.6	~2	1061	100	~87%/200	[125]
NiO/Ni/graphene	LIBs	NA	NA	962	2000	>100%/1000	[128]
Ni–NiO/PCN	SIBs	NA	NA	235.4	1000	84.2%/5000	[129]

GNS, graphene nanosheets.

4.3.7 Other Transition Metal Oxides

Other TMOs such as Cr_2O_3 and MoO_x also offer high theoretical capacity values. Cr_2O_3, for instance, theoretically provides a capacity of 1058 mAh g^{-1} as lithium-ion anode, while simultaneously offering the lowest lithiation voltage (i.e. 0.2 V) among all TMOs described herein. In agreement with the proposed conversion mechanism, experiments revealed the complete reduction of the oxide to metallic chrome and Li_2O. The best charge capacity reported so far for the first cycle, however, has been well below 900 mAh g^{-1}, which led to rather low first cycle CEs. It has been reported that this limitation might be related to the partially irreversible reconversion of Cr_2O_3 to CrO only upon recharge to 3 V [135]. In addition, the cycling stability of powder-based electrodes remained rather poor in the existing literature without applying, e.g. a carbonaceous coating [136, 137]. With regard to the volumetric capacity of such electrodes, though, the overall content of the carbon coating should be well controlled. The use of thin films has been reported as another effective strategy to enhance the cycling stability of Cr_2O_3 [138] thanks to the extensive contact between the electrochemically active component and the current collector. Nevertheless, thin-film electrodes are intrinsically restricted in terms of volumetric and gravimetric capacities when taking into account also the current collector (despite the absence of additional conductive additives, for instance), since for films thicker than 175 nm a dramatic performance fading has been observed.

Molybdenum oxides like MoO_3 have a theoretical specific capacity of 1117 mAh g^{-1}, but have been rarely investigated and accordingly there is little literature available to clearly state that these materials can achieve this value also upon continuous cycling. In fact, commonly, a rapid capacity decay has been observed after several dis-/charge cycles [139, 140]. Another molybdenum oxide that has been investigated is MoO_2. First cycle discharge capacities well above its theoretical value (i.e. 838 mAh g^{-1}) have been reported [141] and stabilized capacities of about 700 mAh g^{-1} when the mesoporosity of the electrode was carefully controlled [142, 143]. Ku et al. [144] developed an approach to further improve the electrochemical performance of MoO_2 by introducing a thermal activation at 120 °C during the first cycle. As a result, they observed capacities of around 800 mAh g^{-1} after 30 cycles. According to their X-ray photoelectron spectroscopy (XPS) and Raman spectroscopy analysis, MoO_2 was indeed reduced to Mo^0 upon lithiation and fully reoxidized to MoO_2 during the subsequent delithiation.

4.4 Summary and Outlook

The large family of conversion materials, i.e. compounds undergoing largely reversible conversion reactions with charge carriers like lithium or sodium, is an attractive class of materials to investigate whether the replacement of lithium by sodium might aid to overcome the previously identified challenges for the reversible charge storage. The reaction mechanisms in SIB electrodes can be, indeed, quite different due to the different phase stabilities. Throughout the sodiation and desodiation processes, sodium can form various stable oxides, like Na_2O, Na_2O_2, and NaO_2, while lithium-based materials are known to form basically Li_2O and Li_2O_2 – mostly Li_2O. Hence, conversion reactions in SIBs may have more intermediate phases, offering a somehow richer electrochemical reactivity and potentially different reaction pathways. By studying and understanding the similarities and differences in crystallography, kinetics, and thermodynamics of the Li- and Na-based conversion reactions, better battery performance may be achieved, including the design of new material chemistries for both Li-ion and Na-ion applications. As such, a comprehensive comparative investigation may also shed further light into the greatest challenge for conversion-type active materials – the voltage hysteresis between the charge and discharge process. In fact, TMO-based conversion-type negative electrode-active materials are, at present, mostly of academic interest until the remaining challenges will be overcome. The rather simple synthesis pathways for (transition) metal oxides allow for the development of advanced material architectures, targeting improved cycling stabilities and reaction kinetics. As such, the corresponding research enables exploring the limits of what might be possible also for other kind of electrode materials and allow for the investigation of, e.g. electron-conducting self-healing binders to overcome the issues related to the low electronic conductivity of the oxides and the pronounced volume variation upon de-/lithiation/sodiation, respectively. Also, the optimization of the electrolyte composition for these highly reactive electrode materials – especially in the lithiated/sodiated state – might pave the way for finding advanced electrolyte compositions also for less catalytically active electrode materials. At least as (if not more) important, though, research efforts will have to focus on the identification of the underlying reaction mechanisms to understand them in detail, which is so far attracting less attention – not least because it is far more complicated. In this regard, this review article shall serve also as motivation for scientists to systematically study their reactivity as a function of the TM cation, its oxidation state, and the reversibly stored alkali metal cation. These aspects are becoming even more important when dealing with binary metal oxides – especially when different storage mechanisms are combined. It appears that the meaningful combination of electrochemical techniques and advanced physicochemical characterization techniques provide a viable way to do so, which, however, requires collaborative efforts of chemists, physicists, theoreticians, and engineers. While great progress has been achieved already since the seminal paper of Tarascon and coworkers, the detailed reaction mechanism and the definite origin of the voltage hysteresis is still unclear – for the electrochemical reaction with lithium as well as with sodium. As such, pure conversion-type anodes (and essentially also cathodes) are not foreseen to play a decisive role for commercial batteries in the near future. As a matter of fact, elements like cobalt, copper, chrome, or ruthenium will most likely not play a great role even if the underlying electrochemical reaction mechanism will be completely understood and methods to overcome these will have been found. However, materials like iron oxide or manganese oxide would be very attractive from an environmental and economic point of view. This certainly ambitious goal may justify the further investigation also of the less attractive ones, in case something valuable can be learnt. More importantly, though, it justifies the search for the major breakthrough(s) in understanding, as this would have a great impact on the technology.

References

1 Etacheri, V., Marom, R., Elazari, R. et al. (2011). Challenges in the development of advanced Li-ion batteries: a review. *Energy & Environmental Science* 4 (9): 3243–3262.

2 Reddy, M.V., Rao, G.V.S., and Chowdari, B.V.R. (2013). Metal oxides and oxysalts as anode materials for Li ion batteries. *Chemical Reviews* 113 (7): 5364–5457.

3 Guo, Y.-G., Hu, J.-S., and Wan, L.-J. (2008). Nanostructured materials for electrochemical energy conversion and storage devices. *Advanced Materials* 20 (15): 2878–2887.

4 Bresser, D., Passerini, S., and Scrosati, B. (2016). Leveraging valuable synergies by combining alloying and conversion for lithium-ion anodes. *Energy & Environmental Science* 9 (11): 3348–3367.

5 Nayak, P.K., Yang, L., Brehm, W. et al. (2018). From lithium-ion to sodium-ion batteries: advantages, challenges, and surprises. *Angewandte Chemie International Edition* 57 (1): 102–120.

6 Nykvist, B. and Nilsson, M. (2015). Rapidly falling costs of battery packs for electric vehicles. *Nature Climate Change* 5: 329.

7 Yamada, Y., Iriyama, Y., Abe, T. et al. (2009). Kinetics of lithium ion transfer at the interface between graphite and liquid electrolytes: effects of solvent and surface film. *Langmuir* 25 (21): 12766–12770.

8 Loeffler, N., Bresser, D., Passerini, S. et al. (2015). Secondary lithium-ion battery anodes: from first commercial batteries to recent research activities. *Platinum Metals Review* 59 (1): 34–44.

9 Tarascon, J.M. (2010). Is lithium the new gold? *Nature Chemistry* 2 (6): 510.

10 Olivetti, E.A., Ceder, G., Gaustad, G.G. et al. (2017). Lithium-ion battery supply chain considerations: analysis of potential bottlenecks in critical metals. *Joule* 1 (2): 229–243.

11 Hwang, J.Y., Myung, S.T., and Sun, Y.K. (2017). Sodium-ion batteries: present and future. *Chemical Society Reviews* 46 (12): 3529–3614.

12 Vaalma, C., Buchholz, D., Weil, M. et al. (2018). A cost and resource analysis of sodium-ion batteries. *Nature Reviews Materials* 3: 18013.

13 Mei, J., Liao, T., Kou, L. et al. (2017). Two-dimensional metal oxide nanomaterials for next-generation rechargeable batteries. *Advanced Materials* 29 (48): 1700176.

14 Huang, S.Y., Kavan, L., Exnar, I. et al. (1995). Rocking chair lithium battery based on nanocrystalline TiO_2 (anatase). *Journal of the Electrochemical Society* 142 (9): L142–L144.

15 Chen, Z., Belharouak, I., Sun, Y.K. et al. (2013). Titanium-based anode materials for safe lithium-ion batteries. *Advanced Functional Materials* 23 (8): 959–969.

16 Ohzuku, T., Ueda, A., and Yamamoto, N. (1995). Zero-strain insertion material of $Li[Li_{1/3}Ti_{5/3}]O_4$ for rechargeable lithium cells. *Journal of the Electrochemical Society* 142 (5): 1431–1435.

17 Bresser, D., Paillard, E., Copley, M. et al. (2012). The importance of "going nano" for high power battery materials. *Journal of Power Sources* 219: 217–222.

18 Senguttuvan, P., Rousse, G., Seznec, V. et al. (2011). $Na_2Ti_3O_7$: lowest voltage ever reported oxide insertion electrode for sodium ion batteries. *Chemistry of Materials* 23 (18): 4109–4111.

19 Pan, H., Lu, X., Yu, X. et al. (2013). Sodium storage and transport properties in layered $Na_2Ti_3O_7$ for room-temperature sodium-ion batteries. *Advanced Energy Materials* 3 (9): 1186–1194.

20 Yan, Z., Liu, L., Shu, H. et al. (2015). A tightly integrated sodium titanate-carbon composite as an anode material for rechargeable sodium ion batteries. *Journal of Power Sources* 274: 8–14.

21 Doeff, M.M., Cabana, J., and Shirpour, M. (2014). Titanate anodes for sodium ion batteries. *Journal of Inorganic and Organometallic Polymers and Materials* 24 (1): 5–14.

22 Xu, T., Wei, W., Gordin, M.L. et al. (2010). Lithium-ion batteries for stationary energy storage. *JOM Journal of the Minerals Metals and Materials Society* 62 (9): 24–30.

23 Pan, F., Yang, J., Huang, Q. et al. (2014). Redox targeting of anatase TiO_2 for redox flow lithium-ion batteries. *Advanced Energy Materials* 4 (15): 1400567.

24 Choi, J.W. and Aurbach, D. (2016). Promise and reality of post-lithium-ion batteries with high energy densities. *Nature Reviews Materials* 1 (4): 16013.

25 Obrovac, M.N. and Chevrier, V.L. (2014). Alloy negative electrodes for Li-ion batteries. *Chemical Reviews* 114 (23): 11444–11502.

26 Idota, Y. (1997). Tin-based amorphous oxide: a high-capacity lithium-ion-storage material. *Science* 276 (5317): 1395–1397.

27 Courtney, I.A. (1997). Key factors controlling the reversibility of the reaction of lithium with SnO_2 and Sn_2BPO_6 glass. *Journal of the Electrochemical Society* 144 (9): 2943.

28 Belliard, F. and Irvine, J. (2001). Electrochemical performance of ball-milled $ZnO–SnO_2$ systems as anodes in lithium-ion battery. *Journal of Power Sources* 97: 219–222.

29 Read, J., Foster, D., Wolfenstine, J. et al. (2001). SnO_2-carbon composites for lithium-ion battery anodes. *Journal of Power Sources* 96 (2): 277–281.

30 Pelliccione, C.J., Ding, Y., Timofeeva, E.V. et al. (2015). In situ XAFS study of the capacity fading mechanisms in ZnO anodes for lithium-ion batteries. *Journal of the Electrochemical Society* 162 (10): A1935–A1939.

31 Poizot, P., Laruelle, S., Grugeon, S. et al. (2000). Nano-sized transition-metal oxides as negative-electrode materials for lithium-ion batteries. *Nature* 407 (6803): 496–499.

32 Zhang, H., Hasa, I., and Passerini, S. (2018). Beyond insertion for Na-ion batteries: nanostructured alloying and conversion anode materials. *Advanced Energy Materials* 8 (17): 1702582.

33 Cabana, J., Monconduit, L., Larcher, D. et al. (2010). Beyond intercalation-based Li-ion batteries: the state of the art and challenges of electrode materials reacting through conversion reactions. *Advanced Materials* 22 (35): E170–E192.

34 Nitta, N. and Yushin, G. (2014). High-capacity anode materials for lithium-ion batteries: choice of elements and structures for active particles. *Particle and Particle Systems Characterization* 31 (3): 317–336.

35 Klein, F., Jache, B., Bhide, A. et al. (2013). Conversion reactions for sodium-ion batteries. *Physical Chemistry Chemical Physics* 15 (38): 15876–15887.

36 Ellis, B.L., Knauth, P., and Djenizian, T. (2014). Three-dimensional self-supported metal oxides for advanced energy storage. *Advanced Materials* 26 (21): 3368–3397.

37 Zhang, J. and Yu, A. (2015). Nanostructured transition metal oxides as advanced anodes for lithium-ion batteries. *Science Bulletin* 60 (9): 823–838.

38 Yuan, C., Wu, H.B., Xie, Y. et al. (2014). Mixed transition-metal oxides: design, synthesis, and energy-related applications. *Angewandte Chemie International Edition* 53 (6): 1488–1504.

39 Mouad, D., Naoaki, Y., Kei, K. et al. (2014). Negative electrodes for Na-ion batteries. *Physical Chemistry Chemical Physics* 16 (29): 15007–15028.

40 Luo, W., Shen, F., Bommier, C. et al. (2016). Na-ion battery anodes: materials and electrochemistry. *Accounts of Chemical Research* 49 (2): 231–240.

41 Bresser, D., Paillard, E., Niehoff, P. et al. (2014). Challenges of "going nano": enhanced electrochemical performance of cobalt oxide nanoparticles by carbothermal reduction and in situ carbon coating. *ChemPhysChem* 15 (10): 2177–2185.

42 Wu, C., Dou, S.X., and Yu, Y. (2018). The state and challenges of anode materials based on conversion reactions for sodium storage. *Small* 14 (22): e1703671.

43 Valvo, M., Lindgren, F., Lafont, U. et al. (2014). Towards more sustainable negative electrodes in Na-ion batteries via nanostructured iron oxide. *Journal of Power Sources* 245: 967–978.

44 Eguia-Barrio, A., Castillo-Martínez, E., Klein, F. et al. (2017). Electrochemical performance of CuNCN for sodium ion batteries and comparison with ZnNCN and lithium ion batteries. *Journal of Power Sources* 367: 130–137.

45 Ong, S.P., Chevrier, V.L., Hautier, G. et al. (2011). Voltage, stability and diffusion barrier differences between sodium-ion and lithium-ion intercalation materials. *Energy & Environmental Science* 4 (9): 3680.

46 Zhu, X.J., Zhu, Y.W., Murali, S. et al. (2011). Nanostructured reduced graphene oxide/Fe_2O_3 composite as a high-performance anode material for lithium ion batteries. *ACS Nano* 5 (4): 3333–3338.

47 Chen, J., Xu, L.N., Li, W.Y. et al. (2005). α-Fe_2O_3 nanotubes in gas sensor and lithium-ion battery applications. *Advanced Materials* 17 (5): 582–586.

48 Reddy, M.V., Yu, T., Sow, C.H. et al. (2007). α-Fe_2O_3 nanoflakes as an anode material for Li-ion batteries. *Advanced Functional Materials* 17 (15): 2792–2799.

49 Xu, X., Cao, R., Jeong, S. et al. (2012). Spindle-like mesoporous α-Fe_2O_3 anode material prepared from MOF template for high-rate lithium batteries. *Nano Letters* 12 (9): 4988–4991.

50 Brandt, A., Winter, F., Klamor, S. et al. (2013). An investigation of the electrochemical delithiation process of carbon coated α-Fe_2O_3 nanoparticles. *Journal of Materials Chemistry A* 1 (37): 11229–11236.

51 Wei, W., Yang, S.B., Zhou, H.X. et al. (2013). 3D graphene foams cross-linked with pre-encapsulated Fe_3O_4 nanospheres for enhanced lithium storage. *Advanced Materials* 25 (21): 2909–2914.

52 He, C.N., Wu, S., Zhao, N.Q. et al. (2013). Carbon-encapsulated Fe_3O_4 nanoparticles as a high-rate lithium ion battery anode material. *ACS Nano* 7 (5): 4459–4469.

53 Backert, G., Oschmann, B., Tahir, M.N. et al. (2016). Facile hybridization of Ni@Fe_2O_3 superparticles with functionalized reduced graphene oxide and its application as anode material in lithium-ion batteries. *Journal of Colloid and Interface Science* 478: 155–163.

54 Zhang, L., Wu, H.B., Madhavi, S. et al. (2012). Formation of Fe_2O_3 microboxes with hierarchical shell structures from metal-organic frameworks and their lithium storage properties. *Journal of the American Chemical Society* 134 (42): 17388–17391.

55 Brandt, A. and Balducci, A. (2013). Ferrocene as precursor for carbon-coated α-Fe_2O_3 nano-particles for rechargeable lithium batteries. *Journal of Power Sources* 230: 44–49.

56 Wang, B., Chen, J.S., Wu, H.B. et al. (2011). Quasiemulsion-templated formation of α-Fe_2O_3 hollow spheres with enhanced lithium storage properties. *Journal of the American Chemical Society* 133 (43): 17146–17148.

57 Aravindan, V., Arun, N., Shubha, N. et al. (2016). Overlithiated $Li_{1+x}Ni_{0.5}Mn_{1.5}O_4$ in all one dimensional architecture with conversion type α-Fe_2O_3: a new approach to eliminate irreversible capacity loss. *Electrochimica Acta* 215: 647–651.

58 Verrelli, R., Brescia, R., Scarpellini, A. et al. (2014). A lithium ion battery exploiting a composite Fe_2O_3 anode and a high voltage $Li_{1.35}Ni_{0.48}Fe_{0.1}Mn_{1.72}O_4$ cathode. *RSC Advances* 4 (106): 61855–61862.

59 Liu, B.Q., Zhang, Q., Jin, Z.S. et al. (2018). Uniform pomegranate-like nanoclusters organized by ultrafine transition metal oxide@nitrogen-doped carbon subunits with enhanced lithium storage properties. *Advanced Energy Materials* 8 (7): 170234.

60 Luo, J.S., Liu, J.L., Zeng, Z.Y. et al. (2013). Three-dimensional graphene foam supported Fe_3O_4 lithium battery anodes with long cycle life and high rate capability. *Nano Letters* 13 (12): 6136–6143.

61 Zhou, G.M., Wang, D.W., Li, F. et al. (2010). Graphene-wrapped Fe_3O_4 anode material with improved reversible capacity and cyclic stability for lithium ion batteries. *Chemistry of Materials* 22 (18): 5306–5313.

62 Ban, C.M., Wu, Z.C., Gillaspie, D.T. et al. (2010). Nanostructured Fe_3O_4/SWNT electrode: binder-free and high-rate Li-ion anode. *Advanced Materials* 22 (20): E145–E149.

63 Zhang, W.M., Wu, X.L., Hu, J.S. et al. (2008). Carbon coated Fe_3O_4 nanospindles as a superior anode material for lithium-ion batteries. *Advanced Functional Materials* 18 (24): 3941–3946.

64 Ming, J., Kwak, W.J., Youn, S.J. et al. (2014). Lithiation of an iron oxide-based anode for stable, high-capacity lithium-ion batteries of porous carbon-Fe_3O_4/$Li[Ni_{0.59}Co_{0.16}Mn_{0.25}]O_2$. *Energy Technology* 2 (9–10): 778–785.

65 Ming, H., Ming, J., Oh, S.M. et al. (2014). Surfactant-assisted synthesis of Fe_2O_3 nanoparticles and F-doped carbon modification toward an improved Fe_3O_4@CF_x/$LiNi_{0.5}Mn_{1.5}O_4$ battery. *ACS Applied Materials & Interfaces* 6 (17): 15499–15509.

66 Zhang, N., Han, X.P., Liu, Y.C. et al. (2015). 3D porous γ-Fe_2O_3@C nanocomposite as high-performance anode material of Na-ion batteries. *Advanced Energy Materials* 5 (5): 1401123.

67 Liu, H., Jia, M.Q., Zhu, Q.Z. et al. (2016). 3D-0D graphene-Fe_3O_4 quantum dot hybrids as high-performance anode materials for sodium-ion batteries. *ACS Applied Materials & Interfaces* 8 (40): 26878–26885.

68 Qin, G.H., Duan, J.Y., Yang, Y.C. et al. (2018). Magnetic field facilitated resilient chain-like Fe_3O_4/C/red P with superior sodium storage performance. *ACS Applied Materials & Interfaces* 10 (7): 6441–6452.

69 Jiang, Y.Z., Hu, M.J., Zhang, D. et al. (2014). Transition metal oxides for high performance sodium ion battery anodes. *Nano Energy* 5: 60–66.

70 Park, D.-Y. and Myung, S.-T. (2014). Carbon-coated magnetite embedded on carbon nanotubes for rechargeable lithium and sodium batteries. *ACS Applied Materials & Interfaces* 6 (14): 11749–11757.

71 Zhou, Y.P., Sun, W.P., Rui, X.H. et al. (2016). Biochemistry-derived porous carbon-encapsulated metal oxide nanocrystals for enhanced sodium storage. *Nano Energy* 21: 71–79.

72 Qi, H., Cao, L., Li, J. et al. (2019). Thin carbon layer coated porous Fe_3O_4 particles supported by rGO sheets for improved stable sodium storage. *ChemistrySelect* 4 (9): 2668–2675.

73 Guo, T., Liao, H., Ge, P. et al. (2018). Fe_2O_3 embedded in the nitrogen-doped carbon matrix with strong C–O–Fe oxygen-bridge bonds for enhanced sodium storages. *Materials Chemistry and Physics* 216: 58–63.

74 Chen, Y., Yuan, X., Yang, C. et al. (2019). γ-Fe_2O_3 nanoparticles embedded in porous carbon fibers as binder-free anodes for high-performance lithium and sodium ion batteries. *Journal of Alloys and Compounds* 777: 127–134.

75 Ni, J., Sun, M., and Li, L. (2019). Highly efficient sodium storage in iron oxide nanotube arrays enabled by built-in electric field. *Advanced Materials* 31 (41): e1902603.

76 Cao, K., Jiao, L., Liu, Y. et al. (2015). Ultra-high capacity lithium-ion batteries with hierarchical CoO nanowire clusters as binder free electrodes. *Advanced Functional Materials* 25 (7): 1082–1089.

77 Rahman, M.M., Glushenkov, A.M., Ramireddy, T. et al. (2014). Electrochemical investigation of sodium reactivity with nanostructured Co_3O_4 for sodium-ion batteries. *Chemical Communications* 50 (39): 5057–5060.

78 Longoni, G., Fiore, M., Kim, J.H. et al. (2016). Co_3O_4 negative electrode material for rechargeable sodium ion batteries: an investigation of conversion reaction mechanism and morphology-performances correlations. *Journal of Power Sources* 332: 42–50.

79 Li, W.Y., Xu, L.N., and Chen, J. (2005). Co_3O_4 nanomaterials in lithium-ion batteries and gas sensors. *Advanced Functional Materials* 15 (5): 851–857.

80 Wu, Z.S., Ren, W.C., Wen, L. et al. (2010). Graphene anchored with Co_3O_4 nanoparticles as anode of lithium ion batteries with enhanced reversible capacity and cyclic performance. *ACS Nano* 4 (6): 3187–3194.

81 Lou, X.W., Deng, D., Lee, J.Y. et al. (2008). Self-supported formation of needle-like Co_3O_4 nanotubes and their application as lithium-ion battery electrodes. *Advanced Materials* 20 (2): 258–262.

82 Li, Y.G., Tan, B., and Wu, Y.Y. (2008). Mesoporous Co_3O_4 nanowire arrays for lithium ion batteries with high capacity and rate capability. *Nano Letters* 8 (1): 265–270.

83 Yan, C., Zhu, Y., Li, Y. et al. (2018). Local built-in electric field enabled in carbon-doped Co_3O_4 nanocrystals for superior lithium-ion storage. *Advanced Functional Materials* 28 (7): 1705951.

84 Jian, Z., Liu, P., Li, F. et al. (2014). Monodispersed hierarchical Co_3O_4 spheres intertwined with carbon nanotubes for use as anode materials in sodium-ion batteries. *Journal of Materials Chemistry A* 2 (34): 13805.

85 Jiang, J., Ma, C., Ma, T. et al. (2019). A novel CoO hierarchical morphologies on carbon nanofiber for improved reversibility as binder-free anodes in lithium/sodium ion batteries. *Journal of Alloys and Compounds* 794: 385–395.

86 Li, J., Wang, D., Zhou, J. et al. (2019). Ti-doped ultra-small CoO nanoparticles embedded in an octahedral carbon matrix with enhanced lithium and sodium storage. *ChemElectroChem* 6 (3): 917–927.

87 Pang, Y., Chen, S., Xiao, C. et al. (2019). MOF derived CoO-NCNTs two-dimensional networks for durable lithium and sodium storage. *Journal of Materials Chemistry A* 7 (8): 4126–4133.

88 Chen, T., Wu, Z., Xiang, W. et al. (2017). Cauliflower-like MnO@C/N composites with multiscale, expanded hierarchical ordered structures as electrode materials for lithium- and sodium-ion batteries. *Electrochimica Acta* 246: 931–940.

89 Luo, W., Hu, X., Sun, Y. et al. (2013). Controlled synthesis of mesoporous MnO/C networks by microwave irradiation and their enhanced lithium-storage properties. *ACS Applied Materials & Interfaces* 5 (6): 1997–2003.

90 Chu, Y., Guo, L., Xi, B. et al. (2018). Embedding MnO@Mn_3O_4 nanoparticles in an N-doped-carbon framework derived from Mn-organic clusters for efficient lithium storage. *Advanced Materials* 30 (6): 1704244.

91 Wang, H.L., Cui, L.F., Yang, Y.A. et al. (2010). Mn_3O_4-graphene hybrid as a high-capacity anode material for lithium ion batteries. *Journal of the American Chemical Society* 132 (40): 13978–13980.

92 Diao, G., Balogun, M.-S., Tong, S.-Y. et al. (2018). Low-valence bicomponent $(FeO)_x(MnO)_{1-x}$ nanocrystals embedded in amorphous carbon as high-performance anode materials for lithium storage. *Journal of Materials Chemistry A* 6 (31): 15274–15283.

93 Reddy, A.L.M., Shaijumon, M.M., Gowda, S.R. et al. (2009). Coaxial MnO_2/carbon nanotube array electrodes for high-performance lithium batteries. *Nano Letters* 9 (3): 1002–1006.

94 Sheng, L., Jiang, H., Liu, S. et al. (2018). Nitrogen-doped carbon-coated MnO nanoparticles anchored on interconnected graphene ribbons for high-performance lithium-ion batteries. *Journal of Power Sources* 397: 325–333.

95 Xiao, Y.-C., Xu, C.-Y., Sun, X.-Y. et al. (2018). Constructing yolk-shell MnO@C nanodiscs through a carbothermal reduction process for highly stable lithium storage. *Chemical Engineering Journal* 336: 427–435.

96 Gao, J., Lowe, M.A., and Abruña, H.D. (2011). Spongelike nanosized Mn_3O_4 as a high-capacity anode material for rechargeable lithium batteries. *Chemistry of Materials* 23 (13): 3223–3227.

97 Weng, Y.T., Huang, T.Y., Lim, C.H. et al. (2015). An unexpected large capacity of ultrafine manganese oxide as a sodium-ion battery anode. *Nanoscale* 7 (47): 20075–20081.

98 He, Y., Xu, P., Zhang, B. et al. (2017). Ultrasmall MnO nanoparticles supported on nitrogen-doped carbon nanotubes as efficient anode materials for sodium ion batteries. *ACS Applied Materials & Interfaces* 9 (44): 38401–38408.

99 Zhang, X., Zhu, G., Yan, D. et al. (2017). MnO@C nanorods derived from metal-organic frameworks as anode for superiorly stable and long-life sodium-ion batteries. *Journal of Alloys and Compounds* 710: 575–580.

100 Li, H., Liu, A., Zhao, S. et al. (2018). In situ growth of a feather-like MnO_2 nanostructure on carbon paper for high-performance rechargeable sodium-ion batteries. *ChemElectroChem* 5 (21): 3266–3272.

101 Banerjee, A., Singh, U., Aravindan, V. et al. (2013). Synthesis of CuO nanostructures from Cu-based metal organic framework (MOF-199) for application as anode for Li-ion batteries. *Nano Energy* 2 (6): 1158–1163.

102 Hu, L., Huang, Y., Zhang, F. et al. (2013). CuO/Cu_2O composite hollow polyhedrons fabricated from metal-organic framework templates for lithium-ion battery anodes with a long cycling life. *Nanoscale* 5 (10): 4186–4190.

103 Park, J.C., Kim, J., Kwon, H. et al. (2009). Gram-scale synthesis of Cu_2O nanocubes and subsequent oxidation to CuO hollow nanostructures for lithium-ion battery anode materials. *Advanced Materials* 21 (7): 803–807.

104 Wang, G., Sui, Y., Zhang, M. et al. (2017). One-pot synthesis of uniform Cu_2O–CuO–TiO_2 hollow nanocages with highly stable lithium storage properties. *Journal of Materials Chemistry A* 5 (35): 18577–18584.

105 Wang, J., Zhang, Q., Li, X. et al. (2014). Three-dimensional hierarchical Co_3O_4/CuO nanowire heterostructure arrays on nickel foam for high-performance lithium ion batteries. *Nano Energy* 6: 19–26.

106 Wu, S., Fu, G., Lv, W. et al. (2018). A single-step hydrothermal route to 3D hierarchical Cu_2O/CuO/rGO nanosheets as high-performance anode of lithium-ion batteries. *Small* 14 (5): 1702667.

107 Zheng, S.F., Hu, J.S., Zhong, L.S. et al. (2008). Introducing dual functional CNT networks into CuO nanomicrospheres toward superior electrode materials for lithium-ion batteries. *Chemistry of Materials* 20 (11): 3617–3622.

108 Ko, S., Lee, J.I., Yang, H.S. et al. (2012). Mesoporous CuO particles threaded with CNTs for high-performance lithium-ion battery anodes. *Advanced Materials* 24 (32): 4451–4456.

109 Verrelli, R., Hassoun, J., Farkas, A. et al. (2013). A new, high performance CuO/$LiNi_{0.5}Mn_{1.5}O_4$ lithium-ion battery. *Journal of Materials Chemistry A* 1 (48): 15329.

110 Verrelli, R., Scrosati, B., Sun, Y.K. et al. (2014). Stable, high voltage $Li_{0.85}Ni_{0.46}Cu_{0.1}Mn_{1.49}O_4$ spinel cathode in a lithium-ion battery using a conversion-type CuO anode. *ACS Applied Materials & Interfaces* 6 (7): 5206–5211.

111 Verrelli, R., Laszczynski, N., Passerini, S. et al. (2016). Electrochemical study of a CuO-carbon conversion anode in ionic liquid electrolyte for application in Li-ion batteries. *Energy Technology: Generation, Conversion, Storage, Distribution* 4 (6): 700–705.

112 Di Lecce, D., Verrelli, R., Campanella, D. et al. (2017). A new CuO-Fe_2O_3-mesocarbon microbeads conversion anode in a high-performance lithium-ion battery with a $Li_{1.35}Ni_{0.48}Fe_{0.1}Mn_{1.72}O_4$ spinel cathode. *ChemSusChem* 10 (7): 1607–1615.

113 Zhang, W., Ma, G., Gu, H. et al. (2015). A new lithium-ion battery: CuO nanorod array anode versus spinel $LiNi_{0.5}Mn_{1.5}O_4$ cathode. *Journal of Power Sources* 273: 561–565.

114 Yuan, S., Huang, X.L., Ma, D.L. et al. (2014). Engraving copper foil to give large-scale binder-free porous CuO arrays for a high-performance sodium-ion battery anode. *Advanced Materials* 26 (14): 2273–2279, 2284.

115 Li, Y., Zhang, M., Qian, J. et al. (2019). Freestanding N-doped carbon coated CuO array anode for lithium-ion and sodium-ion batteries. *Energy Technology: Generation, Conversion, Storage, Distribution* 7 (7): 1900252.

116 Ni, J., Jiang, Y., Wu, F. et al. (2018). Regulation of breathing CuO nanoarray electrodes for enhanced electrochemical sodium storage. *Advanced Functional Materials* 28 (15): 1707179.

117 Liu, H., Zheng, H., Li, L. et al. (2018). Surface-coating-mediated electrochemical performance in CuO nanowires during the sodiation–desodiation cycling. *Advanced Materials Interfaces* 5 (4): 1701255.

118 Wang, L., Zhang, K., Hu, Z. et al. (2013). Porous CuO nanowires as the anode of rechargeable Na-ion batteries. *Nano Research* 7 (2): 199–208.

119 Liu, H., Cao, F., Zheng, H. et al. (2015). In situ observation of the sodiation process in CuO nanowires. *Chemical Communications* 51 (52): 10443–10446.

120 Lu, Y., Zhang, N., Zhao, Q. et al. (2015). Micro-nanostructured CuO/C spheres as high-performance anode materials for Na-ion batteries. *Nanoscale* 7 (6): 2770–2776.

121 Boesenberg, U., Marcus, M.A., Shukla, A.K. et al. (2014). Asymmetric pathways in the electrochemical conversion reaction of NiO as battery electrode with high storage capacity. *Scientific Reports* 4: 7133.

122 Varghese, B., Reddy, M.V., Yanwu, Z. et al. (2008). Fabrication of NiO nanowall electrodes for high performance lithium ion battery. *Chemistry of Materials* 20 (10): 3360–3367.

123 Zou, Y. and Wang, Y. (2011). NiO nanosheets grown on graphene nanosheets as superior anode materials for Li-ion batteries. *Nanoscale* 3 (6): 2615–2620.

124 Zhou, G.M., Wang, D.W., Yin, L.C. et al. (2012). Oxygen bridges between NiO nanosheets and graphene for improvement of lithium storage. *ACS Nano* 6 (4): 3214–3223.

125 Guo, W.X., Sun, W.W., and Wang, Y. (2015). Multilayer CuO@NiO hollow spheres: microwave-assisted metal organic-framework derivation and highly reversible structure-matched stepwise lithium storage. *ACS Nano* 9 (11): 11462–11471.

126 Wang, X., Li, X., Sun, X. et al. (2011). Nanostructured NiO electrode for high rate Li-ion batteries. *Journal of Materials Chemistry* 21 (11): 3571.

127 He, K., Lin, F., Zhu, Y. et al. (2015). Sodiation kinetics of metal oxide conversion electrodes: a comparative study with lithiation. *Nano Letters* 15 (9): 5755–5763.

128 Zou, F., Chen, Y.M., Liu, K. et al. (2016). Metal organic frameworks derived hierarchical hollow NiO/Ni/graphene composites for lithium and sodium storage. *ACS Nano* 10 (1): 377–386.

129 Yang, C.C., Zhang, D.M., Du, L. et al. (2018). Hollow Ni–NiO nanoparticles embedded in porous carbon nanosheets as a hybrid anode for sodium-ion batteries with an ultra-long cycle life. *Journal of Materials Chemistry A* 6 (26): 12663–12671.

130 Balaya, P., Li, H., Kienle, L. et al. (2003). Fully reversible homogeneous and heterogeneous Li storage in RuO_2 with high capacity. *Advanced Functional Materials* 13 (8): 621–625.

131 Li, H., Balaya, P., and Maier, J. (2004). Li-storage via heterogeneous reaction in selected binary metal fluorides and oxides. *Journal of the Electrochemical Society* 151 (11): A1878–A1885.

132 Delmer, O., Balaya, P., Kienle, L. et al. (2008). Enhanced potential of amorphous electrode materials: case study of RuO_2. *Advanced Materials* 20 (3): 501–505.

133 Gregorczyk, K.E., Liu, Y., Sullivan, J.P. et al. (2013). In situ transmission electron microscopy study of electrochemical lithiation and delithiation cycling of the conversion anode RuO_2. *ACS Nano* 7 (7): 6354–6360.

134 Kim, Y., Muhammad, S., Kim, H. et al. (2015). Probing the additional capacity and reaction mechanism of the RuO_2 anode in lithium rechargeable batteries. *ChemSusChem* 8 (14): 2378–2384.

135 Dupont, L., Grugeon, S., Laruelle, S. et al. (2007). Structure, texture and reactivity versus lithium of chromium-based oxides films as revealed by TEM investigations. *Journal of Power Sources* 164 (2): 839–848.

136 Hu, J., Li, H., and Huang, X.J. (2005). Cr_2O_3-based anode materials for Li-ion batteries. *Electrochemical and Solid-State Letters* 8 (1): A66–A69.

137 Hu, J., Li, H., Huang, X.J. et al. (2006). Improve the electrochemical performances of Cr_2O_3 anode for lithium ion batteries. *Solid State Ionics* 177 (26–32): 2791–2799.

138 Grugeon, S., Laruelle, S., Dupont, L. et al. (2005). Combining electrochemistry and metallurgy for new electrode designs in Li-ion batteries. *Chemistry of Materials* 17 (20): 5041–5047.

139 Lee, S.H., Kim, Y.H., Deshpande, R. et al. (2008). Reversible lithium-ion insertion in molybdenum oxide nanoparticles. *Advanced Materials* 20 (19): 3627–3632.

140 Jung, Y.S., Lee, S., Ahn, D. et al. (2009). Electrochemical reactivity of ball-milled $MoO_{(3-y)}$ as anode materials for lithium-ion batteries. *Journal of Power Sources* 188 (1): 286–291.

141 Liang, Y.G., Yang, S.J., Yi, Z.H. et al. (2005). Preparation, characterization and lithium-intercalation performance of different morphological molybdenum dioxide. *Materials Chemistry and Physics* 93 (2–3): 395–398.

142 Ji, X.L., Herle, S., Rho, Y.H. et al. (2007). Carbon/MoO_2 composite based on porous semi-graphitized nanorod assemblies from in situ reaction of tri-block polymers. *Chemistry of Materials* 19 (3): 374–383.

143 Shi, Y.F., Guo, B.K., Corr, S.A. et al. (2009). Ordered mesoporous metallic MoO_2 materials with highly reversible lithium storage capacity. *Nano Letters* 9 (12): 4215–4220.

144 Ku, J.H., Jung, Y.S., Lee, K.T. et al. (2009). Thermoelectrochemically activated MoO_2 powder electrode for lithium secondary batteries. *Journal of the Electrochemical Society* 156 (8): A688–A693.

5

Layered Na-Ion Transition-Metal Oxide Electrodes for Sodium-Ion Batteries

Baskar Senthilkumar[1,2,3], Christopher S. Johnson[4], and Premkumar Senguttuvan[1,2,3]

[1] *New Chemistry Unit, Jawaharlal Nehru Centre for Advanced Scientific Research, Jakkur, Bangalore 560064, India*
[2] *School of Advanced Materials, Jawaharlal Nehru Centre for Advanced Scientific Research, Jakkur, Bangalore 560064, India*
[3] *International Centre for Materials Science, Jawaharlal Nehru Centre for Advanced Scientific Research, Jakkur, Bangalore 560064, India*
[4] *Chemical Sciences and Engineering Division, Argonne National Laboratory, Argonne, IL 60439, United States*

5.1 Introduction

Li-ion batteries (LIBs) have attained their technological maturity and are widely applied for portable consumer electronic devices, hybrid electric vehicles (HEVs), electric vehicles (EVs), and stationary electrical energy-storage (EES) applications [1–3]. However, scarcity, increasing demand, high cost, and uneven distribution of Li (~0.065% in earth's crust) which constrain the successful implementation of LIBs for large-scale energy-storage applications. Indeed, significant research efforts are dedicated to develop new battery chemistries and, among them, sodium-ion batteries have been emerged as the frontrunner due to earth-abundant and low-cost sodium resources [4, 5]. Besides, the operating principle of sodium-ion batteries (SIBs) are akin to its Li-ion counterpart, thus enabling the rapid development of Na-ion electrode and electrolyte materials. In addition, aluminum doesn't alloy with Na metal and, therefore, it can be used as the current collector for anode in the place of expensive copper foil.

Hence, SIBs are a low-cost alternative to LIBs in large-scale EES applications. Recently, SIBs were commercialized by two startup companies, Novasis Energies, Inc. and Faradion Limited [6–8]. Faradion Ltd. developed a 400 Wh battery pack for electric bicycles using the layered cathode material $Na_aNi_{1-x-y-z}Mn_xMg_yTi_zO_2$ and a hard carbon anode. The commercialization of Na-ion batteries by the aforementioned companies demonstrates readiness level of the SIB technology [7].

Performance of the batteries, such as capacity, energy density, operating voltage, and cycle life, mostly depends on the characteristics of the electrode materials [9]. Therefore, finding suitable cathode materials, which can host large

Transition Metal Oxides for Electrochemical Energy Storage, First Edition.
Edited by Jagjit Nanda and Veronica Augustyn.

Na^+-ions (1.02 Å, ~25% larger size of Na^+ relative to Li^+), plays a vital role in realizing high-performance SIBs. So far, various types of cathode materials such as transition-metal oxides, Prussian blue analogues, polyanionic materials, and organic compounds have been studied for SIBs [10–13]. Among these cathode materials, layered metal oxides (Na_xMO_2; M = transition metal) have been studied extensively due to the advantages of high theoretical capacities (up to 250 mAh g^{-1}, 1e^- transfer), simple synthesis process, appropriate operating voltage, and large-scale production. Interestingly, highly reversible Na^+ (de)intercalation is observed in all of the single transition-metal-based layered Na_xMO_2 oxides (M = Fe, Mn, Ni, Co, Cr, Ti, V) in contrast to limited electrochemical reactivities of $LiMO_2$ (M = V, Cr, Fe, Mn) [14, 15]. In recent times, the activity of Cu^{2+}/Cu^{1+} redox couple was also demonstrated in Na_xMO_2 oxides [15]. In this review, recent developments in transition-metal oxide–based cathode materials for SIBs are reviewed and discussed. Specifically, this review will focus on structural classifications of sodium metal oxide (Na_xMO_2), influence of cation compositions, structural changes during battery operation, and anionic redox properties. Several review articles on research progress of SIBs have also been published in the literature [10–13].

5.2 Layered Transition-Metal Oxides

5.2.1 Structural Classification

The general chemical composition of alkali-layered oxides can be written as A_xMO_2 (A = alkali, M = transition metal). This crystal structure is built by sheets of edge-shared MO_6 octahedra and sodium ions, which occupy the interlayer spacing. According to Delmas's notation [16], the structural type of layered oxides can be classified as Xn, wherein X denotes the coordination environment of alkali ions (O: octahedral, P: trigonal prismatic, T: tetrahedral) and n is the number of MO_2 layers present in a single unit cell. Sodium-based layered oxides (Na_xMO_2) mainly crystallize in P- and O-type structures (P2, P3, and O3), while their lithium analogues are found in T- and O-type structures (Figure 5.1) [17]. In the P2- and P3-type structures, oxygen layers are stacked in AB BA and AB BC CA patterns, while O2 and O3 structures are made of AB AC and AB CA BC stackings, respectively (Figure 5.1). Further, monoclinic and orthorhombic distortions in O3- and P2-type structures are observed, and the corresponding phases are denoted as O′3 and P′2, respectively [18]. Further, Na_xMO_2 cathodes undergo a series of phase transformations upon reversible sodium ion (de)insertion during cycling, which are aided by gliding of MO_2 planes: the O3-type structure typically follows O3 ↔ O′3 ↔ P3 ↔ P′3 phase transitions, while P2 ↔ O2 transitions occur in P2-type cathodes. However, the phase transformation between O3- and P2-type phases is not feasible since it requires breaking of transition metal–oxygen bonds, which is energetically more demanding [19]. A schematic illustration of in-plane Na^+-ion diffusion pathways in O- and P-type structures is shown in Figure 5.1e, f. The Na ions in O-type phase diffuse through interstitial tetrahedral sites with high

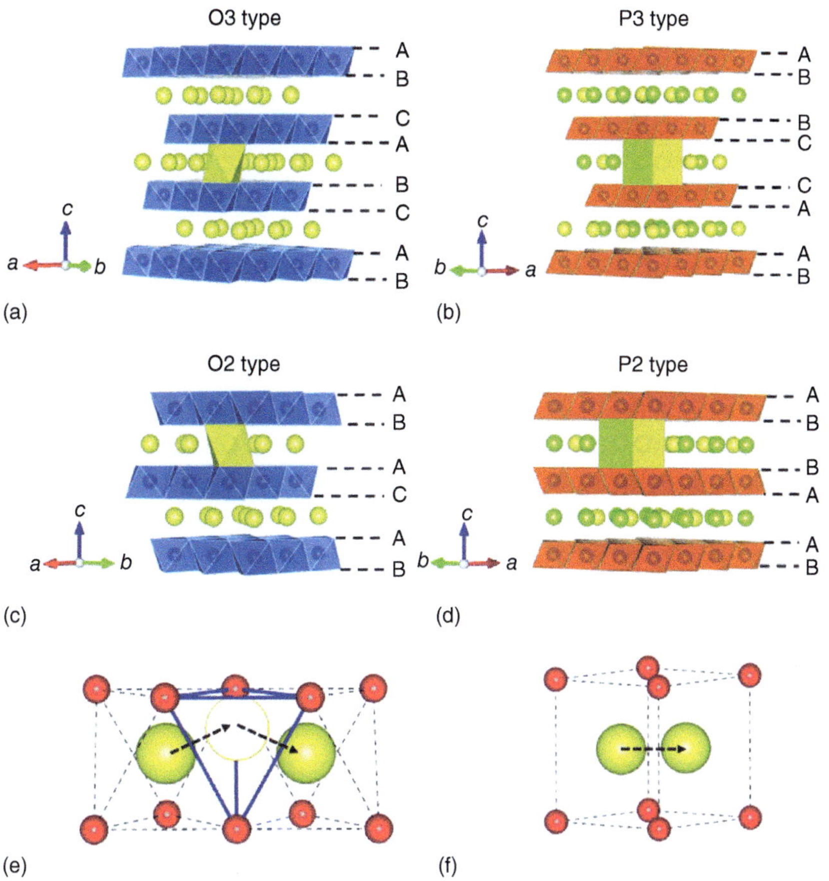

Figure 5.1 Schematics of the structures of layered transition-metal oxide Na_xMO_2 phases. (a) O3-type, (b) P3-type, (c) O2-type, and (d) P2-type. Schematic illustrations of Na^+-ion diffusion pathways through (e) indirect tetrahedral sites in the O-type stacking sequence and (f) direct prismatic sites in the P-type stacking sequence. Source: Guo et al. [17]/with permission of Royal Society of Chemistry.

migration barrier. In contrast, P-type structures have lower diffusion barriers due to direct migration pathways [20].

5.2.2 Single Transition-Metal-Based Layered Transition-Metal Oxides

Electrochemical sodium (de)intercalation studies on single transition-metal (TM)-based layered O3- and P2-type Na_xMO_2 structures were initiated by Delmas et al. [21]. It is worth highlighting that first-row transition-metal cations (except Cu^{3+}) form O3-type sodium-layered oxide cathodes (i.e. $NaMO_2$, M = Sc-Ni), while P2-type oxides are known to exist only for vanadium, manganese, and cobalt. It might be due to the difficulty in stabilizing tetravalent or higher oxidation states of Ti, Cr, Fe, and Ni cations under air or ambient oxygen pressure conditions.

Given the successful implementation of $LiCoO_2$ as the cathode in LIBs by Sony in 1991 [22], it comes as no surprise that its sodium counterpart, Na_xCoO_2, has attracted much attention as Na-ion cathode during the 1980s [21]. Na_xCoO_2 forms in O3-, O′3-, P3, P′3-, and P2-type structures with different stacking sequence of CoO_2 layers, according to the synthesis conditions employed (i.e. sodium content, temperature, and annealing duration) [23]. Initial electrochemical sodium (de)intercalation studies on O3- and P2-type Na_xCoO_2 cathodes were conducted by Delmas et al. [21, 24, 25] A comparison of the electrochemical sodium (de)intercalation properties of O3-, P′3-, and P2-type Na_xCoO_2 is displayed in Figure 5.2a. While O3- and P′3-type $NaCoO_2$ cathodes display almost similar capacities (~140 mAh g^{-1}), the P2-type phase delivered slightly lower reversible capacity of ~120 mAh g^{-1}, which is due to its lower sodium uptake. Their voltage profiles show multiple sets of plateaus irrespective of the CoO_2 stacking sequence, which is due to the multiple phase transitions during electrochemical cycling. The O3 phase transforms into O′3 phase followed by the formation of P3 phase, whereas P2 phase preserves its structure throughout the electrochemical cycling [24]. Recent investigations on P2-Na_xCoO_2 by Delmas and coworkers [25] demonstrated various single and bi-phasic domains due to Na^+/vacancy ordering at different sodium concentrations during the electrochemical cycling (Figure 5.2e), which arise from repulsive interactions between Na–Na and Na–Co. The capacity fading and irreversibility can be suppressed by substituting alkali ions in Na positions [29–32]. Recent reports showed Ca^{2+} substitution into P2- and P3-type Na_xCoO_2 since Na^+ ions (1.02 Å, coordination number [CN] = 6) and Ca^{2+} ions (1.0 Å, CN = 6) have a similar ionic radius. The results showed the improvement in cycling stability for Ca-substituted $Na_xCa_{0.04}CoO_2$ owing to suppression of the irreversible phase transition at high voltage [32].

Layered Na_xMnO_2 cathodes have also received much attention due to their high theoretical capacity (~240 mAh g^{-1}) and low cost. The crystal structures of several Na_xMnO_2 ($0 < x < 1$) phases were summarized by Parant et al. [33]. Na_xMnO_2 phases crystallize into either three-dimensional ($0.2 \leq x \leq 0.44$) or layered (x = 0.7 and 1.0) structures. The layered $NaMnO_2$ is known to exist in two polymorphs, i.e. (i) low-temperature monoclinic α-$NaMnO_2$ (otherwise known as O′3-type $NaMnO_2$) and high-temperature orthorhombic β-$NaMnO_2$. The sodium-deficient phase ($Na_{0.7}MnO_2$) also crystallizes either in hexagonal P2-type structure (*P6₃/mmc*) or distorted orthorhombic P′2-type structure (*Cmcm*). O′3-type $NaMnO_2$ and P′2-type Na_xMnO_2 are thermodynamically stable polymorphs, while P2-type Na_xMnO_2 forms only at controlled synthesis conditions [34, 35]. Electrochemical Na^+-ion (de)intercalation with very narrow reversibility ($x < 0.2$ in $Na_{1-x}MnO_2$) in O′3-type $NaMnO_2$ was first demonstrated by Mendiboure et al. [18]. Later, the same phase was revisited by Ceder and coworkers [36], and the reversibility range was extended up to $x < 0.8$ in $Na_{1-x}MnO_2$, which results in a high discharge capacity of 194 mAh g^{-1}. However, huge capacity fading was observed in subsequent cycles owing to the Jahn–Teller distortion of Mn^{3+} ($t_{2g}^3e_g^1$), which affects the stability of O′3-type $NaMnO_2$ electrodes. The β phase is made of stacks of the edge-shared MnO_6 octahedra, and the sodium ions occupy octahedral sites in the interlayers. It showed a reversible intercalation of 0.82 mol of sodium ions with a good capacity retention (130 mAh g^{-1} after 100 cycles), as shown in

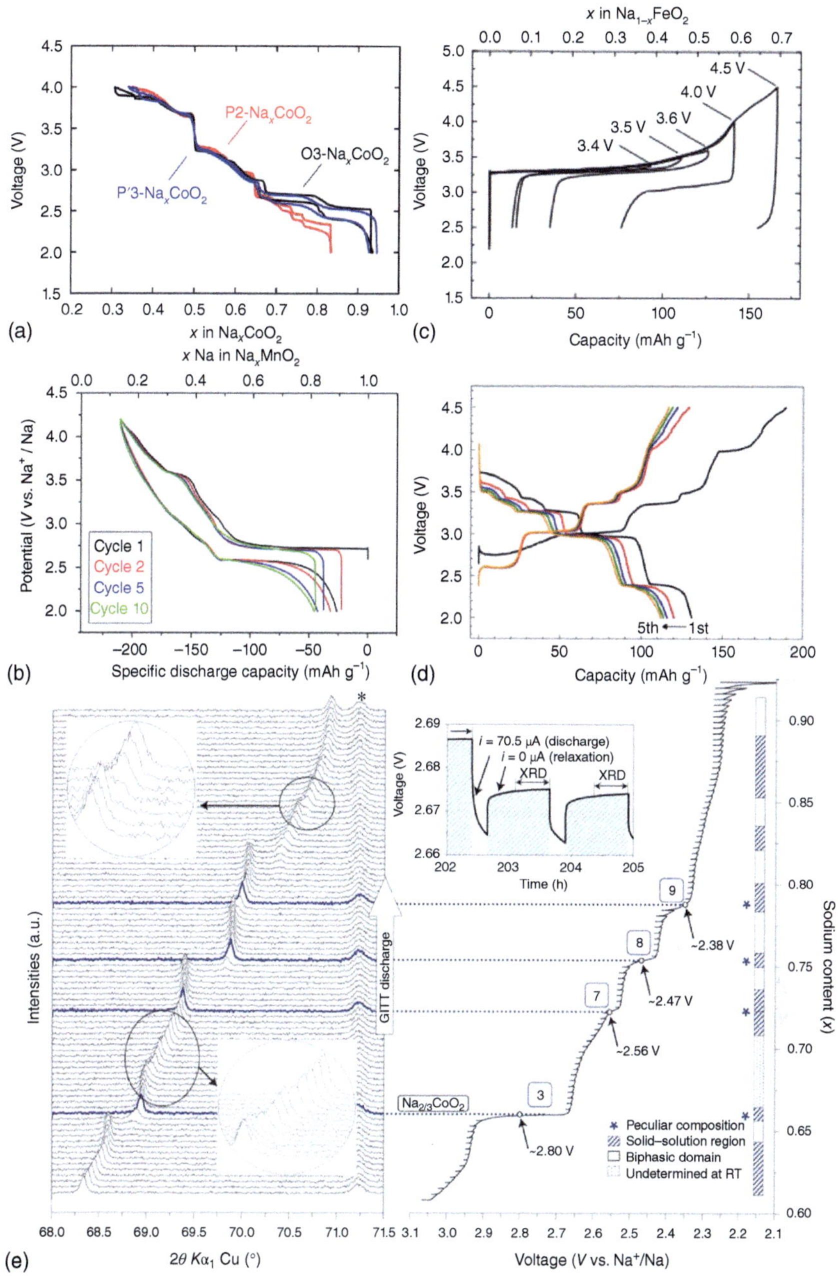

Figure 5.2 The charge–discharge profiles of (a) O3-$NaCoO_2$. Source: Yabuuchi et al. [4]/with permission of American Chemical Society. (b) O3-$NaMnO_2$. Source: Billaud et al. [26]/with permission of American Chemical Society. (c) O3-$NaFeO_2$. Source: Yabuuchi et al. [27]. (d) O3-$NaNiO_2$. Source: Wang et al. [28]/with permission of Elsevier. (e) In situ XRD patterns during sodium ion intercalation in P2-Na_xCoO_2. The galvanostatic intermittent titration technique (GITT) electrochemical battery discharge (right side) enables the equilibrium potential for each given composition to be obtained. Source: Berthelot et al. [25]/with permission of Springer Nature.

Figure 5.2b [26]. P2-Na_xMnO_2 cathodes, which exhibit initial reversible capacities of ~160 mAh g^{-1}, also undergo rapid capacity degradation (67% retention after 50 cycles) [34]. Recently, layered $NaMn_3O_5$ was introduced as a cathode for SIBs. It delivered a high capacity of 219 mAh g^{-1} (energy density: 602 Wh kg^{-1}) in the voltage range of 1.5–4.7 V with an average voltage of 2.7 V [37]. Although Mn-based layered oxides exhibited high capacity, the average voltage is low compared to Fe-based oxides and the Jahn–Teller distortion causes structural irreversibility.

$NaFeO_2$ attained considerable interest as a cathode for SIBs due to its low cost, non-toxicity, abundance of element Fe and stable Fe^{3+}/Fe^{4+} redox at 3.3 V. $NaFeO_2$ can be easily synthesized by solid-state reaction and crystallize in two polymorphs, α-$NaFeO_2$ and β-$NaFeO_2$ [38, 39]. In contrast, an analogous $LiFeO_2$ cannot be prepared as a stable phase by direct synthesis because similar ionic size leads to cation mixing of Li and Fe [40]. α-$NaFeO_2$ is a perfect example of an O3-type structure and is electrochemically active. In contrast, β-$NaFeO_2$, which crystallizes in an orthorhombic structure, is found to be electrochemically inactive. Early reports on reversible extraction and insertion of Na-ions in α-$NaFeO_2$ were demonstrated by Kikkawa et al. and Okada et al. [41, 42]. Solid-state synthesized O3-type $NaFeO_2$ delivered a flat voltage of 3.3 V with a capacity of 80 mAh g^{-1}, as shown in Figure 5.2c [27]. However, extraction of more Na^+ at voltages above 3.4 V reduced reversibility owing to irreversible structural changes and migration of Fe^{3+} ions to face-shared tetrahedral sites, which blocks Na^+ diffusion [27, 43].

Other $NaMO_2$ oxides such as $NaNiO_2$, $NaCrO_2$, and Na_xVO_2 also have been studied as cathodes for SIBs. Albeit direct synthesis of $LiNiO_2$ not being possible, O3-$NaNiO_2$ can be easily prepared by solid-state reaction. The O3-$NaNiO_2$ cathode is attractive due to its high (de)intercalation potential of $Ni^{2+}/Ni^{3+}/Ni^{4+}$ redox couples associated with the multistep voltage profiles and large reversible capacities of ~130 mAh g^{-1} in the voltage range of 4.5–2.0 V (Figure 5.2d) [28, 44–46]. Successive phase transformations (O′3–P′3–P″3–O″3) occur during Na^+ (de)intercalation. A report on in situ-X-ray diffraction (XRD) and galvanostatic intermittent titration technique (GITT) studies revealed that the irreversibility of the $NaNiO_2$ material resulted from the low Na^+ diffusion and slower reaction kinetics [28]. Between the P2 and O3 phases of $NaCrO_2$, the O3 phase was studied extensively owing to better thermal stability and large reversible capacity with small polarization for the Cr^{3+}/Cr^{4+} couple [44, 46–48]. Very recently, synthesis of large grain O3-$NaCrO_2$ has been demonstrated by directly reducing $Na_2Cr_2O_7 \cdot 2H_2O$. The material exhibited a high capacity of 123 mAh g^{-1} and decent capacity retention (88.2% after 500 cycles) [48]. However, $NaCrO_2$ has the drawbacks of low electron conductivity, the toxicity of Cr^{6+}, and cation disorder. Layered Na_xVO_2 exists in two forms: O3-$NaVO_2$ and P2-$Na_{0.7}VO_2$ [49–51]. The reversible capacity observed for O3-$NaVO_2$ (120 mAh g^{-1}) is slightly better than P2-$Na_{0.7}VO_2$ (110 mAh g^{-1}) [52]. Further, Na^+ extraction in $Na_{1-x}VO_2$ above $x = 0.5$ leads to capacity fading, and the low operating voltage (~2.0 V) makes the material less attractive for applications.

5.2.3 Mixed-Metal-Based Layered Transition-Metal Oxides

Although single metal-based layered oxides have large theoretical capacity, the materials suffer from low reversible capacity, multiple voltage plateaus (phase transitions), capacity fading, and limited cycling stability. Presence of mixed metal cations in the layered oxides can provide improved electrochemical performance and structural stability owing to the synergistic effect of different ratios of 3d metal cations. Another motivation of using mixed cations is to reduce the concentration of expensive elements like Ni and Co, limit multiple-phase transitions, decrease the Na^+ diffusion barrier, and increase the electrode performance. In general, Ni increases the capacity and Mn contributes to capacity retention and thermal stability. Co can stabilize the structural framework and increase the voltage. Elements such as Mg, Cu, Ti, or Li can be utilized for structural stability and improved electrochemical performance of the Na_xMO_2 framework.

Layered transition-metal oxides with various binary metal cations can form solid solutions. The first report on an O3-type binary metal cation composition, O3-$NaNi_{0.6}Co_{0.4}O_2$, was made by Saadoune et al. [53]. The material showed a reversible
capacity of ~95 mAh g^{-1} with two voltage plateaus at 2.25 and 2.4 V. The material O3-$NaNi_{1-x}Co_xO_2$ was revisited by Vassilaras et al. [54], who reported a multiple stepwise voltage profile with a capacity of 130 mAh g^{-1} for O3-$NaNi_{0.5}Co_{0.5}O_2$. The multiple-phase transitions are associated with three distinct O3-type (O3, O′3, and O″3) and P3-type (P3, P′3, and P″3) phases. Nevertheless, the O3-$NaNi_{0.5}Co_{0.5}O_2$ delivered good cycling stability and rate performance in the voltage range of 2.0–4.2 V. The O3-$NaFeO_2$ and O3-$NaCoO_2$ can form solid solutions of O3-$NaFe_{1-x}Co_xO_2$ in its complete composition range. O3-type $NaFe_{0.5}Co_{0.5}O_2$ exhibited a high reversible capacity of 160 mAh g^{-1} (Figure 5.3a), which is higher than those of $NaCoO_2$ and $NaFeO_2$. The material exhibited enhanced capacity retention and rate capability [57]. The O3-$NaFe_{0.5}Co_{0.5}O_2$ showed electrochemical reversibility up to 0.7 Na^+ extraction from the structure (only 0.5 Na^+ for $NaFeO_2$) due to the presence of Co, which suppresses Fe migration. Additional extraction of Na^+ beyond 0.7 mol leads to capacity fading and irreversibility. Further, solid solutions of $NaFeO_2$–$NaCoO_2$ synthesized by solid-state reaction were investigated using neutron powder diffraction (NPD) [58]. The NPD analysis of O3-$NaFe_{0.5}Co_{0.5}O_2$ confirmed that there is no ordering of Fe and Co in the transition-metal slab and operando-XRD studies revealed the P3-type phase during charge/discharge over a wide capacity range, and monophasic P′3–O3 transition, which are enhancing the rate performance [58].

Solid solutions of $NaMn_{1-x}Ni_xO_2$ with Ni^{2+} and Mn^{4+} were most extensively studied by several groups. Layered P2-$Na_{2/3}Ni_{1/3}Mn_{2/3}O_2$ was investigated by Lu and Dahn [59] and revisited by Zhao et al. [60]. The P2-type of P2-$Na_{2/3}Ni_{1/3}Mn_{2/3}O_2$ sustains a limited amount of Na exchange (~0.33 mol of sodium ions), while the additional Na^+ ion extraction leads to the phase transition of P2 to O2 due to

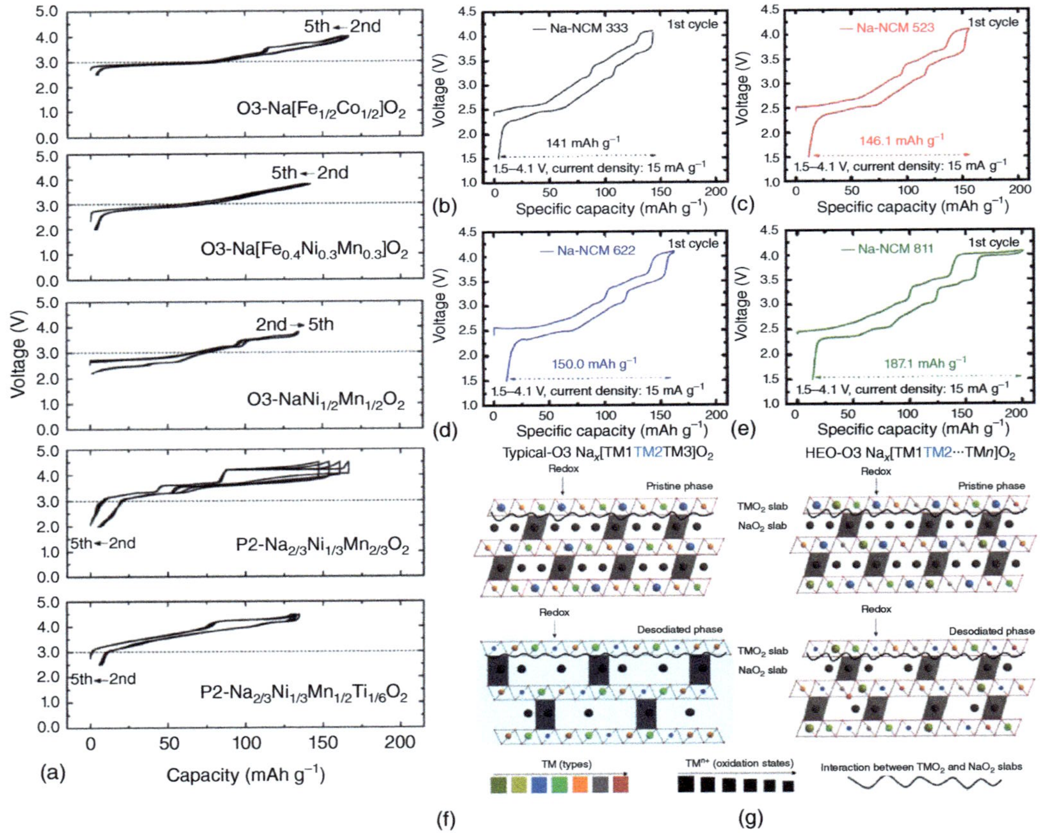

Figure 5.3 (a) Comparison of galvanostatic charge–discharge profiles of various mixed layered oxides: O3-Na[$Co_{1/2}Fe_{1/2}$]O_2, O3-Na[$Fe_{0.4}Ni_{0.3}Mn_{0.3}$]O_2, O3-Na[$Ni_{1/2}Mn_{1/2}$]O_2, P2-$Na_{2/3}$[$Ni_{1/3}Mn_{2/3}$]O_2, and P2-$Na_{2/3}$[$Ni_{1/3}Mn_{1/2}Ti_{1/6}$]O_2. Source: Yabuuchi et al. [4]/with permission of American Chemical Society. Galvanostatic charge–discharge profiles of Na[$Ni_xCo_yMn_z$]O_2: (b) $x = 1/3$ (Na-NCM 333), (c) $x = 0.5$ (Na-NCM 523), (d) $x = 0.6$ (Na-NCM 622), and (e) $x = 0.8$ (Na-NCM 811) in 1.5–4.1 V at 15 mA g^{-1}. Source: Hwang et al. [55]/with permission of Royal Society of Chemistry. Possible mechanism of high-entropy composition in facilitating layered O3-type structure. (f) The conventional O3-type Na-ion cathodes with three different types of transition-metal (TM) elements. (g) The proposed HEO cathodes with multi-transition-metal element components. TM2 is assigned as the redox elements in blue. Source: Zhao et al. [56]/with permission of John Wiley & Sons.

stacking faults. O3-$NaNi_{0.5}Mn_{0.5}O_2$ delivered a large capacity of 185 mAh g^{-1} during charge up to 4.5 V; however, fast capacity fading was observed with more Na^+ extraction (4.5 V) [61]. Very recently, microcrack formation in spherical particles of O3-type $Na[Ni_{0.5}Mn_{0.5}]O_2$ cathode was investigated at different upper cut-off voltages. P′3–O′3 phase transition above 3.6 V causes structural irreversibility and microcrack formation during cycling [62]. Doping of cations such as Zn, Al, Mg, and Cu can suppress the phase changes and microcrack formation by alleviating layer gliding [63–66]. Recent report on Cu^{2+} doping in P2-$Na_{2/3}Ni_{1/3}Mn_{2/3}O_2$ demonstrated improved capacity and cycling performances compared to electrochemically inactive Zn^{2+} and Mg^{2+} doping. The enhanced performances result from structural stabilization (P2/OP4) even at higher voltages and redox activity of both Ni^{2+}/Ni^{4+} and Cu^{2+}/Cu^{3+} [65].

Solid solutions between O3-type $Na(Ni_{1/2}Mn_{1/2})O_2$ and $NaFeO_2$ form O3-type $NaFe_x(Ni_{1/2}Mn_{1/2})_{1-x}O_2$. Ni/Mn substitution for Fe partly suppresses the Fe migration and the $NaFe_{0.4}(Ni_{1/2}Mn_{1/2})_{0.6}O_2$ material delivered a capacity of 130 mAh g^{-1} (Figure 5.3a) and good rate capability [67]. Furthermore, P2-type $Na_{0.67}Mn_{0.65}Fe_{0.2}Ni_{0.15}O_2$ also demonstrated a capacity (208 mAh g^{-1}) as cathode for SIBs [68]. However the P2-type materials and their derivatives suffer from rapid capacity fading due to P2 to O2 phase formation during cycling. Ti substitution in $Na_{2/3}Ni_{1/3}Mn_{2/3-x}Ti_xO_2$ was also studied as cathode material, which exhibited a capacity of 127 mAh g^{-1} with an average voltage of 3.7 V (Figure 5.3a) [69]. Recently, the effect of Sn^{4+} doping in $NaNi_{0.5}Mn_{0.5}O_2$ cathode was investigated. The $NaNi_{0.5}Mn_{0.5}Sn_xO_2$ cathode exhibited a large capacity of 191 mAh g^{-1} and good capacity retention. The Sn^{4+} doping in $NaNi_{0.5}Mn_{0.5}O_2$ inhibits TMO_2 slips and reduces irreversible multiple-phase transitions during electrochemical cycling [70]. Very recently, double substitution of Cu^{2+} for Ni^{2+} and Ti^{4+} for Mn^{4+} in O3-$NaNi_{0.5-y}Cu_yMn_{0.5-z}Ti_zO_2$ (y = 0, 0.05, 0.1; z = 0.1, 0.2) phases was demonstrated by Wang et al. [71], which improve stability against structural phase changes at high potential (4.5 V vs. Na^+/Na^0), and facilitate electrochemically reversible ~0.9 Na^+ (equivalent to ~200 mAh g^{-1}) extraction from the structure. Further, 18 650 Na cells fabricated with $NaNi_{0.4}Cu_{0.1}Mn_{0.4}Ti_{0.1}O_2$ exhibited higher gravimetric and volumetric energy densities (~115 Wh kg^{-1} and ~ 250 Wh l^{-1}) relative to currently available Na-ion $Na_3V_2(PO_4)_2F_3$/C cells (100 Wh kg^{-1}, 175 Wh l^{-1}).

Mn substitution in P2-Na_xCoO_2 can form P2-type $Na_{2/3}Co_{2/3}Mn_{1/3}O_2$ as reported in 2011 by Carlier et al. [72] Reversible Na^+ extraction from the structure involves a solid–solution process except for x = 0.5, which leads to the formation of the ordered P2-$Na_{0.5}Co_{2/3}Mn_{1/3}O_2$ phase. Electrochemical properties of solid–solution of P2-$Na_{2/3}Mn_yCo_{1-y}O_2$ compounds were investigated [73]. The redox potential Co^{4+}/Co^{3+} and Mn^{4+}/Mn^{3+} shifts systematically with Mn substitution. The increase in Mn substitution for Co contributes to an increase in specific capacity while cycling stability degrades. However, P2-$Na_{0.5}Mn_{0.5}Co_{0.5}O_2$ synthesized by a facile mixed hydroxy-carbonate route delivered a high capacity and rate capability [74]. The sodium-containing Mn–Co compounds are moisture sensitive and need to be stored in Ar-filled glovebox, which results in an increase in synthesis and production cost.

Co substitution in the Mn–Ni-based compounds $Na[NiMnCo]O_2$ is attractive and extensively studied with the expectation for high-capacity performance of $Li[NiMnCo]O_2$. Na-ion (de)insertion properties of $NaNi_{1/3}Mn_{1/3}Co_{1/3}O_2$ were first demonstrated by Sathiya et al. The material delivered a capacity of $120\,mAh\,g^{-1}$ and multiphase transitions of O3 → O′3 → P3 → P′3 during charge up to 3.75 V [75]. In their follow-up work, Hwang et al. investigated the role of transition metals in O3-$Na[Ni_xCo_yMn_z]O_2$ (x = 0.33, 0.5, 0.6, and 0.8). Electrochemical performance of the NMC compounds is shown in Figure 5.3b–e, where increase in Ni and Mn content led to improved capacity and retention, respectively [55]. However, higher Ni content (above 60%) results in rapid capacity fading. Furthermore, effects of Co substitution for Ni in $Na_{0.7}Mn_{0.7}Ni_{0.3-x}Co_xO_2$ (x = 0, 0.1, 0.3) were studied by Li et al. [76] The aliovalent Co^{3+} substitution for Ni^{2+} shrinks TMO_6 octahedra and increases electronic conductivity, which results in structural stability, long cycle life, and high-rate capability. Radially aligned hierarchical columnar (RAHC) structured spherical particles with varied chemical compositions from inner end ($Na[Ni_{0.75}Co_{0.02}Mn_{0.23}]O_2$) to the outer end ($Na[Ni_{0.58}Co_{0.06}Mn_{0.36}]O_2$) of the structure were prepared by Hwang et al. [77] The RAHC cathode exhibited enhanced capacity ($157\,mAh\,g^{-1}$ at 0.1 C rate) and rate capability ($132.6\,mAh\,g^{-1}$ at 10 C-rate) compared to bulk cathode. The improved battery performance of the RAHC results from reversible electrochemical reactions ($Ni^{2+}/Ni^{3+}/Ni^{4+}$) of high Ni content in the inner end and suppression of Ni reactivity with electrolyte due to lower Ni and higher Mn content in outer end of RAHC. Further, particle surface coatings (e.g. Al_2O_3, CuO, $NaPO_3$, and ZnO) can enhance the kinetics. Performance improvements and suppressing the damage at the microscopic level of the O3-$Na[Ni_{0.6}Co_{0.2}Mn_{0.2}]O_2$ particle were achieved by Al_2O_3 coating [78].

Multicomponent compounds with various ratios of 3d metal cations present in the layered oxides are gaining considerable attention in the property-oriented new materials and enhancing the performance. Recently, Zhao et al. introduced the high entropy oxide (HEO) concept to find suitable multi-element layered oxides for SIBs [56]. HEO denotes the multi-element metallic systems, which can crystallize in a single phase [79]. Figure 5.3g shows that entropy stabilization of the host structure using a multicomponent system relative to a conventional O3 cathode with three components (Figure 5.3f) facilitates the layered O3-type structure to a larger extent. They have demonstrated the concept with an example of $NaNi_{0.12}Cu_{0.12}Mg_{0.12}Fe_{0.15}Co_{0.15}Mn_{0.1}Ti_{0.1}Sn_{0.1}Sb_{0.04}O_2$, which exhibits excellent rate capability (~80% at 5 C) and longer cycling stability (~83% after 500 cycles). More interestingly, highly reversible phase transitions between O3 and P3 and 60% of capacity were observed in the O3 phase.

5.2.4 Anionic Redox Activity for High Capacity

The anionic redox mechanism concept has emerged as a new paradigm for developing high-capacity positive electrodes [80]. The anionic redox reactions (O^{2-}/O^{n-}) originate from O2p electron unhybridized with transition metals. The oxygen redox process (O^{2-}/O^{n-}) can contribute an extra capacity beyond what is predicted

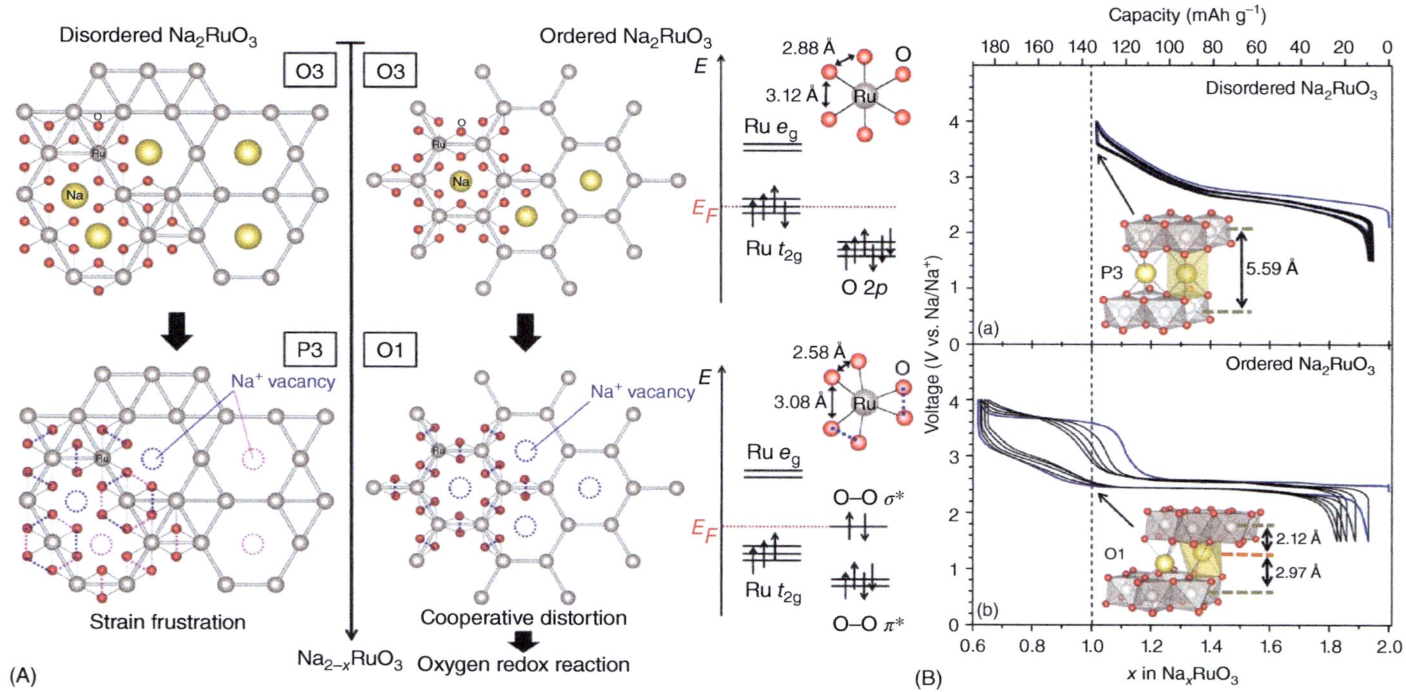

Figure 5.4 (A) Schematic representation of the structural changes for disordered Na_2RuO_3 and ordered Na_2RuO_3. Disordered Na_1RuO_3 cannot accommodate the RuO_6 distortion due to strain frustration, which prevents the oxygen redox reaction. Ordered Na_2RuO_3 can distort cooperatively to raise the energy level of the antibonding σ^* orbital of the O—O bond, leading to the oxygen redox reaction. (B) Galvanostatic cycling curves recorded at 30 mA g^{-1} for disordered and ordered Na_2RuO_3 with the first cycle highlighted in blue. Insets show the coordination environment of Na at $x = 1.0$ for each phase. Source: Mortemard de Boisse et al. [85].

from the transition-metal content alone in layered oxides, which was first reported in Li-excess cathodes such as Li_2RuO_3, $Li_{1.2}Ni_{0.2}Mn_{0.6}O_2$, $Li_2Ru_{0.5}Sn_{0.5}O_2$, and $Li_{1.3}Mn_{0.4}Nb_{0.3}O_2$ [81, 82]. Specifically, anionic redox-active cathodes are needed to overcome the energy-density limitations of SIBs due to lower redox of Na/Na^+ relative to Li/Li^+. Li-rich and Na-rich layered oxides are promising high-capacity cathodes, which can utilize both cationic and anionic redox. Interestingly, anionic redox can be observed in both Na-deficient and Na-rich layered cathode materials in contrast to Li-based layered cathodes. Anionic redox is active only in Li-rich layered materials [83, 84]. Alkali-rich oxides with $A_{2-x}MO_3$ structures are prominent in providing excess capacity due to the anionic redox mechanism. Na-rich Na_2RuO_3 is a famous anionic redox active material, which forms in O3-type structured (*R-3m* space group) polymorphs. In these, one Na_2RuO_3 forms honeycomb-ordered $[Na_{1/3}Ru_{2/3}]O_2$ slabs, while the other one forms disordered $[Na_{1/3}Ru_{2/3}]O_2$ slabs, as shown in Figure 5.4A [85]. The honeycomb-ordered Na_2RuO_3 showed a large capacity of 180 mAh g^{-1}, corresponding to 1.3 Na^+ extraction from the structure, which is higher than the expected capacity from cationic redox of Ru ($1e^-$) (Figure 5.4B). Disordered Na_2RuO_3 showed a capacity of 135 mAh g^{-1} corresponding to the cationic redox of Ru ($1e^-$). As shown in Figure 5.4B, the short O–O distance in distorted RuO_6 octahedra induces orbital reorganization, i.e. the energy level of the anti-bonding σ^* of O—O bond closer to the Fermi level. This phenomenon triggers the anionic redox activity of $Na_{2-x}RuO_3$.

Recent reports showed that excess Na is not necessary for anionic redox activity in Na-based layered oxides. Maitra et al. demonstrated the excess capacity from $Na_{2/3}[Mg_{0.28}Mn_{0.72}]O_2$ is due to oxygen redox activity. They have proved that the excess Na is not required to activate the oxygen redox. In contrast to Na-rich cathodes, $Na_{2/3}[Mg_{0.28}Mn_{0.72}]O_2$ does not lose oxygen owing to the presence of Mg^{2+} ions [65, 66, 84]. Various Mn-based layered oxides $Na_2Mn_3O_7$, O3-$NaNi_{0.5}Mn_{0.5}O_2$, $Na_{0.78}Ni_{0.23}Mn_{0.69}O_2$, and $Na_{0.67}Mn_{0.75}Ni_{0.25}O_2$ also exhibit anionic redox activity [83–87]. Even though oxygen redox provides excess capacity and it can improve the energy density of the SIBs, understanding the mechanism and stabilizing the structure is significant [86, 87].

5.3 Summary and Outlook

SIBs are potential alternative to LIBs for stationary EES applications due to their cost effectiveness, elemental abundance of Na, and having similar (de)intercalating properties as LIBs. Among the various types of cathodes, layered transition-metal oxides are promising in terms of high electrochemical performance. However, the large ionic radius of Na^+ results in complex structural changes during intercalation/de-intercalation process. The multiple-phase transitions lead to stepwise voltage profiles, capacity fading, volume change, large irreversible capacity, and structural instability. To overcome these challenges, cation substitution strategies are utilized to improve the structural stabilities of the oxide framework, which suppress the phase transitions during battery operation. Several P2- and O3-type layered oxides are investigated as cathode materials and are reviewed.

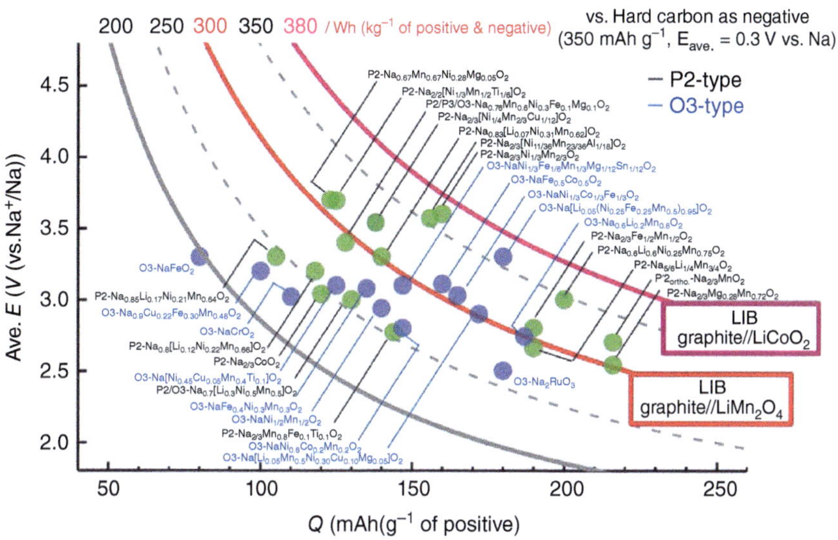

Figure 5.5 Average voltage (V) and energy density (Wh kg^{-1}) versus gravimetric capacity (mAh g^{-1}) for selected O3- and P2-type positive electrode materials for Na-ion batteries. Energy density on the basis of electrode active materials was calculated with the hard carbon (reversible capacity of 350 mAh g^{-1} with E_{ave} = 0.3 V vs. Na metal as a negative electrode material). Source: Kubota et al. [20]/with permission of John Wiley & Sons.

The average voltage, capacity, and energy density of several P2- and O3-type layered oxides are compared in Figure 5.5 [20].

The electrochemical performances of P2- and O3-type layered oxides are comparable to LIB cathodes. Particularly, Ni/Mn-based compounds are high-performing cathodes owing to the full utilization of the double redox couples of Ni^{2+}/Ni^{3+} and Ni^{3+}/Ni^{4+}, which provides high operating voltage and large capacity. Electrochemically inactive Mn^{4+} stabilizes the structure, which results in long cycling stability and retention in reversible capacity. Further, doping of elements (Cu, Mg, Ti, and Li) in Ni/Mn-based compounds enhances the battery performance. Among the reported layered oxide-based cathodes, $NaNi_{0.4}Cu_{0.1}Mn_{0.4}Ti_{0.1}O_2$ and $NaNi_{0.5}Mn_{0.2}Ti_{0.3}O_2$ are found to be the best performing in terms of high capacity, cycling stability, air stability, and high-rate performance. The Mn/Ni-based layered cathodes have the advantages of large-scale synthesis, cost-effective precursors, and facile synthesis methods. Recently, HEO concept was also used to find suitable multi-element layered oxides for high-performance SIBs. To improve the capacity, anionic redox properties of layered cathodes are also investigated. Even though SIBs have the merits of having highly abundant Na, positive electrodes with cost-effective elements, low-cost precursors, and high-performance electrodes, lack of air-stable, high-voltage cathodes and insufficient cell performances of SIBs limits their practical commercialization. Design of new air stable layered metal oxides with proper composition of metal cations could improve the battery performance. Recent developments in layered oxides provide the way to attain the future goals of commercializing SIBs for various applications.

References

1 Whittingham, M.S. (2004). Lithium batteries and cathode materials. *Chemical Reviews* 104 (10): 4271–4301.

2 Li, M., Lu, J., Chen, Z., and Amine, K. (2018). 30 years of lithium-ion batteries. *Advanced Materials* 30 (33): 1–24.

3 Nitta, N., Wu, F., Lee, J.T., and Yushin, G. (2015). Li-ion battery materials: present and future. *Materials Today* 18 (5): 252–264.

4 Yabuuchi, N., Kubota, K., Dahbi, M., and Komaba, S. (2014). Research development on sodium-ion batteries. *Chemical Reviews* 114 (23): 11636–11682.

5 Su, H., Jaffer, S., and Yu, H. (2016). Transition metal oxides for sodium-ion batteries. *Energy Storage Materials* 5: 116–131.

6 Bauer, A., Song, J., Vail, S.Y. et al. (2018). The scale-up and commercialization of nonaqueous Na-ion battery technologies. *Advanced Energy Materials* 8 (17): 1–13.

7 Chen, L., Fiore, M., Wang, J.E. et al. (2018). Readiness level of sodium-ion battery technology: a materials review. *Advanced Sustainable Systems* 2 (3): 1700153.

8 Mauger, A. and Julien, C.M. (2020). State-of-the-art electrode materials for sodium-ion batteries. *Materials (Basel)* 13 (16): 3453.

9 Liu, Q., Hu, Z., Chen, M. et al. (2019). Recent progress of layered transition metal oxide cathodes for sodium-ion batteries. *Small* 15 (32): 1–24.

10 Pan, H., Hu, Y.-S., and Chen, L. (2013). Room-temperature stationary sodium-ion batteries for large-scale electric energy storage. *Energy & Environmental Science* 6 (8): 2338.

11 Senthilkumar, B., Murugesan, C., Sharma, L. et al. (2019). An overview of mixed polyanionic cathode materials for sodium-ion batteries. *Small Methods* 3 (4): 1–23.

12 Barpanda, P., Lander, L., Nishimura, S.I., and Yamada, A. (2018). Polyanionic insertion materials for sodium-ion batteries. *Advanced Energy Materials* 8 (17): 1–26.

13 Rajagopalan, R., Tang, Y., Jia, C. et al. (2020). Understanding the sodium storage mechanisms of organic electrodes in sodium ion batteries: issues and solutions. *Energy & Environmental Science* 13 (6): 1568–1592.

14 Paulsen, J.M. (1999). Layered Li-Mn-oxide with the O_2 structure: a cathode material for Li-ion cells which does not convert to spinel. *Journal of the Electrochemical Society* 146 (10): 3560–3565.

15 Ono, Y. (2018). Structural analysis of $NaCuO_2$ cathode at various charged/discharged stages and its reaction mechanism. *Electrochemistry* 86 (6): 309–314.

16 Delmas, C., Fouassier, C., and Hagenmuller, P. (1980). Structural classification and properties of the layered oxides. *Physica B+C* 99 (1–4): 81–85.

17 Guo, S., Yi, J., Sun, Y., and Zhou, H. (2016). Recent advances in titanium-based electrode materials for stationary sodium-ion batteries. *Energy & Environmental Science* 9 (10): 2978–3006.

18 Mendiboure, A., Delmas, C., and Hagenmuller, P. (1985). Electrochemical intercalation and deintercalation of Na_xMnO_2 bronzes. *Journal of Solid State Chemistry* 57 (3): 323–331.

19 Hwang, J.Y., Myung, S.T., and Sun, Y.K. (2017). Sodium-ion batteries: present and future. *Chemical Society Reviews* 46 (12): 3529–3614.

20 Kubota, K., Kumakura, S., Yoda, Y. et al. (2018). Electrochemistry and solid-state chemistry of $NaMeO_2$ (Me = 3d transition metals). *Advanced Energy Materials* 8 (17): 1703415, 1–30.

21 Delmas, C., Braconnier, J.-J., Fouassier, C., and Hagenmuller, P. (1981). Electrochemical intercalation of sodium in Na_xCoO_2 bronzes. *Solid State Ionics* 3–4: 165–169.

22 Goodenough, J.B. and Kim, Y. (2010). Challenges for rechargeable Li batteries. *Chemistry of Materials* 22 (3): 587–603.

23 Lei, Y., Li, X., Liu, L., and Ceder, G. (2014). Synthesis and stoichiometry of different layered sodium cobalt oxides. *Chemistry of Materials* 26 (18): 5288–5296.

24 Braconnier, J.-J., Delmas, C., Fouassier, C., and Hagenmuller, P. (1980). Comportement electrochimique des phases Na_xCoO_2. *Materials Research Bulletin* 15 (12): 1797–1804.

25 Berthelot, R., Carlier, D., and Delmas, C. (2011). Electrochemical investigation of the P2-Na_xCoO_2 phase diagram. *Nature Materials* 10 (1): 74–80.

26 Billaud, J., Clement, R.J., Armstrong, A.R. et al. (2014). β-$NaMnO_2$: a high-performance cathode for sodium-ion batteries. *Journal of the American Chemical Society* 136 (49): 17243–17248.

27 Yabuuchi, N., Yoshida, H., and Komaba, S. (2012). Crystal structures and electrode performance of alpha-$NaFeO_2$ for rechargeable sodium batteries. *Electrochemistry* 80 (10): 716–719.

28 Wang, L., Wang, J., Zhang, X. et al. (2017). Unravelling the origin of irreversible capacity loss in $NaNiO_2$ for high voltage sodium ion batteries. *Nano Energy* 34 (January): 215–223.

29 Han, S.C., Lim, H., Jeong, J. et al. (2015). Ca-doped $NaxCoO_2$ for improved cyclability in sodium ion batteries. *Journal of Power Sources* 277: 9–16.

30 Sun, L., Xie, Y., Liao, X.Z. et al. (2018). Insight into Ca-substitution effects on O3-type $NaNi_{1/3}Fe_{1/3}Mn_{1/3}O_2$ cathode materials for sodium-ion batteries application. *Small* 14 (21): 1–7.

31 Matsui, M., Mizukoshi, F., and Imanishi, N. (2015). Improved cycling performance of P2-type layered sodium cobalt oxide by calcium substitution. *Journal of Power Sources* 280: 205–209.

32 Ishado, Y., Hasegawa, H., Okada, S. et al. (2020). An experimental and first-principle investigation of the Ca-substitution effect on P3-type layered Na_xCoO_2. *Chemical Communications (Cambridge)* 56 (58): 8107–8110.

33 Parant, J.P., Olazcuaga, R., Devalette, M. et al. (1971). Sur quelques nouvelles phases de formule Na_xMnO_2 ($x \leq 1$). *Journal of Solid State Chemistry* 3 (1): 1–11.

34 Su, D., Wang, C., Ahn, H., and Wang, G. (2013). Single crystalline $Na_{0.7}MnO_2$ nanoplates as cathode materials for sodium-ion batteries with enhanced performance. *Chemistry - A European Journal* 19 (33): 10884–10889.

35 Stoyanova, R., Carlier, D., Sendova-vassileva, M. et al. (2010). Stabilization of over-stoichiometric Mn^{4+} in layered $Na_{2/3}MnO_2$. *Journal of Solid State Chemistry* 183 (6): 1372–1379.

36 Ma, X., Chen, H., and Ceder, G. (2011). Electrochemical properties of monoclinic $NaMnO_2$. *Journal of the Electrochemical Society* 158 (12): A1307–A1312.

37 Guo, S., Yu, H., Jian, Z. et al. (2014). A high-capacity, low-cost layered sodium manganese oxide material as cathode for sodium-ion batteries. *ChemSusChem* 7 (8): 2115–2119.

38 Takeda, Y., Akagi, J., Edagawa, A. et al. (1980). A preparation and polymorphic relations of sodium iron oxide ($NaFeO_2$). *Materials Research Bulletin* 15 (8): 1167–1172.

39 Zhao, J., Zhao, L., Dimov, N. et al. (2013). Electrochemical and thermal properties of α-$NaFeO_2$ cathode for Na-ion batteries. *Journal of the Electrochemical Society* 160 (5): A3077–A3081.

40 Okada, S. and Yamaki, J. (2009). Iron-based rare-metal-free cathodes. In: *Lithium Ion Rechargeable Batteries* (ed. K. Ozawa), 53–66. Weinheim: Wiley VCH Verlag GmbH & Co. KGaA.

41 Kikkawa, S., Miyazaki, S., and Koizumi, M. (1985). Sodium deintercalation from α-$NaFeO_2$. *Materials Research Bulletin* 20 (4): 373–377.

42 Okada, S., Takahashi, Y., Kiyabu, T. et al. (2006). Layered transition metal oxides as cathodes for sodium secondary battery. 210th ECS Meeting Abstracts, MA2006-02, 201.

43 Lee, E., Brown, D.E., Alp, E.E. et al. (2015). New insights into the performance degradation of Fe-based layered oxides in sodium-ion batteries: instability of Fe^{3+}/Fe^{4+} redox in α-$NaFeO_2$. *Chemistry of Materials* 27 (19): 6755–6764.

44 Braconnier, J.J., Delmas, C., and Hagenmuller, P. (1982). Etude par desintercalation electrochimique des systemes Na_xCrO_2 et Na_xNiO_2. *Materials Research Bulletin* 17 (8): 993–1000.

45 Vassilaras, P., Ma, X., Li, X., and Ceder, G. (2013). Electrochemical properties of monoclinic $NaNiO_2$. *Journal of the Electrochemical Society* 160 (2): A207–A211.

46 Kubota, K., Ikeuchi, I., Nakayama, T. et al. (2015). New insight into structural evolution in layered $NaCrO_2$ during electrochemical sodium extraction. *Journal of Physical Chemistry C* 119 (1): 166–175.

47 Tsuchiya, Y., Glushenkov, A.M., and Yabuuchi, N. (2018). Effect of nanosizing on reversible sodium storage in a $NaCrO_2$ electrode. *ACS Applied Nano Materials* 1 (1): 364–370.

48 Wang, Y., Li, W., Hu, G. et al. (2019). Electrochemical performance of large-grained $NaCrO_2$ cathode materials for Na-ion batteries synthesized by decomposition of $Na_2Cr_2O_7{\cdot}2H_2O$. *Chemistry of Materials* 31 (14): 5214–5223.

49 Masashige, O. (2008). Geometrically frustrated triangular lattice system Na_xVO_2: superparamagnetism in $x = 1$ and trimerization in $x \approx 0.7$. *Journal of Physics: Condensed Matter* 20: 145205. (8pp).

50 Mcqueen, T.M., Stephens, P.W., Huang, Q. et al. (2008). Successive orbital ordering transitions in $NaVO_2$. *Physical Review Letters* 101 (October): 166402.

51 Hamani, D., Ati, M., Tarascon, J., and Rozier, P. (2011). Na_xVO_2 as possible electrode for Na-ion batteries. *Electrochemistry Communications* 13: 938–941.

52 Didier, C., Guignard, M., Suchomel, M.R. et al. (2016). Thermally and electrochemically driven topotactical transformations in sodium layered oxides Na_xVO_2. *Chemistry of Materials* 28 (5): 1462–1471.

53 Saadoune, I., Maazaz, A., and Me, M. (1996). On the $Na_xNi_{0.6}Co_{0.4}O_2$ system: physical and electrochemical studies. *Journal of Solid State Chemistry* 122 (1): 111–117.

54 Vassilaras, P., Kwon, D., Dacek, S.T. et al. (2017). Electrochemical properties and structural evolution of O3-type layered sodium mixed transition metal oxides with trivalent nickel. *Journal of Materials Chemistry A* 5: 4596–4606.

55 Hwang, J.-Y., Yoon, C.S., Belharouak, I., and Sun, Y.-K. (2016). A comprehensive study of the role of transition metals in O3-type layered $Na[Ni_xCo_yMn_z]O_2$ ($x = 1/3$, 0.5, 0.6, and 0.8) cathodes for sodium-ion batteries. *Journal of Materials Chemistry A* 4 (46): 17952–17959.

56 Zhao, C., Ding, F., Lu, Y. et al. (2020). High-entropy layered oxide cathodes for sodium-ion batteries. *Angewandte Chemie, International Edition* 59 (1): 264–269.

57 Yoshida, H., Yabuuchi, N., and Komaba, S. (2013). $NaFe_{0.5}Co_{0.5}O_2$ as high energy and power positive electrode for Na-ion batteries. *Electrochemistry Communications* 34: 60–63.

58 Kubota, K., Asari, T., Yoshida, H. et al. (2016). Understanding the structural evolution and redox mechanism of a $NaFeO_2$–$NaCoO_2$ solid solution for sodium-ion batteries. *Advanced Functional Materials* 26 (33): 6047–6059.

59 Lu, Z. and Dahn, J.R. (2001). In situ X-ray diffraction study of P2-$Na_{2/3}[Ni_{1/3}Mn_{2/3}]O_2$. *Journal of the Electrochemical Society* 148 (11): A1225–A1229.

60 Zhao, W., Kirie, H., Tanaka, A. et al. (2014). Synthesis of metal ion substituted P2-$Na_{2/3}Ni_{1/3}Mn_{2/3}O_2$ cathode material with enhanced performance for Na ion batteries. *Materials Letters* 135: 131–134.

61 Komaba, S., Yabuuchi, N., Nakayama, T. et al. (2012). Study on the reversible electrode reaction of $NaNiMnO_2$ for a rechargeable sodium ion battery. *Inorganic Chemistry* 51: 6211–6220.

62 Yu, T., Ryu, H., Han, G., and Sun, Y. (2020). Understanding the capacity fading mechanisms of O3-type $Na[Ni_{0.5}Mn_{0.5}]O_2$ cathode for sodium-ion batteries. *Advanced Energy Materials* 10 (37): 2001609.

63 Singh, G., Tapia-Ruiz, N., Lopez del Amo, J.M. et al. (2016). High voltage mg-doped $Na_{0.67}Ni_{0.3-x}Mg_xMn_{0.7}O_2$ ($x = 0.05$, 0.1) Na-ion cathodes with enhanced stability and rate capability. *Chemistry of Materials* 28 (14): 5087–5094.

64 Hwang, J.-Y., Yu, T.-Y., and Sun, Y.-K. (2018). Simultaneous MgO coating and Mg doping of Na. *Journal of Materials Chemistry A* 6 (35): 16854–16862.

65 Zheng, L., Li, J., and Obrovac, M.N. (2017). Crystal structures and electrochemical performance of air-stable $Na_{2/3}Ni_{1/3-x}Cu_xMn_{2/3}O_2$ in sodium cells. *Chemistry of Materials* 29 (4): 1623–1631.

66 Song, B., Hu, E., Liu, J. et al. (2019). A novel P3-type $Na_{2/3}Mg_{1/3}Mn_{2/3}O_2$ as high capacity sodium-ion cathode using reversible oxygen redox. *Journal of Materials Chemistry A* 7 (4): 1491–1498.

67 Yabuuchi, N., Yano, M., Yoshida, H. et al. (2013). Synthesis and electrode performance of O3-type $NaFeO_2$–$NaNi_{1/2}Mn_{1/2}O_2$ solid solution for rechargeable sodium batteries. *Journal of the Electrochemical Society* 160 (5): A3131–A3137.

68 Yuan, D., Hu, X., Qian, J. et al. (2014). P2-type $Na_{0.67}Mn_{0.65}Fe_{0.2}Ni_{0.15}O_2$ cathode material with high-capacity for sodium-ion battery. *Electrochimica Acta* 116: 300–305.

69 Yoshida, H., Yabuuchi, N., Kubota, K. et al. (2014). P2-type $Na_{2/3}Ni_{1/3}Mn_{2/3-x}Ti_xO_2$ as a new positive electrode for higher energy Na-ion batteries. *Chemical Communications* 50 (28): 3677–3680.

70 Meng, Y., An, J., Chen, L. et al. (2020). A $NaNi_{0.5}Mn_{0.5}Sn_XO_2$ cathode with anti-structural deformation enhancing long lifespan and super power for a sodium ion battery. *Chemical Communications* 56 (58): 8079–8082.

71 Wang, Q., Mariyappan, S., Vergnet, J. et al. (2019). Reaching the energy density limit of layered O3-$NaNi_{0.5}Mn_{0.5}O_2$ electrodes via dual Cu and Ti substitution. *Advanced Energy Materials* 9: 1901785.

72 Carlier, D., Cheng, J.H., Berthelot, R. et al. (2011). The P2-$Na_{2/3}Co_{2/3}Mn_{1/3}O_2$ phase: structure, physical properties and electrochemical behavior as positive electrode in sodium battery. *Dalton Transactions* 40: 9306–9312.

73 Wang, X., Tamaru, M., Okubo, M., and Yamada, A. (2013). Electrode properties of P2–$Na_{2/3}Mn_yCo_{1-y}O_2$ as cathode materials for sodium-ion batteries. *Journal of Physical Chemistry C* 117 (30): 15545–15551.

74 Manikandan, P., Heo, S., Woo, H. et al. (2017). Structural characterization of layered $Na_{0.5}Co_{0.5}Mn_{0.5}O_2$ material as a promising cathode for sodium-ion batteries. *Journal of Power Sources* 363: 442–449.

75 Sathiya, M., Hemalatha, K., Ramesha, K. et al. (2012). Synthesis, structure, and electrochemical properties of the layered sodium insertion cathode material: $NaNi_{1/3}Mn_{1/3}Co_{1/3}O_2$. *Chemistry of Materials* 24 (10): 1846–1853.

76 Li, Z.Y., Zhang, J., Gao, R. et al. (2016). Unveiling the role of Co in improving the high-rate capability and cycling performance of layered $Na_{0.7}Mn_{0.7}Ni_{0.3-x}Co_xO_2$ cathode materials for sodium-ion batteries. *ACS Applied Materials & Interfaces* 8 (24): 15439–15448.

77 Hwang, J., Oh, S., Myung, S. et al. (2015). Radially aligned hierarchical columnar structure as a cathode material for high energy density sodium-ion batteries. *Nature Communications* 6: 6865.

78 Hwang, J.-Y., Myung, S.-T., Choi, J.U. et al. (2017). Resolving the degradation pathways of the O3-type layered oxide cathode surface through the nano-scale aluminum oxide coating for high-energy density sodium-ion batteries. *Journal of Materials Chemistry A* 5 (45): 23671–23680.

79 Rost, C.M., Sachet, E., Borman, T. et al. (2015). Entropy-stabilized oxides. *Nature Communications* 6: 8485.

80 Assat, G. and Tarascon, J. (2018). Fundamental understanding and practical challenges of anionic redox activity in Li-ion batteries. *Nature Energy* 3 (May): 373–386.

81 Sathiya, M., Rousse, G., Ramesha, K. et al. (2013). Reversible anionic redox chemistry in high-capacity layered-oxide electrodes. *Nature Materials* 12 (July): 827–835.

82 Seo, D., Lee, J., Urban, A. et al. (2016). The structural and chemical origin of the oxygen redox activity in layered and cation-disordered Li-excess cathode materials. *Nature Chemistry* 8 (May): 692–697.

83 Xu, H., Guo, S., and Zhou, H. (2019). Review on anionic redox in sodium-ion batteries. *Journal of Materials Chemistry A* 7: 23662–23678.

84 Maitra, U., House, R.A., Somerville, J.W. et al. (2018). Oxygen redox chemistry without excess alkali-metal ions in $Na_{2/3}[Mg_{0.28}Mn_{0.72}]O_2$. *Nature Chemistry* 10: 288–295.

85 Mortemard de Boisse, B., Liu, G., Ma, J., and Yamada, A. (2016). Intermediate honeycomb ordering to trigger oxygen redox chemistry in layered battery electrode. *Nature Communications* 7: 11397.

86 Mortemard de Boisse, B., Nishimura, S., Watanabe, E. et al. (2018). Highly reversible oxygen-redox chemistry at 4.1 V in $Na_{4/7-x}[\square_{1/7}Mn_{6/7}]O_2$ (□: Mn vacancy). *Advanced Energy Materials* 8 (20): 1–7.

87 Song, B., Tang, M., Hu, E. et al. (2019). Understanding the low-voltage hysteresis of anionic redox in $Na_2Mn_3O_7$. *Chemistry of Materials* 31 (10): 3756–3765.

6

Anionic Redox Reaction in Li-Excess High-Capacity Transition-Metal Oxides

Naoaki Yabuuchi[1,2]

[1] *Yokohama National University, Department of Chemistry and Life Science, 79-5 Tokiwadai, Hodogaya-ku, Yokohama, Kanagawa 240-8501, Japan*
[2] *Kyoto University, Elements Strategy Initiative for Catalysts and Batteries, 1-30 Goryo-Ohara, Nishikyo-ku, Kyoto 615-8245, Japan*

6.1 Stoichiometric Layered Oxides for Rechargeable Lithium Batteries

During the past 40 years, rechargeable lithium batteries have been extensively studied for energy-storage applications, and now this technology is highly sophisticated after first commercialization by SONY in 1991. More than a million electric vehicles equipped with an electric motor and lithium battery have been sold per year in the global market. For a long time, the transportation system was dependent on the technology of internal combustion engines. The lithium batteries are starting to substitute clean and sustainable energy resources for fossil fuels. The gravimetric energy density of the lithium batteries is 2–3 times greater compared with batteries with aqueous electrolyte solutions (e.g. metal hydride batteries) [1]. However, the demand for further increase in energy density is still growing to reduce the size of batteries. Lithium batteries are made from two lithium-ion insertion materials, i.e. positive and negative electrode materials, which reversibly store both lithium ions and electrons on charge/discharge processes [2]. The first generation of lithium batteries consisted of $LiCoO_2$ [3] and carbonaceous materials as positive and negative electrode materials, respectively. A theoretical capacity of $LiCoO_2$ is calculated from the following reaction:

$$LiCoO_2 \rightarrow \square CoO_2 + Li^+ + e^- \tag{6.1}$$

Herein, □ denotes a vacant site created by lithium extraction in the host structure. A trivalent Co ion as cationic species in $LiCoO_2$ is oxidized to a tetravalent state, and lithium ions are simultaneously extracted from the host structure to maintain charge neutrality. A theoretical capacity of $LiCoO_2$ is calculated to be 274 mA h g^{-1} from Eq. 6.1 (note that a practical reversible capacity is limited to ~200 mA h g^{-1}). $LiCoO_2$ is still widely used in commercial lithium-ion batteries, especially for portable electronic devices because of its high volumetric energy density.

Transition Metal Oxides for Electrochemical Energy Storage, First Edition.
Edited by Jagjit Nanda and Veronica Augustyn.

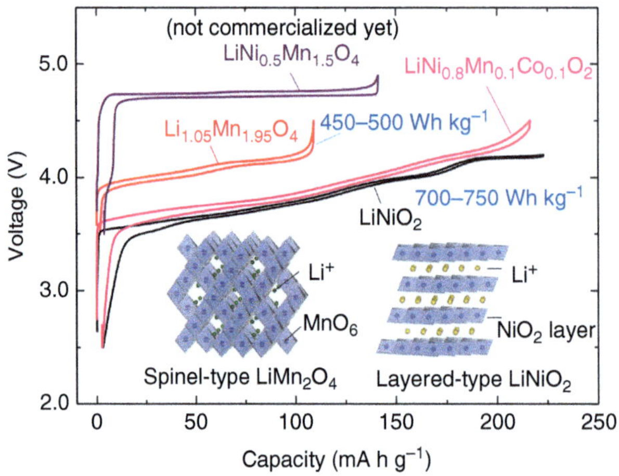

Figure 6.1 Comparison of the charge/discharge curves of representative lithium insertion materials used as positive electrode materials for lithium-ion batteries. Schematic illustrations of spinel-type and layered-type oxides are also compared in the inset. Source: Yabuuchi [4]/with permission of The Chemical Society of Japan & Wiley-VCH Verlag GmbH & Co. KGaA, Weinheim.

For lithium batteries used in electric vehicles, spinel-type $LiMn_2O_4$ and its derivatives were originally used as positive electrode materials. In $LiMn_2O_4$, MnO_6 octahedra share edges, forming a three-dimensional framework structure, and Li ions are located at tetrahedral sites (see Figure 6.1). Nearly all of the Li ions are reversibly extracted from the crystal structure without the destruction of the framework structure. $LiMn_2O_4$ is used as a 4-V class positive electrode material. However, the reversible capacity is limited to ~120 mA h g^{-1} as shown in Figure 6.1, and the available energy density is limited to 450–500 Wh kg^{-1} based on metallic lithium. To increase the energy density of lithium batteries, Ni-based layered materials are now used for electric vehicle applications. Galvanostatic charge/discharge curves of stoichiometric $LiNiO_2$ are also shown in Figure 6.1. The crystal structure of stoichiometric $LiMO_2$ (M = transition-metal ions) is classified as a rocksalt-related layered structure with a cubic close-packed (ccp) array of oxide ions. Li and M ions are ordered into alternate layers perpendicular to [111] of the original cubic lattice, forming a layered structure with a rhombohedral lattice. $LiCoO_2$ and $LiNiO_2$ are isostructural, and formation of a solid solution is possible in the entire composition range of $LiCo_xNi_{1-x}O_2$ ($0 \leq x \leq 1$) [5]. The reversible capacity of stoichiometric $LiNiO_2$ exceeds 200 mA h g^{-1} even though the operating average voltage as the electrode material is slightly lowered compared with spinel-type $LiMn_2O_4$. The energy density with it as the positive electrode material reaches 700–750 Wh kg^{-1}. Because stoichiometric $LiNiO_2$ is thermally unstable after delithiation and shows detrimental phase transitions on delithiation [6–8], other transition-metal ions are partially substituted for nickel ions. The Co/Al-substituted system, for instance, $LiNi_{0.8}Co_{0.15}Al_{0.05}O_2$ [9], is currently used for practical batteries. The Mn/Co-substituted system was also extensively studied for positive electrode materials [10–13]. Galvanostatic charge/discharge curves of $LiNi_{0.8}Mn_{0.1}Co_{0.1}O_2$ are also shown and compared with stoichiometric $LiNiO_2$ in Figure 6.1. Average operating voltage is slightly higher than the pure Ni system, and voltage plateaus associated with phase transitions are less pronounced for $LiNi_{0.8}Mn_{0.1}Co_{0.1}O_2$. Lithium and charge ordering in the layered structure is disturbed by metal substitution [14]. Although the Ni-based layered materials are currently used for electric vehicle applications, the available energy density is restricted below 800 Wh kg^{-1} on the basis of metallic lithium. Therefore, the development of high-capacity positive electrode materials, which outperform the Ni-based layered materials, is necessary to further increase energy density of commercial lithium batteries.

6.2 Li-Excess Rocksalt Oxides as High-Capacity Positive Electrode Materials

Because the theoretical capacities of positive electrode materials depend on lithium contents in the host structures, substitution of lithium ions for transition-metal ions in $LiMO_2$ ($Li_{1+x}M_{1-x}O_2$) is indispensable to design high-capacity positive electrode materials. In the past decade, Li-enriched materials, Li_2MO_3-type layered materials ($M = Mn^{4+}$, Ru^{4+}, and other tetravalent transition-metal ions), which are also classified based on the cation-ordered rocksalt-type structure [15], have been extensively studied as high-capacity positive electrode materials [16–24]. Both compositions, $LiMO_2$ and Li_2MO_3, are found in a Li-M-O triangular phase diagram on a rocksalt tie-line shown in Figure 6.2. Li_2MO_3 is also reformulated as $Li_{4/3}M_{2/3}O_2$ when the oxygen content in the chemical formula is normalized to 2. These oxides are often

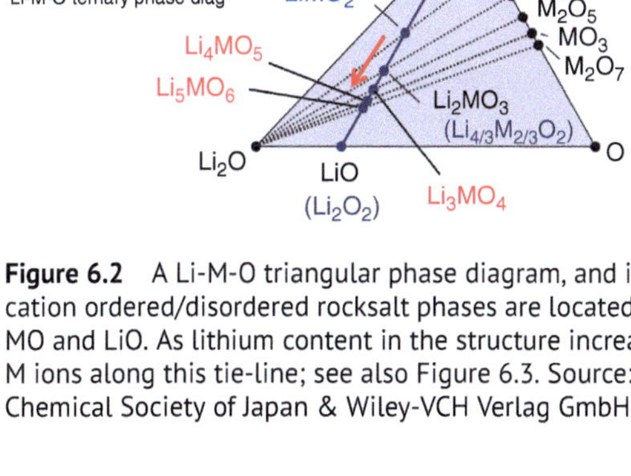

Figure 6.2 A Li-M-O triangular phase diagram, and its expanded view is also shown. Many cation ordered/disordered rocksalt phases are located on a tie-line (solid blue line) between MO and LiO. As lithium content in the structure increases, so too do the oxidation states of M ions along this tie-line; see also Figure 6.3. Source: Yabuuchi [4]/with permission of The Chemical Society of Japan & Wiley-VCH Verlag GmbH & Co. KGaA, Weinheim.

called "lithium-excess" and "lithium-rich" materials because the lithium content is enriched as compared with conventional layered materials. Among the Li_2MO_3-type oxides, a Mn-system, Li_2MnO_3, and its derivatives, have been the most widely studied as electrode materials. The theoretical capacity of Li_2MnO_3 is calculated from the following reaction:

$$Li_2MnO_3 \rightarrow \square_2MnO_3 + 2Li^+ + 2e^- \quad (6.2)$$

Herein, □ denotes a vacant site created by lithium extraction in the host structure.

The theoretical capacity is 459 mA h g^{-1} if two electrons and lithium ions are successfully and reversibly extracted from the host structure. This value is much higher than those of stoichiometric layered oxides. However, Li_2MnO_3 was regarded as an electrochemically inactive material because of the difficulty of oxidizing tetravalent Mn ions into higher oxidation states. Nevertheless, the fact is that Li_2MnO_3 is electrochemically active associated with the contribution of anions (oxide ions) for the charge compensation process [19, 21]. Detailed reaction mechanisms of Li_2MnO_3 as the electrode material are found in the literature [26]. Historically, charge compensation by noncationic species had been known in a sulfide-based compound before 1990 [27]. Sulfide ions are relatively soft and easily polarizable compared with oxide ions, and oxidation of sulfide ions (S^{2-}) and the formation of disulfide (persulfide) ions ($2S^{2-}$) is a well-known process in many chemical/biological processes. Ligand hole stabilization has also been known. It is associated with charge transfer from the oxide ion to the transition-metal ions in a heavily hybridized system with oxygen 2p orbitals, for example, Fe^{4+} in $SrFeO_3$ [28]. Similarly, the possibility of charge transfer from oxygen 2p to Co^{4+} in Li_xCoO_2 was also discussed in 1999 [29]. If such ligand holes are directly created by electrochemical oxidation in the lithium-excess system, this process is classified as "anionic" redox.

The crystal structure of Li_2MnO_3 is essentially the same as $LiCoO_2$, the only difference being in the presence of additional lithium ions in the transition-metal layers as shown in Figure 6.3. Further enrichment of lithium content in the structure is also achieved with the presence of transition-metals ions with higher oxidation states [4]. Li_3NbO_4 with pentavalent niobium ions [30], Li_4MoO_5 with hexavalent molybdenum ions [31], and Li_5ReO_6 with heptavalent rhenium ions [32] are, indeed, found in the literature. These lithium-excess oxides are also classified as cation-ordered rocksalt-type structures (Figure 6.3) with the common ccp array of oxide ions, and found in the rocksalt tie-line shown in Figure 6.2. Although niobium and molybdenum as 4d-transition-metal ions and rhenium as a 5d-transition-metal ion are heavier ions, the theoretical capacities of these oxides are attractive as electrode materials as compared in Table 6.1. These transition-metal as well as lithium ions are located in octahedral sites in the ccp array of oxide ions, but the arrangements of these ions are different from each other, as shown in Figure 6.3. The theoretical capacity reaches 526 mA h g^{-1} for Li_4MoO_5 when anionic redox is effectively utilized. However, these lithium-excess compounds consist of transition-metal

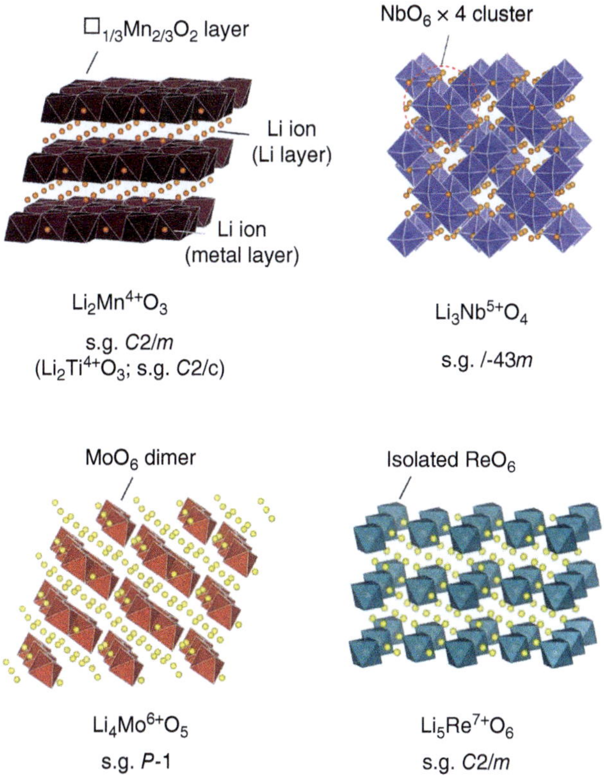

Figure 6.3 Schematic illustrations of the crystal structures of different lithium-excess oxides: Li_2MnO_3, Li_3NbO_4, Li_4MoO_5, and Li_5ReO_6. These illustrations were drawn using the VESTA program [25]. In artwork, □ denotes a vacant site created by lithium extraction in the host structure. Source: Yabuuchi [4]/with permission of John Wiley & Sons.

Table 6.1 Comparison of the theoretical capacity on the basis of lithium content in the structures of different lithium-excess materials.

$LiCoO_2$	Li_2MnO_3	Li_3NbO_4	Li_4MoO_5	Li_5ReO_6
$274\,mA\,h\,g^{-1}$	$459\,mA\,h\,g^{-1}$	$452\,mA\,h\,g^{-1}$	$526\,mA\,h\,g^{-1}$	$423\,mA\,h\,g^{-1}$

ions without conductive d-electrons and are essentially insulators and/or semiconductors. Therefore, the anionic redox reaction is kinetically hindered. Lithium insertion materials as electrode materials require both high ionic and electronic conductivity. To induce electronic conductivity, partial substitution of transition metals for these oxides is an effective strategy to design high-capacity positive electrode materials.

6.3 Reversible and Irreversible Anionic Redox for Li_3NbO_4- and Li_2TiO_3-Based Oxides

Although pure Li_3NbO_4 is electrochemically inactive, the substitution of 3d transition-metal ions for Nb/Li ions provides conductive electrons to the structure [33, 34]. Such 3d transition-metal ions can accept electrons from oxide ions, leading to the activation of anionic redox coupled with partial cationic redox. Cation ordering observed for pure Li_3NbO_4 is lost by metal substitution, and a cation-disordered rocksalt-type structure is formed. Historically, cation-disordered rocksalt oxides were regarded as electrochemically inactive as electrode materials because of the absence of Li migration paths in bulk structures [35–37]. Nevertheless, facile lithium migration is possible for lithium-excess cation-disordered rocksalt $Li_{1+x}Me_{1-x}O_2$, which is associated with the formation of a percolative network for lithium migration in the host structure [38–40].

Charge/discharge curves of Mn^{3+}- and Fe^{3+}-substituted Li_3NbO_4, i.e. $Li_{1.3}Nb_{0.3}Mn_{0.4}O_2$ and $Li_{1.3}Nb_{0.3}Fe_{0.4}O_2$, are compared in Figure 6.4. Both samples deliver initial charge capacities of ~350 mA h g^{-1} with a voltage plateau at 4.1–4.3 V. However, a clear difference is noted for the discharge process. Polarization on discharge is much larger for the Fe-substituted sample, and the second charge curve is completely different from the initial charge. The voltage plateau is also observed only for the initial charge. In contrast, profiles of the initial and second charge are similar for the Mn-substituted sample with a sloping region (3–4 V) followed by a voltage plateau at 4.3 V. A large reversible capacity of 300 mA h g^{-1} is observed for the Mn-substituted sample with relatively high voltage on discharge. The observed discharge capacities are much larger than the theoretical capacity calculated based on the Mn^{3+}/Mn^{4+} cationic redox. This observation clearly indicates that charge compensation is realized by anionic redox. Similar to Li_3NbO_4, Li_2TiO_3 is also used as a host structure for anionic redox, and galvanostatic charge/discharge curves of $Li_{1.2}Ti_{0.4}M_{0.4}O_2$ (M = Mn^{3+} and Fe^{3+}) are also shown in Figure 6.4. Niobium- and tatanium-based oxides show similar voltage profiles, and similar charge compensation processes for both systems are expected.

To further examine charge compensation mechanisms and the contribution of oxide ions on the charge compensation process, soft X-ray absorption spectroscopy (XAS) was utilized for $Li_{1.2-x}Ti_{0.4}Mn_{0.4}O_2$ (Figure 6.5) [41]. On the charge process, no change is observed for Ti L-edge XAS spectra, suggesting that Ti is not responsible for charge compensation. In the sloping region (3–4 V on charge), the shift of Mn L-edge XAS spectra to the higher-energy region indicates oxidation of Mn from a trivalent to tetravalent state, while no change is found on charge in the voltage plateau region. In contrast, a new peak appears at 530 eV for O K-edge XAS spectra, and systematic changes as a function of charge capacities are observed during charge in the plateau region. Moreover, this process is reversible and comes back to its original position after discharge. Such a new peak after charge is also evidenced for Li_2MnO_3-based electrode materials [42, 43], but more pronounced changes are noted for $Li_{1.3-y}Nb_{0.3}Mn_{0.4}O_2$ and $Li_{1.2-y}Ti_{0.4}Mn_{0.4}O_2$. When it is assumed that charge compensation is realized by anionic redox, formal oxidation states change

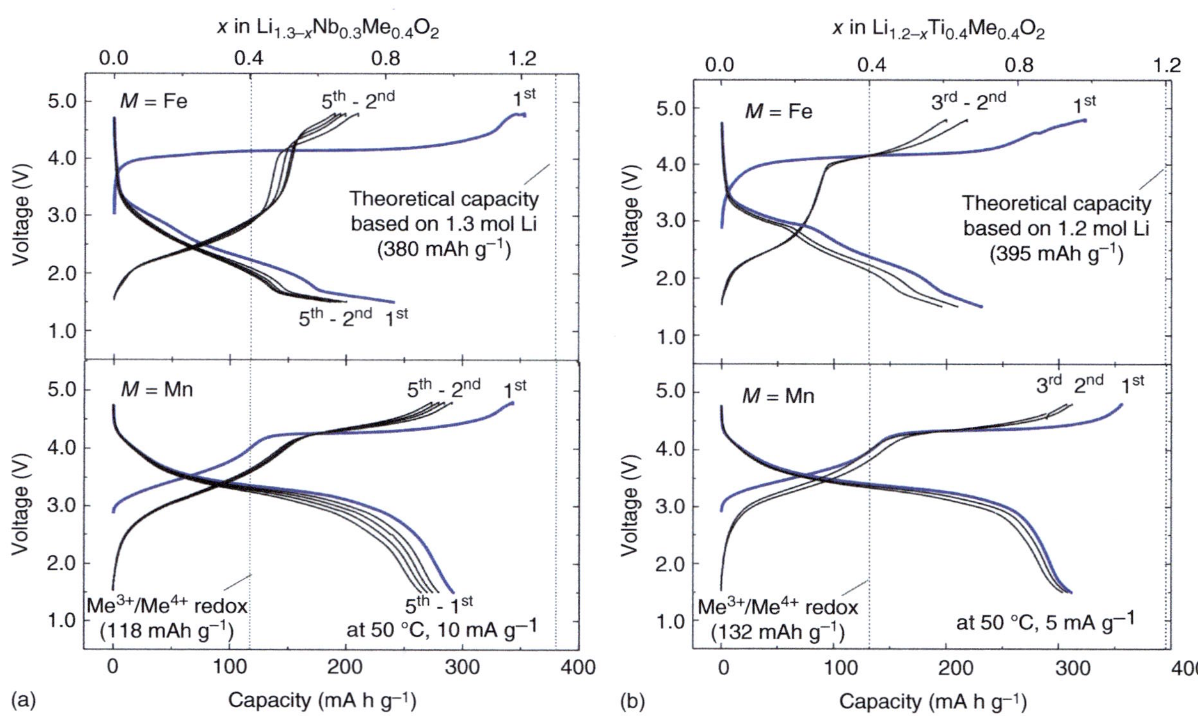

Figure 6.4 Comparison of the charge/discharge curves of (a) $Li_{1.3}Nb_{0.3}M_{0.4}O_2$ and (b) $Li_{1.2}Ti_{0.4}M_{0.4}O_2$ ($M = Mn^{3+}$ and Fe^{3+}). Source: Yabuuchi [26]/ with permission of The Chemical Society of Japan.

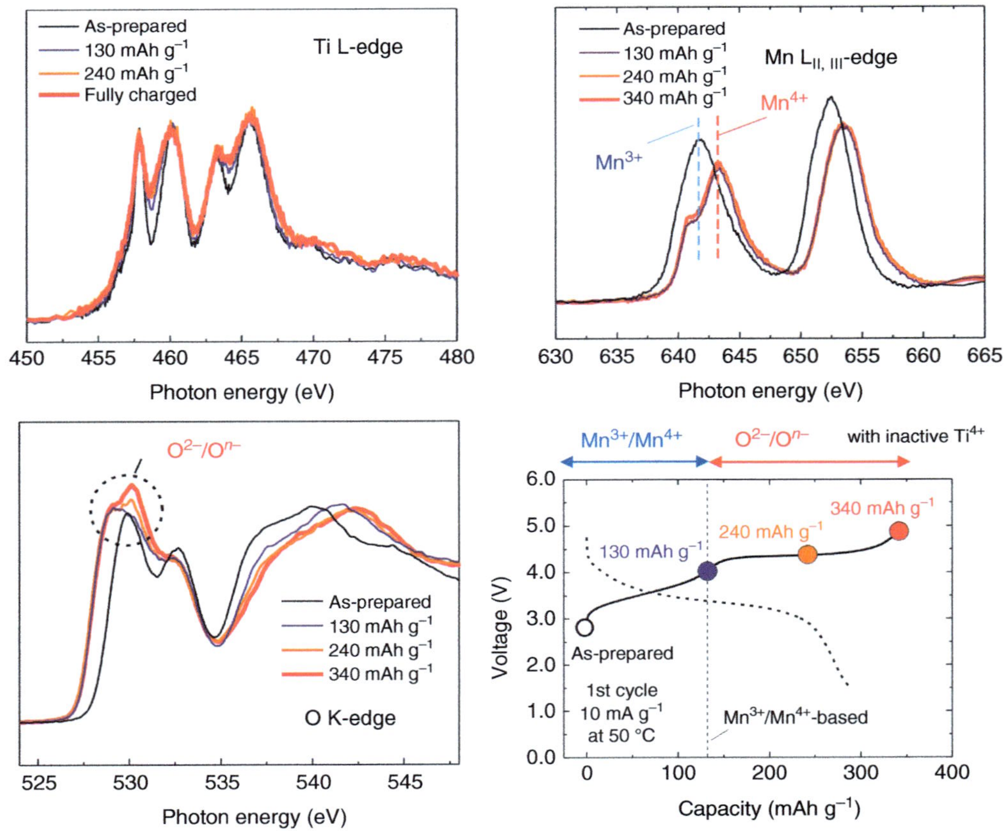

Figure 6.5 Changes in the electronic structures for $Li_{1.2-x}Ti_{0.4}Mn_{0.4}O_2$ on initial charge/discharge: Ti L-edge, Mn L-edge, and O K-edge XAS spectra. The points where XAS spectra have been collected are also shown.

from $Li_{1.2}Ti_{0.4}Mn^{3+}{}_{0.4}O_2{}^{2-}$ for the as-prepared sample to $Li_0Ti_{0.4}Mn^{4+}{}_{0.4}O_2{}^{1.6-}$ for the fully charged state. The origin of the new peak has been debatable for a long time [44]. Recently, this peak has been explained as oxidation of π-like bonds between Mn^{4+} (t_{2g}^3) and O 2p orbitals [45]. After the oxidation of oxygen, holes are energetically stabilized by π-type interaction with Mn^{4+}, which is possibly accompanied by rehybridization of molecular orbitals, and probably accounts for the trend observed in the XAS spectra.

As shown in Figure 6.4, the Fe-substituted samples, $Li_{1.3-y}Nb_{0.3}Fe_{0.4}O_2$ and $Li_{1.2-y}Ti_{0.4}Fe_{0.4}O_2$, show a completely different trend. These results suggest that an irreversible phase transition occurs on the initial charge process. Soft XAS spectra of the Fe-substituted Li_3NbO_4 sample, $Li_{1.3-y}Nb_{0.3}Fe_{0.4}O_2$, on initial charge/discharge are shown in Figure 6.6 [41]. In the O K-edge XAS spectra, a new peak at 527.5 eV appears after charge to point b (120 mA h g^{-1}) in Figure 6.6. The energy position is clearly different from the Mn-substituted systems. However, because no change is observed from Fe L-edge XAS spectra, oxygen contributes

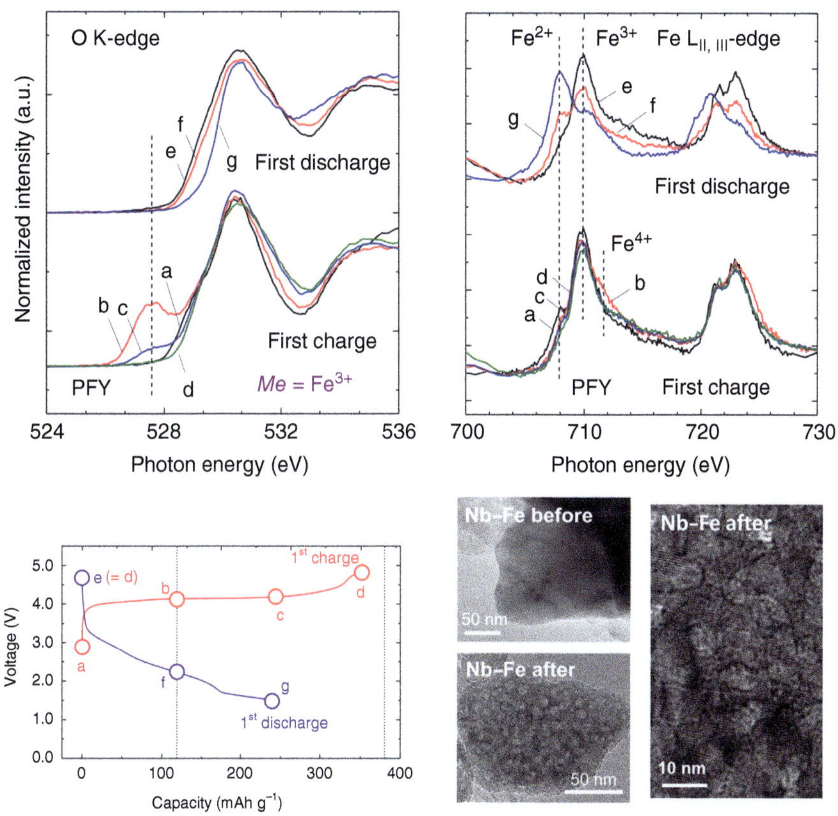

Figure 6.6 Changes in the electronic structures for $Li_{1.3-x}Nb_{0.3}Fe_{0.4}O_2$ on initial charge/discharge: Fe L-edge and O K-edge XAS spectra. The points where XAS spectra have been collected are also plotted. TEM images of $Li_{1.3}Nb_{0.3}Fe_{0.4}O_2$ particles before and after electrochemical cycling at 50 °C. Oxygen loss for the Fe system results in the formation of nanosized grains in the single particle.

to the charge compensation process. A similar change was also reported for $Li_{1.19}Ti_{0.38}Fe_{0.42}O_2$ [46]. In addition, this new peak disappears on further charge to 4.8 V. This fact indicates that this intermediate species is energetically unstable, leading to oxygen loss as an irreversible process by electrochemical oxidation. Such irreversible oxygen loss results in the reconstruction of particle morphology and formation of nanosized particles, and this fact is clearly evidenced in TEM images of Figure 6.6. After oxygen loss on charge, Fe^{3+} is electrochemically reduced to Fe^{2+} on initial "discharge," as clearly observed in the Fe L-edge XAS spectra [41]. Instability of anionic redox is also responsible in the huge polarization observed in Figure 6.4 for the Fe systems. Similarly, oxygen loss was found to be the dominative process for $Li_{1.33}Sb_{0.33}Fe_{0.33}O_2$ [47].

For the case of oxides with Fe^{3+}, covalency is relatively high, and therefore, the Fermi level consists of both iron and oxygen [41]. After charge, both iron and oxygen are, therefore, oxidized. However, Fe^{4+} is chemically less stable, and therefore oxygen donates electrons to reduce iron ions. As a result, oxygen is further oxidized, leading to oxygen release. In contrast, the Mn—O bond has a relatively high ionic character, and Mn^{4+} is chemically more stable. The holes created by oxidation are isolated in the structure (small interaction with manganese ions), and therefore high reversibility is achieved for the manganese system.

6.4 Activation of Anionic Redox by Chemical Bonds with High Ionic Characters

Each oxide ion in conventional and stoichiometric layered $LiMO_2$ is coordinated by three transition-metal ions and three lithium ions. For the Li-excess layered oxides, Li_2MO_3, oxide ions are coordinated by two transition-metal ions and by four lithium ions. Coordination numbers of lithium ions are increased by Li enrichment. Since the bonding character of Li–O is essentially ionic, the increase in the coordination numbers of lithium ions influences the electronic structures of oxide ions. Oxide ions gain more electrons from Li ions, and a net oxidation number approaches its formal oxidation number of −2. Recently, a rational explanation about such considerations has been provided by a DFT study of lithium-excess oxides [48]. The lithium-excess oxide contains the Li–O–Li configuration, and the Li–O–Li configuration is essentially unhybridized with transition-metal ions. The energy of this unhybridized state is lower than those of hybridized O 2p states with transition-metal ions, and thus, oxide ions are more easily oxidized for the lithium-excess oxides.

Additionally, Nb^{5+} and Ti^{4+} ions are highly ionized ions compared with the late transition-metal ions. Mixing between metal d orbitals and oxygen 2p orbitals is less pronounced, and thus similar to the lithium enrichment in the structure, the character of oxide ions becomes more ionic. This fact is beneficial to activate anionic redox, leading to a lowering of the electrochemical potential. Initial charge/discharge profiles of the conventional lithium-excess oxide, $Li_{1.2}Co_{0.13}Ni_{0.13}Mn_{0.54}O_2$ [23], and $Li_{1.2}Ti_{0.4}Mn_{0.4}O_2$ [41] in Li cells are compared in Figure 6.7. Well-defined voltage plateaus are observed for both samples, but the voltage related to anionic

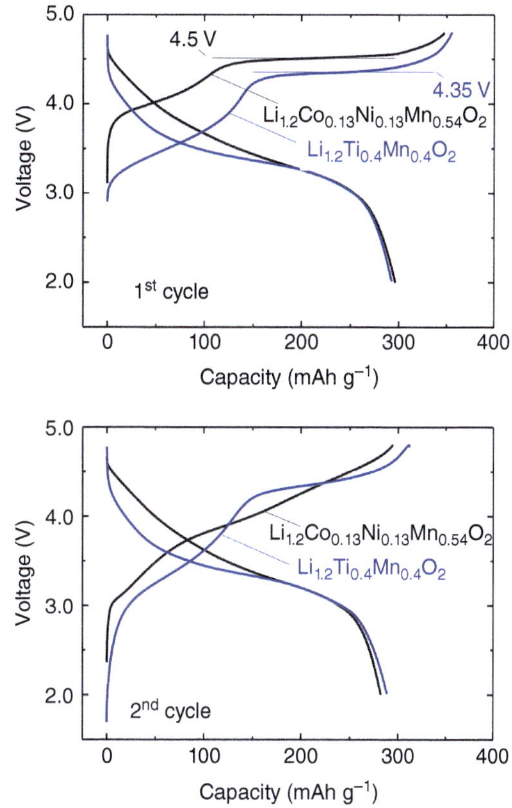

Figure 6.7 Comparison of the first and second charge/discharge curves of the conventional Li-excess layered oxide, $Li_{1.2}Ni_{0.13}Ni_{0.13}Mn_{0.54}O_2$, and the cation disordered rocksalt oxide, $Li_{1.2}Ti_{0.4}Mn_{0.4}O_2$. Source: Yabuuchi [26]/with permission of The Chemical Society of Japan.

redox is 0.15 V lower for $Li_{1.2-y}Ti_{0.4}Mn_{0.4}O_2$ than that of $Li_{1.2-x}Co_{0.13}Ni_{0.13}Mn_{0.54}O_2$. Oxide ions are more easily oxidized with a lower electrochemical potential than for $Li_{1.2-y}Ti_{0.4}Mn_{0.4}O_2$. Due to the presence of Nb^{5+} and Ti^{4+}, anionic redox is more easily accessed when compared with the lithium-excess oxides containing late transition-metal ions. Furthermore, for $Li_{1.2-y}Co_{0.13}Ni_{0.13}Mn_{0.54}O_2$, the voltage plateau is lost for the second charge process, as shown in Figure 6.7, because the partial oxygen loss and structural reconstruction processes cannot be avoided on initial charge [23, 49–52]. Nevertheless, as pointed out by a recent article [45], the nature of chemical bonds between oxygen and ions without valence electrons, like Nb^{5+} and Ti^{4+}, is essentially nonbonding, and therefore these ions cannot energetically stabilize anionic redox. Instead, anionic redox would be stabilized by tetravalent manganese ions through π-type interaction with oxygen.

6.5 Li_4MoO_5 as a Host Structure for Lithium-Excess Oxides

Li_4MoO_5 has also been studied as a host structure for high-capacity positive electrode materials. To extract lithium ions from Li_4MoO_5, divalent nickel ions are in part substituted for lithium and molybdenum ions on the basis of the

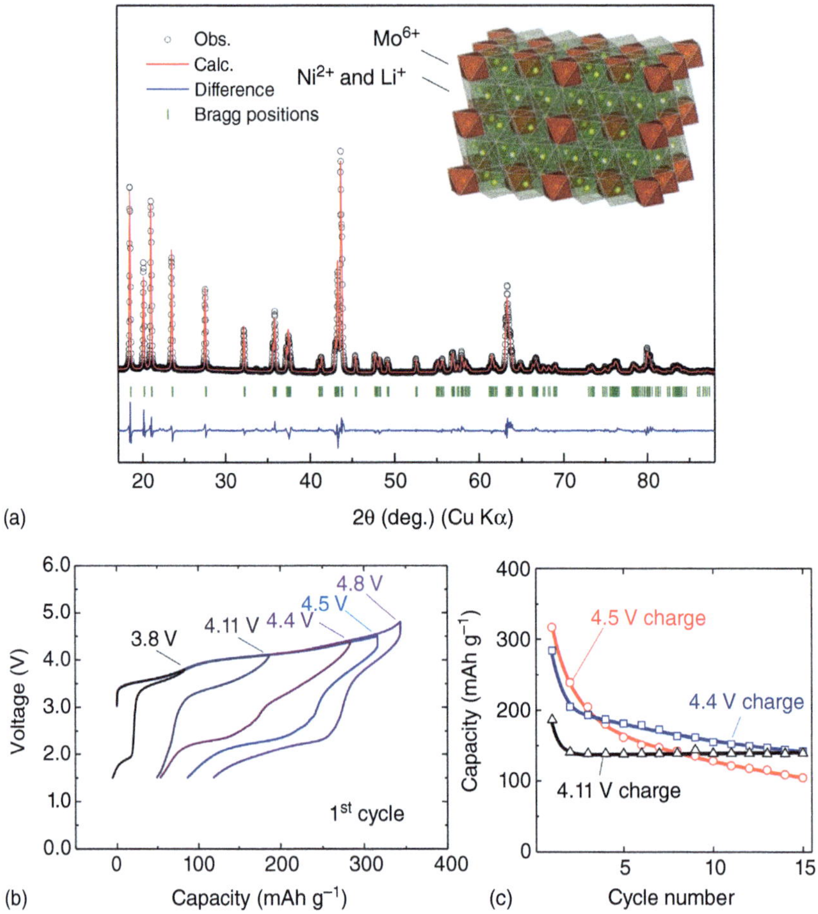

Figure 6.8 (a) Rietveld refinement on $Li_{4/3}Ni_{1/3}Mo_{1/3}O_2$, and a schematic illustration of the crystal structure. (b) Galvanostatic oxidation/reduction curves of $Li_{4/3}Ni_{1/3}Mo_{1/3}O_2$ with different cutoff voltages at 10 mA g^{-1}. Charge capacity retention with 4.11, 4.5, and 4.8 V cutoff voltages is also compared in (c). Source: Yabuuchi [53]/with permission of American Chemical Society.

x Li_4MoO_5–(1–x) NiO binary system with two different rocksalt-type related structures. Synthesis of single-phase samples is achieved in the entire range in this binary system [53]. An XRD pattern of the 1 : 1 mixture of Li_4MoO_5 and NiO, i.e. Li_4NiMoO_6 is shown in Figure 6.8a. Li_4NiMoO_6 is also reformulated as $Li_{4/3}Ni_{1/3}Mo_{1/3}O_2$. The crystal structure of Li_4NiMoO_6 was analyzed by Rietveld analysis, and it is found that the crystal structure is related to that of Li_5ReO_6. Rhenium ions at 2a sites in Li_5ReO_6 are replaced by molybdenum ions. Additionally, one nickel ion is substituted for one Li ion, resulting in the chemical formula of Li_4NiMoO_6. Nickel and lithium ions are essentially randomly distributed at the same octahedral sites, whereas perfect ordering is observed for Mo ions, as shown in a schematic illustration of the crystal structure in the inset of Figure 6.8a [53].

The electrochemical properties of $Li_{4/3}Ni_{1/3}Mo_{1/3}O_2$ with different cut-off voltages for the charge (oxidation) process are shown in Figure 6.8b. Although a large initial charge capacity is observed up to 4.8 V, the polarization on charge/discharge is large as the electrode material. Electrode reversibility is improved by lowering the cut-off voltage even though reversible capacity is inevitably lowered. No capacity loss was observed for 15 continuous cycles in the range of 1.5–4.11 V (Figure 6.8c). When the cut-off voltage is increased from 4.11 to 4.5 V, the discharge (reduction) voltage profile changes. A clear voltage plateau at around 2 V appears after charge above 4.5 V, which is clearly different from the voltage profile observed for 4.11 V. To examine the difference in voltage profiles for 4.11 and 4.5 V cutoffs, *ex situ* XRD patterns were collected after electrochemical cycling. Diffraction lines related to Mo ordering, observed in a two-theta range of 20°–35°, have disappeared after the 4.5-V cycle, and major diffraction lines in the XRD pattern can be assigned to a cation-disordered rocksalt structure [53]. This observation suggests that an irreversible phase transition, including Mo migration in the bulk structure, occurs upon cycling to 4.5 V. Clear difference is observed in the XRD pattern for the 4.11-V cycle. Diffraction lines related to Mo ordering are still observed after charge to 4.11 V, suggesting partial oxygen loss on oxidation above 4.11 V [53].

Similarly, a binary system, x Li_4MoO_5–$(1-x)$ $LiFeO_2$, is studied as electrode materials for rechargeable lithium batteries [54]. A single-phase sample is obtained at $x = 0.5$, and it is found that $Li_{1.42}Mo_{0.29}Fe_{0.29}O_2$ ($x = 0.5$) also crystallizes into a Li_5ReO_6-type structure. Although an initial charge capacity of $Li_{1.42}Mo_{0.29}Fe_{0.29}O_2$ reaches 350 mA h g^{-1}, irreversible phase transition associated with oxygen loss and molybdenum migration on charge is evidenced by X-ray diffraction and X-ray absorption spectroscopy [54]. The irreversible phase transition inevitably results in large polarization on discharge. For late transition metals, such as iron, cobalt, and nickel ions, competition of cationic and anionic redox reactions cannot be avoided and is associated with relatively high covalent natures. The holes created in oxygen are destabilized by charge transfer to chemically unstable cationic species, leading to oxygen loss as the irreversible process.

6.6 Extremely Reversible Anionic Redox for Li_2RuO_3 System

Although irreversible anionic redox was evidenced for the late 3d-transition metal ions, an exception is found in Li_2RuO_3 ($Li_{4/3}Ru_{2/3}O_2$) [4]. Ruthenium as the 4d-transition-metal ion has a much more covalent nature with oxide ions. However, chemical stability of ruthenium ions with higher oxidation states is relatively high when compared with 3d transition-metal ions, and therefore unfavorable charge transfer from oxygen on charge is effectively suppressed. Moreover, anionic redox is easily activated associated with the high electronic conductivity of Li_2RuO_3 [16], leading to highly reversible anionic redox. Typical charge/discharge curves of Li_2RuO_3 are shown in Figure 6.9 collected at 50 °C. A reversible capacity of 300 mA h g^{-1}, which corresponds to >90% of the theoretical capacity, is observed,

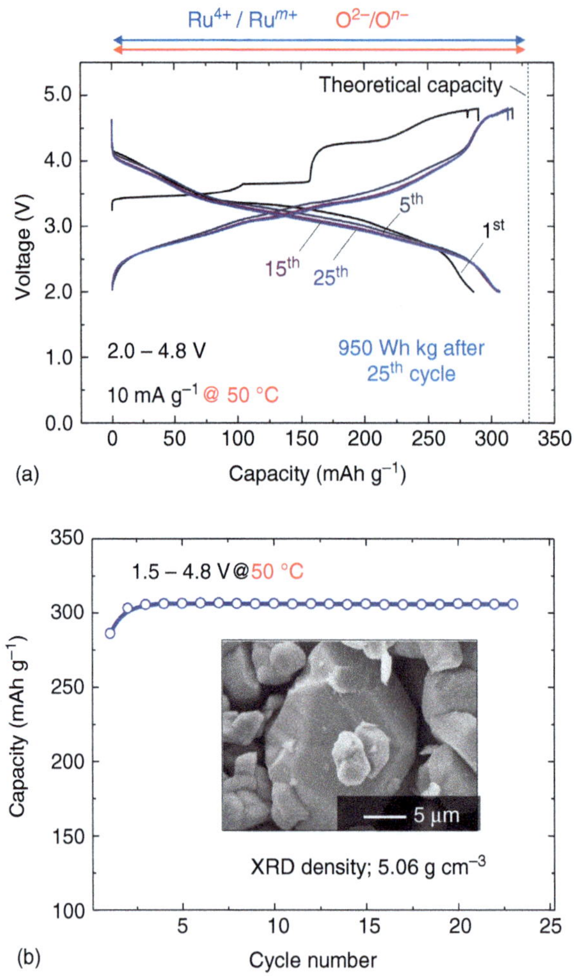

Figure 6.9 (a) Charge/discharge curves of a Li/Li_2RuO_3 ($Li_{4/3}Ru_{2/3}O_2$) cell at a rate of 10 mA g^{-1} at 50 °C, and (b) its capacity retention. No deterioration is evidenced for 25 cycles. Particle morphology of the sample observed by SEM is also shown in the inset. Source: Yabuuchi [4]/with permission of The Chemical Society of Japan & Wiley-VCH Verlag GmbH & Co. KGaA, Weinheim.

and surprisingly nearly 2 M of lithium ions are reversibly extracted from Li_2RuO_3 with the cationic/anionic redox, forming $\square_2RuO_3$ ($\square$ denotes a vacant site) after charge to 4.8 V. Moreover, such electrode properties are realized from highly dense particles with large size of ~15 μm, as shown in the SEM image of Figure 6.9 [4]. No capacity fading is observed even at 50 °C. Because the density (g cm^{-3}) estimated from the crystallographic parameters of Li_2RuO_3 is also higher than for Ni-based layered materials, both gravimetric/volumetric energy density clearly outperform stoichiometric layered materials. The only problem is found in the material cost of ruthenium.

For conventional layered oxides, e.g. $LiCoO_2$ and $LiNiO_2$, extraction of 1.0 M of Li ions from the host structures results in the formation of the O1 phase [55] (additionally, some stacking faults are noted for Li_xNiO_2 [56]), and thus drastic shrinkage of interlayer distances cannot be avoided. This fact is generally used to account for insufficient cyclability for $LiCoO_2$ and $LiNiO_2$ when all Li ions are extracted from the host structures. As a layered material, a similar drawback was anticipated for Li_2RuO_3. Nevertheless, no degradation during electrochemical cycling is evidenced in this system. Further studies on reaction mechanisms for the Ru system as model materials of the anionic redox are needed to explain the outstanding experimental observation. Similar to Li_2RuO_3, if 2 M of lithium ions are reversibly extracted from Li_2MO_3 (4/3 M of lithium ions from $Li_{4/3}M_{2/3}O_2$) with the chemistry of 3d transition-metal ions, available reversible capacity possibly exceeds 400 mA h g^{-1} as the electrode materials.

As shown in this chapter, reversible capacities of 300 mA h g^{-1} are realized for many chemistries with anionic redox. For the case of the ruthenium system, Li_2RuO_3, the observed reversible capacity is very close to that of the theoretical capacity. In contrast, for the manganese-based oxides, observed reversible capacities are smaller than those of theoretical capacities, and reversible capacities are possibly further increased by further optimization of electrode materials. However, the average operating voltage of these compounds is relatively low (c. 3.3 V vs. lithium metal) as positive electrode materials. In contrast, in the nickel-based system, a high average operating voltage is attractive even though observed reversible capacity is currently limited. If reversibility of anionic redox is improved for the nickel-based system, the available energy density would be significantly improved.

6.7 Anionic Redox for Sodium-Storage Applications

The use of anionic redox is not limited to the lithium system and is easily extended to systems with different mobile ions. Indeed, anion redox has been extended for sodium-storage applications. Na_2MnO_3, which is a hypothetical Na counterpart of Li_2MnO_3, is considered to be a host structure for high-capacity sodium insertion materials. Nevertheless, synthesis of Na_2MnO_3 ($Na(Na_{1/3}Mn_{2/3})O_2$) with an isostructural phase of Li_2MnO_3 is difficult by conventional synthesis routes because of a large difference in size between Na^+ and Mn^{4+} ions. Therefore, instead of $Na(Na_{1/3}Mn_{2/3})O_2$, the use of $Na(Li_{1/3}Mn_{2/3})O_2$ has been proposed [57]. Although synthesis of the stoichiometric phase is difficult, a nonstoichiometric Na-deficient phase, $Na_{5/6}(Li_{1/4}Mn_{3/4})O_2$, has been successfully obtained [57]. Similar to Li_2MnO_3-based electrode materials, a well-defined voltage plateau is observed for $Na_{5/6}(Li_{1/4}Mn_{3/4})O_2$ in a Na cell. Partial oxygen loss results in the disappearance of the plateau from the second cycle, but the sample delivers a large reversible capacity of 200 mA h g^{-1} in a Na cell [57]. Similarly, a large reversible capacity was reported for $Na_{0.6}Li_{0.6}Ni_{0.25}Mn_{0.75}O_y$ [58], and such large capacity cannot be obtained without the contribution of anion redox. A reversible voltage plateau at 4.2 V vs. Na metal (4.5 V vs. Li metal), presumably associated with anionic redox, has also been reported for $Na_{0.6}(Li_{0.2}Mn_{0.8})O_2$ [59]. Mg ions are substituted for Li ions in transition-metal layers. $Na_{2/3}(Mg_{0.28}Mn_{0.72})O_2$ delivers a large reversible capacity of more than 200 mA h g^{-1} [60], and the contribution from reversible anionic redox has been recently reported [61].

For the case of larger cationic species, like ruthenium ions, direct synthesis of a sodium-excess phase is possible. Similar to Li_2RuO_3, Na_2RuO_3 crystallizes into the α-$NaFeO_2$-type layered structure. Two layered polymorphs, in-plane Na/Ru ordered and disordered phases, exist and the anionic redox is activated only for the ordered phase [62]. The origin of the activation has been proposed to be the formation and stabilization of the intermediate phase, ilmenite-type $NaRuO_3$, associated with glide of layers. A similar phase is also found in the Li counterpart [16]. This process involves a phase transition from ccp to hexagonal closed packing (hcp) for the oxygen arrangement. It has also been proposed that the large distortion of RuO_6 octahedra for the ilmenite phase triggers the formation of O–O sigma antibonding state, and thus the anion redox is also activated [62].

Similar to Li_3NbO_4, Na_3NbO_4 has been also studied as a model host structure for a high-capacity positive electrode material. Na and Nb ions are located at octahedral sites in a ccp array of oxide ions, and clusters consisting of four Nb ions are surrounded by sodium ions, forming a layered structure (see Figure 6.10a). If all Na ions are extracted/reinserted from/into the crystal lattice, the theoretical capacity reaches 356 mA h g^{-1}. A binary system, Na_3NbO_4–$NaMnO_2$, has been examined as potential high-capacity positive electrode materials for sodium battery applications [63]. However, synthesis of samples by conventional calcination is impossible. Phase segregation into Na_3NbO_4 and $NaMnO_2$ was evidenced, and a narrow solid solution range is expected in this binary system. Therefore, mechanical milling was used to synthesize metastable phases. X-ray diffraction patterns of a mixture of Na_3NbO_4 and $NaMnO_2$ before and after the mechanical milling are shown in Figure 6.10a. The mixture gradually changes into a cation-disordered rocksalt-type structure, and a color change of the powder is observed after mechanical milling. Although the size of Na^+ is much larger than those of Nb^{5+} and Mn^{3+}, all cations are located at the same octahedral sites in the ccp lattice. The sample segregates into a mixture of Na_3NbO_4 and $NaMnO_2$ after heating, indicating that this phase is metastable [63]. From the scientific point of view, such a metastable phase is quite interesting and important as an insertion material. However, from the practical point of view, the difficulty in the synthesis of these materials hinders the use for battery applications because of the cost limitation. Therefore, the research progress on synthesis of metastable phases, possibly the development of a continuous flow process with low cost, is needed to solve this practical problem in the future.

Electrochemical properties of the samples prepared by conventional calcination and mechanical milling are compared in Figure 6.10b. A theoretical capacity reaches 311 mA h g^{-1} when all Na ions are extracted from $Na_{1.3}Nb_{0.3}Mn_{0.4}O_2$. However, the sample prepared by conventional calcination delivers only 95 mA h g^{-1}, corresponding to nearly the theoretical capacity based on Mn^{3+}/Mn^{4+} redox. This result indicates that anionic redox is not activated in this phase. In contrast, the

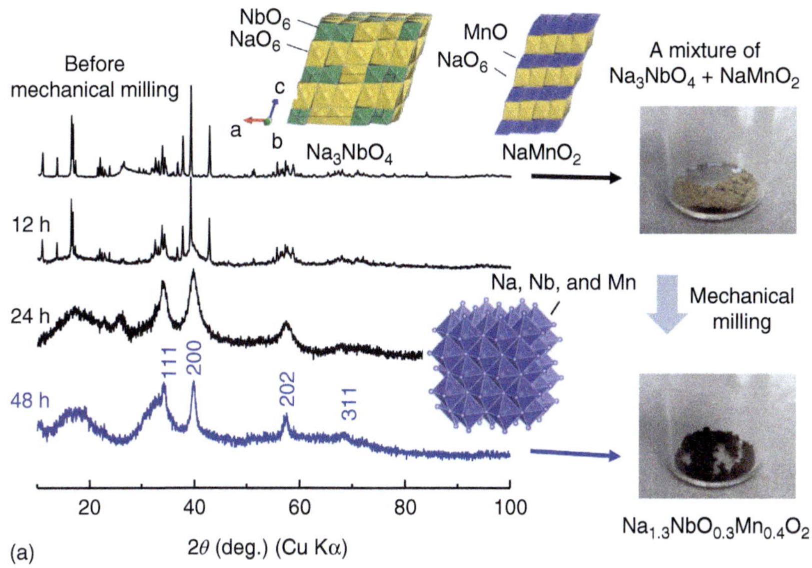

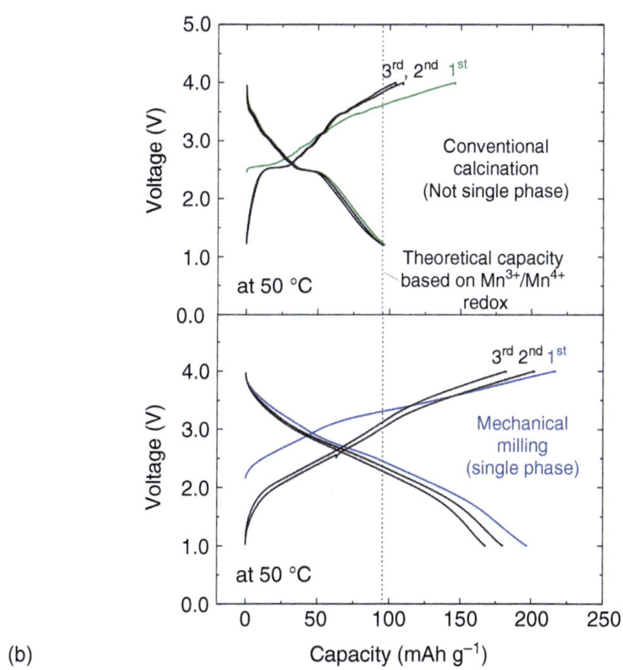

Figure 6.10 (top) Synthesis of $Na_{1.3}Nb_{0.3}Mn_{0.4}O_2$ by mechanical milling; (a) X-ray diffraction patterns of a mixture of Na_3NbO_4 and $NaMnO_2$ before/after the mechanical milling. Photographs of powders are also shown. (b) Electrochemical properties of $Na_{1.3}Nb_{0.3}Mn_{0.4}O_2$ at a rate of 10 mA g^{-1}; comparison of electrode performance for the sample prepared by the conventional calcination and mechanical milling. Source: Sato et al. [63]/with permission of American Chemical Society.

sample obtained by mechanical milling delivers a large reversible capacity of approximately 200 mA h g^{-1} at 50 °C, indicating that anionic redox is effectively activated for the sodium-excess condition. Electrochemical voltage of the sample increases almost linearly with two different slope angles, which change at 3.2 V. The potential profile of the sample is similar to the case of Li_2MnO_3-based electrode materials. Nevertheless, a clear voltage plateau, as observed for $Li_{1.3}Nb_{0.3}Mn_{0.4}O_2$ [33, 41], is not found for the Na counterpart. Moreover, cyclability of the sample in the Na cell is not acceptable for battery applications. Reversibility is completely lost after only 20 cycles [63]. Although reversibility of anionic redox seems to be relatively low for the Na system, $Na_{1.14}Ti_{0.29}Mn_{0.57}O_2$ prepared by mechanical milling with the less Na-excess composition shows much better cyclability with anionic redox [64]. In addition, good reversibility as electrode materials is realized for stoichiometric and nanosize $NaMnO_2$ prepared by mechanical milling [65]. Therefore, further optimization of chemical compositions for sodium-excess oxides possibly improves the reversibility of anionic redox chemistry for sodium-storage applications in the future.

6.8 Future Perspectives of Anionic Redox for Energy-Storage Applications

Lithium-excess compounds with rocksalt-related structures are attractive as positive electrode materials for rechargeable lithium/sodium batteries. Higher lithium (sodium) contents with fewer transition-metal ions in framework structures effectively increase theoretical capacities as positive electrode materials. A wide variety of material designs are possible with different chemistry on the basis of anionic redox coupled with cationic redox. In the past several years, many new positive electrode materials appeared. Moreover, the concept of anion redox, not only for the oxide system, is extended to the sulfide system [66] and oxyfluoride system [67–69]. These concepts are also extended to the non-rocksalt system, e.g. $Li_4Mn_2O_5$ [70, 71] and Li_2O system [72–74]. Experimental and theoretical understanding of anionic redox is also further accelerated from the research progress on oxygen electrocatalysts as another important anionic redox system [75], and a possible synergy between battery materials and electrocatalysts is anticipated [76]. However, cyclability of anionic redox in electrode materials is currently not enough for practical battery applications. The surface of electrode materials can be chemically reactive after the activation of anionic redox, probably leading to unfavorable side reactions with the electrolyte and growth of impedance on electrochemical cycling, especially at elevated temperatures [77]. Nevertheless, as shown in Figure 6.9, Li_2RuO_3 shows excellent cyclability with partial anionic redox and will be used for battery applications. Further studies of degradation mechanisms for these materials are, therefore, encouraged to solve the cyclability problem of anionic redox. Indeed, stabilization of anionic redox has been recently proposed by partial fluorination [69, 78]. Fluorination to metal oxides results in lowering oxidation state of transition-metal ions and thus effectively modifying the balance between cationic and anionic redox on electrochemical reactions [4, 79]. On charge, excess enrichment of holes in oxygen results in irreversible oxygen dimerization [80], leading to the unavailable oxygen loss [77]. In addition, coordination numbers of transition-metal ions with d electrons to oxygen also significantly influence the reversibility of anionic redox. Further experimental and theoretical understanding results in the development of high-energy electrode materials with highly reversible anionic redox in the future.

References

1 Dunn, B., Kamath, H., and Tarascon, J.-M. (2011). Electrical energy storage for the grid: a battery of choices. *Science* 334: 928–935.

2 Ohzuku, T. and Ueda, A. (1994). Why transition metal (di) oxides are the most attractive materials for batteries. *Solid State Ionics* 69: 201–211.

3 Mizushima, K., Jones, P.C., Wiseman, P.J. et al. (1980). Li_xCoO_2 ($0<x\leq1$): a new cathode material for batteries of high energy density. *Materials Research Bulletin* 15: 783–789.

4 Yabuuchi, N. (2019). Material design concept of lithium-excess electrode materials with Rocksalt-related structures for rechargeable non-aqueous batteries. *The Chemical Record* 19: 690–707.

5 Ueda, A. and Ohzuku, T. (1994). Solid-state redox reactions of $LINI_{1/2}CO_{1/2}O_2$ (*R*-3*M*) for 4-volt secondary lithium cells. *Journal of the Electrochemical Society* 141: 2010–2014.

6 Arai, H., Okada, S., Sakurai, Y. et al. (1998). Thermal behavior of $Li_{1-y}NiO_2$ and the decomposition mechanism. *Solid State Ionics* 109: 295–302.

7 Guilmard, M., Croguennec, L., Denux, D. et al. (2003). Thermal stability of lithium nickel oxide derivatives. Part I: $Li_xNi_{1.02}O_2$ and $Li_xNi_{0.89}Al_{0.16}O_2$ ($x = 0.50$ and 0.30). *Chemistry of Materials* 15: 4476–4483.

8 Guilmard, M., Croguennec, L., and Delmas, C. (2003). Thermal stability of lithium nickel oxide derivatives. Part II: $Li_xNi_{0.70}Co_{0.15}Al_{0.15}O_2$ and $Li_xNi_{0.90}Mn_{0.10}O_2$ ($x = 0.50$ and 0.30). Comparison with $Li_xNi_{1.02}O_2$ and $Li_xNi_{0.89}Al_{0.16}O_2$. *Chemistry of Materials* 15: 4484–4493.

9 Makimura, Y., Sasaki, T., Nonaka, T. et al. (2016). Factors affecting cycling life of $LiNi_{0.8}Co_{0.15}Al_{0.05}O_2$ for lithium-ion batteries. *Journal of Materials Chemistry A* 4: 8350–8358.

10 Ohzuku, T. and Makimura, Y. (2001). Layered lithium insertion material of $LiCo_{1/3}Ni_{1/3}Mn_{1/3}O_2$ for lithium-ion batteries. *Chemistry Letters* 30 (7): 642–643.

11 Lu, Z.H., MacNeil, D.D., and Dahn, J.R. (2001). Layered $LiNi_xCo_{1-2x}Mn_xO_2$ cathode materials for lithium-ion batteries. *Electrochemical and Solid-State Letters* 4: A200–A203.

12 Noh, H.-J., Youn, S., Yoon, C.S. et al. (2013). Comparison of the structural and electrochemical properties of layered $Li[Ni_xCo_yMn_z]O_2$ ($x = 1/3$, 0.5, 0.6, 0.7, 0.8 and 0.85) cathode material for lithium-ion batteries. *Journal of Power Sources* 233: 121–130.

13 Yabuuchi, N., Koyama, Y., Nakayama, N. et al. (2005). Solid-state chemistry and electrochemistry of $LiCo_{1/3}Ni_{1/3}Mn_{1/3}O_2$ for advanced lithium-ion batteries. *Journal of the Electrochemical Society* 152: A1434–A1440.

14 Reimers, J.N. and Dahn, J.R. (1992). Electrochemical and in situ X-ray diffraction studies of lithium intercalation in $LixCoO_2$. *Journal of the Electrochemical Society* 139: 2091–2097.

15 Mather, G.C., Dussarrat, C., Etourneau, J. et al. (2000). A review of cation-ordered rock salt superstructure oxides. *Journal of Materials Chemistry* 10: 2219–2230.

16 Kobayashi, H., Kanno, R., Kawamoto, Y. et al. (1995). Structure and lithium deintercalation of $Li_{2-x}RuO_3$. *Solid State Ionics* 82: 25–31.

17 Lu, Z.H., Beaulieu, L.Y., Donaberger, R.A. et al. (2002). Synthesis, structure, and electrochemical behavior of $Li[Ni_xLi_{1/3-2x/3}Mn_{2/3-x/3}]O_2$. *Journal of the Electrochemical Society* 149: A778–A791.

18 Lu, Z.H. and Dahn, J.R. (2002). Understanding the anomalous capacity of $Li/Li[Ni_xLi_{(1/3-2x/3)}Mn_{(2/3-x/3)}]O_2$ cells using in situ X-ray diffraction and electrochemical studies. *Journal of the Electrochemical Society* 149: A815–A822.

19 Robertson, A.D. and Bruce, P.G. (2003). Mechanism of electrochemical activity in Li_2MnO_3. *Chemistry of Materials* 15: 1984–1992.

20 Johnson, C.S., Kim, J.S., Lefief, C. et al. (2004). The significance of the Li_2MnO_3 component in 'composite' $xLi_2MnO_3{\cdot}(1-x)LiMn_{0.5}N_{i0.5}O_2$ electrodes. *Electrochemistry Communications* 6: 1085–1091.

21 Yu, D.Y.W., Yanagida, K., Kato, Y. et al. (2009). Electrochemical activities in Li_2MnO_3. *Journal of the Electrochemical Society* 156: A417–A424.

22 Sathiya, M., Ramesha, K., Rousse, G. et al. (2013). High performance $Li_2Ru_{1-y}Mn_yO_3$ ($0.2 \leq y \leq 0.8$) cathode materials for rechargeable lithium-ion batteries: their understanding. *Chemistry of Materials* 25: 1121–1131.

23 Yabuuchi, N., Kubota, K., Aoki, Y. et al. (2016). Understanding particle-size-dependent electrochemical properties of Li_2MnO_3-based positive electrode materials for rechargeable lithium batteries. *The Journal of Physical Chemistry C* 120: 875–885.

24 Thackeray, M.M., Johnson, C.S., Vaughey, J.T. et al. (2005). Advances in manganese-oxide 'composite' electrodes for lithium-ion batteries. *Journal of Materials Chemistry* 15: 2257–2267.

25 Momma, K. and Izumi, F. (2011). VESTA 3 for three-dimensional visualization of crystal, volumetric and morphology data. *Journal of Applied Crystallography* 44: 1272–1276.

26 Yabuuchi, N. (2017). Solid-state redox reaction of oxide ions for rechargeable batteries. *Chemistry Letters* 46: 412–422.

27 Blandeau, L., Ouvrard, G., Calage, Y. et al. (1987). Transition-metal dichalcogenides from disintercalation processes. Crystal structure determination and Mossbauer study of Li_2FeS_2 and its disintercalates Li_xFeS_2 ($0.2 \leq x \leq 2$). *Journal of Physics C: Solid State Physics* 20: 4271.

28 Bocquet, A.E., Fujimori, A., Mizokawa, T. et al. (1992). Electronic structure of $SrFe^{4+}O_3$ and related Fe perovskite oxides. *Physical Review B* 45: 1561–1570.

29 Tarascon, J.M., Vaughan, G., Chabre, Y. et al. (1999). In situ structural and electrochemical study of $Ni_{1-x}Co_xO_2$ metastable oxides prepared by soft chemistry. *Journal of Solid State Chemistry* 147: 410–420.

30 Ukei, K., Suzuki, H., Shishido, T. et al. (1994). Li_3NbO_4. *Acta Crystallographica Section C* 50: 655–656.

31 Hoffmann, R. and Hoppe, R. (1989). Zwei neue Ordnungs-Varianten des NaCl-Typs: Li_4MoO_5 und Li_4WO_5. *Zeitschrift für Anorganische und Allgemeine Chemie* 573: 157–169.

32 Betz, T. and Hoppe, R. (1984). Über Perrhenate. 2. Zur Kenntnis von Li_5ReO_6 und Na_5ReO_6 – mit einer Bemerkung über Na_5IO_6. *Zeitschrift für Anorganische und Allgemeine Chemie* 512: 19–33.

33 Yabuuchi, N., Takeuchi, M., Nakayama, M. et al. (2015). High-capacity electrode materials for rechargeable lithium batteries: Li_3NbO_4-based system with cation-disordered rocksalt structure. *Proceedings of the National Academy of Sciences* 112: 7650–7655.

34 Yabuuchi, N., Takeuchi, M., Komaba, S. et al. (2016). Synthesis and electrochemical properties of $Li_{1.3}Nb_{0.3}V_{0.4}O_2$ as a positive electrode material for rechargeable lithium batteries. *Chemical Communications* 52: 2051–2054.

35 de Picciotto, L.A., Thackeray, M.M., David, W.I.F. et al. (1984). Structural characterization of delithiated $LiVO_2$. *Materials Research Bulletin* 19: 1497–1506.

36 Ado, K., Tabuchi, M., Kobayashi, H. et al. (1997). Preparation of $LiFeO_2$ with alpha-$NaFeO_2$-type structure using a mixed-alkaline hydrothermal method. *Journal of the Electrochemical Society* 144: L177–L180.

37 Obrovac, M.N., Mao, O., and Dahn, J.R. (1998). Structure and electrochemistry of $LiMO_2$, (M = Ti, Mn, Fe, Co, Ni) prepared by mechanochemical synthesis. *Solid State Ionics* 112: 9–19.

38 Lee, J., Urban, A., Li, X. et al. (2014). Unlocking the potential of Cation-disordered oxides for rechargeable lithium batteries. *Science* 343: 519–522.

39 Urban, A., Lee, J., and Ceder, G. (2014). The configurational space of Rocksalt-type oxides for high-capacity lithium battery electrodes. *Advanced Energy Materials* 4: 1400478. -n/a.

40 Lee, J., Seo, D.-H., Balasubramanian, M. et al. (2015). A new class of high capacity cation-disordered oxides for rechargeable lithium batteries: Li–Ni–Ti–Mo oxides. *Energy & Environmental Science* 8: 3255–3265.

41 Yabuuchi, N., Nakayama, M., Takeuchi, M. et al. (2016). Origin of stabilization and destabilization in solid-state redox reaction of oxide ions for Li batteries. *Nature Communications* 7: 13814.

42 Oishi, M., Fujimoto, T., Takanashi, Y. et al. (2013). Charge compensation mechanisms in $Li_{1.16}Ni_{0.15}Co_{0.19}Mn_{0.50}O_2$ positive electrode material for Li-ion batteries analyzed by a combination of hard and soft X-ray absorption near edge structure. *Journal of Power Sources* 222: 45–51.

43 Oishi, M., Yamanaka, K., Watanabe, I. et al. (2016). Direct observation of reversible oxygen anion redox reaction in Li-rich manganese oxide, Li_2MnO_3, studied by soft X-ray absorption spectroscopy. *Journal of Materials Chemistry A* 4: 9293–9302.

44 Gent, W.E., Lim, K., Liang, Y. et al. (2017). Coupling between oxygen redox and cation migration explains unusual electrochemistry in lithium-rich layered oxides. *Nature Communications* 8: 2091.

45 Okubo, M. and Yamada, A. (2017). Molecular orbital principles of oxygen-redox battery electrodes. *ACS Applied Materials & Interfaces* 9: 36463–36472.

46 Glazier, S.L., Li, J., Zhou, J. et al. (2015). Characterization of disordered $Li_{(1+x)}Ti_{2x}Fe_{(1-3x)}O_2$ as positive electrode materials in Li-ion batteries using percolation theory. *Chemistry of Materials* 27: 7751–7756.

47 McCalla, E., Sougrati, M.T., Rousse, G. et al. (2015). Understanding the roles of anionic redox and oxygen release during electrochemical cycling of lithium-rich layered Li_4FeSbO_6. *Journal of the American Chemical Society* 137: 4804–4814.

48 Seo, D.-H., Lee, J., Urban, A. et al. (2016). The structural and chemical origin of the oxygen redox activity in layered and cation-disordered Li-excess cathode materials. *Nature Chemistry* 8: 692–697.

49 Armstrong, A.R., Holzapfel, M., Novak, P. et al. (2006). Demonstrating oxygen loss and associated structural reorganization in the lithium battery cathode $Li[Ni_{0.2}Li_{0.2}Mn_{0.6}]O^2$. *Journal of the American Chemical Society* 128: 8694–8698.

50 Jiang, M., Key, B., Meng, Y.S. et al. (2009). Electrochemical and structural study of the layered, "Li-excess" lithium-ion battery electrode material $Li[Li_{1/9}Ni_{1/3}Mn_{5/9}]O_2$. *Chemistry of Materials* 21: 2733–2745.

51 Yabuuchi, N., Yoshii, K., Myung, S.-T. et al. (2011). Detailed studies of a high-capacity electrode material for rechargeable batteries, $Li_2MnO_3-LiCo_{1/3}Ni_{1/3}Mn_{1/3}O_2$. *Journal of the American Chemical Society* 133: 4404–4419.

52 Koga, H., Croguennec, L., Mannessiez, P. et al. (2012). $Li_{1.20}Mn_{0.54}Co_{0.13}Ni_{0.13}O_2$ with different particle sizes as attractive positive electrode materials for lithium-ion batteries: insights into their structure. *The Journal of Physical Chemistry C* 116: 13497–13506.

53 Yabuuchi, N., Tahara, Y., Komaba, S. et al. (2016). Synthesis and electrochemical properties of Li_4MoO_5–NiO binary system as positive electrode materials for rechargeable lithium batteries. *Chemistry of Materials* 28: 416–419.

54 Matsuhara, T., Tsuchiya, Y., Yamanaka, K. et al. (2016). Synthesis and electrode performance of Li_4MoO_5-$LiFeO_2$ binary system as positive electrode materials for rechargeable lithium batteries. *Electrochemistry* 84: 797–801.

55 Amatucci, G.G., Tarascon, J.M., and Klein, L.C. (1996). CoO_2, the end member of the Li_xCoO_2 solid solution. *Journal of the Electrochemical Society* 143: 1114–1123.

56 Croguennec, L., Pouillerie, C., Mansour, A.N. et al. (2001). Structural characterisation of the highly deintercalated $Li_xNi_{1.02}O$ phases (with $x \leq 0.30$). *Journal of Materials Chemistry* 11: 131–141.

57 Yabuuchi, N., Hara, R., Kajiyama, M. et al. (2014). New O_2/P_2-type Li-excess layered manganese oxides as promising multi-functional electrode materials for rechargeable Li/Na batteries. *Advanced Energy Materials* 4: 1301453.

58 Slater, M.D., Kim, D., Lee, E. et al. (2013). Sodium-ion batteries. *Advanced Functional Materials* 23: 947–958.

59 Du, K., Zhu, J., Hu, G. et al. (2016). Exploring reversible oxidation of oxygen in a manganese oxide. *Energy & Environmental Science* 9: 2575–2577.

60 Yabuuchi, N., Hara, R., Kubota, K. et al. (2014). A new electrode material for rechargeable sodium batteries: P2-type $Na_{2/3}[Mg_{0.28}Mn_{0.72}]O_2$ with anomalously high reversible capacity. *Journal of Materials Chemistry A* 2: 16851–16855.

61 Maitra, U., House, R.A., Somerville, J.W. et al. (2018). Oxygen redox chemistry without excess alkali-metal ions in $Na_{2/3}[Mg_{0.28}Mn_{0.72}]O_2$. *Nature Chemistry* 10 (3): 288.

62 Mortemard de Boisse, B., Liu, G., Ma, J. et al. (2016). Intermediate honeycomb ordering to trigger oxygen redox chemistry in layered battery electrode. *Nature Communications* 7: 11397.

63 Sato, K., Nakayama, M., Glushenkov, A.M. et al. (2017). Na-excess cation-disordered Rocksalt oxide: $Na_{1.3}Nb_{0.3}Mn_{0.4}O_2$. *Chemistry of Materials* 29: 5043–5047.

64 Kobayashi, T., Zhao, W., Rajendra, H.B. et al. (2019). Nanosize cation-disordered Rocksalt oxides: Na_2TiO_3–$NaMnO_2$ binary system. *Small* 15: 1902462.

65 Sato, T., Sato, K., Zhao, W. et al. (2018). Metastable and nanosize cation-disordered rocksalt-type oxides: revisit of stoichiometric $LiMnO_2$ and $NaMnO_2$. *Journal of Materials Chemistry A* 6: 13943–13951.

66 Sakuda, A., Takeuchi, T., Okamura, K. et al. (2014). Rock-salt-type lithium metal sulphides as novel positive-electrode materials. *Scientific Reports* 4: 4883.

67 House, R.A., Jin, L., Maitra, U. et al. (2018). Lithium manganese oxyfluoride as a new cathode material exhibiting oxygen redox. *Energy & Environmental Science* 11: 926–932.

68 Lee, J., Kitchaev, D.A., Kwon, D.-H. et al. (2018). Reversible Mn^{2+}/Mn^{4+} double redox in lithium-excess cathode materials. *Nature* 556: 185–190.

69 Ji, H., Wu, J., Cai, Z. et al. (2020). Ultrahigh power and energy density in partially ordered lithium-ion cathode materials. *Nature Energy* 5: 213–221.

70 Freire, M., Kosova, N.V., Jordy, C. et al. (2016). A new active Li–Mn–O compound for high energy density Li-ion batteries. *Nature Materials* 15: 173–177.

71 Diaz-Lopez, M., Freire, M., Joly, Y. et al. (2018). Local structure and lithium diffusion pathways in $Li_4Mn_2O_5$ high capacity cathode probed by Total scattering and XANES. *Chemistry of Materials* 30: 3060–3070.

72 Okuoka, S., Ogasawara, Y., Suga, Y. et al. (2014). A new sealed lithium-peroxide battery with a co-doped Li_2O cathode in a superconcentrated lithium bis(fluorosulfonyl)amide electrolyte. *Scientific Reports* 4: 5684.

73 Kobayashi, H., Hibino, M., Ogasawara, Y. et al. (2016). Improved performance of co-doped Li_2O cathodes for lithium-peroxide batteries using $LiCoO_2$ as a dopant source. *Journal of Power Sources* 306: 567–572.

74 Harada, K., Hibino, M., Kobayashi, H. et al. (2016). Electrochemical reactions and cathode properties of Fe-doped Li_2O for the hermetically sealed lithium peroxide battery. *Journal of Power Sources* 322: 49–56.

75 Mueller, D.N., Machala, M.L., Bluhm, H. et al. (2015). Redox activity of surface oxygen anions in oxygen-deficient perovskite oxides during electrochemical reactions. *Nature Communications* 6: 6097.

76 Grimaud, A., Hong, W.T., Shao-Horn, Y. et al. (2016). Anionic redox processes for electrochemical devices. *Nature Materials* 15: 121–126.

77 Kobayashi, Y., Sawamura, M., Kondo, S. et al. (2020). Activation and stabilization mechanisms of anionic redox for Li storage applications: joint experimental and theoretical study on Li_2TiO_3–$LiMnO_2$ binary system. *Materials Today* 37: 43–55.

78 Lee, J., Papp, J.K., Clément, R.J. et al. (2017). Mitigating oxygen loss to improve the cycling performance of high capacity cation-disordered cathode materials. *Nature Communications* 8: 981.

79 Lun, Z., Ouyang, B., Cai, Z. et al. (2020). Design principles for high-capacity Mn-based cation-disordered Rocksalt cathodes. *Chem* 6: 153–168.

80 Ben Yahia, M., Vergnet, J., Saubanère, M. et al. (2019). Unified picture of anionic redox in Li/Na-ion batteries. *Nature Materials* 18: 496–502.

7

Transition Metal Oxides in Aqueous Electrolytes

Xiaoqiang Shan and Xiaowei Teng

University of New Hampshire, Department of Chemical Engineering, 33 Academic Way, Durham, NH 03824, USA

7.1 Introduction: Opportunities and Challenges of Aqueous Batteries

Rechargeable electrochemical energy storage (EES) is playing an increasingly important role in large-scale stationary energy storage, emerging as one of the key solutions to effectively integrate high shares of renewable energy generation from solar panels and wind turbines into power systems [1, 2]. The utility-scale EES could provide a greater feed-in of electricity from renewable sources into the grid through load shifting (storing excess electricity generation during peak production for later consumption) and frequency regulation (second-by-second adjustment of power to maintain grid frequency). Therefore, EES systems can provide more reliable and cheaper electricity in isolated grids and off-grid communities when paired with renewable generators, in sharp contrast to relying on expensive diesel fuel for electricity generation. The global energy storage deployment is expected to increase by over 40% each year until 2025. Currently, the global energy storage sector is dominated by the utility-scale stationary battery and hydroelectric dams. But it is expected that small-scale battery storage, especially so-called behind-the-meter (BTM) batteries, is expected to increase, complementing utility-scale applications (Figure 7.1) significantly (https://www.irena.org/Newsroom/Articles/2020/Mar/Battery-Storage-Paves-Way-for-a-Renewable-Powered-Future). This trend is driven by the rapidly falling prices of battery packs, along with a growing consumer market for electric vehicles (EVs) and plug-in hybrid EVs and increasing deployment of distributed renewable energy generation and smart grids [3, 4].

Essential criteria required for stationary EES include low construction/maintenance cost, low safety hazards, high round-trip efficiency, long cycle life, and environmental friendliness [5, 6]. While non-aqueous lithium-ion batteries (LIBs) based on intercalation chemistry have clearly dominated current applications with high-energy density, there have been concerns regarding their high cost, supply risk, and safety hazards primarily residing in the usage of highly

Transition Metal Oxides for Electrochemical Energy Storage, First Edition.
Edited by Jagjit Nanda and Veronica Augustyn.

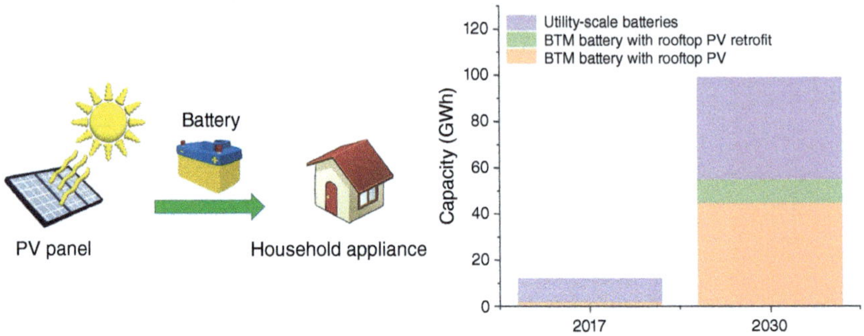

Figure 7.1 Stationary battery storage capacity growth from 2017–2030, where BTM batteries stand for behind-the-meter batteries, which are connected behind the utility meter of commercial, industrial, or residential customers. Source: Based on https://www.irena.org/Newsroom/Articles/2020/Mar/Battery-Storage-Paves-Way-for-a-Renewable-Powered-Future.

inflammable organic electrolyte [7]. Especially, there are concerns for large-scale implementation of non-aqueous LIBs in utility storage, where the cost, safety, cycling life, and power capability become more imperative than mere energy density [8]. Replacing organic solvent with water-based electrolytes has significant implications: besides their intrinsic non-flammability, aqueous electrolytes offer 2–3 orders of magnitude higher ionic conductivity compared to organic solvents and thus contribute to high rate performance [9, 10]. Meanwhile, manufacturing costs incurred from moisture control can be dramatically mitigated during the fabrication of aqueous batteries [11, 12].

Notwithstanding, current aqueous EES systems still cannot rival the performance of non-aqueous LIBs due to limited thermodynamically stable potential window, beyond which hydrogen evolution reaction (HER) and oxygen evolution reaction (OER) may occur [13]. Moreover, capacitive charge storage processes (e.g. electrical double-layer [EDL] capacitance or pseudocapacitance) often dominate the charge storage mechanism in an aqueous EES system, which significantly improves the power performance and cycle life and also displays insufficient storage capacity and energy performance due to the inactive diffusion-limited redox charge storage [14].

7.2 Electrochemistry of Aqueous Batteries

7.2.1 Potential Window

In an electrochemical cell, the open-circuit voltage (OCV), namely the cell voltage when the current of a cell is kept infinitely small, is limited by the electrochemical potentials between the cathode and anode also by the electrochemical window of the electrolyte [15]. For a non-aqueous electrolyte, its E_g is usually stable over a wide potential window (>3.5 V) during redox processes. However, for an aqueous electrolyte, the electrochemical window of water is equal to 1.23 V, much smaller than the potential between anode and cathode ($\mu_c - \mu_a$). Therefore, the working

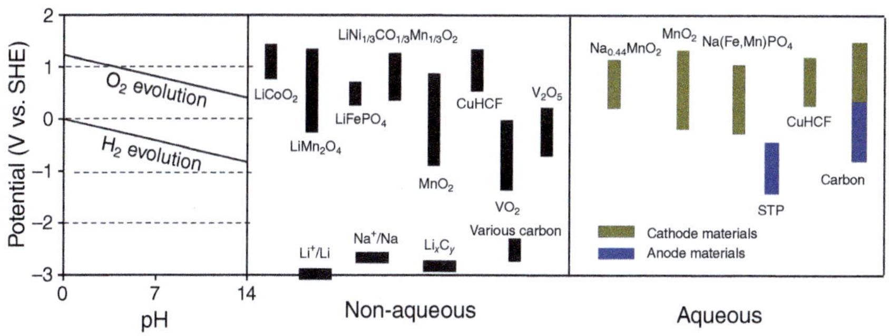

Figure 7.2 Comparison of water potential window (Pourbaix diagram) with commonly used electrode materials for non-aqueous and aqueous batteries (STP stands for sodium titanium phosphate; CuHCF stands for copper hexacyanoferrate with a Prussian blue crystal structure).

voltage window for an aqueous EES cell is often determined by the thermodynamically stable voltage window of water instead of the electrochemical potentials of the electrodes (μ_c or μ_a).

The Pourbaix diagram shown in Figure 7.2 describes the dependence of an electrochemical cell voltage on the pH value of the electrolyte. Primarily, the electrode potential of HER and OER can be described via Nernst equation. It is clearly seen that the thermodynamically stable voltage window for an aqueous EES device is always equal to 1.23 V in any given pH value. Since the energy density of an EES system is determined by its operating voltage window and storage capacity, this voltage window is too small to achieve a competitive energy performance compared to its non-aqueous counterparts.

Extensive efforts have been made to break through the 1.23 V potential window limit in aqueous EES. Recent exploration of "water-in-salt" electrolyte can significantly expand the kinetically stable voltage window of such a concentrated aqueous electrolyte up to 3.0 V by dissolving a high concentration of lithium bis(trifluoromethane sulfonyl)imide (LiTFSI) at extremely high concentrations (molality > 20 M) in water [16]. In a "water-in-salt" electrolyte, most of the water molecules are confined within the solvation sheath of the cation and thus are unable to react with the water-sensitive metal or metal alloy anode. Furthermore, the Li-containing electrolyte also reacts with electrode materials, and a passivation layer is formed on the surface. Such a solid-electrolyte interphase (SEI) as a protective film on both anode and cathode surfaces further increases the energy barrier of water decomposition during high-voltage (3.0 V) cycling [13, 17, 18]. Meanwhile, the resulting SEI maintains good ionic conductivity for the transport of charge carriers between the electrolyte and electrode.

In comparison with SEI formed via the reduction of Li-containing concentrated electrolyte, interphase formation via the interaction between metal oxides and water has been investigated less extensively. Not long before, it was generally accepted that in conventional aqueous electrolytes containing water and strongly dissociated inorganic salts, protective interphase resulting from the decomposition of water is very unlikely to form. This is because the decomposition products from water such as H_2 and OH^- (the decomposition products from water reduction) or O_2 and H^+ (the decomposition products from water oxidation) cannot deposit on the surface of the electrode materials in a dense solid state. However, Shan et al. showed that interaction between the monoclinic manganese oxide ($Mn^{2+}{}_2Mn^{4+}{}_3O_8$), the only layered and bivalent manganese oxide, and water during the electrochemical cycling results in a new form of hydroxylated interphase on the surface of Mn_5O_8 (Figure 7.3) [19]. In such a bivalent manganese oxide, Mn^{2+}-terminated surface has a lower surface energy (0.87 J m^{-2}) than Mn^{4+}-terminated surface (1.80 J m^{-2}), and thus the Mn_5O_8 material is enclosed by Mn^{2+}-rich surface, favoring the formation of hydroxylated interphase upon the interaction between Mn^{2+}-rich surface and water. The resulting hydroxylated interphase on the surface of Mn_5O_8 materials inhibits the HER and OER and thus enables a kinetically stable potential window of 2.5 V for aqueous Na-ion storage in a three-electrode half-cell and of 3.0 V in asymmetric full cell.

7.2.2 Diverse Charge Transfer and Storage Processes in Aqueous Batteries

The overall electrochemistry of an aqueous battery is very similar to that of the conventional non-aqueous system, where charge carriers (e.g. alkali) transfer through the electrolyte and the electron through the external circuit between the anode and cathode. However, the electro-kinetics of aqueous batteries are very different from non-aqueous counterparts, primarily due to the high ionic mobilities of charge carriers in an aqueous solution. Moreover, due to the high solubility of acids and bases in the aqueous solution, H^+ and OH^- as charge carriers have been much more often observed in aqueous batteries than in non-aqueous batteries, bringing more complexities to the charge transfer and storage mechanisms.

7.2.2.1 Overview of Various Storage Mechanisms

The EDL capacitive process relies on the charge separation through two parallel layers of charge at near electrode surface utilizing static charge (non-faradaic). As seen in Figure 7.4, the first layer (Helmholtz layer) consists of ions chemically adsorbed on the electrode surface, namely between the inner Helmholtz plane (IHP) and outer Helmholtz plane (OHP). On the other hand, the second layer consists of ions attracted to the surface charge via the coulombic force, electrically screening the first layer. This second layer is loosely associated with the electrode and also named as "diffuse layer." The storage capacity of an EDL is determined by the electrode/electrolyte interface, in which EDL capacitances usually range from 10 to 20 μF cm^{-2}. Carbon is an ideal EDL electrode material for its high electrical conductivity and large specific surface area, delivering a storage capacity of up to 150 F g^{-1} in ionic liquid electrolytes.

The pseudocapacitive process relies on the charge transfer process primarily at electrode/electrolyte interface utilizing three faradaic mechanisms, including underpotential deposition, redox pseudocapacitance, and intercalation pseudocapacitance [20]. Underpotential deposition involves the electrodeposition of a

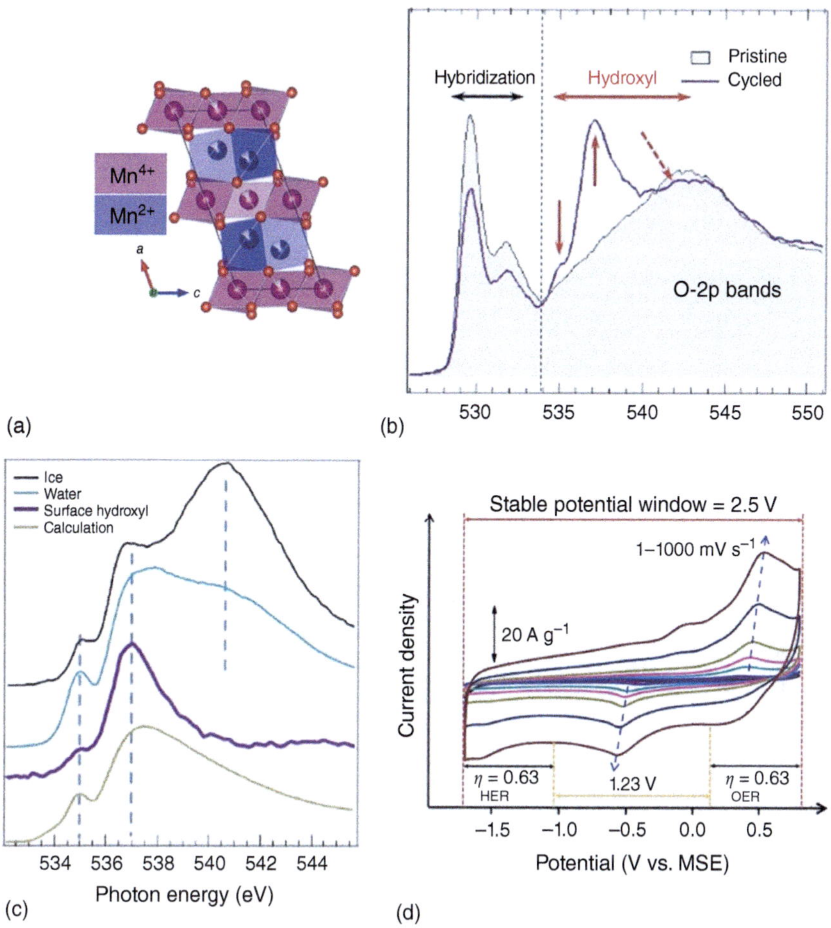

Figure 7.3 (a) Crystal structure of Mn_5O_8 with Mn^{2+} (in blue) and Mn^{4+} (in purple); (b) Oxygen K-edge soft X-ray absorption spectroscopy data of Mn_5O_8 (pristine) and Mn_5O_8 electrode after two CV cycles between −1.7 and 0.8 V (vs. mercury surface electrode [MSE]). Features at 535 and 537 eV peaks are fingerprints of the water; (c) A comparison of the oxygen feature of the surface hydroxyl layer of Mn_5O_8 with water, ice, and one of the calculations with aligned H-bonds but lengthened O–O (3.50 and 3.00 Å) distances; (c) CVs of Mn_5O_8 in 0.1 M Na_2SO_4 aqueous electrolyte showing a stable 2.5 V potential window. Source: Shan et al. [19]. CC BY 4.0.

species, significantly reducing a metal cation or a proton to a solid metal at a potential above its electrode potential (equilibrium potential described by Nernst equation). Examples of underpotential deposition include the deposition of less noble metal cation on the surface of another metal or the deposition of the proton on the surface of a noble metal such as platinum. Redox pseudocapacitance occurs at or near the surface of an electrode via the adsorption of ions. Intercalation pseudocapacitance occurs when ions intercalate into the channels or layers of a redox-active material, accompanied by a faradaic charge transfer with no

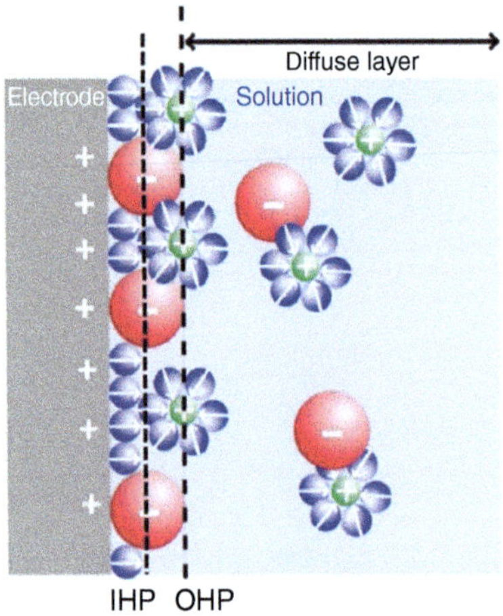

Figure 7.4 Scheme of EDL.

crystallographic phase change. Examples of redox pseudocapacitance include protonation of hydrous RuO_2 [21], or interaction between alkali with layered metal oxides such as birnessite (δ-MnO_2) [22]. Besides metal oxides, metal carbides or nitrides are also ideal pseudocapacitive electrode materials for their reversible redox activity, wide electrochemical potential, and high chemical stability [20, 23]. The storage capacity of a pseudocapacitive material is around $\sim 100\,\mu F\,cm^{-2}$, nearly 1 order of magnitude higher than that of an EDL material (Figure 7.5).

Diffusion-limited redox process has been found in most battery materials, which relies on the kinetically limited insertion of alkali. In this process, ion intercalation and de-intercalation are intrinsically tied to the slow kinetics of solid-phase transition between the intercalated and non-intercalated phases (e.g. nucleation barrier) and/or slow solid-state diffusion, as shown in the following equation:

$$MO + A^+ + e^- \leftrightarrow AMO \tag{7.1}$$

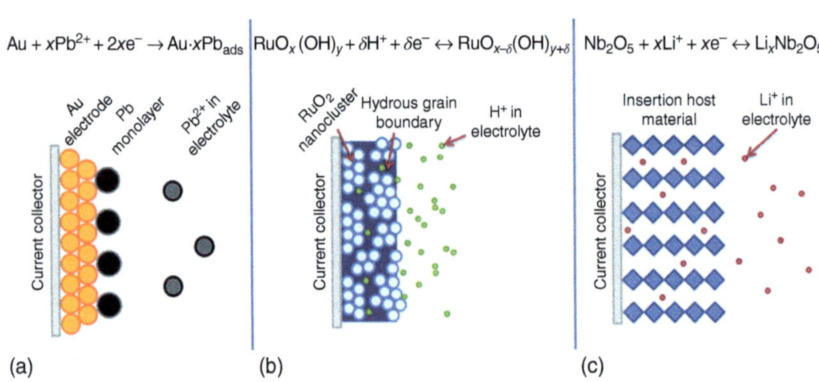

Figure 7.5 Scheme of three faradaic pseudocapacitance mechanisms. (a) Underpotential deposition; (b) redox pseudocapacitance; (c) intercalation psuedocapacitance. Source: Augustyn et al. [20]/with permission of Royal Society of Chemistry.

where MO is the transition metal oxide and A^+ is the alkali (charge carrier). Upon the insertion of A^+ into MO, M cations will be reduced to balance the charges and vice versa. The Gibbs phase rule of thermodynamics can be applied to battery reaction when the system is under thermodynamic equilibrium. The number of degrees of freedom (F) is given by

$$F = C - P + 2 \tag{7.2}$$

where C is the number of components and P is the number of coexisting phases. The intercalation process involves a two-component system ($C = 2$), namely the intercalant A^+ and the host electrode (MO). For a single-phase redox process ($P = 1$), especially a solid-solution reaction between A^+ and MO, the insertion of A^+ doesn't cause the generation of a new phase, resulting in a degree of freedom of 3 ($F = 2 - 1 + 2$). When the two intensive parameters (temperature and pressure) are taken into count, there is only one more degree of freedom ($F = 1$). Thus, the chemical potential (OCV) has to function as temperature, pressure, and composition (concentration of A in the MO). Since battery reaction often takes place under an isobaric (constant pressure) and isothermal (constant temperature) condition, when composition changes, the OCV has to change accordingly, as shown in Figure 7.6a. On the other hand, in the case of a two-phase mixture, namely the MO and AMO are in different phases, the degree of freedom is 2 ($F = 2 - 2 + 2$). Then, there is no more independent degree of freedom besides temperature and pressure, so all thermodynamic functions such as OCV should remain constant when the composition changes under an isobaric and isothermal condition, as shown in Figure 7.6b.

7.2.2.2 Semi-quantitative Analysis of Storage Mechanism from Sweeping Voltammetry Analysis

Understanding the electro-kinetics of charge storage inside the metal oxides can be obtained by analyzing the current–voltage curves at various scan rates obtained from cyclic voltammetry (CV) measurements in the half-cell [23]. For a strictly

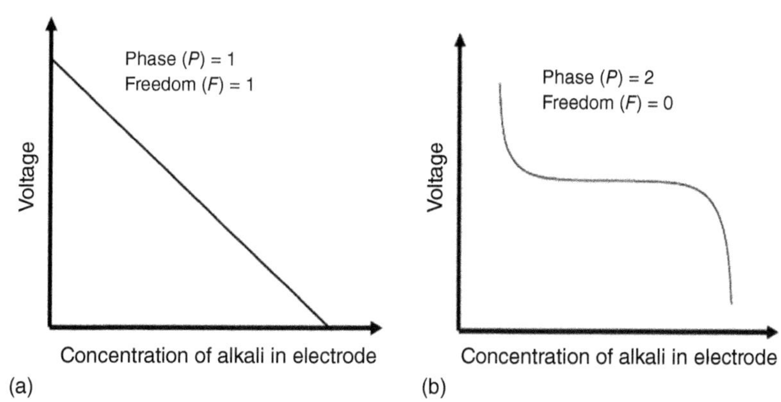

Figure 7.6 Phase rule in battery reaction in a system involving (a) single-phase process and (b) two-phase process.

diffusion-limited redox reaction, the rate of charge transfer reactions, namely the current (i_d), can be expressed in Eq. (7.3) or a simplified Eq. (7.4):

$$i_d = 0.495FCA\left(\frac{D\alpha nFv}{RT}\right)^{\frac{1}{2}} \tag{7.3}$$

$$i_d = k_d v^{0.5} \tag{7.4}$$

where C is the concentration of charge carriers in the accumulation layer, α is the charge transfer coefficient, D is the diffusion coefficient of the charge carrier inside the electrode materials, n is the number of electrons involved in the Faradaic reaction, A is the surface area of the electrode materials, F is Faraday's constant, R is the molar gas constant, and T is the temperature.

On the other hand, the capacitive current (i_c) from an EDL capacitance or a pseudocapacitance has a linear dependence on the scan rate according to Eq. (7.5) or in a simplified Eq. (7.6):

$$i_c = AC_c v \tag{7.5}$$

$$i_c = k_c v \tag{7.6}$$

where C_c is the capacitance from the capacitive process and A is a constant. Assuming no other kinetic limitations present (e.g. ohmic drop; mass transport in electrode pores; nucleation of new phase), the overall current at a given potential can be expressed as the sum of two separate charge storage mechanisms, that is capacitive current and kinetic current as shown in Eq. (7.7). At higher scan rates, the overall current is dominated by capacitive current (i_c), due to its stronger linear dependence on scan rates shown in Eq. (7.7), whereas the overall current is dominated by diffusion-limited kinetic current (i_d) at lower scan rates. In this context, the overall current (i_{total}) is usually described by a simple power law, as shown in Eq. (7.8).

$$i_{\text{total}} = i_c + i_d = k_c v + k_d v^{0.5} \tag{7.7}$$

$$i_{\text{total}} = a v^b \tag{7.8}$$

where a is an adjustable parameter and b is a variable heavily dependent on the relative contribution from i_c or i_d. It is apparent that the value of b is equal to either 0.5 or 1 when the overall currents are strictly dominated by capacitive (i_c) or kinetic (i_d) current, respectively. Moreover, rearrangement of Eq. (7.7) forms:

$$i_{\text{total}}/v^{0.5} = k_c v^{0.5} + k_d \tag{7.9}$$

Plotting $i/v^{1/2}$ vs. $v^{1/2}$ curves at a given potential, k_c and k_d can be calculated. Therefore, contributions from capacitive charge storage i_c and diffusion-limited redox charge storage i_d during the CVs were extracted quantitatively (Figure 7.7) [19].

7.2.2.3 Storage Mechanisms in Electrolyte with Different pH Values

Aqueous electrolytes generally include acidic (e.g. HCl, H_2SO_4), neutral (Na_2SO_4, KCl), and alkaline (e.g. NaOH, KOH) solutions. Table 7.1 shows the physical properties of several typical charge carriers, including bare ionic radius, hydrated ionic radius, and ionic mobility [24–26]. The sizes of charge carriers (e.g. H^+, Li^+, Na^+,

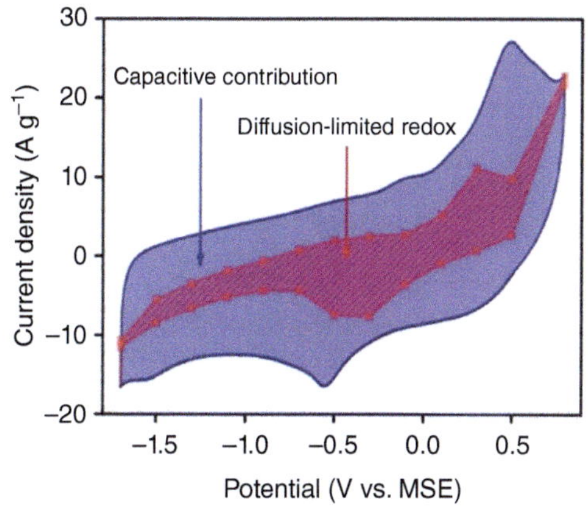

Figure 7.7 Kinetic analysis of CVs. Source: Shan et al. [19]. CC BY 4.0.

Table 7.1 Physical properties of ions with bare ionic radius, hydrated radius, water number in shell, hydrated enthalpy, and ionic mobility [24–26].

Ions	Bare ionic radius (pm)	Hydrated radius (pm)	Water number in shell	Hydration enthalpy ($kJ\ mol^{-1}$)	Ionic mobility ($\Omega^{-1}\ cm^2\ mol^{-1}$)
H	1.15	280	12	−1050	349.8
Li	94	382	5.2	−475	38.7
Na	117	358	3.5	−365	50.1
K	149	331	2.6	−295	73.5
Mg	72	428	10	−2010	53.1
Zn	74	430	9.6	−1955	56.6
Al	53	480	20.4	−4525	63

Source: Based on Jaffrezic-Renault and Dzyadevych [24], Marcus [25], Volkov et al. [26].

K^+) in bare or hydrated states, the ion mobility, and concentration can affect the charge storage reaction. Various hydrated metal ions such as Na^+ can be stored in the host electrode in a neutral electrolyte for energy storage, while protons play a role as charge carrier media in acidic and alkaline (generated proton from the break of OH bonds during OH^- insertion) solutions.

Pseudocapacitive Storage in Acidic Electrolytes RuO_2 is one of the first well-known transition metal oxides showing faradic surface redox pseudocapacitive storage mechanism, utilizing protons as charge carriers in an acidic electrolyte (Figure 7.8a) [27]. A theoretical capacitance of $1450\ F\ g^{-1}$ can be achieved in H_xRuO_2 when x reaches 2 for a potential window of 1 V. Meanwhile, H_2SO_4 is the most widely used acid electrolyte for aqueous storage with a high ionic conductivity due to the lowest hydration radius and fastest ionic mobility of hydronium ions (Table 7.1). Although no very distinct electrochemical features are observed (Figure 7.8e,h), RuO_2 can achieve a very high redox pseudocapacitive storage capacitance around $700–1000\ F\ g^{-1}$ from an easy proton insertion process with its intrinsic metallic

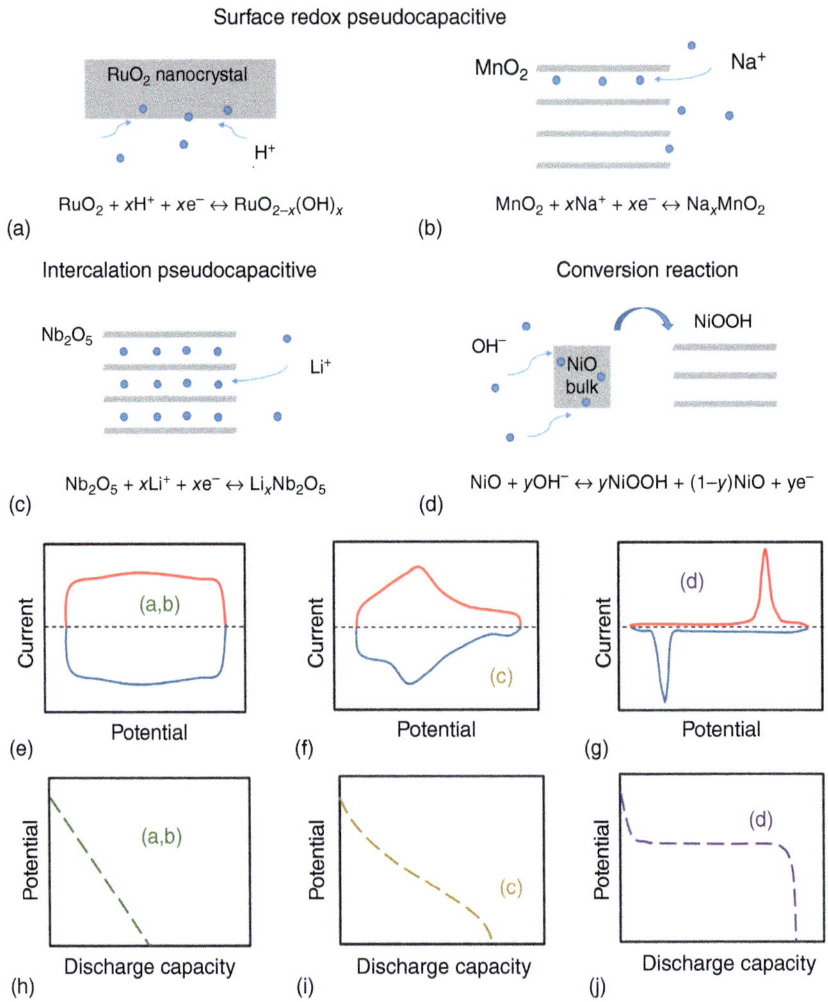

Figure 7.8 Schematic of storage reaction mechanisms with (a, b) surface redox pseudocapacitive behaviors of RuO_2 in acidic electrolyte and MnO_2 in the neutral electrolyte, (c) intercalation pseudocapacitive behaviors of Nb_2O_5, (d) conversion reaction between NiO and NiOOH in alkaline electrolyte with (e, h) surface redox pseudocapacitive, (f, i) intercalation pseudocapacitive, and (g, j) conversion reaction features.

conductive property as well as the introduced amorphous structure or structural water [20, 24]. However, due to the scarcity and high cost of RuO_2 for EES, transition metal oxides such as MoO_3 and WO_3 are also studied with the observation of pseudocapacitive redox reactions in acid electrolytes with large proton storage capacity [28–31]. Proton insertion with active redox process could be very promising especially for its high rate capability, which is encouraged from the smallest H^+ hydrated radius or Grotthuss conduction mechanism [32, 33].

Pseudocapacitive Storage Mechanism in Neutral Electrolytes A neutral electrolyte is often used for battery reaction since the salt solution with different types of charge carriers affords more charge storage chemistries. For example, an alkali such as Na^+ can function as a charge carrier to insert into the host electrode layered MnO_2 (Figure 7.8b). Goodenough and coworker are one of the first groups studying amorphous $MnO_2 \cdot H_2O$ in a 2 M KCl aqueous electrolyte, showing a faradic pseudocapacitive behavior with a specific capacitance of $200\,F\,g^{-1}$ [34]. Similar to RuO_2, amorphous $MnO_2 \cdot H_2O$ shows a redox pseudocapacitive electrochemical features with close to rectangular CV curves. However, due to the low electron conductivity of MnO_2 relative to RuO_2, the obtained capacitance is much smaller than its theoretical value $1110\,F\,g^{-1}$, assuming one-electron charge transfer in a 1 V potential window. Na_2SO_4 neutral electrolyte is most widely studied in aqueous storage, possibly due to its low cost and good ionic conductivity. The pseudocapacitive charge storage mechanism of MnO_2 birnessite in a Na_2SO_4 electrolyte is widely studied using various spectroscopic tools to confirm the redox process and electrochemically activity. Variation of the MnO_2 birnessite structure upon charge and discharge from X-ray diffraction (XRD) is observed due to the intercalation and de-intercalation of Na^+ cations between the layered sheets [35]. Further studies using X-ray absorption near-edge structure (XANES) and X-ray photoelectron spectroscopy (XPS) techniques show the evolutions of Mn valence states and Na concentrations in MnO_2 electrode during the potential-dependent faradaic redox reaction in Na_2SO_4 electrolyte [22, 36]. Comprehensive studies of alkali (e.g. Li^+, Na^+, K^+) effects on pseudocapacitive storage of MnO_2 are conducted, electrochemical redox activities of MnO_2 all can be observed in alkali salts solutions [37], and Raman spectroscopy further shows the quantitative cation-size effect of Li^+, Na^+, and K^+ on spectral features (e.g. band position, intensity, and width) during cation insertion into MnO_2 at different potentials [38]. It suggests the properties of alkali metal ions (e.g. hydrated radii) in the neutral electrolyte play a role in pseudocapacitive storage.

The pseudocapacitive storage could occur on the surface or bulk with the insertion or extraction of cations (e.g. H^+, Na^+). However, often the reactions happen on the surface of electrodes unless the size domain of metal oxides reduces to the nanoscale. Otherwise, the cation diffusion only occurs on the surface of the bulk electrode with surface redox pseudocapacitive behaviors. Augustyn et al. first report the intercalation pseudocapacitive charge storage mechanism [39]. It shows the Li^+ insertion into Nb_2O_5 layered bulk electrode without solid-state diffusion limitations (Figure 7.8c), and such rapid Li^+ diffusion encourages the high-rate EES. As illustrated in Figure 7.8f,i, distinct electrochemical redox features can be observed from CV, while no distinct crystal structure lattice changes upon charge and discharge. Intercalation pseudocapacitive storage has also been observed in VO_2(B) oxides with highly decreased interaction energy and Li^+ diffusion kinetic barrier [40]. The application of intercalation pseudocapacitive behavior in $(NH_4)_2SO_4$ aqueous electrolyte is also reported in layered V_2O_5 with good rate performance [41, 42]. Further understanding of intercalation pseudocapacitance is highly anticipated because it not only offers a high-rate electrochemical performance but also increases the storage capacity relative to that from a surface redox pseudocapacitive process.

Mixed Storage Mechanism in Neutral Electrolytes Although there are surface-redox and intercalation pseudocapacitive reactions of transition metal oxides are observed in an aqueous neutral electrolyte, the mixed charge storage mechanism of these two could coincide during the electrochemical reactions, possibly due to the special design of oxide material and structure. For example, in a neutral Na_2SO_4 electrolyte, cation-intercalated $Na_{0.55}Mn_2O_4 \cdot 1.5H_2O$ shows a surface-controlled pseudocapacitive behavior in the low potential region (0–0.8 V), while an intercalation pseudocapacitive behavior at the higher potential [43]. With structural water-mediated cation insertion process, Shan et al. reported the solid-state formed MnO_2 birnessite with Na-rich and disordered property shows mixed storage mechanisms of both pseudocapacitive and diffusion-limited redox in a Na_2SO_4 electrolyte [44]. Therefore, the storage mechanism in neutral aqueous could be influenced by the interplay between the electrolytes and the oxides, which necessitates the concurrent analyses of the electro-kinetics and the evolution of crystal and electronic structures during the redox reactions.

Conversion Storage Mechanism in Alkaline Electrolytes Battery-like conversion reaction of transition metal oxides for aqueous storage majorly occurs in alkaline electrolytes. For example, compact rock-salt-type NiO bulk electrode in alkaline storage leads to the oxidation state variation between Ni^{2+} and Ni^{3+} and the structural transformation into layered NiOOH during charge and discharge (Figure 7.8d). The breaking/formation of O—H bonds for the faradic storage processes shows distinct redox features in CV and a plateau in chronopotentiometry (CP) from the conversion reaction during discharge (Figure 7.8g,j). Highly concentrated KOH (6 M) is often used as the electrolyte since it is low cost (vs. NaOH) and generates higher phase conversion ability with large storage capacity, although the faradic redox reactions could be limited by solid-state diffusion of charge carrier. NiO has a very high theoretical specific capacity (714 mAh g^{-1}), and the obtained electrochemical value is above 278 mAh g^{-1} [45–51]. More efforts are needed to improve the reaction kinetics of the conversion storage and address the corrosion issues arising from using an electrolyte with high alkalinity.

7.3 Transition Metal Oxides for Aqueous EES

Among numerous electrode materials for aqueous batteries, transition metal oxides have been extensively studied, along with other compounds, including metal hydroxide and polyanionic metal materials. Transition metal oxides for aqueous energy storage are categorized into cathodes and anodes depending on the potential window for their redox activities. The cathodes materials (e.g. MnO_2, V_2O_5, and NiO) have an averaged working Augustyn potential above 0 V vs. saturated calomel electrode (SCE), while anode materials (e.g. Fe_3O_4, FeOOH, and Bi_2O_3) have an

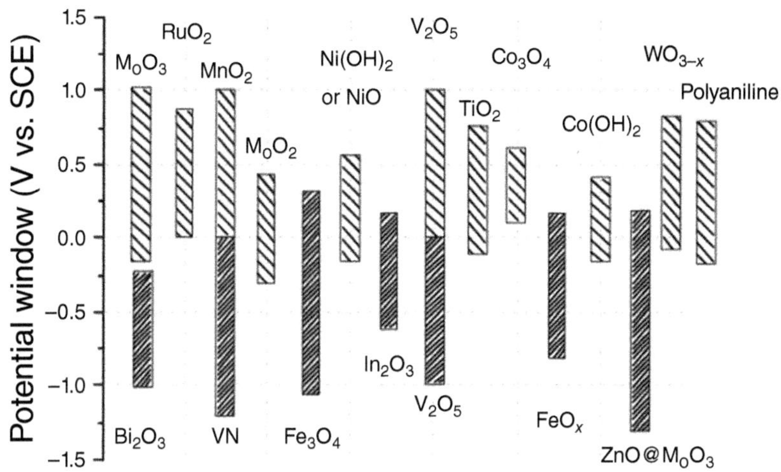

Figure 7.9 The working potentials for various transition metal oxides in aqueous electrolytes for the cathodes and anodes. Source: Qu et al. [52]/with permission of John Wiley & Sons.

average working potential below 0 V (Figure 7.9) [52]. In the section, structural and electrochemical properties of various metal oxide electrode materials in aqueous batteries are discussed, with a primary focus on the influences of material design and functionalization (e.g. crystal structure engineering, doping, and defect chemistry, composting with other additives, and electrode surface and structural water modifications) on improving their charge storage reactions.

7.3.1 Manganese Compounds

Manganese oxide is one of the transition metal oxides widely studied for aqueous energy storage due to its variety of crystal structures, redox-active Mn valence between Mn^{2+} and Mn^{4+}, low cost, and environment friendlessness. In this section, various strategies to improve the electrochemical performance of manganese oxides in aqueous energy storage have been summarized, and the perspective with further research direction for manganese oxides has been provided. Notably, this book chapter is more concentrated on the manganese oxides as electrode materials for aqueous storage. However, other research about manganese oxides (e.g. MnO_2) or other transition metal oxides could function as electro-catalysts in metal–air batteries [53, 54], which is beyond the scope of this work.

7.3.1.1 Crystal Structures of Manganese Oxides for Aqueous Storage

Due to the arrangement of $[MnO_6]$ octahedral building units, there are various polymorphic manganese oxides materials, including MnO, Mn_3O_4, Mn_2O_3, α-MnO_2, β-MnO_2, λ-MnO_2, and δ-MnO_2 (Figure 7.10) [55, 56]. Rock-salt MnO, tetragonal Mn_3O_4, bixbyite Mn_2O_3 are close-packed oxides with very limited vacant sites available for intercalation of charge carriers. They are often used as aqueous cathodes with only capacitive storage. It is worth mentioning that these tunnel oxides are

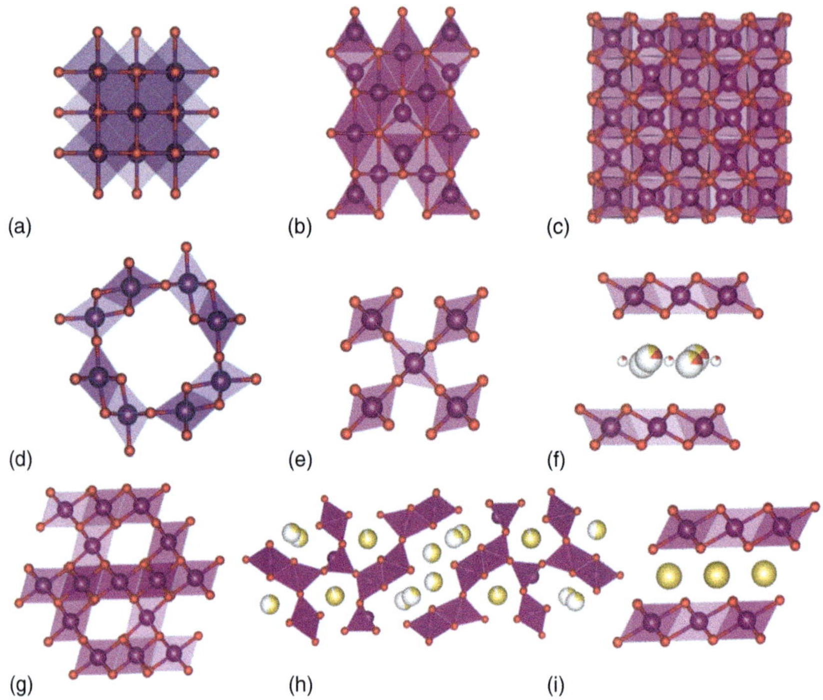

Figure 7.10 Crystal lattice structures of manganese-based oxides materials drawn with VESTA software (Mn: magenta; O: red; Na^+: yellow). (a) MnO; (b) Mn_3O_4; (c) Mn_2O_3; (d) α-MnO_2; (e) β-MnO_2; (f) δ-MnO_2; (g) λ-MnO_2; (h) $Na_{0.44}MnO_2$; (i) O_3-$NaMnO_2$.

usually labeled by two numbers corresponding to the number of octahedra forming the tunnel in each direction. Hollandite α-MnO_2 (2 × 2), pyrolusite β-MnO_2 (1 × 1), spinel λ-MnO_2 (a mixture of 2 × 1 and 1 × 1, made by removal the lithium from spinel $LiMn_2O_4$) are typical tunnel structures, which can make storage charge carriers (e.g. H^+, Na^+) in bulk [57, 58], while birnessite δ-MnO_2 storages charges between the layered frameworks [59]. α-MnO_2 and δ-MnO_2 are more attractive as cathodes compared to the other manganese dioxides majorly due to higher storage capacity from their large site openness and higher stability in aqueous electrolyte [60].

Comprehensive electrochemical studies on different crystallographic manganese dioxides are conducted in 0.1 M Na_2SO_4, and all show the rectangular shape voltammetry with a fingerprint for the pseudocapacitive redox behaviors (Figure 7.11). The charge storage values are quantitatively evaluated from galvanostatic charge and discharge cycling, which has the following order: $\alpha \approx \delta > \gamma > \lambda > \beta$ with a maximum specific capacitance 200–300 F g^{-1} (Figure 7.12b,c) [60, 70]. With pre-intercalated metal ions, some other manganese oxides such as tunnel structural $Na_{0.44}MnO_2$ and layered O3-$NaMnO_2$ are also considered for potentially large storage capacity in aqueous (Figure 7.11d–f).

$Na_{0.44}MnO_2$ has an orthorhombic tunnel-structured (mixture of 2 × 2 and 1 × 3 tunnels) with pre-intercalated Na-ions. Whitacre et al. first reported this material

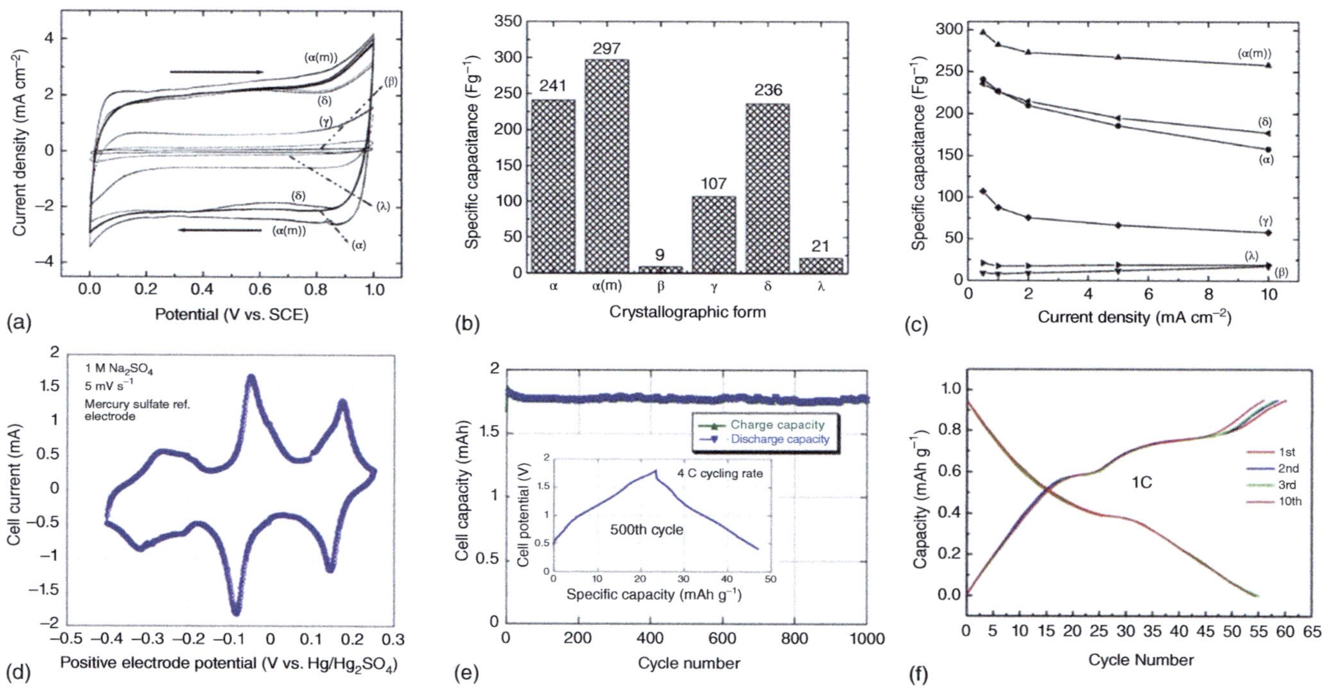

Figure 7.11 (a) Cyclic voltammetry of α-, α(m)-, β-, γ-, δ-, and λ- MnO_2 between 0 and 1.0 V (vs. SCE) in aqueous 0.1 M Na_2SO_4 at 20 mV s^{-1} [60]; (b) specific capacitances of MnO_2 electrodes with various crystallographic forms at current density of 0.5 mA cm^{-2} [60]; (c) specific capacitances at various current densities of charge–discharge cycling [60]; (d) CV of $Na_{0.44}MnO_2$ as working electrode tested at 5 mV s^{-1} in 1 M Na_2SO_4 in a three-electrode half-cell (vs. Hg/Hg_2SO_4) [61]; (e) cell charge and discharge capacities as a function of cycle number for 4 C galvanostatic cycling a potential window of 0.4–1.8 V [61]; (f) discharge and charge curves of $NaMnO_2$ cathode at 1 C rate [62]. Source (a–c): Devaraj and Munichandraiah [60]/with permission of American Chemical Society, (d, e) Whitacre et al. [61]/with permission of Elsevier, (f) Hou et al. [62]/with permission of Royal Society of Chemistry.

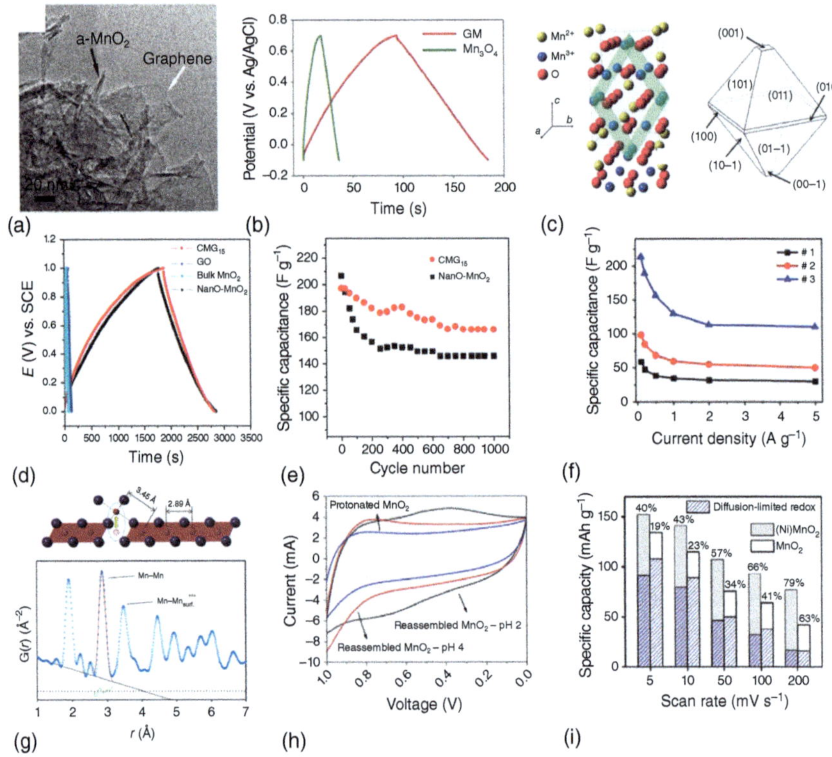

Figure 7.12 (a) TEM image of α-MnO_2 nanowires in graphene sheets [63]; (b) galvanostatic charging and discharging curves of Mn_3O_4 and graphene/Mn_3O_4 composite (GM) at $1\,A\,g^{-1}$ [64]; (c) crystal structure of Mn_3O_4 with various exposed facets [65]; (d) charge/discharge curves of CMG_{15}, nano-MnO_2, GO, and bulk MnO_2 at $0.2\,A\,g^{-1}$ [66]; (e) cycle life of CMG_{15} and nano-MnO_2 at $0.2\,A\,g^{-1}$ in 1 M Na_2SO_4 solution [66]; (f) CV curves of sample 1–3 at $50\,mV\,s^{-1}$ in a three-electrode configuration [67]; (g) PDF of δ-MnO_2 equilibrated at pH = 2 showing Mn surface defects [68]; (h) CV curves for δ-MnO_2 reassembled in various pHs [68]; (i) calculated diffusion-limited redox and capacitive charge storage for MnO_2 and Ni-doped (Ni)MnO_2 birnessite [69]. Source: (a) Wu et al. [63]/with permission of American Chemical Society, (b) Lee et al. [64]/with permission of American Chemical Society, (c) Yeager et al. [65]/with permission of John Wiley & Sons, (d, e) Chen et al. [66]/With permission of American Chemical Society, (f) Hu et al. [67]/with permission of Elsevier, (g, h) Gao et al. [68]/Springer Nature/CC BY 4.0, (i) Shan et al. [69]/with permission of American Chemical Society.

as cathode used in aqueous energy storage. $Na_{0.44}MnO_2$ has a specific capacity of $45\,mAh\,g^{-1}$ within 0.5 V potential window (over $300\,F\,g^{-1}$ in specific capacitance) in 1 M Na_2SO_4 electrolyte, and CV shows multiple distinct redox peaks with many phase transformations unlike previously described MnO_2 with typically pseudocapacitive behaviors (Figure 7.12d) [61, 71]. Although it exhibits stable performance after 1000 discharge cycles without loss in capacity by coupling with activated carbon as the negative electrode (Figure 7.12e), the achieved capacity is still unsatisfactory, which is far from its theoretical capacity of about $120\,mAh\,g^{-1}$. Layered $NaMnO_2$ materials (e.g. O3 $NaMnO_2$ shown in Figure 7.11d–f) with even higher Na-rich concentrations are very popular cathodes used in non-aqueous Li- or Na-ion batteries with a high theoretical capacity above 150 mAh g^{-1}, which can be classified into four common crystal structures (O3, P3, O2, P2) depending on construction and stacking of edge-sharing MnO_6 octahedral and alkali metal ions [72, 73]. There are few studies on such layered $NaMnO_2$ for aqueous energy storage, except that an aqueous Na-ion battery based on $NaMnO_2$-$NaTi_2(PO_4)_3$ hybrid system in 2 M CH_3COONa aqueous electrolyte has been reported with an energy density of 30 Wh kg^{-1} [62]. However, the resulting capacity of $NaMnO_2$ in aqueous half-cell is low with only about 55 mAh g^{-1} at 1 C (Figure 7.12f), and whether there is any phase evolution of $NaMnO_2$ into other types of MnO_2 is unknown because $NaMnO_2$ is water- and air-sensitive electrode [74, 75]. Although air-stable P2-type cathodes are reported [76], the water intercalation into layered $NaMnO_2$ could be a major issue especially for aqueous energy storage, which triggers the direct reaction between $NaMnO_2$ and water–electrolyte with the formation of insulators such as NaOH and Na_2CO_3 [75].

The polymorphic structures of manganese oxides play a vital role in their intrinsic charge storage ability as the electrodes, and they point out that layered δ-MnO_2 as well as tunnel-structured α-MnO_2 and $Na_{0.44}MnO_2$ could be promising Mn-based oxides for aqueous storage. There are indeed more effects focusing on improving the electrochemical performance on these types of materials by the various strategies such as compositing with other additives, surface engineering crystal facets, edge sites, and domain size, doping and defect chemistry, and controlling intercalated species, as the following discussion shows.

7.3.1.2 Compositing Manganese Oxides with Other Additives

Mixing carbon-based materials (e.g. graphene, carbon black) with manganese oxides is the most straightforward approach to increase the electrochemical performance since the electron conductivity of manganese oxides can be enhanced [63, 64]. Figure 7.12a shows that α-MnO_2 nanowires mixed with graphene sheets can be used as the positive electrode in an aqueous Na_2SO_4-based asymmetric electrochemical capacitors, providing an energy density of 30.4 Wh kg^{-1} and a power density of 5000 W kg^{-1} at 7.0 Wh kg^{-1} [63]. In addition, graphene/Mn_3O_4 composite also shows higher capacitance as the supercapacitor electrode than that of Mn_3O_4 (Figure 7.12b) [64].

7.3.1.3 Surface Engineering Crystal Facets, Edge Sites, and Bulk/Nano Size Domain

Increasing the number of active sites for cation storage is another approach to achieving a high Mn oxide capacity [65, 66, 77, 78]. Yeager et al. report the synthesis of Mn_3O_4 particles with (101) facets as a pseudocapacitive cathode material with a specific mass capacitance of 261 F g^{-1} (Figure 7.12c), and the stable spinel structure, as well as the redox activity of Mn, was revealed by *in situ* XRD and XANES measurements. Moreover, the density functional theory (DFT) calculations demonstrate that the enhanced redox activity may be attributed to the easier process of absorption and desorption of Na-ions on the exposed active (101) crystal facet [65]. Similar research work shows that the higher surface adsorption pseudocapacitance is observed for β-MnO_2 with a large percentage of (001) surface compared to those of other surfaces, which resulted from a lower ion diffusion barrier in the electrode [77]. Reducing the size of the bulk electrode into a nanoscale or increasing the surface area by modifying the morphology from particles into ribbons/sheets or electrode architecture (e.g. 1D, 2D) certainly can largely enhance the specific capacity majorly all due to the increased active sites for accessible cation storage [66, 78]. For example, nano-MnO_2 with the nanocrystal and more exposure active sites show a much longer charge–discharge time than that of bulk MnO_2 (Figure 7.12d) [66]. In addition, it is observed that the CMG_{15} (chemically synthesized graphene oxide and MnO_2 nanocomposite when the feeding ratio of MnO_2/graphene oxide [GO] is 15/1) exhibits better cycling stability compared to that of pure nano-MnO_2. Although carbon additive provides no significant enhancement of specific capacitance of MnO_2/GO composite, it suggests that the carbon may also stabilize the pseudocapacitive reaction. Although carbon additives and surface engineering are certainly promising for increasing the storage capacity, some issues still need to be addressed, including (i) the increased volume with carbon additive could potentially decrease volume density of manganese-based electrode; (ii) the stability and cycling performance of nanoscaled manganese oxides may not be as good as the bulk electrode since more active sites could potentially cause more side reactions.

7.3.1.4 Doping and Defect Chemistry

Doping manganese oxides with other redox-active metal species and/or introducing defects to manganese oxides are also attractive for improving storage capacity without compromising the volumetric energy density. Instead of introducing carbon into the electrode as a mixture, other transition metal ions (e.g. Co, Ni) are incorporated into the framework of the manganese oxides during synthesis or possibly with certain site defects generated [47, 67–69, 79–81], which may increase the electron conductivity of the electrode. For example, series Al-doped α-MnO_2 samples with various Al^{3+}:Mn^{2+} molar ratio have been synthesized (samples 1, 2, 3 with ratio of 0.25 : 1, 0.5 : 1, and 1 : 1), and it shows an enhanced current response from CV curves with an improved specific capacitance as the Al doping amount increases (Figure 7.12e,f) [67]. X-ray pair distribution function (PDF) data show exfoliation of crystalline K_xMnO_2 and reassembly at different pH solutions not only can increase the surface area of resulting δ-MnO_2 but also introduces surface Mn vacancies (Figure 7.12g) [68]. As shown in Figure 7.12h, the CV response increases as the used pH decreases to pH (=2). The δ-MnO_2 equilibrated at low pH (=2) has about 26.5% of Mn vacancies, and such generated point defects can increase pseudocapacitance to over 300 $F\,g^{-1}$ with a reduced charge transfer resistance. Shan et al. report that the framework doping of Ni atoms into δ-MnO_2 birnessite, verified by neutron total scattering and PDF analysis, demonstrates a large enhancement of pseudocapacitive charge storage (Figure 7.12i) [69]. Moreover, the Co-doped manganese oxides have also been studied, in which neutron and X-ray PDF show bi-phases with layered $MnO_2 \cdot H_2O$ birnessite phase and a $(Co_{0.83}Mn_{0.13}Va_{0.04})_{tetra}(Co_{0.38}Mn_{1.62})_{octa}O_{3.72}$ (Va: vacancy; tetra: tetrahedral sites; octa: octahedral sites) spinel phase. Such electrode has an increased capacity compared to the pure Mn_3O_4, majorly due to the enhanced diffusion-limited redox process from doped redox-active Co atoms as well as the cation vacancy of the tetrahedral sites in spinel phase [81]. As discussed previously, tunnel-structured $Na_{0.44}MnO_2$ is limited by its low capacity around $50\,mAh\,g^{-1}$, further studies show that $Na_{0.44}MnO_2/NaTi_2(PO_4)_3$ system without using activated carbon as negative electrode in aqueous Na_2SO_4 or $NaClO_4$ can result in a full-cell capacity above $100\,mAh\,g^{-1}$ [82, 83]. Doping approach (e.g. Ti) is also applied in this oxide material possibly to boost its capacity. Although the doped Ti serves redox center (Ti^{3+}/Ti^{4+}) similar to Mn atoms with superior cyclability, the $Na_{0.44}[Mn_{1-x}Ti_x]O_2$ shows a lower capacity only close to $40\,mAh\,g^{-1}$ in aqueous, much smaller than that in a non-aqueous electrolyte (above $100\,mAh\,g^{-1}$) [84, 85]. Therefore, the selection of doped transition metal ions into manganese oxides could be crucial to their electrochemical performance in aqueous storage because some (e.g. Ni, Co) may increase the storage capacity, whereas others could only improve the cycling stability.

7.3.1.5 Pre-intercalated Species

δ-MnO_2 birnessite is one of the most important manganese oxides for its layered structure and potential to store charges in aqueous electrolyte stably. Despite tremendous research effects on improving electrochemical performance, pre-intercalation of ions in the interlayer is considered as an effective strategy to tune the structure of the host electrode, and thus the electrochemical properties. Various alkalis (e.g. Na^+, K^+, Mg^{2+}) or conductive polymers (e.g. polyaniline PANI) have been reported to occupy between the interlayers of MnO_2 birnessite [86–89]. For example, Figure 7.13a shows that Na or K incorporated δ-MnO_2 has a superior capacity of $74.6\,mAh\,g^{-1}$ as the cathode coupling with $NaTi_2(PO_4)_3$ anode compared to the Na-MnO_2 counterpart [86]. PANI-MnO_2 nanocomposite shows a large basal spacing of about 1.47 nm, benefiting an enhanced Na-ion storage with a specific capacitance of $330\,F\,g^{-1}$ at $1\,A\,g^{-1}$, compared to those of PANI ($187\,F\,g^{-1}$) and MnO_2 ($208\,F\,g^{-1}$) [86].

Recently, some new strategies with the water chemistry regarding the intercalated structural water involved in the electrochemical reactions of δ-MnO_2 have been interesting and raising attention, which may modify storage reaction kinetics or capacity [44, 90–93]. The introduction of structural water during the electrochemical process enables the phase transition from spinel to a layered Birnessite structure in an aqueous $MgSO_4$ electrolyte after four charge–discharge cycles (Figure 7.13b), and electrochemical CP results show an increased specific capacity from around $100\,mAh\,g^{-1}$ to above $200\,mAh\,g^{-1}$ (Figure 7.13c) [90]. In addition, further studies of such spinel-to-Birnessite phase transition can also be observed in other aqueous electrolytes such as Li_2SO_4 and Na_2SO_4, and it exhibits similar electrochemical behaviors although best performance is achieved in $MgSO_4$ solution (Figure 7.13d) [91]. Structural water contained in the MnO_2 interlayers could also help the cycling performance. For example, the layered Na-birnessite (Na-Bir), as well as de-hydrated Na-Bir synthesized by heat treatment, has been reported, and hydrated layered MnO_2 manifests a specific capacity of $80\,mAh\,g^{-1}$ at 1 C without obvious capacity loss after 150 cycles while the de-hydrated one only retains 60% of the initial capacity (Figure 7.13e) [92]. The promotional effects of structural water on

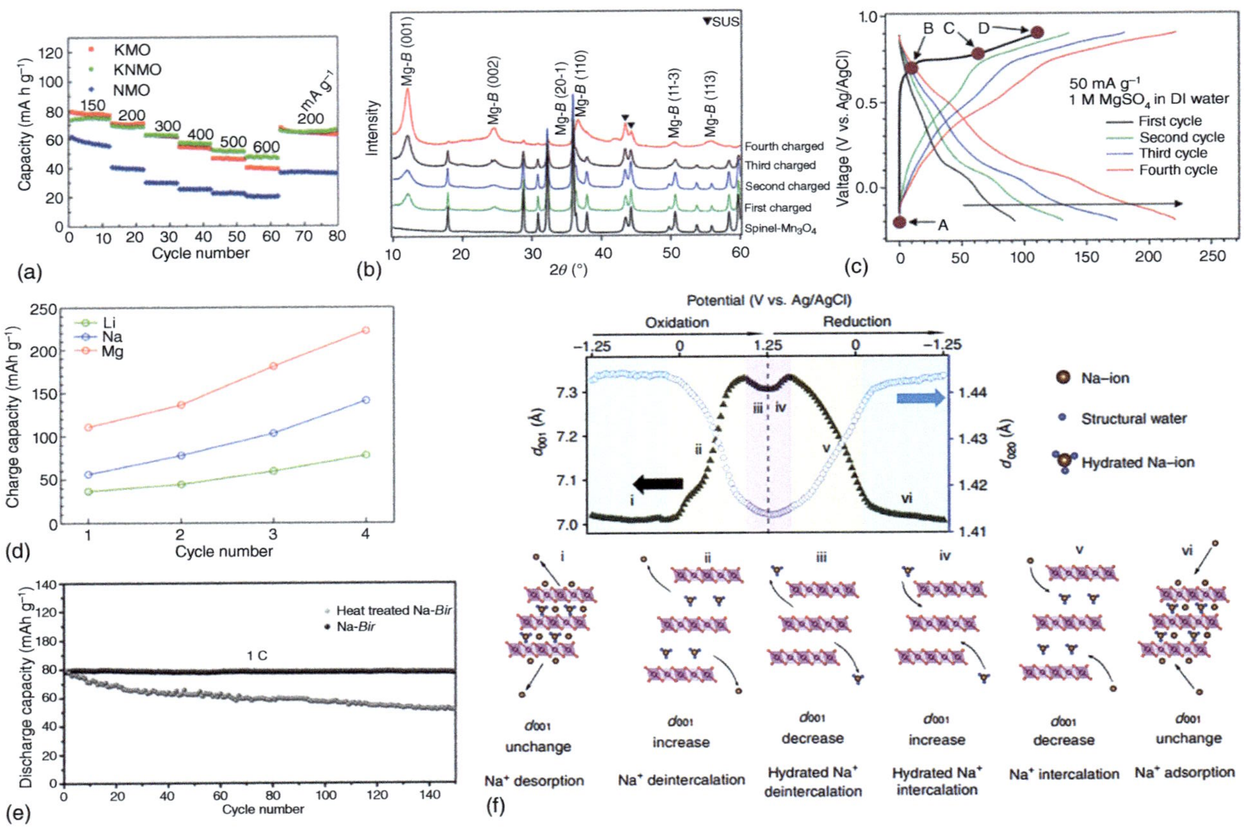

Figure 7.13 (a) Charge/discharge profiles of $NaTi_2(PO_4)_3$/A(Na,K)-δ-MnO_2 hybrid full cell at 0.2 A g^{-1} [86]; (b) XRD patterns with spinel-to-layer phase transition at different cycling states in 1 M $MgSO_4$ aqueous electrolyte [90]; (c) charge–discharge profiles at 0.05 A g^{-1} [90]; (d) specific capacity evolution at 0.05 A g^{-1} using different aqueous electrolyte 1 M Li_2SO_4, 1 M Na_2SO_4, and 1 M $MgSO_4$ [91]; (e) cycling performance of Na-Bir and heat-treated Na-Bir at 1 C in Na_2SO_4 [92]; (f) Schematic of Na-ions and water motion during the redox process of MnO_2 birnessite in aqueous Na_2SO_4 electrolyte [44]. Source: (a) Liu et al. [86]/with permission of Royal Society of Chemistry, (b, c) Kim et al. [90]/with permission of American Chemical Society, (d) Kim et al. [91]/with permission of American Chemical Society, (e) Zhang et al. [92]/with permission of Royal Society of Chemistry, (f) Shan et al. [44]/Springer Nature/CC BY 4.0.

the electrochemical performance for the hydrated δ-MnO_2 can also be observed in non-aqueous electrolytes [93]. As illustrated in Figure 7.13f, Shan et al. report that the co-intercalation of the hydrated water and sodium ions from layered δ-MnO_2 birnessite electrode during the high potential charging process results in the shrinkage of interlayer distance and thus stabilizes the layered structure [44]. The role of crystal structure water in MnO_2 definitely is vital to the electrochemical reactions in aqueous storage, and more importantly, it could be a generic method for improving the capacity or stability in aqueous because Mitchell et al. report similar results that such water chemistry could be involved in other transition metal oxides such as tungsten oxides and titanium oxides [94, 95].

7.3.2 Ni Compounds

Due to the Jahn–Teller effect of Ni^{3+} in $LiNiO_2$ spinel, as well as the hydrolysis reaction between $LiNiO_2$ and water, NiO has been seen as the only stable nickel oxide, which can have the electrochemical storage reactions in aqueous electrolyte. Particularly, the conversion reaction ($NiO + OH^- \leftrightarrow NiOOH$) in the alkaline electrolyte has been widely reported, and Yeager et al. further demonstrate the redox-active center of Ni atoms by *in situ* XANES tests [96]. Further improvements of the electrochemical performance of NiO in an alkaline medium by various strategies have been summarized [45–50, 97]. For example, honeycomb-like mesoporous NiO microspheres have been synthesized (Figure 7.14a), and such three-dimensional (3D) hierarchical nanostructure provides a high specific capacitance 1250 F g^{-1} at 1 A g^{-1} in 6 M KOH aqueous solution [49].

It is reported that a lithiation process can induce a rescaling effect on NiO nanoparticles by decreasing their sizes from ~10 nm to a sub-nanoscale on reduced graphene oxides (rGOs; Figure 7.14b,c). This downsizing increases NiO capacitance while maintaining a 100% capacity retention over 100 000 redox cycles [48]. It is speculated that the Ni defects from the local distortion of [NiO_6] octahedra can tailor electronic structures of the electrode and thus potentially boost their pseudocapacitance. For example, Choi and coworkers report a distortion of [NiO_6] in Mix(NiMnCo)O solid solution structure. Such distortion breaks the degenerate e_g level of Ni^{2+} (Figure 7.14d), so that the Jahn–Teller lattice instability necessary for the $Ni^{2+}/^{3+}$ redox flip can be effectively diminished during charge–discharge. As a result, MixO-rGO (rGO stands for reduced graphene oxide) shows a significantly increased capacitance (Figure 7.14e) relative to NiO-rGO and CoO-rGO (MnO is inactive in alkaline solution) [47]. In addition, surface modification shows the incorporation of hydroxyl groups on the surface of NiO (Figure 7.14f). It is claimed

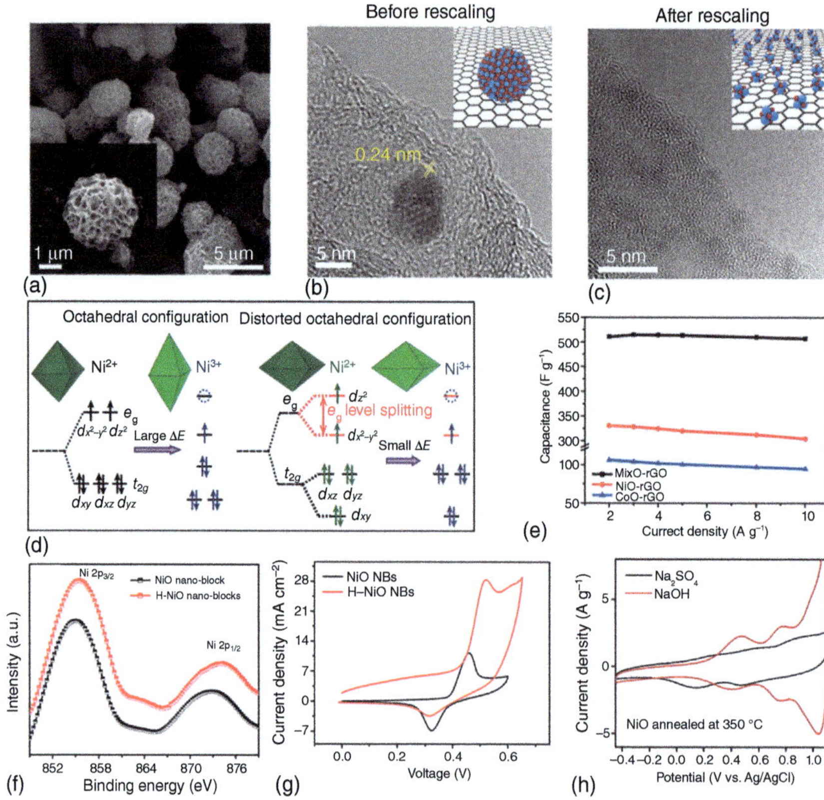

Figure 7.14 (a) Scanning electron microscopy (SEM) of obtained honeycomb-like NiO after an annealing treatment of $Ni(OH)_2$ [49]; (b) TEM of single NiO nanoparticle on rGO [48]; (c) NiO particles after lithiation-induced rescaling [48]; (d) Schematic description of 3d orbital configurations for octahedral and distorted octahedral fields of Ni atoms in Mix(NiMnCo)O solid solution [47]; (e) capacitance of MixO-rGO compared to NiO and CoO [47]; (f) XPS of NiO nano-block and hydrogenated H-NiO nano-block [50]; (g) CVs of NiO and H-NiO at 100 mV s^{-1} [50]; (h) 350 °C annealed NiO tested at 0.1 M Na_2SO_4 and NaOH aqueous electrolyte between −0.45 and 1.15 V (all potentials converted vs. Ag/AgCl) at the scan rate of 50 mV s^{-1}. Source: (a) Ren et al. [49]/with permission of American Chemical Society, (b, c) Ock et al. [48]/with permission of John Wiley & Sons, (d, e) Lee et al. [47]/ with permission of John Wiley & Sons, (f, g) Singh et al. [50]/with permission of American Chemical Society, (h) Unpublished work.

that the hydrogenated NiO nanoblock architecture has a better pseudocapacitive performance compared to the regular NiO nanoblocks (Figure 7.14g) [50]. Improving the electrochemical performance of NiO in alkaline by modifying the architecture, electronic, or surface structures is vital. Although there are few studies of NiO electrodes in an aqueous neutral electrolyte, it is expected that the charge storage mechanism of NiO in a neutral electrolyte could be significantly different from that in a basic solution. Unlike the phase conversion reaction in an alkaline solution, recent work shows that NiO has only a surface intercalation redox process in a neutral solution (Figure 7.14h), suggesting such storage reaction could be possibly controlled by the crystal lattice spacing of NiO.

7.3.3 Vanadium Compounds

Vanadium-based compounds can exist in a variety of crystal structures. This is due to the multiple oxidation states and multiple anion-coordination polyhedra possible for vanadium. Among various polymorphs, VO_2, V_2O_5, metal vanadates, and vanadium-based NASICON (which are isostructural with Na super ionic conductor) have been of immense interest for aqueous batteries.

Monoclinic VO_2 possesses a tunnel structure (Figure 7.15a), allowing rapid and reversible insertion and extraction of alkali. In fact, it is one of the first transition metal oxides used as the anode in aqueous batteries. Li et al. reported a spinel $LiMn_2O_4//VO_2$ cell using 5 M $LiNO_3$ and 1 mM LiOH aqueous electrolyte [98]. This prototype aqueous battery cell operates at an average voltage of 1.5 V with an energy density of 75 Wh kg^{-1}. Despite its success as anode materials in the early generation of aqueous batteries, there are a few challenges for the VO_2 anode system: first, VO_2 synthesis involves a rather complicated preparation procedure due to the metastable nature of VO_2 under ambient condition; second, the electrode potential of $VO_2/LiVO_2$ redox couple is around −0.43 V vs. standard hydrogen electrode (SHE), very similar to the potential of HER at a neutral pH (−0.42 V vs. SHE). Therefore, final control over electrolyte pH value is very important for VO_2-based aqueous batteries, whereas unwanted HER occurring at a lower pH electrolyte would reduce the Coulombic efficiency of the anode. Third, the chemical stability of VO_2 in the presence of oxygen remains problematic primarily upon

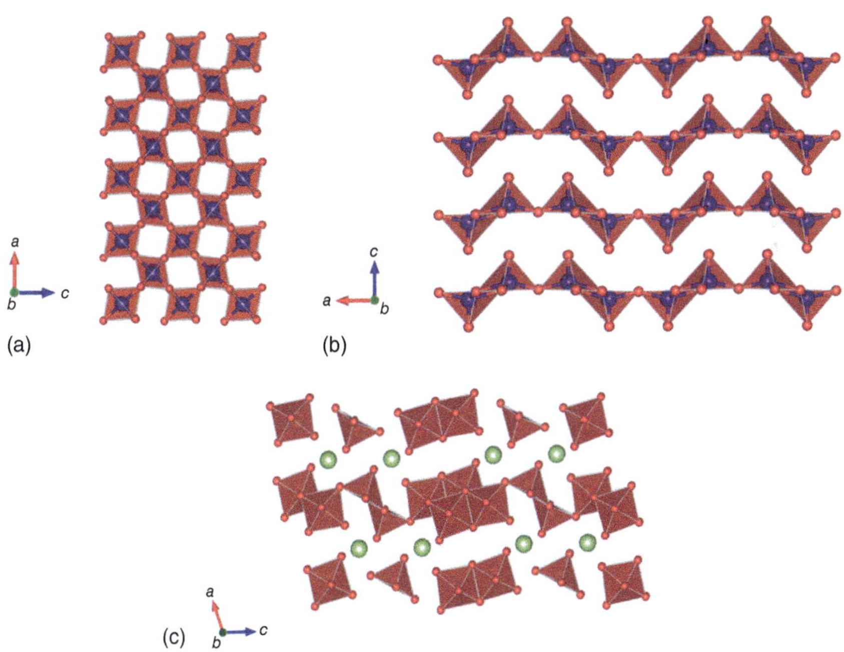

Figure 7.15 Crystalline structures of (a) VO_2, (b) V_2O_5, and (c) Li-V_3O_8.

prolonged cycling. The oxidation of VO_2 over oxygen exposure inevitably increased the electrode potential of the $VO_2/LiVO_2$ redox couple and resulted in voltage loss.

V_2O_5 contains distorted [VO_6] octahedra and forms a bilayered structure (Figure 7.15b), where water molecules, alkali, or even large organic molecules can insert between the layers to form an intercalated structure. These intercalants between the V_2O_5 bilayers play a crucial role in facilitating insertion and extraction of charge carrier into V_2O_5 host lattice and often improve electrode cyclability. For example, the intercalated water molecule provides an electrostatic screening mechanism that reduces the coulombic interactions between the charge carrier and the host V_2O_5 electrode, improving the mobility of the charge carrier during insertion, especially for the aqueous multivalent charge storage (e.g. Zn^{2+} or Mg^{2+}). Yeager et al. report highly crystalline V_2O_5 layered nanofibers with pre-intercalated K-ions as cathode materials for aqueous K-ion storage, in which the V_2O_5 nanofibers show an average diameter of ~120 nm and in length up to a few micrometers (Figure 7.16) [99]. A "rolling-up" formation mechanism has been proposed, where [VO_6] crystallites are exfoliated into layered $K_{0.33}V_2O_5$ crystalline sheets at mild temperature (200 °C), followed by bending and rolling-up of thin layers at high temperatures (450 °C) to form 1D nanostructures. Since the largest periodicity of

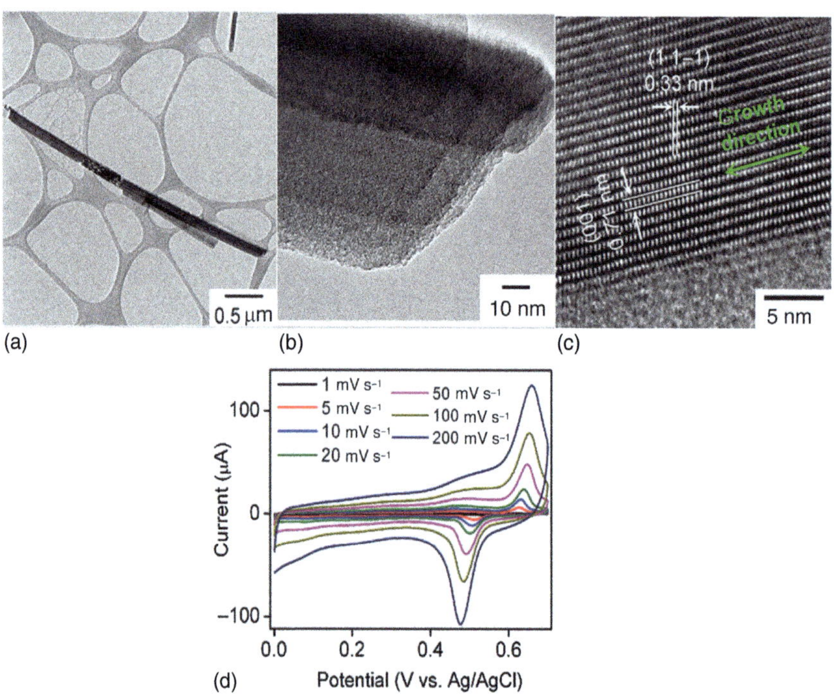

Figure 7.16 (a–c) TEM images of $K_{0.33}V_2O_5$ layered nanofibers after being annealed at 450 °C; (d) analyses: (a) CVs of $K_{0.33}V_2O_5$ layered nanofibers/poly(3,4-ethylenedioxythiophene) (PEDOT) as loading of 5 and 1.25 μg measured in 1 M KCl with a three-electrode half-cell. Source: (a–c) Yeager et al. [99]/with permission of John Wiley & Sons, (d) Yeager et al. [99]/with permission of John Wiley & Sons.

nanolayers is along the <001> (0.96 nm), it is speculated that the rolling-up along <001> is the most energetically favorable mechanism. Figure 7.16d shows the CVs of $K_{0.33}V_2O_5$ layered nanofibers, of which well-defined anodic and cathodic peaks appear and represent characteristics of diffusion-limited K^+ extraction and insertion behavior. The calculated gravimetric capacitance (C_{MS}) is strongly dependent on the scan rates, showing a maximum value of 136 F g^{-1}.

Besides highly crystalline counterparts, disordered V_2O_5 has also drawn much research interest recently. Charles et al. report the preparation of disordered $K_{0.2}V_2O_5$ nanosheets in an aqueous reaction at room temperature by reacting potassium hydroxide with vanadyl(IV) sulfate hydrate [100]. Transmission electron microscopy (TEM) shows that the sheet dimensions vary from sheet to sheet but are on the order of hundreds of nanometers. The $K_{0.2}V_2O_5$ nanosheets exhibit excellent capacitance at low scan rates, 661 F g^{-1} (178 mAh g^{-1}) at a scan rate of 5 mV s^{-1} in a 1 V potential window. A significant amount of capacitance is retained even at higher scan rates, 334 F g^{-1} (93 mAh g^{-1}) at 200 mV s^{-1}. Assuming one charge transfer per vanadium atom, the theoretical capacitance of V_2O_5 is 294 mAh g^{-1}, 0.6 electron transfer per vanadium atom. Moreover, the roles of structural water on K-ion storage of disordered V_2O_5 have also been studied using neutron total scattering and PDF analysis, as shown in Figure 7.17a. The fully hydrated V_2O_5 material (after soaking disordered V_2O_5 in water for two weeks) shows a much larger coherence length, the rearrangement of the local structure, and larger occupancy of the water intercalated between the layers, compared with partially hydrated V_2O_5 (the as-made disordered V_2O_5 without water soaking). These results indicate that disordered V_2O_5 could be stabilized via a strong interaction between O (from structural water) and V atoms. V_2O_5 engaged by structural water exhibits superior capacity (0.89 electrons transfer per vanadium atom) for aqueous K-ion storage in a half-cell. *In situ* XRD has been conducted to examine the evolution of (001) diffraction peak upon the three cycles of CVs. The results demonstrate that disordered V_2O_5 layered materials engaged with structural water show more continuous expansion/contraction of (001) diffraction planes and a wider redox-active potential window during electrochemical cycling in a contour plot (Figure 7.17b), compared with highly crystalline V_2O_5 materials (Figure 7.17c). These results are important for designing a new type of disordered electrode materials stabilized by structural water for aqueous energy storage with a large storage capacity.

7.3.3.1 Li or Na Vanadates

Vanadium can coordinate with oxygen atoms in many possible arrangements of tetrahedral or octahedral building units such as VO_4^{3-} (orthovanadates), $V_2O_7^{4-}$ (pyrovanadates), or $[V_3O_8]^-$ polymorphs. The frameworks are charge-balanced with cations of appropriate numbers, sizes, and charges (along with water molecules in some cases), resulting in a metal–vanadate family in either a layered- or a tunnel-type structure. In the case of a layered framework, for instance, in LiV_3O_8, the layers consist of negatively charged $[V_3O_8]^-$ units (Figure 7.15c). Notably, alkali can play an important role in opening and stabilizing the crystal structures, where

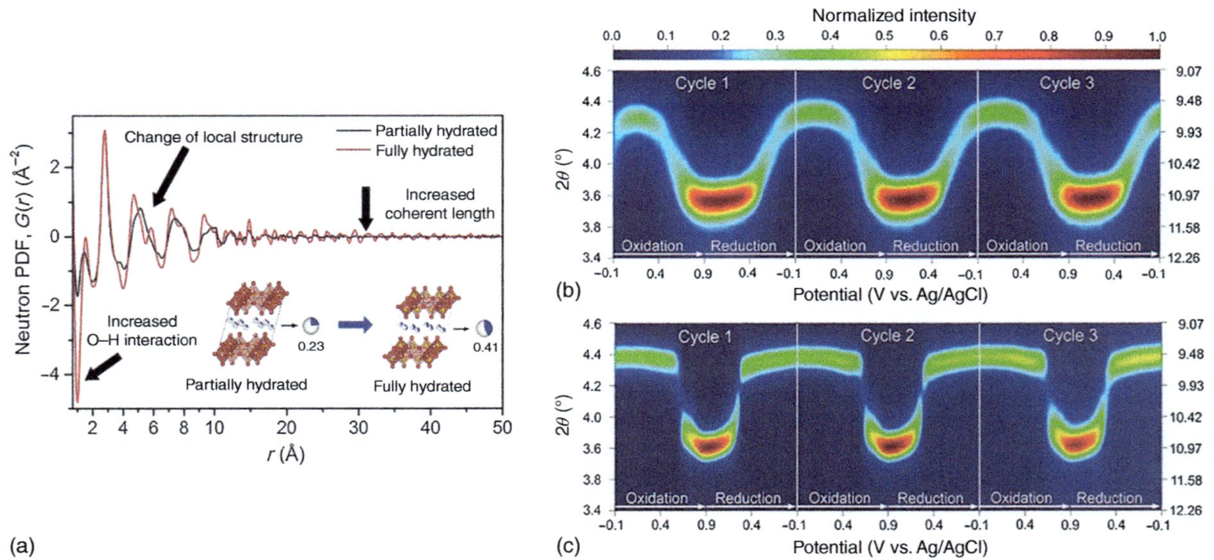

Figure 7.17 (a) Neutron PDFs of partially and fully hydrated disordered V_2O_5 nanosheets. The contour plot of *in situ* XRD spectra of the (001) plane during CVs for (b) disordered V_2O_5 and (c) ordered V_2O_5 materials. Source: Charles et al. [100]/Springer Nature/CC BY 4.0.

the negatively charged V–O framework layers are held together electrostatically via the alkali "pillars" in between the layers.

7.3.4 Iron Compounds

As the fourth richest element in earth's crust, iron (Fe) potentially can satisfy nearly all criteria for "green" battery electrode materials, such as a wide variety of oxidation states, geographic accessibility, low cost, and environmental beneficence. Moreover, active forms of Fe such as hydroxide [$Fe(OH)_2$] or oxyhydroxide (FeOOH) have a low solubility in alkaline solutions, avoiding the formation of Fe dendrite and the need for ion-selective separator in a battery cell.

7.3.4.1 Fe/Fe_3O_4

Fe has been successfully used as anode materials in Ni/Fe alkaline battery, originally developed in 1901 by Thomas Edison. However, this century-old system faces challenges in modern energy supply systems due to historic low energy and power density, especially at Fe anode. The charge–discharge process of an alkaline iron electrode is complex, owing to solid-state transformation within the surface films. On charging, $Fe(OH)_2$ is reduced to Fe at a −0.88 V (Eq. (7.1)), which is always negative to the HER under the same condition (Eq. (7.2)). This gassing issue not only decreases the charging efficiency but also impairs the discharge potential of the full cell due to the ohmic drop imposed by H_2 bubbles formed near the Fe electrode. Moreover, metallic iron is a thermodynamically unstable phase and suffers from corrosion through prolonged cycling. Several sulfide additives, such as FeS, PbS, and Bi_2S_3, have increased the hydrogen overpotential. On discharge, utilization of $Fe \rightarrow Fe^{3+}$ redox couple is difficult, where a more stable magnetite Fe_3O_4 usually forms (Eq. (7.12)), though $Fe(OH)_3$ is a more desirable discharged phase for achieving a high-capacity Fe electrode. Therefore, the actual charge storage mechanism in alkaline Fe anode is usually considered to occur during the two main stages of charge–discharge described in Eqs. (7.10) and (7.12), involving the redox couples of $Fe \leftrightarrow Fe(OH)_2 \leftrightarrow Fe_3O_4$.

$$Fe + 2OH \rightarrow Fe(OH)_2 + 2e^- \quad E^\theta = -0.88\ \text{V vs.SHE} \tag{7.10}$$

$$2H_2O + 2e^- \rightarrow H_2 + 2OH^- \quad E^\theta = -0.83\ \text{V vs.SHE} \tag{7.11}$$

$$3Fe(OH)_2 + 2OH^- \leftrightarrow Fe_3O_4 + 4H_2O + 2e^- \quad E^\theta = -0.76\ \text{V vs.SHE} \tag{7.12}$$

One potential issue with Fe anode material is the insulating nature of the $Fe(OH)_2$ phase, which could passivate the electrode. Therefore, conductive carbon-based materials have been used as additives to improve electrode conductivity. Rajan et al. report preparation of a mixture of metallic iron (α-Fe) and magnetite (Fe_3O_4) grafted with carbon through high-temperature decomposition of ferrous oxalate (FeC_2O_4) and polyvinyl alcohol (PVA) composites [101]. With Bi_2S_3 additives, the mixed-phase Fe/Fe_3O_4 anode undergoes a deep charging to −1.05 V vs. SHE, showing $Fe_3O_4 \rightarrow Fe(OH)_2 \rightarrow Fe$ reaction process without causing significant

HER. The authors suggest that the formation of Bi metal upon charging Bi_2S_3 ($E^{\theta} = -0.82$ V vs. SHE) on the surface of the anode materials could impose a great barrier for HER. A specific discharge capacity over 400 mAh g^{-1} at a current density of 0.1 A g^{-1} and a faradaic efficiency of 80% for the iron electrode is reported, attributed to the increased amount of metallic iron in the active material and its concomitant *in situ* carbon grafting during the synthesis. A more detailed study of the role of Bi_2S_3 additive on Fe anode reaction has been reported by Manohar et al. [102], showing that the presence of sulfide derivatives decreases the detrimental effect of iron passivation on the electrode capacity by forming a conductive FeS layer with surface iron atoms, thereby decreasing the insulating effect caused by the formation of passivating $Fe(OH)_2$ phase. In sharp contrast, when the Bi_2O_3 additive was employed, there was no significant increase in the capacity of the iron electrode observed, indicating that it is the sulfur atoms that are important rather than the presence of bismuth alone.

7.3.4.2 Fe_2O_3/FeOOH

Recently, there have been increasing research interests in nanosized ferric oxide (Fe_2O_3) and ferro oxyhydroxide (FeOOH), owing to their high availability, environmental friendliness, and, more significantly, a potential of three-electron transfer redox behavior ($Fe^{3+} \leftrightarrow Fe$). Like vanadium, iron can coordinate with oxygen and/or hydrogen atoms with various arrangements of Fe–O tetrahedra or octahedra building units, forming various iron oxide and hydroxide structures. The formation of FeOOH from Fe^{2+} and/or Fe^{3+} aqueous solutions has been studied for decades. Hydrolysis of iron(III) solutions at ambient temperatures results in an amorphous precipitate of $Fe(OH)_3$. This amorphous compound is thermodynamically unstable and may gradually transform to various polymorphs of FeOOH through controlling the reaction time, temperature, and pH value of the solution. The latter factor is particularly important in governing the precipitation of FeOOH in solution, follow-up crystallization (aging) process, as well as the crystalline phases. For example, at ambient conditions, FeOOH often formed at high pH solution, while Fe_2O_3 formed in a solution with medium to low pH values.

FeOOH exists in several polymorphs forms such as α- , β-, γ-, and δ-FeOOH (Figure 7.18), where α- and γ-FeOOH form a stable orthorhombic structure, and δ-FeOOH is thermodynamically unstable under an ambient atmosphere. Interestingly, β-FeOOH has a [2×2] hollandite-type tunnel structure (space group: $I4/m$), in which two edge-sharing $[FeO_6]$ octahedral units are twisted by 180°, forming an $[Fe_2O_{10}]$ unit. Each $[Fe_2O_{10}]$ building block shares one corner with two neighboring $[Fe_2O_{10}]$ units to form a framework of tunnel structures. β-FeOOH has been considered as battery electrode for allowing the intercalation of alkali (owing to the unique tunnel structure) and the above-mentioned conversion reaction between Fe^3 and Fe (owing to weaker ionic bonding between Fe^{3+} and OH), while the actual mechanism by which this process occurs is still a matter of speculation.

Zhu et al. reported an environmentally benign process employing the iron rust scraped from corrosives as initiating materials [103]. The collected iron rust was

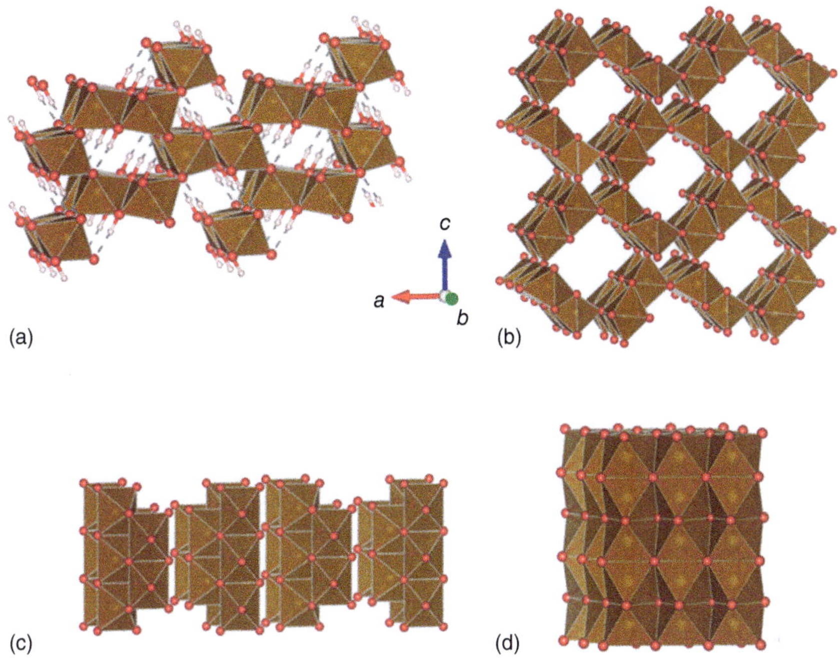

Figure 7.18 Crystalline structures of different polymorphs of FeOOH such as (a) α- , (b) β-, (c) γ-, and (d) δ-FeOOH.

ground and dissolved in nitric acid (HNO_3), followed by conventional hydrothermal treatment, to finally form sphere-like α-Fe_2O_3 nanocrystals with a central size of ~30 nm (Figure 7.19a). The as-made Fe_2O_3 materials can be directly used as actives for the anode of alkaline Fe batteries without further electrochemical activation. The α-Fe_2O_3 nanosphere exhibits an outstanding anodic performance with a high capacity (with a maximum value of ~269 mAh g^{-1}), good cyclability (nearly no capacity decay after 500 cycles), and excellent rate capabilities. CV measurement conducted in a three-electrode system in 3 M KOH solution (Figure 7.19b,c) showed multiple redox peaks, claiming the realization of full valence state transformation of elemental Fe ($Fe^{3+} \leftrightarrow Fe^{2+}$) within a potential window between −1.45 and −0.5 V (vs. Ag/AgCl), equivalent to −1.25 and −0.30 V (vs. SHE). On charging, a strong reduction peak at ~−1.14 V can be attributed to an electrochemical conversion reaction from Fe^{3+} to Fe^{2+}, while an oxidation peak at −0.78 V can be attributed to its reverse reaction. However, it is highly possible that these two redox features resulted from $Fe_3O_4 \leftrightarrow Fe(OH)_2$.

7.4 Conclusion

A variety of strategies to improve the electrochemical performance of transition metal oxides in aqueous EES have been discussed, with a primary focus on the commonly seen electrode materials such as Mn-, Ni-, V-, and Fe-based oxides. Moreover,

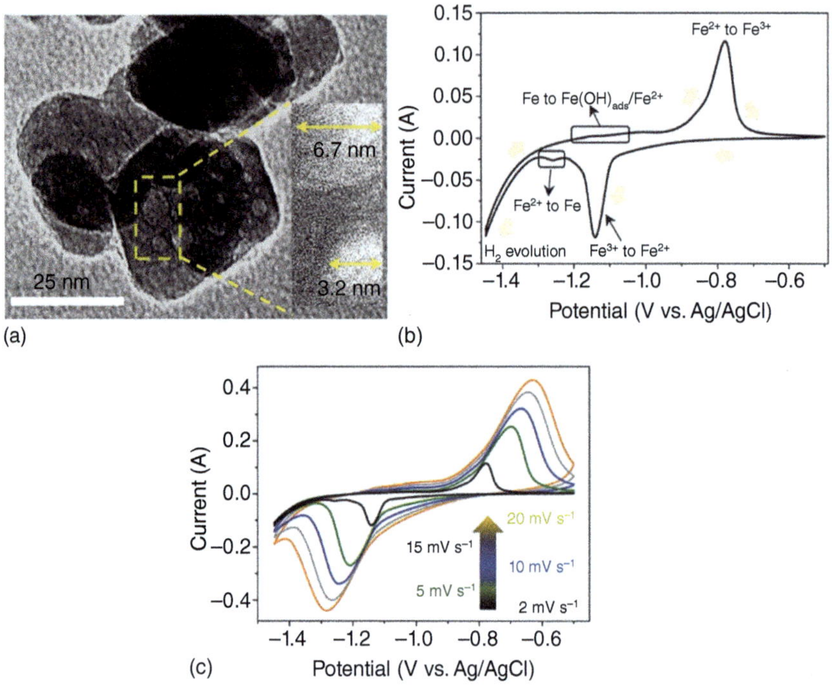

Figure 7.19 (a) TEM image of α-Fe_2O_3 nanosphere electrode; (b) CV scan of α-Fe_2O_3 nanosphere electrode in 6 M KOH at a current rate of 2 mV s^{-1}; (c) CV plots at various sweeping rates. (a) Source: (a) Zhu et al. [103]/with permission of American Chemical Society.

the perspectives regarding the future research opportunities and challenges of these metal oxides toward aqueous EES have also been provided. It is mentioned that, besides the materials mentioned in this work, other research on many transition metal oxides could also function as electrode materials for aqueous EES, which is beyond the scope of this work.

Acknowledgments

This work was supported by the US Department of Energy (DOE), Office of Science, Basic Energy Sciences under Award DE-SC0018922.

References

1 Divya, K.C. and Ostergaard, J. (2009). Battery energy storage technology for power systems: an overview. *Electric Power Systems Research* 79 (4): 511–520.

2 Dunn, B., Kamath, H., and Tarascon, J.M. (2011). Electrical energy storage for the grid: a battery of choices. *Science* 334 (6058): 928–935.

3 Das, H.S., Rahman, M.M., Li, S., and Tan, C.W. (2020). Electric vehicles standards, charging infrastructure, and impact on grid integration: a technological review. *Renewable and Sustainable Energy Reviews* 120: 109618.

4 Yong, J.Y., Ramachandaramurthy, V.K., Tan, K.M., and Mithulananthan, N. (2015). A review on the state-of-the-art technologies of electric vehicle, its impacts and prospects. *Renewable and Sustainable Energy Reviews* 49: 365–385.

5 Larcher, D. and Tarascon, J.M. (2015). Towards greener and more sustainable batteries for electrical energy storage. *Nature Chemistry* 7 (1): 19–29.

6 Maddukuri, S., Malka, D., Chae, M.S. et al. (2020). On the challenge of large energy storage by electrochemical devices. *Electrochimica Acta* 354: 136771.

7 Goodenough, J.B. and Kim, Y. (2010). Challenges for rechargeable Li batteries. *Chemistry of Materials* 22 (3): 587–603.

8 Liu, J., Zhang, J.-G., Yang, Z. et al. (2013). Materials science and materials chemistry for large scale electrochemical energy storage: from transportation to electrical grid. *Advanced Functional Materials* 23 (8): 929–946.

9 Jiang, L., Lu, Y., Zhao, C. et al. (2019). Building aqueous K-ion batteries for energy storage. *Nature Energy* 4 (6): 495–503.

10 Wang, Y., Yi, J., and Xia, Y. (2012). Recent progress in aqueous lithium-ion batteries. *Advanced Energy Materials* 2 (7): 830–840.

11 Kim, H., Hong, J., Park, K.Y. et al. (2014). Aqueous rechargeable Li and Na ion batteries. *Chemical Reviews* 114 (23): 11788–11827.

12 Chao, D., Zhou, W., Xie, F. et al. (2020). Roadmap for advanced aqueous batteries: from design of materials to applications. *Science Advances* 6 (21): 4098.

13 Xu, K. and Wang, C.S. (2016). Widening voltage windows. *Nature Energy* 1: 16161.

14 Cook, J.B., Kim, H.-S., Lin, T.C. et al. (2017). Pseudocapacitive charge storage in thick composite MoS_2 nanocrystal-based electrodes. *Advanced Energy Materials* 7 (2): 1601283.

15 Peljo, P. and Girault, H.H. (2018). Electrochemical potential window of battery electrolytes: the HOMO–LUMO misconception. *Energy & Environmental Science* 11 (9): 2306–2309.

16 Suo, L., Borodin, O., Gao, T. et al. (2015). "Water-in-salt" electrolyte enables high-voltage aqueous lithium-ion chemistries. *Science* 350: 938–943.

17 Suo, L.M., Borodin, O., Sun, W. et al. (2016). Advanced high-voltage aqueous lithium-ion battery enabled by "water-in-bisalt" electrolyte. *Angewandte Chemie International Edition* 55 (25): 7136–7141.

18 Suo, L.M., Borodin, O., Wang, Y.S. et al. (2017). "Water-in-salt" electrolyte makes aqueous sodium-ion battery safe, green, and long-lasting. *Advanced Energy Materials* 7 (21): 1701189.

19 Shan, X.Q., Charles, D.S., Lei, Y.K. et al. (2016). Bivalence Mn_5O_8 with hydroxylated interphase for high-voltage aqueous sodium-ion storage. *Nature Communications* 7: 13379.

20 Augustyn, V., Simon, P., and Dunn, B. (2014). Pseudocapacitive oxide materials for high-rate electrochemical energy storage. *Energy & Environmental Science* 7 (5): 1597–1614.

21 Hadzijordanov, S., Angersteinkozlowska, H., Vukovic, M., and Conway, B.E. (1978). Reversibility and growth-behavior of surface oxide-films at ruthenium electrodes. *Journal of the Electrochemical Society* 125 (9): 1471–1480.

22 Toupin, M., Brousse, T., and Bélanger, D. (2004). Charge storage mechanism of MnO_2 electrode used in aqueous electrochemical capacitor. *Chemistry of Materials* 16 (16): 3184–3190.

23 Liu, T.C., Pell, W.G., Conway, B.E., and Roberson, S.L. (1998). Behavior of molybdenum nitrides as materials for electrochemical capacitors – comparison with ruthenium oxide. *Journal of the Electrochemical Society* 145 (6): 1882–1888.

24 Jaffrezic-Renault, N. and Dzyadevych, S.V. (2008). Conductometric microbiosensors for environmental monitoring. *Sensors* 8 (4): 2569–2588.

25 Marcus, Y. (1991). Thermodynamics of solvation of ions. Part 5. – Gibbs free energy of hydration at 298.15 K. *Journal of the Chemical Society, Faraday Transactions* 87 (18): 2995–2999.

26 Volkov, A.G., Paula, S., and Deamer, D.W. (1997). Two mechanisms of permeation of small neutral molecules and hydrated ions across phospholipid bilayers. *Bioelectrochemistry and Bioenergetics* 42 (2): 153–160.

27 Sugimoto, W., Iwata, H., Yokoshima, K. et al. (2005). Proton and electron conductivity in hydrous ruthenium oxides evaluated by electrochemical impedance spectroscopy: the origin of large capacitance. *Journal of Physical Chemistry B* 109 (15): 7330–7338.

28 Jo, C., Hwang, I., Lee, J. et al. (2011). Investigation of pseudocapacitive charge-storage behavior in highly conductive ordered mesoporous tungsten oxide electrodes. *Journal of Physical Chemistry C* 115 (23): 11880–11886.

29 Mendoza-Sánchez, B., Brousse, T., Ramirez-Castro, C. et al. (2013). An investigation of nanostructured thin film α-MoO_3 based supercapacitor electrodes in an aqueous electrolyte. *Electrochimica Acta* 91: 253–260.

30 Mitchell, J.B., Geise, N.R., Paterson, A.R. et al. (2019). Confined interlayer water promotes structural stability for high-rate electrochemical proton intercalation in tungsten oxide hydrates. *ACS Energy Letters* 4 (12): 2805–2812.

31 Wang, X., Xie, Y., Tang, K. et al. (2018). Redox chemistry of molybdenum trioxide for ultrafast hydrogen-ion storage. *Angewandte Chemie International Edition* 57 (36): 11569–11573.

32 Whittingham, M.S. (2004). Hydrogen motion in oxides: from insulators to bronzes. *Solid State Ionics* 168 (3): 255–263.

33 Wu, X., Hong, J.J., Shin, W. et al. (2019). Diffusion-free grotthuss topochemistry for high-rate and long-life proton batteries. *Nature Energy* 4 (2): 123–130.

34 Lee, H.Y. and Goodenough, J.B. (1999). Supercapacitor behavior with KCl electrolyte. *Journal of Solid State Chemistry* 144 (1): 220–223.

35 Athouël, L., Moser, F., Dugas, R. et al. (2008). Variation of the MnO_2 birnessite structure upon charge/discharge in an electrochemical supercapacitor electrode in aqueous Na_2SO_4 electrolyte. *Journal of Physical Chemistry C* 112 (18): 7270–7277.

36 Yeager, M., Du, W., Si, R. et al. (2012). Highly efficient $K_{0.15}MnO_2$ birnessite nanosheets for stable pseudocapacitive cathodes. *Journal of Physical Chemistry C* 116 (38): 20173–20181.

37 Kuo, S.-L. and Wu, N.-L. (2006). Investigation of pseudocapacitive charge-storage reaction of $MnO_2{\cdot}nH_2O$ supercapacitors in aqueous electrolytes. *Journal of the Electrochemical Society* 153 (7): A1317.

38 Chen, D., Ding, D., Li, X. et al. (2015). Probing the charge storage mechanism of a pseudocapacitive MnO_2 electrode using *in operando* Raman spectroscopy. *Chemistry of Materials* 27 (19): 6608–6619.

39 Augustyn, V., Come, J., Lowe, M.A. et al. (2013). High-rate electrochemical energy storage through Li^+ intercalation pseudocapacitance. *Nature Materials* 12 (6): 518–522.

40 Xia, C., Lin, Z., Zhou, Y. et al. (2018). Large intercalation pseudocapacitance in 2D VO_2(B): breaking through the kinetic barrier. *Advanced Materials* 30 (40): 1803594.

41 Chao, D. and Fan, H.J. (2019). Intercalation pseudocapacitive behavior powers aqueous batteries. *Chem* 5 (6): 1359–1361.

42 Dong, S., Shin, W., Jiang, H. et al. (2019). Ultra-fast NH_4^+ storage: strong H bonding between NH_4^+ and bi-layered V_2O_5. *Chem* 5 (6): 1537–1551.

43 Zhang, Q., Levi, M.D., Dou, Q. et al. (2019). The charge storage mechanisms of 2D cation-intercalated manganese oxide in different electrolytes. *Advanced Energy Materials* 9 (3): 1802707.

44 Shan, X., Guo, F., Charles, D.S. et al. (2019). Structural water and disordered structure promote aqueous sodium-ion energy storage in sodium-birnessite. *Nature Communications* 10 (1): 4975.

45 Jeong, H.M., Choi, K.M., Cheng, T. et al. (2015). Rescaling of metal oxide nanocrystals for energy storage having high capacitance and energy density with robust cycle life. *Proceedings of the National Academy of Sciences of the United States of America* 112 (26): 7914–7919.

46 Lai, H., Wu, Q., Zhao, J. et al. (2016). Mesostructured NiO/Ni composites for high-performance electrochemical energy storage. *Energy & Environmental Science* 9 (6): 2053–2060.

47 Lee, H.J., Lee, J.H., Chung, S.-Y., and Choi, J.W. (2016). Enhanced pseudocapacitance in multicomponent transition-metal oxides by local distortion of oxygen octahedra. *Angewandte Chemie International Edition* 55 (12): 3958–3962.

48 Ock, I.W., Choi, J.W., Jeong, H.M., and Kang, J.K. (2018). Synthesis of pseudocapacitive polymer chain anode and subnanoscale metal oxide cathode for aqueous hybrid capacitors enabling high energy and power densities along with long cycle life. *Advanced Energy Materials* 8 (10): 1702895.

49 Ren, X., Guo, C., Xu, L. et al. (2015). Facile synthesis of hierarchical mesoporous honeycomb-like NiO for aqueous asymmetric supercapacitors. *ACS Applied Materials & Interfaces* 7 (36): 19930–19940.

50 Singh, A.K., Sarkar, D., Khan, G.G., and Mandal, K. (2014). Hydrogenated NiO nanoblock architecture for high performance pseudocapacitor. *ACS Applied Materials & Interfaces* 6 (7): 4684–4692.

51 Zhang, M., Huang, Z., Shen, Z. et al. (2017). High-performance aqueous rechargeable Li–Ni battery based on $Ni(OH)_2$/NiOOH redox couple with high voltage. *Advanced Energy Materials* 7 (17): 1700155.

52 Qu, Q., Yang, S., and Feng, X. (2011). 2D sandwich-like sheets of iron oxide grown on graphene as high energy anode material for supercapacitors. *Advanced Materials* 23 (46): 5574–5580.

53 Cheng, F., Shen, J., Ji, W. et al. (2009). Selective synthesis of manganese oxide nanostructures for electrocatalytic oxygen reduction. *ACS Applied Materials & Interfaces* 1 (2): 460–466.

54 An, L., Huang, B., Zhang, Y. et al. (2019). Interfacial defect engineering for improved portable zinc–air batteries with a broad working temperature. *Angewandte Chemie International Edition* 58 (28): 9459–9463.

55 Wei, W., Cui, X., Chen, W., and Ivey, D.G. (2011). Manganese oxide-based materials as electrochemical supercapacitor electrodes. *Chemical Society Reviews* 40 (3): 1697–1721.

56 Kitchaev, D.A., Dacek, S.T., Sun, W., and Ceder, G. (2017). Thermodynamics of phase selection in MnO_2 framework structures through alkali intercalation and hydration. *Journal of the American Chemical Society* 139 (7): 2672–2681.

57 Tompsett, D.A. and Islam, M.S. (2013). Electrochemistry of hollandite α-MnO_2: Li-ion and Na-ion insertion and Li_2O incorporation. *Chemistry of Materials* 25 (12): 2515–2526.

58 Hunter, J.C. (1981). Preparation of a new crystal form of manganese dioxide: λ-MnO_2. *Journal of Solid State Chemistry* 39 (2): 142–147.

59 Abou-El-Sherbini, K.S., Askar, M.H., and Schöllhorn, R. (2002). Hydrated layered manganese dioxide: Part I. Synthesis and characterization of some hydrated layered manganese dioxides from α-$NaMnO_2$. *Solid State Ionics* 150 (3): 407–415.

60 Devaraj, S. and Munichandraiah, N. (2008). Effect of crystallographic structure of MnO_2 on its electrochemical capacitance properties. *Journal of Physical Chemistry C* 112 (11): 4406–4417.

61 Whitacre, J.F., Tevar, A., and Sharma, S. (2010). $Na_4Mn_9O_{18}$ as a positive electrode material for an aqueous electrolyte sodium-ion energy storage device. *Electrochemistry Communications* 12 (3): 463–466.

62 Hou, Z., Li, X., Liang, J. et al. (2015). An aqueous rechargeable sodium ion battery based on a $NaMnO_2$–$NaTi_2(PO_4)_3$ hybrid system for stationary energy storage. *Journal of Materials Chemistry A* 3 (4): 1400–1404.

63 Wu, Z.-S., Ren, W., Wang, D.-W. et al. (2010). High-energy MnO_2 nanowire/graphene and graphene asymmetric electrochemical capacitors. *ACS Nano* 4 (10): 5835–5842.

64 Lee, J.W., Hall, A.S., Kim, J.-D., and Mallouk, T.E. (2012). A facile and template-free hydrothermal synthesis of Mn_3O_4 nanorods on graphene sheets for supercapacitor electrodes with long cycle stability. *Chemistry of Materials* 24 (6): 1158–1164.

65 Yeager, M.P., Du, W., Wang, Q. et al. (2013). Pseudocapacitive hausmannite nanoparticles with (101) facets: synthesis, characterization, and charge-transfer mechanism. *ChemSusChem* 6 (10): 1983–1992.

66 Chen, S., Zhu, J., Wu, X. et al. (2010). Graphene oxide–MnO_2 nanocomposites for supercapacitors. *ACS Nano* 4 (5): 2822–2830.

67 Hu, Z., Xiao, X., Chen, C. et al. (2015). Al-doped α-MnO_2 for high mass-loading pseudocapacitor with excellent cycling stability. *Nano Energy* 11: 226–234.

68 Gao, P., Metz, P., Hey, T. et al. (2017). The critical role of point defects in improving the specific capacitance of δ-MnO_2 nanosheets. *Nature Communications* 8 (1): 14559.

69 Shan, X., Guo, F., Page, K. et al. (2019). Framework doping of Ni enhances pseudocapacitive Na-ion storage of (Ni)MnO_2 layered birnessite. *Chemistry of Materials* 31 (21): 8774–8786.

70 Yin, B., Zhang, S., Jiang, H. et al. (2015). Phase-controlled synthesis of polymorphic MnO_2 structures for electrochemical energy storage. *Journal of Materials Chemistry A* 3 (10): 5722–5729.

71 Tevar, A.D. and Whitacre, J.F. (2010). Relating synthesis conditions and electrochemical performance for the sodium intercalation compound $Na_4Mn_9O_{18}$ in aqueous electrolyte. *Journal of the Electrochemical Society* 157 (7): A870.

72 Delmas, C., Fouassier, C., and Hagenmuller, P. (1980). Structural classification and properties of the layered oxides. *Physica B+C* 99 (1): 81–85.

73 Kubota, K., Yabuuchi, N., Yoshida, H. et al. (2014). Layered oxides as positive electrode materials for Na-ion batteries. *MRS Bulletin* 39 (5): 416–422.

74 Lu, Z. and Dahn, J.R. (2001). Intercalation of water in P2, T2 and O2 structure $A_z[Co_xNi_{1/3-x}Mn_{2/3}]O_2$. *Chemistry of Materials* 13 (4): 1252–1257.

75 Sun, Y.-K. (2020). Direction for commercialization of O_3-type layered cathodes for sodium-ion batteries. *ACS Energy Letters* 5 (4): 1278–1280.

76 Mu, L., Xu, S., Li, Y. et al. (2015). Prototype sodium-ion batteries using an air-stable and Co/Ni-free O_3-layered metal oxide cathode. *Advanced Materials* 27 (43): 6928–6933.

77 Chen, C., Xu, K., Ji, X. et al. (2017). Promoted electrochemical performance of β-MnO_2 through surface engineering. *ACS Applied Materials & Interfaces* 9 (17): 15176–15181.

78 Gu, J., Fan, X., Liu, X. et al. (2017). Mesoporous manganese oxide with large specific surface area for high-performance asymmetric supercapacitor with enhanced cycling stability. *Chemical Engineering Journal* 324: 35–43.

79 Cai, J., Liu, J., and Suib, S.L. (2002). Preparative parameters and framework dopant effects in the synthesis of layer-structure birnessite by air oxidation. *Chemistry of Materials* 14 (5): 2071–2077.

80 Kang, J., Hirata, A., Kang, L. et al. (2013). Enhanced supercapacitor performance of MnO_2 by atomic doping. *Angewandte Chemie International Edition* 52 (6): 1664–1667.

81 Shan, X., Charles, D.S., Xu, W. et al. (2018). Biphase cobalt–manganese oxide with high capacity and rate performance for aqueous sodium-ion electrochemical energy storage. *Advanced Functional Materials* 28 (3): 1703266.

82 Li, Z., Young, D., Xiang, K. et al. (2013). Towards high power high energy aqueous sodium-ion batteries: the $NaTi_2(PO_4)_3/Na_{0.44}MnO_2$ system. *Advanced Energy Materials* 3 (3): 290–294.

83 Wu, W., Shabhag, S., Chang, J. et al. (2015). Relating electrolyte concentration to performance and stability for $NaTi_2(PO_4)_3/Na_{0.44}MnO_2$ aqueous sodium-ion batteries. *Journal of the Electrochemical Society* 162 (6): A803–A808.

84 Wang, Y., Liu, J., Lee, B. et al. (2015). Ti-substituted tunnel-type $Na_{0.44}MnO_2$ oxide as a negative electrode for aqueous sodium-ion batteries. *Nature Communications* 6 (1): 6401.

85 Wang, Y., Mu, L., Liu, J. et al. (2015). A novel high capacity positive electrode material with tunnel-type structure for aqueous sodium-ion batteries. *Advanced Energy Materials* 5 (22): 1501005.

86 Liu, Y., Qiao, Y., Zhang, W. et al. (2015). Nanostructured alkali cation incorporated δ-MnO_2 cathode materials for aqueous sodium-ion batteries. *Journal of Materials Chemistry A* 3 (15): 7780–7785.

87 Post, J.E. and Veblen, D.R. (1990). Crystal structure determinations of synthetic sodium, magnesium, and potassium birnessite using TEM and the Rietveld method. *American Mineralogist* 75 (5–6): 477–489.

88 Chen, R., Zavalij, P., and Whittingham, M.S. (1996). Hydrothermal synthesis and characterization of $K_xMnO_2{\cdot}yH_2O$. *Chemistry of Materials* 8 (6): 1275–1280.

89 Zhang, X., Ji, L., Zhang, S., and Yang, W. (2007). Synthesis of a novel polyaniline-intercalated layered manganese oxide nanocomposite as electrode material for electrochemical capacitor. *Journal of Power Sources* 173 (2): 1017–1023.

90 Kim, S., Nam, K.W., Lee, S. et al. (2015). Direct observation of an anomalous spinel-to-layered phase transition mediated by crystal water intercalation. *Angewandte Chemie International Edition* 54 (50): 15094–15099.

91 Kim, S., Lee, S., Nam, K.W. et al. (2016). On the mechanism of crystal water insertion during anomalous spinel-to-birnessite phase transition. *Chemistry of Materials* 28 (15): 5488–5494.

92 Zhang, X., Hou, Z., Li, X. et al. (2016). Na-birnessite with high capacity and long cycle life for rechargeable aqueous sodium-ion battery cathode electrodes. *Journal of Materials Chemistry A* 4 (3): 856–860.

93 Nam, K.W., Kim, S., Yang, E. et al. (2015). Critical role of crystal water for a layered cathode material in sodium ion batteries. *Chemistry of Materials* 27 (10): 3721–3725.

94 Mitchell, J.B., Lo, W.C., Genc, A. et al. (2017). Transition from battery to pseudocapacitor behavior via structural water in tungsten oxide. *Chemistry of Materials* 29 (9): 3928–3937.

95 Fleischmann, S., Sun, Y., Osti, N.C. et al. (2020). Interlayer separation in hydrogen titanates enables electrochemical proton intercalation. *Journal of Materials Chemistry A* 8 (1): 412–421.

96 Yeager, M.P., Su, D., Marinković, N.S., and Teng, X. (2012). Pseudocapacitive NiO fine nanoparticles for supercapacitor reactions. *Journal of the Electrochemical Society* 159 (10): A1598–A1603.

97 Nam, K.-W., Yoon, W.-S., and Kim, K.-B. (2002). X-ray absorption spectroscopy studies of nickel oxide thin film electrodes for supercapacitors. *Electrochimica Acta* 47 (19): 3201–3209.

98 Li, W., Dahn, J.R., and Wainwright, D.S. (1994). Rechargeable lithium batteries with aqueous-electrolytes. *Science* 264 (5162): 1115–1118.

99 Yeager, M.P., Du, W.X., Bishop, B. et al. (2013). Storage of potassium ions in layered vanadium pentoxide nanofiber electrodes for aqueous pseudocapacitors. *ChemSusChem* 6 (12): 2231–2235.

100 Charles, D.S., Feygenson, M., Page, K. et al. (2017). Structural water engaged disordered vanadium oxide nanosheets for high capacity aqueous potassium-ion storage. *Nature Communications* 8: 15520.

101 Rajan, A.S., Sampath, S., and Shukla, A.K. (2014). An in situ carbon-grafted alkaline iron electrode for iron-based accumulators. *Energy & Environmental Science* 7 (3): 1110–1116.

102 Manohar, A.K., Yang, C.G., Malkhandi, S. et al. (2013). Enhancing the performance of the rechargeable iron electrode in alkaline batteries with bismuth oxide and iron sulfide additives. *Journal of the Electrochemical Society* 160 (11): A2078–A2084.

103 Zhu, J.H., Li, L.P., Xiong, Z.H. et al. (2017). Evolution of useless iron rust into uniform α-Fe_2O_3 nanospheres: a smart way to make sustainable anodes for hybrid Ni–Fe cell devices. *ACS Sustainable Chemistry & Engineering* 5 (1): 269–276.

8

Nanostructured Transition Metal Oxides for Electrochemical Energy Storage

Simon Fleischmann, Ishita Kamboj, and Veronica Augustyn

North Carolina State University, Department of Materials Science & Engineering, Raleigh, NC 27695, USA

8.1 Fundamental Electrochemistry of Nanostructured TMOs

Transition metal oxides (TMOs) are commonly used in electrochemical energy storage (EES) devices, such as lithium-ion batteries (LIBs), since they can undergo highly reversible redox reactions with ions from the electrolyte. When a typical battery is discharged, electrons flow through the external circuit from the negative electrode (anode, electrode that is being oxidized) to the positive electrode (cathode, electrode that is being reduced), while ions are transported through the electrolyte to compensate the charge. To recharge the battery, the redox reaction is reversed by inducing a current flow in the opposite direction via an external voltage.

In LIBs, the most common cathode material is based on lithium cobalt oxide ($LiCoO_2$), which has been used as a host for reversible Li^+ intercalation since 1980 [1]. While today's ubiquitous LIB anode material is graphite, TMO anodes can offer significant advantages over graphite in terms of power and safety. It is not only the elemental composition of the electrode materials that plays a crucial role in determining the electrochemical properties: the size and shape of the particles are also important. In particular, nanostructuring has emerged as a common strategy to influence the electrochemical properties of TMO electrodes over the past two decades [2–4].

Given the prevalence of TMO electrodes in EES, it is important to have a fundamental understanding of the impacts of nanostructuring on the electrochemical properties. This chapter will describe the effects of TMO nanostructuring on the electrochemical performance and explore the origins from both thermodynamic and kinetic perspectives.

8.1.1 Thermodynamics of Charge Storage in Nanostructured TMOs

To discuss how nanostructuring of TMOs influences the behavior of electrochemical reactions, it is necessary to understand the thermodynamic principles governing

Transition Metal Oxides for Electrochemical Energy Storage, First Edition.
Edited by Jagjit Nanda and Veronica Augustyn.

electrochemical processes. Generally, in an EES device, electrical energy is converted into chemical energy during the charging step and vice versa during discharge. In this sense, an EES device is both an energy conversion and storage device. From a thermodynamic perspective, charging of a LIB means an increase in the chemical potential of lithium, μ_{Li}. This occurs by removing the amount n_{Li} of lithium from the cathode and inserting it into the anode. In this case, the chemical potential can be defined as a function of the change in Gibbs free energy δG according to [5]

$$\mu_{\mathrm{Li}} = \frac{\delta G}{\delta n_{\mathrm{Li}}}. \tag{8.1}$$

According to Maier, the chemical potential of lithium can be described as "a measure of how much a component [lithium] is 'disliked' in a given phase" [5]. This means that lithium is more "disliked" when intercalated into the anode, and according to Eq. (8.1), this leads to an increase in the Gibbs free energy during charging. Consequently, the Gibbs free energy of the lithiation reaction R, $\Delta_R G$, results in a driving force (also called electromotive force), which is the equilibrium cell voltage E of the charged LIB according to

$$E = \frac{-\Delta_R G}{z_R F}, \tag{8.2}$$

where z_R is the number of electrons transferred and F is Faraday's constant [5].

The total chemical potential of a component α, μ_α, can be described as the sum of its chemical potential in the bulk form, μ_α^∞ (infinite particle size), and its excess chemical potential μ_α^{ex}. This excess chemical potential is caused by capillary pressure on the particle due to its surface curvature [6]. Excess chemical potentials are an inverse function of the particle size; they usually become no longer negligible for very small particle radii in the range of a few nanometers, when significant surface curvature is observed. In nanostructured electrode materials, they can have a significant impact on thermodynamic properties like the redox potential. Assuming a particle is in its equilibrium shape, the so-called Wulff form [5, 7], the excess chemical potential of the component α can be described by [8]

$$\mu_\alpha^{\mathrm{ex}} = \nu_\alpha 2\frac{\overline{\gamma}}{\overline{r}}, \tag{8.3}$$

where ν_α is the partial molar volume of component α, $\overline{\gamma}$ the area-averaged surface tension, and $\overline{r}$ the area-averaged distance of the crystal plane j to the particle center (a_j is the area of surface j and a is the total surface), with [8]

$$\overline{\gamma} = \sum_j \frac{a_j \gamma_j}{a}, \tag{8.4}$$

$$\overline{r} = \sum_j \frac{a_j r_j}{a}. \tag{8.5}$$

This leads to an additional term of the Gibbs free energy of reaction as a function of effective particle size $\overline{r}$ (with the stoichiometry coefficient υ_α) [8]:

$$\Delta_R G = \Delta_R G(r \to \infty) + \sum_\alpha 2\upsilon_\alpha \frac{\overline{\gamma_\alpha}}{\overline{r_\alpha}} \nu_\alpha. \tag{8.6}$$

Going back to the specific case of the LIB, the chemical potential of lithium is increased by an excess contribution:

$$\mu_{Li}(\overline{r}) = \mu_{Li}^{\infty}(r \rightarrow \infty) + 2\frac{\overline{\gamma}(\overline{r})}{\overline{r}} v_{Li}. \tag{8.7}$$

The dependence of electrochemical properties like the cell voltage (or from a half-cell point of view: the redox potential) on the particle size has been demonstrated experimentally for a number of model systems, such as a pair of silver electrodes. In an electrochemical cell consisting of two silver metal electrodes, where one electrode consists of nanocrystalline Ag (ca. 50 nm) and the other of annealed bulk Ag, an electromotive force, i.e. a cell voltage, of several millivolts can be measured between the electrodes [9]. This is a direct consequence of the increased total chemical potential of silver in its nanocrystalline form.

In the case of many nanosized TMOs that exhibit a sharp redox potential plateau in their bulk form, the excess chemical potential is responsible for a shift to a more sloping potential curve during electrochemical cycling. This effect is amplified by the dispersion of crystallite sizes and shapes in many nanosized TMO electrodes, which in turn leads to a broad distribution in thermodynamic redox potentials from a macroscopic point of view [5].

These thermodynamic considerations about the chemical potentials of nanosized materials are based on the assumption of ideal crystals, as well as an immobile, frozen-in host lattice. In reality, however, the presence of crystal defects, the deformation of the host lattice during ion (de-)intercalation, the changed atomic bonding environment at the surface compared to the bulk, and the local disordering in the near-surface region have a significant impact on the redox site energies and hence on the observed potential development during electrochemical cycling [5, 10].

Examples for this behavior have been shown in many studies of lithium intercalation behavior of nanosized TMOs operating in non-aqueous electrolytes. Instead of a specific redox potential for the (de-)lithiation of $LiCoO_2$ (Figure 8.1) [10], or V_2O_5 [11], sloping potential developments were reported. This behavior is not limited to typical LIB systems. Boukhalfa et al. reported on the sloping redox potential curve of nanoscopic, amorphous VO_x coated on carbon nanotubes (CNTs) via atomic layer deposition (ALD) in an aqueous 8 M LiCl electrolyte [12], and Fleischmann et al. showed sodium insertion in nanoscopic V_2O_5 [13] from organic 1 M $NaClO_4$ in acetonitrile electrolyte with a sloping potential development.

Thermodynamic considerations into the electrochemistry of nanostructured materials not only apply to the redox potential, but also impact the thermodynamics of phase transformations that often accompany redox reactions in TMO electrodes. In general, the driving force of a phase transformation from α to β is given by the free energy difference between the two phases, $\Delta G_{\alpha \rightarrow \beta}$, by

$$\Delta G_{\alpha \rightarrow \beta} = G_{total,\beta} - G_{total,\alpha}. \tag{8.8}$$

The total free energy of a phase is determined by the sum of its volume (bulk) free energy, G_v, and its surface energy per unit area of the material, γ_s [14]

$$G_{total}\,(\text{J mol}^{-1}) = G_v + \gamma_s\,(\text{N m}^{-1}) \cdot \text{surface area}\,(\text{m}^2\,\text{mol}^{-1}). \tag{8.9}$$

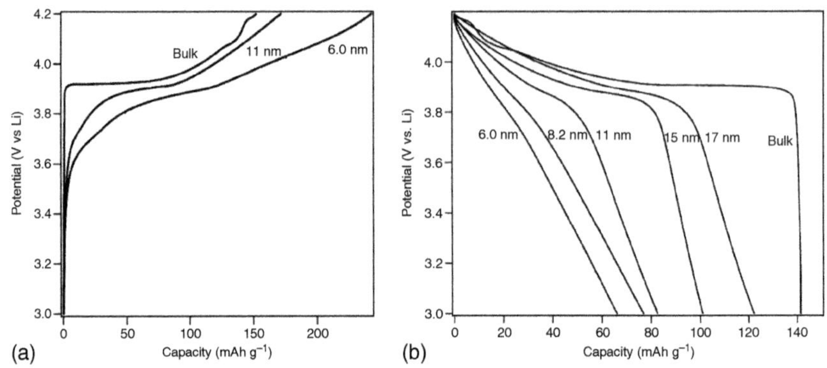

Figure 8.1 Comparison of the voltage profiles of $LiCoO_2$ as a function of crystallite size: (a) first anodic cycle and (b) second cathodic cycle. The oxide was cycled vs. lithium metal in 1 M $LiClO_4$ in ethylene carbonate and diethyl carbonate electrolyte. Source: Okubo et al. [10]/with permission of American Chemical Society.

Clearly, the surface area is a function of the particle size, where smaller particle sizes will lead to an increased specific surface area per mole of material. As a consequence, the total free energy (per mole) of a phase will increase for smaller particle sizes due to the increased surface energy term of the phase, affecting the driving force of the phase transformation according to Eq. (8.8) [14].

The nucleation of a new phase β within the matrix of α is further influenced by a crystallite size-dependent strain energy that is associated with differences in the density between α and β. The surface tension, which was already described as a function of crystallite size in Eq. (8.3), will increase the strain energy for the formation of a phase if it causes a positive volume change, and vice versa [14].

Most TMO electrodes undergo a phase transformation during redox reactions to accommodate the bonding changes induced by the intercalation of cations. The phase transformations that include nucleation and growth of a new phase are – together with solid-state diffusion of ions – the rate-limiting steps in most EES processes. Kim et al. demonstrated that the lithium intercalation kinetics in MoO_2 were significantly improved by nanostructuring because the phase transformation was suppressed in the nanostructured material. *Ex situ* X-ray diffraction (XRD) patterns show that microsized MoO_2 undergoes a transformation from the monoclinic to the orthorhombic phase at around 1.25 V vs. Li/Li^+ (Figure 8.2a), while the monoclinic phase was maintained in nanosized MoO_2 in the same voltage range while exhibiting the same degree of lithiation (Figure 8.2b) [15]. This resulted not only in improved lithiation kinetics, but also led to a decrease in the potential hysteresis during charge/discharge, yielding improved reversibility and energy efficiency.

8.1.2 Kinetics of Charge Storage in Nanostructured TMOs

Among the main motivations for utilizing nanostructured TMO electrodes in EES is the fact that these materials often show improved charge storage kinetics compared to their bulk (micron-scale) equivalent. Compared to larger-sized particles,

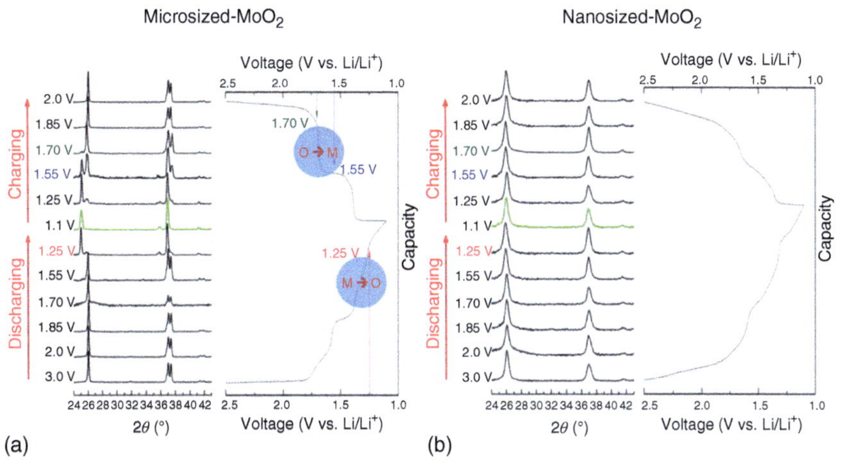

Figure 8.2 *Ex situ* XRD patterns of MoO_2 galvanostatically cycled in 1 M $LiClO_4$ in propylene carbonate (PC) electrolyte at a rate of 21 mA g^{-1}. (a) Microsized-MoO_2 shows a phase transition from monoclinic to orthorhombic at 1.25 V during the cathodic cycle and a return to monoclinic during the anodic cycle at c. 1.85 V. (b) Nanosized-MoO_2 shows no electrochemically induced phase transition during cycling, but a slight shift to lower diffraction angles supports a solid-solution intercalation mechanism. Both materials showed the same specific capacity of around 210 mAh g^{-1} at this rate.
Source: Kim et al. [15]/with permission of IOP Publishing.

nanostructured materials exhibit a larger ratio of surface and near-surface atoms with respect to bulk atoms, and a larger contact area with the electrolyte (electrochemical interface). This leads to shorter transport paths in nanosized particles for both the reacting ionic species (e.g. Li^+) and the electrons, enabling higher cycling rates and/or the use of materials exhibiting low ionic or electronic conductivities [2].

Assuming a TMO electrode exhibits sufficient electronic conductivity and porosity for electrolyte transport, the rate-limiting step for lithium intercalation is usually the solid-state diffusion of lithium ions within the TMO lattice. According to Fick's laws of diffusion, the diffusion length L is related to the flux of lithium ions, j_{Li}, and to the diffusion time t_d by [6]

$$j_{\mathrm{Li}} \propto L^{-1}, \tag{8.10}$$

$$t_d \propto L^2. \tag{8.11}$$

For electrodes consisting of micron-sized TMO particles, at a chosen cycling rate ($\propto t_d^{-1}$) the redox reaction is under semi-infinite diffusion control, when

$$L > \sqrt{Dt_d}, \tag{8.12}$$

where D is the diffusion coefficient. In electrochemical cyclic voltammetry experiments, the Randles–Ševčík equation describes semi-infinite diffusion of an electrolyte species:

$$i = 0.4958n \cdot FAC^* \cdot D^{1/2}\left(\frac{\alpha nF}{RT}\right)^{\frac{1}{2}} \nu^{1/2}, \tag{8.13}$$

where i is the peak current, n is the number of electrons transferred, F is Faraday's constant, A is the surface area, C^* is the maximum concentration of the reduced species in the structure, α is the transfer coefficient, R is the ideal gas constant, and T is the temperature [16]. It has also been applied to understand semi-infinite diffusion of electrochemically intercalated ions in solid-state materials [17].

Sufficient nanostructuring can lead to the redox reaction no longer being diffusion controlled for a chosen cycling rate, when

$$L \ll \sqrt{Dt_d}. \tag{8.14}$$

This situation is also described in some electrochemical literature as the "modified thin-film electrode," and the redox peak current i of such an electrode in a cyclic voltammetry experiment with the sweep rate v is given by

$$i = \frac{nF^2}{4RT} A\Gamma^* v, \tag{8.15}$$

where the amount of redox-active adsorbed species at the surface is Γ [16, 18].

The dependence of the redox peak current on the sweep rate is often used to characterize the extent of diffusion limitation in a given electrode material. It can be used to deconvolute surface-controlled current ($i \propto v$) and semi-infinite diffusion-controlled current ($i \propto \sqrt{v}$) from the total current at a given cycling rate. A detailed description on how to apply this method, often referred to as the k_1, k_2 analysis, can be found in literature [16, 19].

A comparison between the lithium intercalation properties of anatase TiO_2 films consisting of several crystalline sizes (7, 10, and 30 nm) demonstrated the increased diffusion limitations for increased particle sizes. Using the k_1, k_2 analysis, Wang et al. demonstrate that, at a rate of 0.5 mV s^{-1}, the smaller the TiO_2 particle size, the greater the proportion of surface-controlled contributions to the total current (shaded area, Figure 8.3a). Moreover, smaller particles retained more of the initial charge storage capacity at short timescales (Figure 8.3b). While the maximum charge was nearly independent of the particle size, the film with 30 nm particles only exhibited 63% of its initial capacity at a charging time of 300 seconds, while the film with 7 nm particles maintained almost 100% of the initial charge. This demonstrates how diffusion length and diffusion time, as presented in Eqs. (8.12) and (8.14), and here equivalent to particle size and charging time, affect the kinetic limitations of redox reactions in TMO electrodes.

Finally, decreasing particle size led to decreased redox potential hysteresis (Figure 8.3c). A larger peak separation, i.e. an increased overpotential of the lithiation reaction, is unfavorable as it will cause decreased energy efficiency of the energy storage device employing the TiO_2 electrode. It should be noted that decreased particle sizes will not always lead to decreased overpotentials and a higher energy efficiency. As hypothesized by Eftekhari, nanostructuring may lead to an increased interfacial resistance, which in turn could increase the overpotential of the lithiation reaction [20].

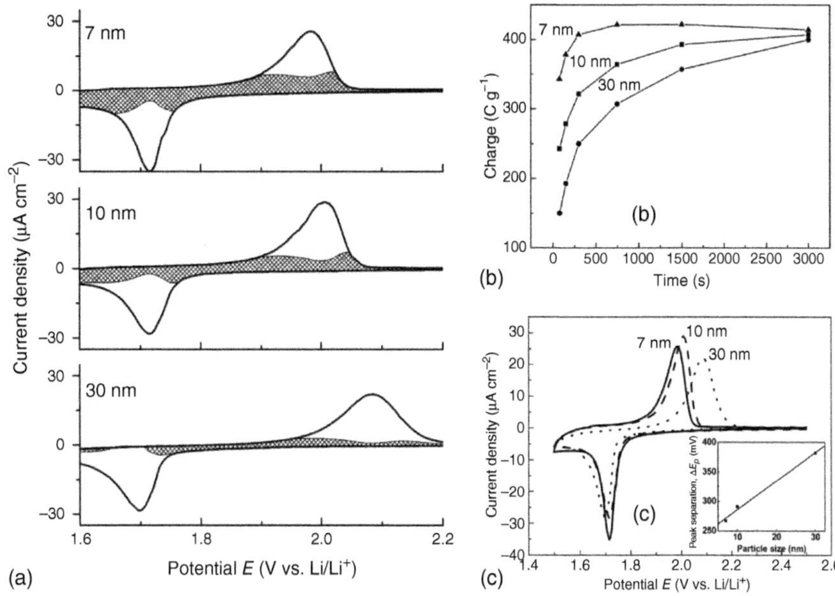

Figure 8.3 Comparison of electrochemical behavior of nanocrystalline anatase TiO_2 films as a function of particle size (7, 10, and 30 nm) cycled vs. lithium metal in 1 M $LiClO_4$ in propylene carbonate: (a) cyclic voltammograms of the TiO_2 films with the surface-controlled current (shaded) deconvoluted via the k_1, k_2 analysis, (b) rate behavior based on calculated charge of the anodic voltammetric cycle, and (c) overlay of voltammetric responses at 0.5 mV s^{-1} with corresponding separation between anodic and cathodic peak in the inset. Source: Wang et al. [19]/with permission of American Chemical Society.

8.2 Emerging Nanostructured TMOs

Efforts to utilize nanostructured TMOs as electrodes in EES devices are mainly driven by the desire to increase their specific power by reducing solid-state diffusion distances and potentially suppress phase transformations during redox. They can roughly be divided into three groups depending on their composition and charge storage mechanism: (i) lithium-containing TMOs that can be used as cathodes in LIBs; (ii) binary TMOs converting into metallic nanoparticles and Li_2O upon reduction that are used as anodes for conversion charge storage; and (iii) binary TMOs maintaining their structure to host guest ions that are used for intercalation charge storage. This chapter describes the most commonly used TMO materials in EES, explains their charge storage mechanisms, and showcases with specific examples how nanostructuring impacts their charge storage properties.

8.2.1 Nanostructured TMO Cathodes for LIBs

Nanostructured lithium-containing TMOs are desirable for LIB cathodes with high rate capability. The three main groups are spinel oxides such as LiM_2O_4, layered oxides of the $LiMO_2$ type, and olivine structures such as $LiMPO_4$ (Figure 8.4),

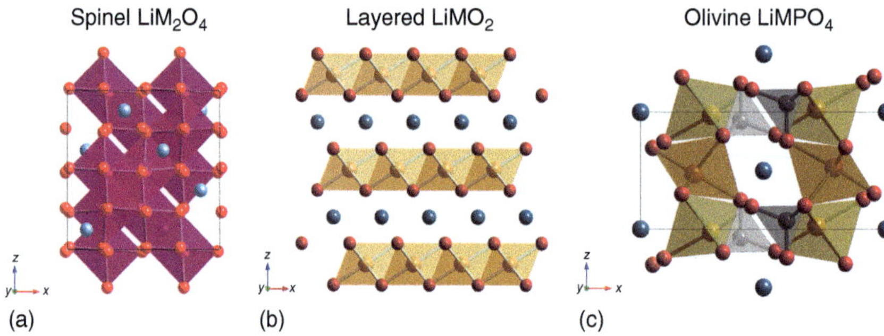

Figure 8.4 Crystal structures of (a) the cubic spinel $LiMn_2O_4$, (b) the layered $LiCoO_2$, and (c) the olivine $LiFePO_4$ viewed along the *b*-axis.

where M is typically a third-row transition metal. The synthesis of nanostructured lithium-containing TMOs from solid-state synthesis is challenging, because high temperatures (>600 °C) are required for crystallization, and this can readily lead to the growth of micron-scale grains.

The lithium spinel $LiMn_2O_4$ has been employed as a cathode in LIBs due to its suitable redox potential around 4 V vs. Li/Li^+ and structural stability over the entire lithiation range ($0 \leq x \leq 1$). Lithium ions are stored in tetrahedral sites and can reversibly (de-)intercalate with maintenance of the cubic spinel phase [21]. Further lithiation ($1 < x \leq 2$) occurs at around 3 V and leads to a Jahn–Teller distortion of the Mn^{3+} and transition to the tetragonal phase above $x \approx 1.2$, which leads to poor cycling stability [22]. For high-power operation, maintaining the spinel structure in the lithiation range between $0 \leq x \leq 1$ is advantageous and intercalation reactions are limited by solid-state diffusion of lithium in the spinel. Nanostructuring $LiMn_2O_4$ is one strategy to increase the kinetics via reduced lithium diffusion distance. The synthesis of nanocrystalline $LiMn_2O_4$ via solid-state synthesis is facilitated by its comparatively low formation temperature around 550 °C, where a high degree of grain growth can be avoided to some extent [23].

Cyclic voltammograms of $LiMn_2O_4$ in organic lithium-containing electrolyte show two distinct redox peaks at around 3.9 and 4.1 V, which correspond to the stepwise oxidation from $LiMn_2O_4$ to $Li_{0.5}Mn_2O_4$ and to Mn_2O_4, respectively, during the anodic cycle (Figure 8.5a) [23]. A comparison of nanoporous $LiMn_2O_4$ with different particle sizes (70, 40, 20, and 10 nm) showed that smaller particle sizes gave rise to a higher rate capability due to decreased lithium diffusion distances, but the specific capacity of nanostructured $LiMn_2O_4$ significantly decreased and was only a fraction of the theoretical capacity of 148 mAh g^{-1} for a one-electron redox process (Figure 8.5b,c) [24].

The lithium intercalation process involves the Mn^{3+}/Mn^{4+} redox couple for $Li_xMn_2O_4$ with $0 \leq x \leq 1$. A transmission electron microscopy study found that the surface tetrahedral sites of the spinel structure contained Mn^{2+} instead of Li^+, forming Mn_3O_4 (Figure 8.5d) [25], and X-ray photoelectron spectroscopy confirmed the presence of up to 40% of Mn^{2+} in nanostructured mesoporous thin films of $LiMn_2O_4$ [23]. The strongly reduced Mn^{2+} is not accessible for redox in the studied

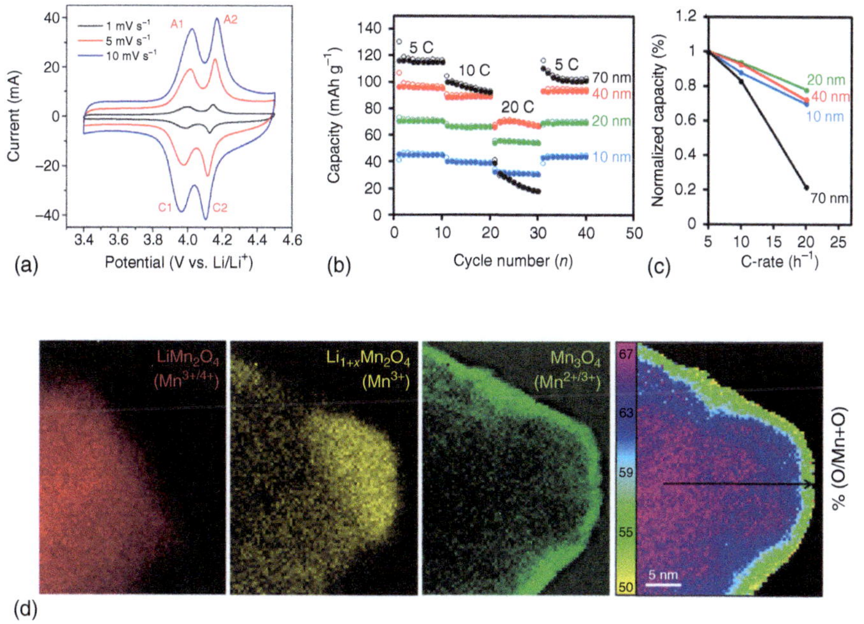

Figure 8.5 (a) Cyclic voltammograms of a nanostructured, mesoporous thin film of $LiMn_2O_4$ at sweep rates of 1, 5, and 10 mV s^{-1}. Source: Lesel et al. [23]/with permission of American Chemical Society. (b) Specific capacity as a function of galvanostatic cycling rate and particle size of nanoporous $LiMn_2O_4$ powders with (c) a rate comparison of normalized capacities. Source: Lesel et al. [24]/with permission of American Chemical Society. (d) High-angle annular dark-field scanning transmission electron micrographs with a mapping of the manganese oxidation state obtained from electron energy loss spectroscopy of the $LiMn_2O_4$ surface region. Source: Amos et al. [25] / with permission of American Chemical Society.

potential range and thus cannot participate in the lithium intercalation reaction, explaining the reduced capacity for nanostructured $LiMn_2O_4$ spinel.

Layered lithium-TMO materials with the structural formula $LiMO_2$ are widely employed as cathodes in LIBs. $LiCoO_2$ was the first layered oxide used in a commercial LIB, and it shows a reversible theoretical capacity of 140 mAh g^{-1} in the range of Li_xCoO_2 with $0.5 \leq x \leq 1$. Deeper delithiation below $x = 0.5$ causes structural and chemical instabilities that lead to poor cyclability. $LiCoO_2$ synthesized via a hydrothermal reaction between CoOOH and LiOH enabled control over the particle size, with high LiOH concentration and short reaction times favoring small nanoparticles of about 6 nm crystallite size [10]. Decreasing the $LiCoO_2$ particle size led to improved rate handling due to shortened diffusion paths but lower specific capacity (see Figure 8.1) that could be linked to significant amounts of non-redox active (within the studied potential range) Co^{2+} in the near-surface region [10].

The toxicity, scarcity, and human rights concerns associated with cobalt motivate the search for suitable replacements. There are many $LiMO_2$ compositions available where cobalt is partially or fully substituted by metals such as Ni, Mn, Fe, or Al. Nickel-rich layered oxides are attractive due to their higher theoretical capacity and lower cost than $LiCoO_2$. $LiNiO_2$ suffers from a number of issues, including thermal instability in the charged (delithiated) state and the migration of Ni^{2+} into lithium sites, which impedes lithium diffusion and decreases the reversible capacity [26]. Substituting nickel with manganese to form $LiNi_{1-y}Mn_yO_2$ results in a layered oxide structure for $y \leq 0.5$, but transforms to a spinel for structures rich in manganese ($y > 0.5$) [27, 28]. In the manganese-poor structure ($y \leq 0.5$), manganese is present in its Mn^{4+} state and does not change valency during cycling. This leads to the formation of Ni^{2+} during reduction and an increased mixing of Ni^{2+} with Li^+ sites over prolonged cycling, which limits the capacity retention, but a significant increase in thermal stability can be achieved [28, 29]. To mitigate these issues, internal nanostructuring of micron-sized particles has been explored, such as gradients of the Ni/Mn/Co concentrations throughout the particle [30], or the addition of internal macrovoids [31]. The sole use of nanoparticles for nickel-rich cathodes without further mitigation strategies, however, can be detrimental due to increased instabilities and surface side reactions.

The olivines based on $LiMPO_4$ (where M is a transition metal, mostly Fe or Mn) exhibit an orthorhombic structure with lithium sites located in chains of edge-sharing LiO_6 octahedra [32, 33]. The most commonly used phase, lithium iron phosphate ($LiFePO_4$, LFP), was discovered in 1997 as a candidate for high-rate LIB cathodes [33]. It has a redox plateau around 3.5 V vs. Li/Li^+ and a theoretical capacity of up to 170 mAh g^{-1}. Lithium diffusion occurs in a highly anisotropic manner along the *b*-axis, and, in *bulk* particles, (de-)lithiation exhibits a two-phase mechanism resulting in a separation of lithium-rich ($LiFePO_4$) and lithium-poor ($FePO_4$) phases with a moving phase separation boundary [33]. It was found that the rate capability of the material could be improved by carbon coating the particles for improved electrical conductivity and by decreasing the particle size [34]. However, the observed rate behavior could not fully be explained by lower diffusion distances in accordance with Fick's laws, challenging the theory that solid-state diffusion of lithium was always the rate-limiting step for lithium (de-)intercalation in LFP. Experimental and theoretical studies found a smaller miscibility gap between the two phases for decreased particle size and that phase separation could be suppressed: Nanoparticular Li_xFePO_4 exhibits a solid-solution lithiation regime (single-phase process instead of two-phase process) for certain ranges of x as a function of the particle size [35–39]. Gibot et al. demonstrated a transition from a flat plateau to a more sloping potential when the LFP particle size was decreased from 140 to 40 nm (Figure 8.6a), while *in situ* XRD showed a continuous shift of the diffraction peaks, indicative of a solid-solution intercalation process for the 40 nm LFP (Figure 8.6b) [39]. It was further demonstrated that the solid-solution regime becomes larger at higher cycling rates, indicative of the formation of metastable Li_xFePO_4 structures, which will separate into thermodynamically stable $LiFePO_4$ and $FePO_4$ upon prolonged relaxation [40]. Observation of the spatial Fe oxidation state distribution within a particle was possible via *operando* transmission X-ray microscopy (TXM). It revealed a transformation from phase-separating to solid-solution intercalation due to the formation of nonuniform solid-solution domains at moderate rates, while uniform solid-solutions formed at high currents [41]. This example illustrates how particle size and current can affect the thermodynamics and kinetics of intercalation reactions in TMO electrodes.

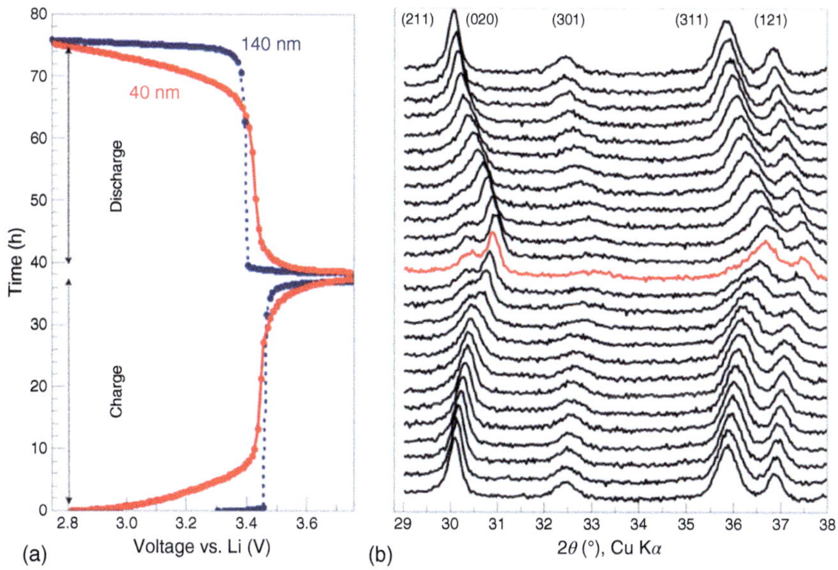

Figure 8.6 Size- and rate-dependence of the (de-)lithiation mechanism in $LiFePO_4$. (a) Galvanostatic charge/discharge for 40 nm (red) and 140 nm (blue) LFP particles, and (b) the *in situ* XRD for the 40 nm particles that indicates a solid-solution intercalation mechanism. The red diffraction pattern aligns closely with $FePO_4$. Source: Gibot et al. [39]/with permission of Springer Nature.

The understanding gained from advanced characterization tools, particularly *in operando* conditions, can help guide materials design for improved EES properties.

8.2.2 Nanostructured Binary TMOs for Conversion-Type Charge Storage

TMOs with the structural formula MO or M_3O_4, where M is a transition metal like Co, Ni, Cu, or Fe, exhibit high capacities at low potentials vs. Li/Li^+, making them possible candidates for LIB anodes. They can undergo reversible conversion reactions with lithium ions that lead to the formation of a metal and Li_2O [42]:

$$M_aO_b + 2nLi^+ + 2ne^- \leftrightarrow aM + bLi_2O$$

The full reduction of the transition metal in the first cycle involves at least two electrons, significantly greater than typical single-electron transfer in intercalation redox reactions [2, 43].

Poizot et al. reported the conversion reaction of CoO with lithium and found a large voltage plateau around 0.8 V vs. Li/Li^+ during the first cathodic cycle that led to the formation of Co nanoparticles and Li_2O. The authors hypothesized that the reversibility of the reaction was due to the high reactivity of Co nanoparticles (1–5 nm crystallite size) with the Li_2O matrix [43]. It is noteworthy that this internal nanostructuring of the electrode material components is not a consequence of the synthesis, but forms *in situ* during the first reduction cycle of the micron-sized TMO particles. The high capacity delivered by conversion-type charge storage reactions, however, comes at the price of a large voltage hysteresis between the reduction and oxidation steps. This is in large part due to the energy penalty of nucleation and interface formation between the nanosized metal and Li_2O crystallites and leads to drastically decreased energy efficiencies compared to intercalation-type charge storage reactions [42].

Among the most popular TMOs with a conversion reaction mechanism is magnetite, Fe_3O_4, which has high theoretical capacity and is low cost [44, 45]. Fe_3O_4 has an inverse spinel crystal structure with Fe in both tetrahedral and octahedral sites (Figure 8.7a). It undergoes sequential lithiation reactions during the first cathodic cycle. First, a rock-salt-phase $Li_xFe_3O_4$ is formed via lithium intercalation, with $0 < x < 2$, followed by the cation-segregated phase $Li_2O{\cdot}FeO$, until the formation of the final products, Li_2O and nanoparticular, metallic Fe [46]. This multi-step lithiation reaction involves up to eight electron transfers per Fe_3O_4 with a theoretical capacity of 926 mAh g^{-1}. The subsequent anodic (delithiation) cycle forms FeO with the rock-salt structure (Figure 8.7b). The special feature of these reaction sequences is that the cubic-close-packed O^{2-} anion network is retained throughout the whole process, while only cations (Li^+, Fe^{2+}, Fe^{3+}) are displaced and reordered, which is hypothesized to be essential for maintaining the high reversibility of the conversion reaction [46].

The use of nanostructured Fe_3O_4 can increase the rate capability of the electrode [44], which is generally hindered by the slow ion diffusion in the densely packed crystal structure. It was found that the fraction of Fe^{3+} in the overall material increases for smaller Fe_3O_4 particles, as demonstrated by X-ray adsorption spectra for a range of crystallite sizes (Figure 8.7c). Consequently, a higher discharge capacity can be obtained in the cathodic cycle [47]. While this effect was demonstrated as favorable for the intercalation reaction ($Li_xFe_3O_4$), it was unclear whether a higher fraction of Fe^{3+} will also increase the reversible capacity of the conversion reaction.

The use of advanced characterization tools like *ex situ* X-ray absorption spectroscopy (XAS) and scanning transmission electron microscopy (STEM), as well as TXM, gave important insights into the effect of nanostructuring on the electrochemical performance of Fe_3O_4 conversion electrodes in lithium cells. Electrodes containing smaller nanocrystallites (8 nm) discharge more evenly and show a more homogeneous distribution of FeO in the particle agglomerates, whereas larger particles (28 nm) mainly formed FeO at the agglomerate edges (Figure 8.8) [48]. Moreover, due to the increased diffusion length in larger particles, inhomogeneous lithiation leads to regions with both metallic Fe and intercalated $Li_xFe_3O_4$. The inhomogeneity led to fractured and misaligned FeO nanograins in the subsequent anodic cycle [49].

Another obstacle for maintaining the high reversibility of the conversion reaction is the formation of aggregates and agglomerates of nanoparticles during cycling, which impede ion transport to the active particles [50]. The use of an appropriate electrode architecture that embeds the active particles in a 3D network can mitigate these issues and improve the charge storage performance [51]. This topic will be further explored in Section 8.3.

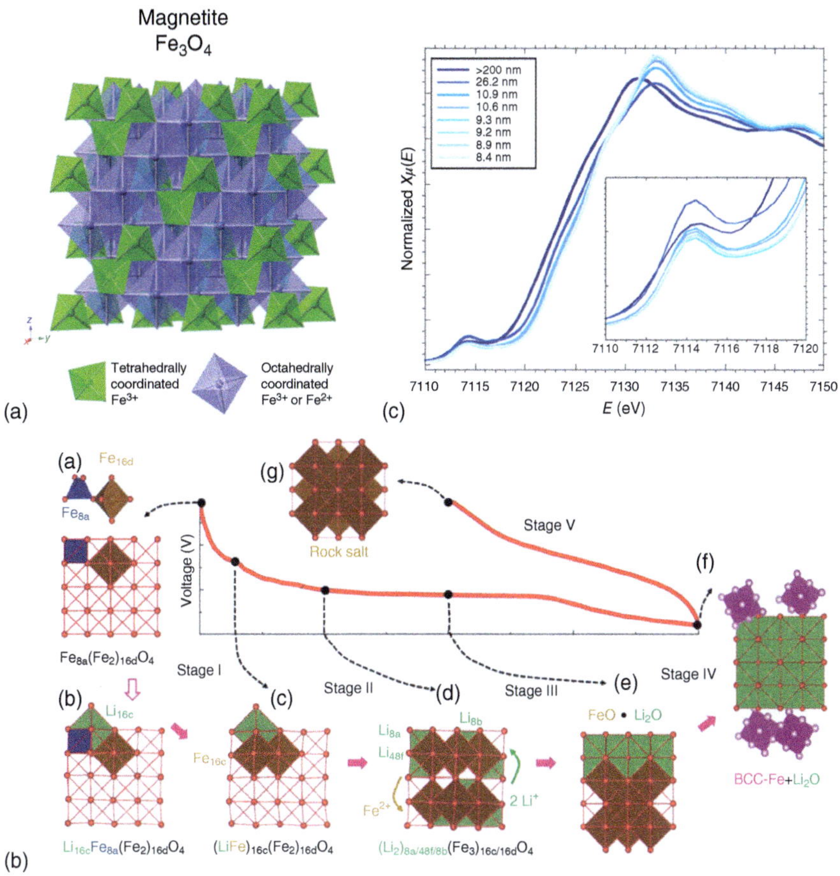

Figure 8.7 (a) Inverse spinel cubic crystal structure of magnetite Fe_3O_4 with tetrahedrally coordinated Fe^{3+} and octahedrally coordinated Fe^{3+} and Fe^{2+}. Source: Bruck et al. [45]/ with permission of Royal Society of Chemistry. (b) Sequential lithiation reaction including intercalation and conversion mechanism. Source: Zhang et al. [46]/with permission of John Wiley & Sons. (c) XAS shows an increased fraction of Fe^{3+} in Fe_3O_4 with smaller particle sizes. Source: Menard et al. [47]/with permission of Elsevier.

8.2.3 Nanostructured Binary TMOs for Intercalation-Type Charge Storage

The intercalation of lithium into binary TMOs is typically above 1 V vs. Li/Li^+ and therefore within the electrochemical stability window of most organic, carbonate-based electrolytes. In one sense, this is an advantage for the use of nanostructured binary TMOs as anodes in LIBs or electrochemical capacitors for high power and safety. On the other hand, it decreases the energy storage of the device. There are numerous examples of nanostructured binary TMOs undergoing intercalation, and this has been the topic of several reviews [2–4, 52–54].

One of the most studied nanostructured binary TMOs for intercalation charge storage is titanium oxide (TiO_2), since it is low cost, abundant, and exhibits a high theoretical capacity. It has many polymorphs, but anatase [19] and especially

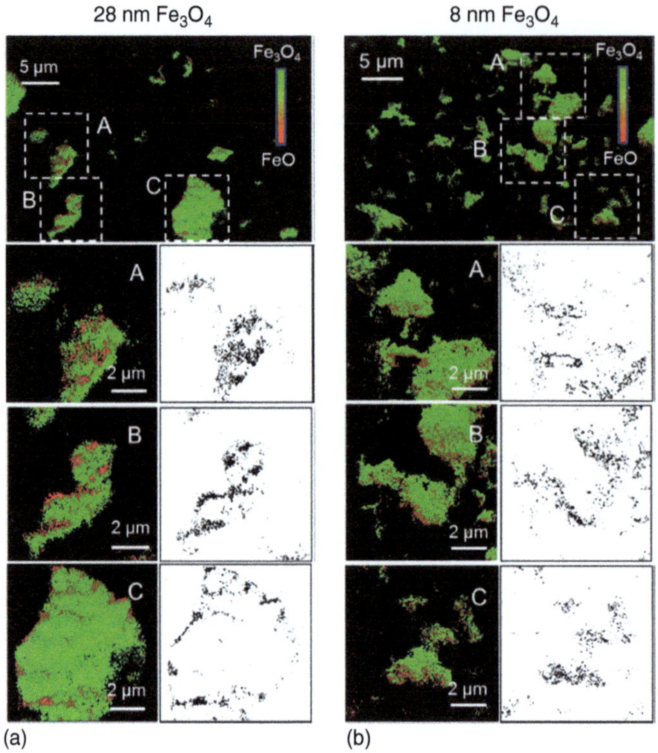

Figure 8.8 Transmission X-ray micrographs with X-ray absorption near-edge structure (TXM-XANES) maps of the distributions of Fe_3O_4 and FeO in lithiated Fe_3O_4 nanoparticles agglomerates of (a) 28 nm and (b) 8 nm primary particle size. The black pixels in the black and white images correspond to reduced FeO. Source: Bock et al. [48] / with permission of American Chemical Society.

TiO_2(B) (Figure 8.9a) [58] have shown the most attractive performance for lithium intercalation. The intercalation of lithium into TiO_2(B) at around 1.6 V has a theoretical capacity of 335 mAh g^{-1} for one-electron transfer per Ti (Li_xTiO_2, $x = 1$), and experimentally, this capacity can only be reached in nanostructured materials [59, 60]. In contrast, bulk TiO_2(B) exhibits a maximum lithiation capacity up to c. $x = 0.7$ [61]. In the cyclic voltammogram, TiO_2(B) nanowires exhibit redox peaks that are broader than the bulk phase (Figure 8.9b), likely caused by capillary effects on the free energy described in Section 8.1.1. The fast kinetics in nanostructured TiO_2(B) are attributed to the preferred 1D lithium diffusion path along the *b*-axis, leading to very short diffusion paths in nanowire, nanotube, or nanosheet structures, where the *b*-axis is radial to the walls or perpendicular to the sheets [55, 60]. A drawback of TiO_2(B) nanostructures is that they exhibit a first cycle irreversibility, which was traced back to side reactions of lithium with surface functionalities (–OH) and surface water. Surface treatment of nanotubes with lithium ethoxide (C_2H_5OLi) to remove such functionalities is one strategy to mitigate the high first cycle loss (Figure 8.9c) [57]. The example of TiO_2(B)

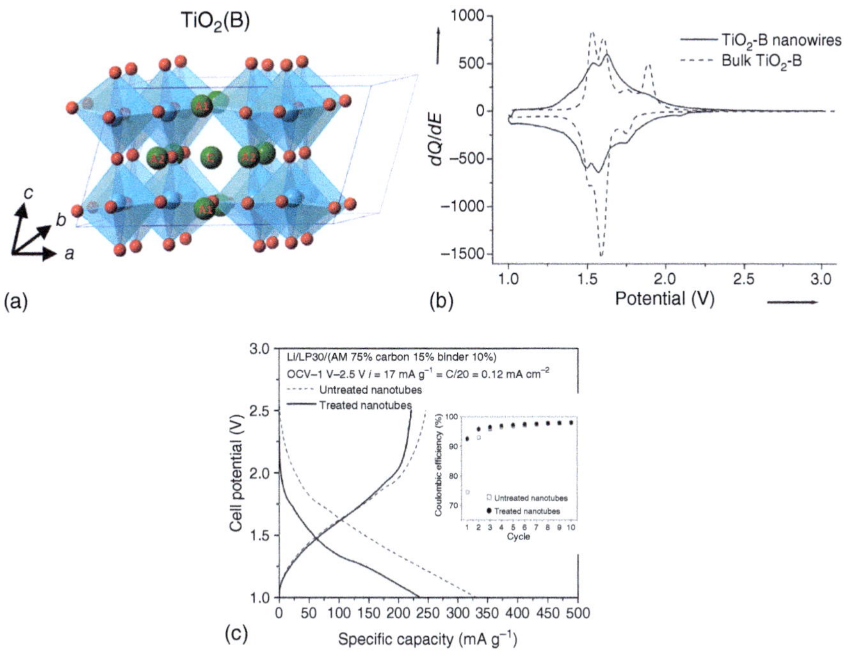

Figure 8.9 Crystal structure of TiO_2(B) showing the tunnel along the *b*-axis and different lithium storage sites (C, A1, A2). Source: Dylla et al. [55]/with permission of American Chemical Society. (b) Incremental capacity plots from galvanostatic cycling comparing the lithiation behavior of TiO_2(B) nanowires and bulk TiO_2(B) at 10 mA g^{-1}. Source: Armstrong et al. [56]/with permission of John Wiley & Sons. (c) First galvanostatic cycle (inset: coulombic efficiency vs. cycle number) of surface-treated (with C_2H_5OLi) and untreated TiO_2(B) nanotubes showing how first cycle losses can be minimized. Source: Brutti et al. [57]/with permission of John Wiley & Sons.

demonstrates how nanostructuring can be a powerful tool to not only increase the rate behavior of materials, but also the maximum specific capacity.

Nanostructured vanadium oxides (mainly V_2O_5, VO_2, and amorphous VO_x) have been developed for a wide range of energy storage applications due to their ability to reversibly intercalate several types of cations from aqueous and organic electrolytes. Compared to its bulk form, nanostructured and amorphous V_2O_5 does not undergo an irreversible phase transition upon lithium intercalation and exhibits a pseudocapacitive, sloping charge–discharge behavior [62, 63]. Due to its poor electrical conductivity, hybridization strategies of vanadia with highly conductive carbon nanoparticles have been developed that were successful at delivering combined high power and energy with good cycling stability. These include deposition of nanosized or amorphous V_2O_5 layers on porous carbons or CNTs via ALD [12, 13, 64] and hydrothermal syntheses where nanocrystalline V_2O_5 was directly grown onto carbon nanomaterials [65, 66]. The nanohybrid materials exhibited pseudocapacitive intercalation behavior of Li^+ or Na^+ from organic or aqueous electrolytes demonstrating the versatility of nanostructured vanadia nanohybrids.

8.3 Implementation of Nanostructured TMOs in Electrode Architectures

While the implementation of nanostructured TMOs as electrode materials in EES devices can offer numerous opportunities concerning specific capacity, power, and even cycling stability, they also face challenges. In summary, there are several main challenges [3, 52]:

(1) Agglomeration of TMO nanoparticles and tendency to segregate from conductive carbon phase.
(2) Difficulty to maintain electrical conductivity in thick electrodes throughout the entire volume [54].
(3) An increased tendency for parasitic side reactions between electrode and electrolyte, especially in the first cycle.
(4) Generally lower density than bulk materials, leading to decreased volumetric capacity [67].

These challenges have necessitated the research and development of advanced electrode architectures to rival the standard slurry architecture commonly used in commercial technologies (such as LIBs). To understand why this shift has occurred, we must first understand the basic architectural requirements of a well-performing electrode. These include [68, 69]:

(1) High electronic conductivity throughout the electrode. Internal resistance must be minimized to optimize electrochemical activity (i.e. must ensure electrons reach all electrochemically active materials to enable electrochemical reactions).
(2) High ionic conductivity throughout the electrode, and within the electrolyte that permeates its pores. Electrodes must have sufficient ion transport pathways such that ions can readily access active material throughout the electrode.
(3) The electrode must have the ability to handle any strains caused by electrochemical cycling without incurring significant structural degradation.

A slurry electrode is made by first preparing a homogeneous mixture of active material, conductive additive, and polymeric binder particles dispersed in a solvent [69]. This mixture is cast onto a flat metal foil current collector, dried, and pressed to maximize density and adhesion. Figure 8.10a illustrates the nanoscale morphology of a model slurry electrode: the active material has maximum surface area exposed to the electrolyte with conductive additive and binder particles homogeneously dispersed around it. In this ideal slurry morphology, all three of the criteria listed above for a well-performing electrode are met. However, this model morphology has proven to be quite difficult to achieve in practice, especially for nanosized active materials. Slurry architectures are less effective for nanosized materials primarily because smaller particles are more susceptible to agglomeration and segregation in dispersions, resulting in inhomogeneous slurries and the clustering of active material particles in the slurry electrode [52, 71, 72]. As a consequence of segregation, components are not well dispersed and instead form a hierarchical structure where the active material forms clusters around which the conductive additive and binder

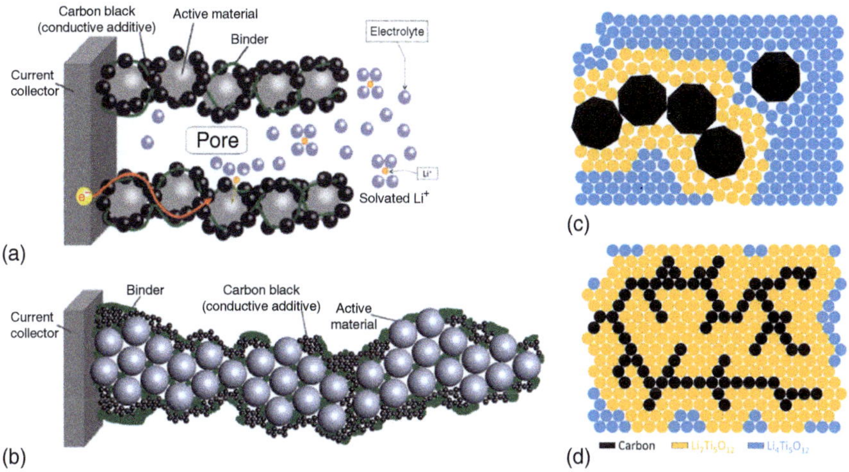

Figure 8.10 Schematics of slurry electrodes comparing (a) a model electrode, with continuous conductive pathways and ample pore space to maximize interfacial area between the active material and electrolyte vs. (b) a realistic electrode with aggregation of components and minimal pore space. Source: Kraytsberg et al. [69]/with permission of John Wiley & Sons. Illustration of improved electrical percolation networks when (c) micron-scale carbon particles are replaced with (d) nano-sized carbon particles for use in slurries with high concentrations of nano-sized TMOs. Source: Widmaier et al. [70]/with permission of Elsevier.

coagulate, as illustrated in Figure 8.10b. The resulting electrode lacks continuous electrical percolation networks and avenues for electrolyte to access active material surfaces, and fails to achieve both high electronic and ionic conductivity. This compromises the achievable rate capability and capacity. Widmaier et al. investigated the electrochemical performance of carbon and nanostructured lithium titanate (LTO) composite electrodes as a function of microstructural composition [70]. At high LTO concentrations, the LTO particles formed aggregates, trapping the micron-sized activated carbon in an insulating matrix and disrupting electrical percolation. The authors concluded that for high LTO concentrations, sub-micron sized carbon particles must be used to create continuous electrical percolation networks throughout a composite electrode as shown in Figure 8.10c,d.

The morphology of realistic slurry electrodes features clusters of densely packed active material particles, as shown in Figure 8.10b. These clusters are often disrupted by void formation and separation, compromising cluster strength [69]. Certain nanostructured metal oxides can be susceptible to these disruptions due to their propensity for volume change during charge/discharge [73]. This volume change is more concerning for TMOs undergoing conversion reactions, but has also been observed for some intercalation materials [2, 74]. Volume change causes particle clusters to split up into aggregates that are weakly connected to nearby aggregates, but strongly interconnected within. On the macroscale, this causes the electrode to be increasingly susceptible to cracking during electrochemical cycling [72].

There have been attempts to adapt existing slurry-making protocols to nanosized materials, primarily through incorporating advanced mixing techniques and tailoring slurry viscosity, but slurry inhomogeneity and resulting poor electrochemical performance have continued to present issues for nanoscale active materials [69]. This has prompted increased interest in exploring advanced architectures for deterministic electrode design.

The following discussion will cover different classes of electrode architectures for use with nanosized TMO active materials. These architectures can broadly be classified using different design variations. In this context, the term "hybrid" describes electrodes in which several components (electrochemically active components, conductive components, structural components, etc.) are synergistically combined with binding on a molecular level to enhance the electronic and/or ionic conductivity, electrochemical reactivity, or cycling stability [54]. These components can range from conductive polymers such as polyaniline (PANI) or poly(3,4-ethylenedioxythiophene) (PEDOT) to carbons such as CNTs or graphene [73, 75]. Independently, nanostructured current collectors can themselves serve as scaffolds for TMO or hybrid films [76]. In many cases, the nanostructuring of the current collector enables a streamlined manufacturing process with a more repeatable output [77]. The subsequent analysis of emerging electrode architectures will be organized according to the geometry of conductive phase in the architecture (i.e. how many dimensions exist for electron conduction). Figure 8.11 presents common examples of the various architectural geometries.

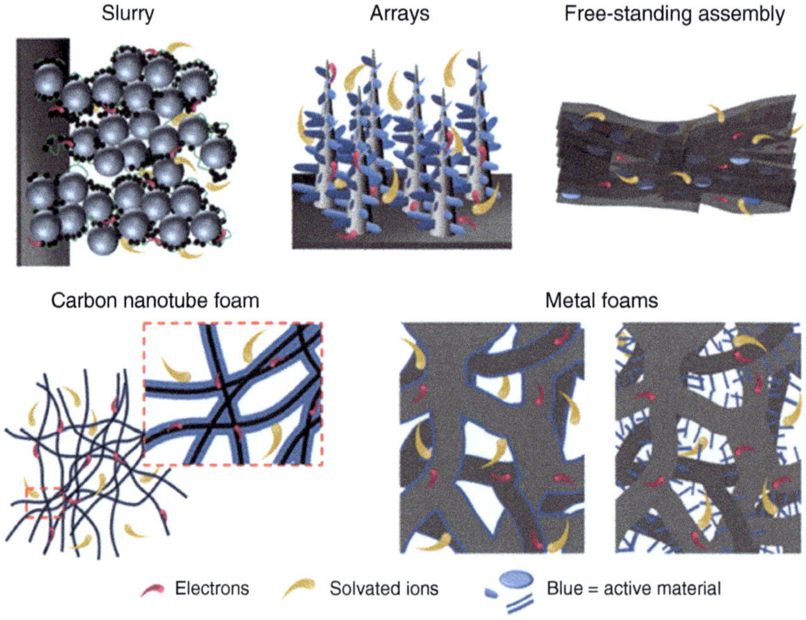

Figure 8.11 Examples of emerging electrode architectures that provide well-defined electron transport pathways. Arrays are an example of one-dimensional architectures, whereas assemblies and foams are three-dimensional architectures. Source: Spencer and Augustyn [68]/with permission of Springer Nature.

8.3.1 One-Dimensional and Two-Dimensional Architectures

8.3.1.1 Nanowires and Nanotubes

One type of electrode architecture uses aligned nanowires (also termed nanorods or nanopillars) and nanotubes either as current collectors themselves, or grown from a current collector to act as its conductive extensions. In either case, there is typically an increase in available current collector surface area for the efficient incorporation of higher mass loadings of active material, as well as increased electronic conductivity throughout the electrode. The space between nanowires facilitates electrolyte access and helps accommodate volume changes that may accompany charge storage [73].

The most employed metals for nanowire current collectors are copper or nickel, fabricated by electrodeposition onto templates or through reduction of metal oxides [68, 73]. Pure metal nanowire arrays have high electronic conductivity and high surface area for deposition of the active material [78]. This has made them attractive substrates for nanostructured LIB electrodes. Taberna et al. developed such an architecture via electrodeposition of nanostructured Fe_3O_4 onto a Cu nanorod scaffold, as shown in Figure 8.12 [44]. The Cu nanorods were grown with a template on a flat current collector. Compared to a planar electrode of the same material, this architecture saw a sixfold improvement in power density, attributed to the electrical conductivity of the copper scaffold.

Though insulative in bulk, TMO nanowires can exhibit high electron mobility if they are single crystals (due to decreased electron scattering from grain boundaries) [73]. This has made them of interest as negative electrodes in hybrid supercapacitors, as well as for high-rate LIB applications.

Li et al. studied a Ni nanotube architecture where large Ni nanotubes (~250 nm in diameter) were grown from a flat substrate via alternating electrodeposition and electrochemical dealloying techniques [79]. Fe_3O_4 nanoparticles were grown on the internal and external nanotube walls, and coated with birnessite-type δ-MnO_2 films creating a conversion-type electrode. The resulting architecture, shown in Figure 8.13, has a morphology that is maintained even after the electrode undergoes electrochemical conversion. The δ-MnO_2 film is hypothesized to hold the Fe_3O_4 nanoparticles intact and prevent large volume changes during cycling, and the hollow structure maximizes the active material/electrolyte interfacial area. Both of these factors were proposed to enable higher capacity by accommodating additional Li^+ ions compared to an architecture of Ni/Fe_3O_4 alone. After 1000 cycles, the $Ni/Fe_3O_4/MnO_2$ architecture appeared to have almost the same morphology as before cycling, with 96% capacity retention.

Electrode architectures with aligned, 1D nanostructured current collectors can exhibit good electrochemical performance, but they may lack mechanical integrity at high active mass loadings [73]. Smaller nanowire or nanotube diameters lead to higher overall surface area for the deposition of active material, but high mass loadings become limited by the strength of smaller nanowires. Conversely, larger nanowires lead to a higher volume fraction of the electrode being the current collector, also limiting high mass loadings [80]. A second obstacle has been the

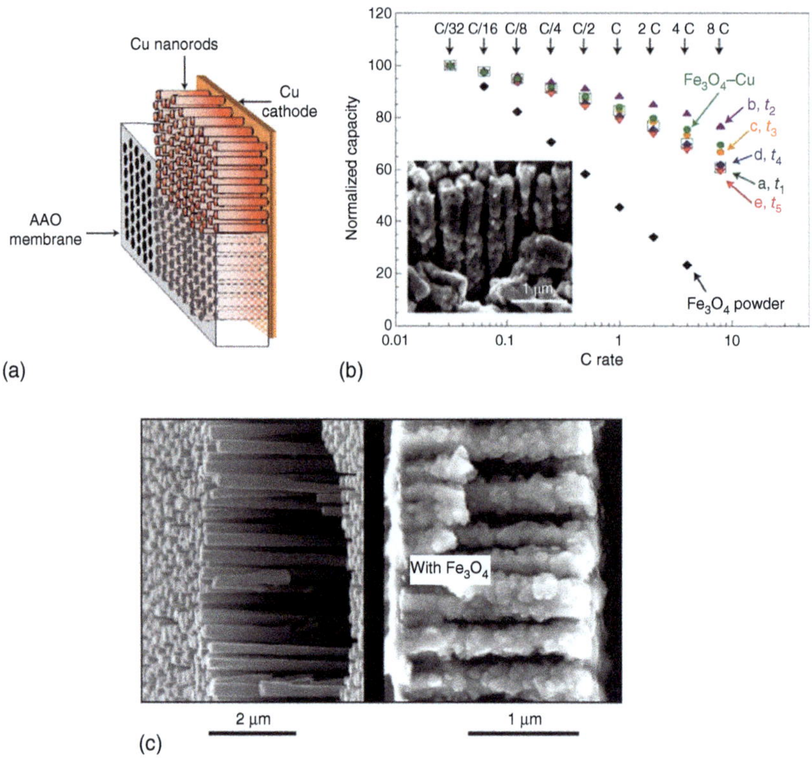

Figure 8.12 (a) Template-assisted growth of copper nanorods on a flat substrate, (b) capacity retention of the Fe_3O_4–Cu electrode compared to an Fe_3O_4 powder electrode, and (c) side view of the Fe_3O_4–Cu architecture before (left) and after (right) Fe_3O_4 deposition. Source: Taberna et al. [44] / with permission of Springer Nature.

lack of control in engineering the length of nanowires, specifically those made via template-assisted electrodeposition techniques [68]. While increasing the length of the wires would provide more surface area, longer aligned nanowires tend to form clusters when released from templates [81]. Optimizing the diameter and length of 1D nanowires for a balance between mechanical integrity and high volumetric capacity has presented an ongoing challenge for this type of architecture.

A prominent example of a hybrid 1D nanostructure is core/shell nanowires. A core/shell architecture is one where the "core" (or center) of the structure is a different material than the "shell" (or outer layer) [82]. These wires are engineered with the goal of reaping the benefits of different materials by combining them in multiple layers [83]. Core/shell structures have been explored with various combinations of metal, metal oxide, and polymer materials. In the case of heterostructures of different TMOs, such as V_2O_5 coated on SnO_2 wires, a single-crystalline nanowire core was proposed to enable high electrical conductivity [84]. In the last decade, research efforts have focused on optimizing the manufacturing processes of these nanowire core/shell structures to gain precise control over structure and morphology. Xia et al. proposed a combination of hydrothermal synthesis and chemical bath deposition to

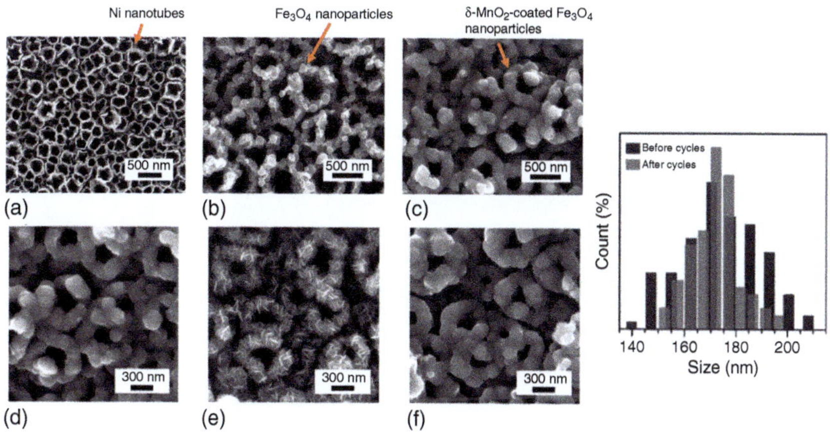

Figure 8.13 Morphology of a nanotube electrode architecture before and after electrochemical cycling: (a) the initial Ni nanotube array, (b) after Fe_3O_4 nanoparticle growth, and (c) after coating δ-MnO_2. (d) Architecture before electrochemical cycling, (e) discharged, and (f) charged. (g) Size distribution of features before and after 1000 electrochemical cycles, indicating minimal size change due to cycling. Source: Li et al. [79] / with permission of John Wiley & Sons.

fabricate two different metal oxide core/shell nanowire arrays for conversion-type electrodes, Co_3O_4/NiO and ZnO/NiO, as shown in Figure 8.14 [85]. The Co_3O_4/NiO core/shell structure exhibited improved electrochemical performance over arrays of the individual components alone. The areal capacity of this core/shell structure was almost double that of the Co_3O_4 nanowire arrays, and over 16 times higher than the NiO nanoflake arrays. The rate capability was also significantly higher for the core/shell structure, which maintained 95.1% of its highest capacity after 6000 cycles compared to 85.5% for Co_3O_4 and 56.7% for NiO. The hierarchical structure provided by the 3D microporous Ni foam current collector also improved the energy conversion efficiency compared to electrodes using flat fluorine-doped tin oxide (FTO) substrates.

The exploration of electrode architectures has thus overwhelmingly evolved in recent years to focus on hierarchical architectures that combine nanostructured TMO materials with 1D, 2D, and 3D electronic conduction pathways. Section 8.3.2 will discuss common 3D substrate choices and architectures.

8.3.2 Three-Dimensional Architectures

8.3.2.1 Assemblies

Nanostructured materials have been used in free-standing assemblies due to advances in synthesis techniques like electrodeposition, hydrothermal synthesis, and ALD that enable their deposition onto conductive scaffolds to form hierarchical structures that exhibit good electronic and ionic conductivity [68]. Carbon materials, such as CNTs and graphene, are the most widely explored for use as scaffolds in hierarchical assemblies due to their high electronic conductivities [12, 86, 87].

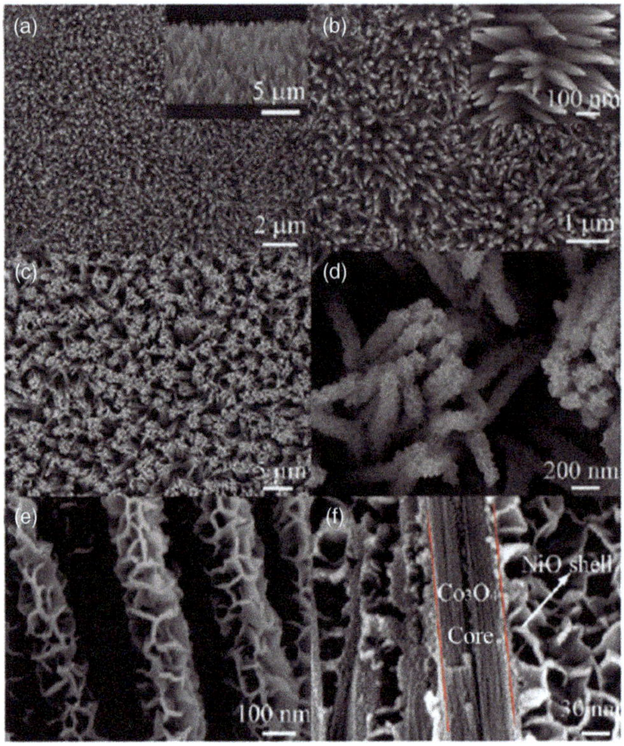

Figure 8.14 Scanning electron microscopy (SEM) images of (a, b) Co_3O_4 nanowires alone and (c, d) Co_3O_4 nanowires with a NiO shell. (e, f) High magnification of the Co_3O_4/NiO heterostructure shows the flaky, porous NiO shell. Source: Xia et al. [85] / with permission of American Chemical Society.

These conductive materials can self-assemble into layered or disordered structures that provide ample surface area for active material loading. However, graphene sheets tend to restack because of strong π–π interactions, leading to a decreased accessible surface area. This can be mitigated by, for example, etching nanopores into the restacked graphene assemblies using hydrogen peroxide [88, 89]. Such a free-standing, 3D holey graphene architecture can serve as a conductive substrate for nanostructured TMOs.

Sun et al. employed the holey graphene architecture as a conductive scaffold to incorporate Nb_2O_5 nanoparticles, creating electrodes with high loading of active material [87]. Nb_2O_5 was deposited onto graphene oxide sheets in solution and then underwent a reduction/crystallization step to form orthorhombic Nb_2O_5 on graphene sheets. These sheets were mixed with holey graphene oxide and then reduced and annealed. During this step, the sheets self-assembled into a structure that is porous at multiple length scales, allowing electrolyte to permeate throughout the volume. The highly interconnected network was proposed to provide channels for ion and electron transport, enabling high areal capacities at high mass loadings (11 mg cm^{-2}), even at high rates (10–50 C).

Figure 8.15 SEM images of V_2O_5 nanoparticles suspended in a graphene aerogel matrix at (a) low and (b) high resolution. Source: Gao et al. [93] / with permission of Elsevier.

Just like the nanowire arrays from Section 8.3.1, these assemblies are examples of how materials with 1D and 2D pathways for electronic conduction (such as CNTs and graphene flakes, respectively) can be combined or grown on a flat or porous substrate to form 3D architectures that serve as scaffolds for TMO deposition.

8.3.2.2 Foams

Metallically conductive foams have been considered ideal scaffolds for nanostructured TMOs to enable high areal capacity energy storage electrodes [68]. This is because they have high surface area for active material, macropores for electrolyte to permeate through the electrode's thickness, and continuous percolation for electronic conduction. Foams also allow for free-standing structures and are amenable to many synthesis approaches that facilitate control over morphology, size, and active mass loading such as electrodeposition and hydrothermal synthesis.

The most common foam materials are metals and carbon. Metal foams such as those of nickel and copper are mechanically robust and available at low costs but typically add a higher inactive mass to the electrode than carbon-based foams [68, 90]. Therefore, carbon foams have been explored to achieve simultaneous high gravimetric energy and power densities. The most explored carbon foams are made from graphene or CNTs. In some studies, CNTs were grown from graphene foam to increase the overall surface area [91].

8.3.2.3 Aerogels

TMOs can be prepared as low-density aerogels with high porosity to facilitate electrolyte diffusion and high surface area, as well as good electronic conductivity [68]. Aerogels have been widely studied for electrode architectures [11, 92]. In one study, a graphene aerogel was used as scaffold for the hydrothermal deposition of V_2O_5 nanoparticles, shown in Figure 8.15 [93]. The nanoparticles adhered to the graphene matrix well and accommodated the volume expansion of V_2O_5 during Li^+ intercalation in non-aqueous electrolyte. Although overall additional research is necessary to create binder-free aerogels that can accommodate high active mass loading, this study suggests that aerogels are suited to provide sufficient electronic and ionic conductivity for simultaneous high energy and power electrodes.

8.4 Conclusions

Due to the contribution of surface sites and short diffusion lengths, nanostructured TMOs can exhibit EES performance that is better than their bulk counterparts. There are examples of the commercialization of nanostructured TMOs, most prominently in the case of $LiFePO_4$ cathodes for LIBs. To fully harness the performance of these materials, future efforts are needed to enable their integration into advanced electrode architectures, including the need for high volumetric capacity, cycling stability, and low cost of synthesis.

References

1 Mizushima, K., Jones, P.C., Wiseman, P.J., and Goodenough, J.B. (1980). Li_xCoO_2 ($0 < x < -1$): a new cathode material for batteries of high energy density. *Materials Research Bulletin* 15 (6): 783–789.

2 Aricò, A.S., Bruce, P., Scrosati, B. et al. (2005). Nanostructured materials for advanced energy conversion and storage devices. *Nature Materials* 4 (5): 366–377.

3 Bruce, P.G., Scrosati, B., and Tarascon, J.M. (2008). Nanomaterials for rechargeable lithium batteries. *Angewandte Chemie International Edition* 47 (16): 2930–2946.

4 Wu, H.B., Chen, J.S., Hng, H.H., and Lou, X.W. (2012). Nanostructured metal oxide-based materials as advanced anodes for lithium-ion batteries. *Nanoscale* 4 (8): 2526–2542.

5 Maier, J. (2013). Thermodynamics of electrochemical lithium storage. *Angewandte Chemie International Edition* 52 (19): 4998–5026.

6 Maier, J. (2007). Size effects on mass transport and storage in lithium batteries. *Journal of Power Sources* 174 (2): 569–574.

7 Landau, L.D. and Lifshitz, E.M. (2013). *Course of Theoretical Physics*. Elsevier.

8 Jamnik, J. and Maier, J. (2003). Nanocrystallinity effects in lithium battery materials: aspects of nano-ionics. Part IV. *Physical Chemistry Chemical Physics* 5 (23): 5215–5220.

9 Schroeder, A., Fleig, J., Drings, H. et al. (2004). Excess free enthalpy of nanocrystalline silver, determined in a solid electrolyte cell. *Solid State Ionics* 173 (1–4 SPEC. ISS.): 95–101.

10 Okubo, M., Hosono, E., Kim, J. et al. (2007). Nanosize effect on high-rate Li-ion intercalation in $LiCoO_2$ electrode. *Journal of the American Chemical Society* 129 (23): 7444–7452.

11 Dong, W., Rolison, D.R., and Dunn, B. (2000). Electrochemical properties of high surface area vanadium oxide aerogels. *Electrochemical and Solid-State Letters* 3 (10): 457–459.

12 Boukhalfa, S., Evanoff, K., and Yushin, G. (2012). Atomic layer deposition of vanadium oxide on carbon nanotubes for high-power supercapacitor electrodes. *Energy & Environmental Science* 5 (5): 6872–6879.

13 Fleischmann, S., Leistenschneider, D., Lemkova, V. et al. (2017). Tailored mesoporous carbon/vanadium pentoxide hybrid electrodes for high power pseudocapacitive lithium and sodium intercalation. *Chemistry of Materials* 29 (20): 8653–8662.

14 Gribb, A.A. and Banfield, J.F. (1997). Particle size effects on transformation kinetics and phase stability in nanocrystalline TiO_2. *American Mineralogist* 82 (7–8): 717–728.

15 Kim, H.-S., Cook, J.B., Tolbert, S.H., and Dunn, B. (2015). The development of pseudocapacitive properties in nanosized-MoO_2. *Journal of the Electrochemical Society* 162 (5): A5083–A5090.

16 Fleischmann, S., Mitchell, J.B., Wang, R. et al. (2020). Pseudocapacitance: from fundamental understanding to high power energy storage materials. *Chemical Reviews* 120 (14): 6738–6782.

17 Lindström, H., Södergren, S., Solbrand, A. et al. (1997). Li^+ ion insertion in TiO_2 (anatase). 2. Voltammetry on nanoporous films. *Journal of Physical Chemistry B* 101 (39): 7717–7722.

18 Bard, A.J. and Faulkner, L.R. (2000). *Electrochemical Methods: Fundamentals and Applications*. Wiley.

19 Wang, J., Polleux, J., Lim, J., and Dunn, B. (2007). Pseudocapacitive contributions to electrochemical energy storage in TiO_2 (anatase) nanoparticles. *Journal of Physical Chemistry C* 111 (40): 14925–14931.

20 Eftekhari, A. (2017). Energy efficiency: a critically important but neglected factor in battery research. *Sustainable Energy & Fuels* 1 (10): 2053–2060.

21 Thackeray, M.M. (1995). Structural considerations of layered and spinel lithiated oxides for lithium ion batteries. *Journal of the Electrochemical Society* 142 (8): 2558.

22 Thackeray, M.M., David, W.I.F.F., Bruce, P.G., and Goodenough, J.B. (1983). Lithium insertion into manganese spinels. *Materials Research Bulletin* 18 (4): 461–472.

23 Lesel, B.K., Ko, J.S., Dunn, B., and Tolbert, S.H. (2016). Mesoporous $Li_xMn_2O_4$ thin film cathodes for lithium-ion pseudocapacitors. *ACS Nano* 10 (8): 7572–7581.

24 Lesel, B.K., Cook, J.B., Yan, Y. et al. (2017). Using nanoscale domain size to control charge storage kinetics in pseudocapacitive nanoporous $LiMn_2O_4$ powders. *ACS Energy Letters* 2 (10): 2293–2298.

25 Amos, C.D., Roldan, M.A., Varela, M. et al. (2016). Revealing the reconstructed surface of $Li[Mn_2]O_4$. *Nano Letters* 16 (5): 2899–2906.

26 Manthiram, A., Knight, J.C., Myung, S.T. et al. (2016). Nickel-rich and lithium-rich layered oxide cathodes: progress and perspectives. *Advanced Energy Materials* 6 (1): 1501010.

27 Rossen, E., Jones, C.D.W., and Dahn, J.R. (1992). Structure and electrochemistry of $Li_xMn_yNi_{1-y}O_2$. *Solid State Ionics* 57 (3–4): 311–318.

28 Sun, Y.K., Lee, D.J., Lee, Y.J. et al. (2013). Cobalt-free nickel rich layered oxide cathodes for lithium-ion batteries. *ACS Applied Materials & Interfaces* 5 (21): 11434–11440.

29 Manthiram, A., Song, B., and Li, W. (2017). A perspective on nickel-rich layered oxide cathodes for lithium-ion batteries. *Energy Storage Materials* 6 (October 2016): 125–139.

30 Sun, Y.K., Chen, Z., Noh, H.J. et al. (2012). Nanostructured high-energy cathode materials for advanced lithium batteries. *Nature Materials* 11 (11): 942–947.

31 Kalluri, S., Cha, H., Kim, J. et al. (2020). Building high-rate nickel-rich cathodes by self-organization of structurally stable macrovoid. *Advancement of Science* 7 (7): 1902844.

32 Zhang, W.J. (2011). Structure and performance of $LiFePO_4$ cathode materials: a review. *Journal of Power Sources* 196 (6): 2962–2970.

33 Padhi, A.K., Nanjundaswamy, K.S., and Goodenough, J.B. (1997). Phospho-olivines as positive-electrode materials for rechargeable lithium batteries. *Journal of the Electrochemical Society* 144 (4): 1188.

34 Striebel, K., Shim, J., Srinivasan, V., and Newman, J. (2005). Comparison of $LiFePO_4$ from different sources. *Journal of the Electrochemical Society* 152 (4): A664–A670.

35 Meethong, N., Huang, H.Y.S., Carter, W.C., and Chiang, Y.M. (2007). Size-dependent lithium miscibility gap in nanoscale $Li_{1-x}FePO_4$. *Electrochemical and Solid-State Letters* 10 (5): 134–138.

36 Meethong, N., Huang, H.Y.S., Speakman, S.A. et al. (2007). Strain accommodation during phase transformations in olivine-based cathodes as a materials selection criterion for high-power rechargeable batteries. *Advanced Functional Materials* 17 (7): 1115–1123.

37 Meethong, N., Kao, Y.H., Tang, M. et al. (2008). Electrochemically induced phase transformation in nanoscale olivines $Li_{1-x}MPO_4$ (M = Fe, Mn). *Chemistry of Materials* 20 (19): 6189–6198.

38 Burch, D. and Bazant, M.Z. (2009). Size-dependent spinodal and miscibility gaps for intercalation in nanoparticles. *Nano Letters* 9 (11): 3795–3800.

39 Gibot, P., Casas-Cabanas, M., Laffont, L. et al. (2008). Room-temperature single-phase Li insertion/extraction in nanoscale $Li_{(x)}FePO_4$. *Nature Materials* 7 (9): 741–747.

40 Zhang, X., Van Hulzen, M., Singh, D.P. et al. (2014). Rate-induced solubility and suppression of the first-order phase transition in olivine $LiFePO_4$. *Nano Letters* 14 (5): 2279–2285.

41 Lim, J., Li, Y., Alsem, D.H. et al. (2016). Origin and hysteresis of lithium compositional spatiodynamics within battery primary particles. *Science* 353 (6299): 566–571.

42 Cabana, J., Monconduit, L., Larcher, D., and Palacín, M.R. (2010). Beyond intercalation-based Li-ion batteries: the state of the art and challenges of electrode materials reacting through conversion reactions. *Advanced Materials* 22 (35): 170–192.

43 Poizot, P., Laruelle, S., Grugeon, S. et al. (2000). Nano-sized transition-metal oxides as negative-electrode materials for lithium-ion batteries. *Nature* 407 (6803): 496–499.

44 Taberna, P.L., Mitra, S., Poizot, P. et al. (2006). High rate capabilities Fe_3O_4-based Cu nano-architectured electrodes for lithium-ion battery applications. *Nature Materials* 5 (7): 567–573.

45 Bruck, A.M., Cama, C.A., Gannett, C.N. et al. (2016). Nanocrystalline iron oxide based electroactive materials in lithium ion batteries: the critical role of crystallite size, morphology, and electrode heterostructure on battery relevant electrochemistry. *Inorganic Chemistry Frontiers* 3 (1): 26–40.

46 Zhang, W., Bock, D.C., Pelliccione, C.J. et al. (2016). Insights into ionic transport and structural changes in magnetite during multiple-electron transfer reactions. *Advanced Energy Materials* 6 (10): 1–11.

47 Menard, M.C., Marschilok, A.C., Takeuchi, K.J., and Takeuchi, E.S. (2013). Variation in the iron oxidation states of magnetite nanocrystals as a function of crystallite size: the impact on electrochemical capacity. *Electrochimica Acta* 94: 320–326.

48 Bock, D.C., Kirshenbaum, K.C., Wang, J. et al. (2015). 2D cross sectional analysis and associated electrochemistry of composite electrodes containing dispersed agglomerates of nanocrystalline magnetite, Fe_3O_4. *ACS Applied Materials & Interfaces* 7 (24): 13457–13466.

49 Bock, D.C., Pelliccione, C.J., Zhang, W. et al. (2017). Size dependent behavior of Fe_3O_4 crystals during electrochemical (de)lithiation: an in situ X-ray diffraction, ex situ X-ray absorption spectroscopy, transmission electron microscopy and theoretical investigation. *Physical Chemistry Chemical Physics* 19 (31): 20867–20880.

50 Knehr, K.W., Brady, N.W., Cama, C.A. et al. (2015). Modeling the mesoscale transport of lithium-magnetite electrodes using insight from discharge and voltage recovery experiments. *Journal of the Electrochemical Society* 162 (14): A2817–A2826.

51 Shi, Y., Zhang, J., Bruck, A.M. et al. (2017). A tunable 3D nanostructured conductive gel framework electrode for high-performance lithium ion batteries. *Advanced Materials* 29 (22): 1–8.

52 Jung, S.-K., Hwang, I., Chang, D. et al. (2020). Nanoscale phenomena in lithium-ion batteries. *Chemical Reviews* 120 (14): 6684–6737.

53 Guo, Y.G., Hu, J.S., and Wan, L.J. (2008). Nanostructured materials for electrochemical energy conversion and storage devices. *Advanced Materials* 20 (15): 2877–2887.

54 Fleischmann, S., Tolosa, A., and Presser, V. (2018). Design of carbon/metal oxide hybrids for electrochemical energy storage. *Chemistry A European Journal* 24 (47): 12143–12153.

55 Dylla, A.G., Henkelman, G., and Stevenson, K.J. (2013). Lithium insertion in nanostructured TiO_2(B) architectures. *Accounts of Chemical Research* 46 (5): 1104–1112.

56 Armstrong, A.R., Armstrong, G., Canales, J. et al. (2005). Lithium-ion intercalation into TiO_2-B nanowires. *Advanced Materials* 17 (7): 862–865.

57 Brutti, S., Gentili, V., Menard, H. et al. (2012). TiO_2-(B) nanotubes as anodes for lithium batteries: origin and mitigation of irreversible capacity. *Advanced Energy Materials* 2 (3): 322–327.

58 Marchand, R., Brohan, L., and Tournoux, M. (1980). TiO_2(B) a new form of titanium dioxide and the potassium octatitanate $K_2Ti_8O_{17}$. *Materials Research Bulletin* 15 (8): 1129–1133.

59 Armstrong, A.R., Armstrong, G., Canales, J., and Bruce, P.G. (2005). TiO_2-B nanowires as negative electrodes for rechargeable lithium batteries. *Journal of Power Sources* 146 (1–2): 501–506.

60 Armstrong, G., Armstrong, A.R., Canales, J., and Bruce, P.G. (2006). TiO_2(B) nanotubes as negative electrodes for rechargeable lithium batteries. *Electrochemical and Solid-State Letters* 9 (3): A139–A143.

61 Zukalová, M., Kalbáč, M., Kavan, L. et al. (2005). Pseudocapacitive lithium storage in TiO_2(B). *Chemistry of Materials* 17 (5): 1248–1255.

62 Chernova, N.A., Roppolo, M., Dillon, A.C., and Whittingham, M.S. (2009). Layered vanadium and molybdenum oxides: batteries and electrochromics. *Journal of Materials Chemistry* 19 (17): 2526–2552.

63 Augustyn, V., Simon, P., and Dunn, B. (2014). Pseudocapacitive oxide materials for high-rate electrochemical energy storage. *Energy & Environmental Science* 7 (5): 1597.

64 Fleischmann, S., Jäckel, N., Zeiger, M. et al. (2016). Enhanced electrochemical energy storage by nanoscopic decoration of endohedral and exohedral carbon with vanadium oxide via atomic layer deposition. *Chemistry of Materials* 28 (8): 2802–2813.

65 Chen, Z., Augustyn, V., Wen, J. et al. (2011). High-performance supercapacitors based on intertwined CNT/V_2O_5 nanowire nanocomposites. *Advanced Materials* 23 (6): 791–795.

66 Fleischmann, S., Zeiger, M., Jäckel, N. et al. (2017). Tuning pseudocapacitive and battery-like lithium intercalation in vanadium dioxide/carbon onion hybrids for asymmetric supercapacitor anodes. *Journal of Materials Chemistry A* 5 (25): 13039–13051.

67 Sun, H., Zhu, J., Baumann, D. et al. (2019). Hierarchical 3D electrodes for electrochemical energy storage. *Nature Reviews Materials* 4 (1): 45–60.

68 Spencer, M.A. and Augustyn, V. (2019). Free-standing transition metal oxide electrode architectures for electrochemical energy storage. *Journal of Materials Science* 54: 13045–13069.

69 Kraytsberg, A. and Ein-Eli, Y. (2016). Conveying advanced Li-ion battery materials into practice the impact of electrode slurry preparation skills. *Advanced Energy Materials* 6 (21): 1–23.

70 Widmaier, M., Jäckel, N., Zeiger, M. et al. (2017). Influence of carbon distribution on the electrochemical performance and stability of lithium titanate based energy storage devices. *Electrochimica Acta* 247: 1006–1018.

71 Nakajima, M. and Yabuuchi, N. (2017). Lithium-excess cation-disordered rocksalt-type oxide with nanoscale phase segregation: $Li_{1.25}Nb_{0.25}V_{0.5}O_2$. *Chemistry of Materials* 29 (16): 6927–6935.

72 Bauer, W. and Nötzel, D. (2014). Rheological properties and stability of NMP based cathode slurries for lithium ion batteries. *Ceramics International* 40 (3): 4591–4598.

73 Jiang, J., Li, Y., Liu, J. et al. (2012). Recent advances in metal oxide-based electrode architecture design for electrochemical energy storage. *Advanced Materials* 24 (38): 5166–5180.

74 Li, H., Wang, Z., Chen, L., and Huang, X. (2009). Research on advanced materials for Li-ion batteries. *Advanced Materials* 21 (45): 4593–4607.

75 Tang, P., Han, L., and Zhang, L. (2014). Facile synthesis of graphite/PEDOT/MnO_2 composites on commercial supercapacitor separator membranes as flexible and high-performance supercapacitor electrodes. *ACS Applied Materials & Interfaces* 6 (13): 10506–10515.

76 Arthur, T.S., Bates, D.J., Cirigliano, N. et al. (2011). Three-dimensional electrodes and battery architectures. *MRS Bulletin* 36 (7): 523–531.

77 Candelaria, S.L., Shao, Y., Zhou, W. et al. (2012). Nanostructured carbon for energy storage and conversion. *Nano Energy* 1 (2): 195–220.

78 Jiang, J., Li, Y., Liu, J., and Huang, X. (2011). Building one-dimensional oxide nanostructure arrays on conductive metal substrates for lithium-ion battery anodes. *Nanoscale* 3 (1): 45–58.

79 Li, Y.-Q., Li, J., Lang, X. et al. (2017). Lithium ion breathable electrodes with 3D hierarchical architecture for ultrastable and high-capacity lithium storage. *Advanced Functional Materials* 27 (29): 1700447.

80 Lukatskaya, M.R., Dunn, B., and Gogotsi, Y. (2016). Multidimensional materials and device architectures for future hybrid energy storage. *Nature Communications* 7: 1–13.

81 Wang, W., Tian, M., Abdulagatov, A. et al. (2012). Three-dimensional Ni/TiO_2 nanowire network for high areal capacity lithium ion microbattery applications. *Nano Letters* 12 (2): 655–660.

82 Wang, J., Zhang, Q., Li, X. et al. (2014). Three-dimensional hierarchical Co_3O_4/CuO nanowire heterostructure arrays on nickel foam for high-performance lithium ion batteries. *Nano Energy* 6: 19–26.

83 Wang, J., Zhang, Q., Li, X. et al. (2015). Smart construction of three-dimensional hierarchical tubular transition metal oxide core/shell heterostructures with high-capacity and long-cycle-life lithium storage. *Nano Energy* 12: 437–446.

84 Yan, J., Sumboja, A., Khoo, E., and Lee, P.S. (2011). V_2O_5 loaded on SnO_2 nanowires for high-rate Li ion batteries. *Advanced Materials* 23 (6): 746–750.

85 Xia, X., Tu, J., Zhang, Y. et al. (2012). High-quality metal oxide core/shell nanowire arrays on conductive substrates for electrochemical energy storage. *ACS Nano* 6 (6): 5531–5538.

86 Fleischmann, S., Zeiger, M., Quade, A. et al. (2018). Atomic layer deposited molybdenum oxide/carbon nanotube hybrid electrodes: influence of crystal structure on lithium-ion capacitor performance. *ACS Applied Materials & Interfaces* 10 (22): 18675–18684.

87 Sun, H., Mei, L., Liang, J. et al. (2017). Three-dimensional holey-graphene/niobia composite architectures for ultrahigh-rate energy storage. *Science (80–.)* 356 (6338): 599–604.

88 Xu, Y., Lin, Z., Zhong, X. et al. (2014). Holey graphene frameworks for highly efficient capacitive energy storage. *Nature Communications* 5: 1–8.

89 Xu, Y., Chen, C.Y., Zhao, Z. et al. (2015). Solution processable holey graphene oxide and its derived macrostructures for high-performance supercapacitors. *Nano Letters* 15 (7): 4605–4610.

90 Chao, D., Xia, X., Liu, J. et al. (2014). A V_2O_5/conductive-polymer core/shell nanobelt array on three-dimensional graphite foam: a high-rate, ultrastable, and freestanding cathode for lithium-ion batteries. *Advanced Materials* 26 (33): 5794–5800.

91 Guan, C., Liu, J., Wang, Y. et al. (2015). Iron oxide-decorated carbon for supercapacitor anodes with ultrahigh energy density and outstanding cycling stability. *ACS Nano* 9 (5): 5198–5207.

92 Long, J.W., Swider, K.E., Merzbacher, C.I., and Rolison, D.R. (1999). Voltammetric characterization of ruthenium oxide-based aerogels and other RuO_2 solids: the nature of capacitance in nanostructured materials. *Langmuir* 15 (3): 780–784.

93 Gao, X.T., Liu, Y.T., Zhu, X.D. et al. (2018). V_2O_5 nanoparticles confined in three-dimensionally organized, porous nitrogen-doped graphene frameworks: flexible and free-standing cathodes for high performance lithium storage. *Carbon* 140: 218–226.

9

Interfaces in Oxide-Based Li Metal Batteries*

Moran Balaish[1], Kun Joong Kim[1], Masaki Wadaguchi[2], Lingping Kong[1], and Jennifer L.M. Rupp[1,3]

[1] *Massachusetts Institute of Technology, Department of Materials Science and Engineering, 77 Massachusetts Ave, Cambridge, MA 02139, USA*
[2] *NGK Spark Plug Co., Ltd, 2808 Iwasaki, Komaki-shi, Aichi 485-8510, Japan*
[3] *Massachusetts Institute of Technology, Department of Electrical Engineering and Computer Science, 77 Massachusetts Ave, Cambridge, MA 02139, USA*

9.1 Introduction

The drive toward a safer alternative to conventional liquid-based batteries has motivated academic and industrial pursuits toward high-energy-density solid-state batteries, and in particular all-solid-state Li metal batteries (ASSLBs), which are believed to be one of the most promising candidates to attain the desired energy densities (>500 Wh kg^{-1}, >700 Wh L^{-1}) [1] and power densities (>10 kW kg^{-1}) [2]. The high-energy density is achieved through the adoption of high-capacity (1000 mA h g^{-1}) [3] and high-voltage (5 V) [4] cathode materials and the optimal anode material, lithium (Li) metal. *Solid electrolytes* are incombustible, non-volatile, non-flammable, and stable at elevated temperatures; they also exhibit for some solid Li-ion conductors higher transference number (t_{Li+}) close to unity when compared to their liquid counterparts, high elastic modulus, and wide electrochemical stability without any leakage problems. In addition, solid electrolytes (i) have a negligible self-discharge due to their low electronic conductivity; (ii) allow high current densities without concentration polarization due to the immobility of the anionic framework, potentially leading to higher power density and energy density (via coupling with thick electrodes); (iii) possess larger thermal conductivity than liquids, thereby mitigating to some extent heat-dissipation issues linked to batteries; (iv) may prevent unwanted electrode "cross-talk" due to immiscibility of the decomposition products in the solid electrolyte; and (v) are less prone to aging mechanisms and are expected to support longer lifetime devices owing to the slower

Moran Balaish and Kun Joong Kim contributed equally to this work.

*This chapter was originally published in Advanced Energy Materials, 2021, Volume 11, Issue 1; DOI: 10.1002/aenm.202002689. Reproduced with permission of WILEY-VCH GmbH.

Transition Metal Oxides for Electrochemical Energy Storage, First Edition.
Edited by Jagjit Nanda and Veronica Augustyn.

reactivity of solids compared to liquids. The incorporation of solid electrolytes in Li-ion batteries (LIBs), either Li^+-conductive polymers (e.g. polyethylene oxide [PEO], polyacrylonitrile [PAN], polymethyl methacrylate [PMMA], polyvinylidene fluoride [PVDF]), or inorganic conductive ceramics (e.g. amorphous lithium phosphorus oxynitride [LiPON], perovskite $Li_xLa_{(2-x)/3}TiO_3$ [LLTO], Na super ionic conductor (NASICON)-type $Li_{1.3}Al_{0.3}Ti_{1.7}(PO_4)_3$ [LATP], lithium super ionic conductor (LISICON)-type $Li_{14}Zn(GeO_4)_4$, garnet-type $Li_7La_3Zr_2O_{12}$ [LLZO], and sulfides [thio-LISICON Li_2S–P_2S_5 types]), began for many of these materials in the 1980s and 1990s, respectively [5, 6]. Replacing the liquid and/or gel electrolyte and separator with an intrinsically non-flammable, non-volatile solid electrolyte shows great promise and is perhaps the ultimate solution toward safer high-energy-density ASSLBs. Moreover, the high density of solid electrolytes compared to liquid electrolytes necessitates the use of lithium metal as the anode material in an ASSLB configuration to increase the volumetric and gravimetric energy density by roughly 70% and 40%, respectively, as opposed to the 10% decrease in gravimetric energy density if the traditional graphite anode were used (Figure 9.1) [7]. On the cathode side, the wider electrochemical stability of solid electrolytes should support an increase in the cell voltage from 4.2 to 5 V (with the adoption of high-voltage cathodes), without significant electrolyte decomposition, resulting in a potential increase of more than 20% in energy density [7].

Nonetheless, commercialization of solid electrolytes, especially for electric propulsion and load-leveling applications, requires the resolution of the following critical issues: (i) *interfaces*: reduction of active surface area for reactions and decrease of charge-transfer resistance at electrode/electrolyte interfaces, (ii) *safety*: overall safety and possibly that with lithium metal, and (iii) *manufacturing costs*:

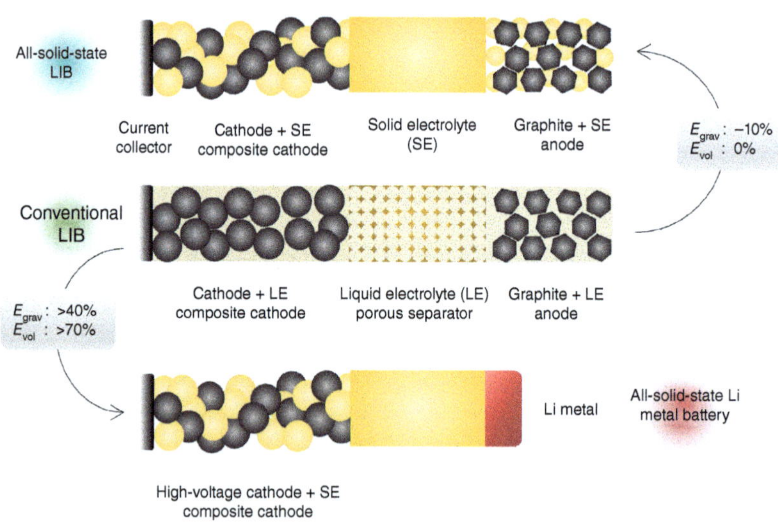

Figure 9.1 Change in energy density upon replacing liquid electrolyte in conventional LIB with solid electrolyte (all-solid-state LIB) and Li metal anode (all-solid-state Li metal batteries). The volumetric and gravimetric energy densities are represented by E_{vol} and E_{grav}, respectively. Source: Janek and Zeier [7]/with permission of Springer Nature.

continuous efforts have been focused on the realization of high-energy-density ASSLBs. Recently, a sulfide-based bulk-type cell with minimalistic configuration consisting of a Li-metal anode, β-Li_3PS_4 solid electrolyte, and $Li(Ni_{0.6}Co_{0.2}Mn_{0.2})O_2$ cathode active material exhibited a total cell resistance (summation of ohmic and polarization resistances) of 350–450 Ω cm^2 at room-temperature cycling at 0.16–0.19C [8]. In addition, the state-of-the-art garnet-based ASSLB with Li_2CO_3-coated LCO, $(Li,C)B_{0.3}O_{0.3}$, and LLZO as a cathode composite, LLZO as the solid electrolyte, and a Li-metal anode exhibited a total interfacial resistance of approximately 270 Ω cm^2 tested at 100 °C under 0.05C, which predominantly originated from the cathode interface [9]. Overall, the total cell resistances for both cases were still much higher than those of commercial LIBs (10–20 Ω cm^2) and low current density (< 0.2C) still to be overcome [2, 10]. Similarly, a recent holistic analysis of state-of-the-art solid-state battery performance suggests that a practical and high-specific-energy ASSLB cell requires internal resistance of less than ~40 Ω cm^2 to allow cycling at 1C with more than 90% energy efficiency [8], implying that further efforts are needed to reduce the cell impedance. Generally, the high impedance originates from both thick electrolytes with lowered conductivity and limited electrochemical reaction at the electrolyte/electrode interfaces. Therefore, reduction of the solid–electrolyte thickness and interfacial resistances are important tasks. As the key redox reactions during battery operation occur at the interfaces, controlling the interface quality is imperative, lending further credence to the significant efforts being placed on engineering cathode and cathode/electrolyte architectures.

This review brings to the forefront the predominant interfacial challenges facing high-energy-density Li-metal solid-state batteries by highlighting the oxide-based promising inorganic solid electrolytes. First, the key properties of promising Li-oxide solid electrolytes are briefly discussed. Next, the origins of one of the paramount issues leading to poor performance of oxide-based solid-state batteries, the interfacial resistance at the electrolyte/anode and electrolyte/cathode interfaces, are addressed and possible strategies to mitigate these issues are presented. Finally, future perspectives, guidelines, and selective interface engineering strategies toward the resolution of these interfacial challenges are outlined. More research details and comparative discussions of oxide- and sulfide-based electrolyte and solid-state batteries can be found in recent review [11].

9.2 Solid Oxide Electrolytes

The realization of solid electrolytes is a critical step toward an intrinsically safe ASSLB, which with improved packing density through polar stacking [12] could, in principle, result in high-energy-density solid-state batteries by coupling with a lithium-metal anode and high-voltage cathode. Key properties of solid electrolytes include high ionic conductivity, which has recently been linked to the critical current density (CCD) for Li dendrite initiation at the Li/electrolyte interface [13], negligible electronic conductivity, wide electrochemical stability window, and chemical compatibility with the cathode (and anode) material. The electrolyte should not participate in the lithium insertion/extraction process and should only act as an ion-conducting solid. Other stringent requirements for battery design and engineering include wide thermal stability, adequate mechanical properties, good electrode/electrolyte adhesion, and capability of being manufactured into ideally thin electrolytes (10–25 μm) with lowered area-specific resistance (ASR). One major prerequisite for solid electrolytes is satisfactory room-temperature ionic conductivity. However, room-temperature conductivity greater than 1 mS cm^{-1} has only been exhibited by a handful of solid lithium-conductors, including oxides such as LLTO, LATP, LLZO, and the recently discovered $LiTa_2PO_8$ [14, 15]. Moreover, among these materials, only LLZO has demonstrated total conductivity greater than 1 mS cm^{-1} with good stability against Li metal, receiving much attention as next-generation electrolyte for ASSLB [16–20]; that of the others is hindered by high resistance at grain boundaries or instability against Li-metal anodes. First-principles calculation [21] has shown that the wide electrochemical window observed in many experiments (0–5 V) is not thermodynamically intrinsic to the solid material but is the result of sluggish kinetics of the decomposition reaction leading to the formation of a passivation layer (interphase), similar to the solid–electrolyte interphase (SEI), which inhibits further decomposition of the bulk material. Currently, only one solid electrolyte, i.e. a LiPON thin-film electrolyte, has exhibited excellent stability for 10 000 cycles when paired with a high-voltage cathode of $LiMn_{1.5}Ni_{0.5}O_4$ and others are still to be proven [4]. Oxide solid electrolytes (e.g. the perovskite $Li_{3.3}La_{0.56}TiO_3$, NASICON-type $LiTi_2(PO_4)_3$, LISICON-type $Li_{14}Zn(GeO_4)_4$, and garnet-type $Li_7La_3Zr_2O_{12}$) exhibit generally better resistance to oxidation than other types of material classes such as sulfides and nitrides. They have higher oxidation potential, with the NASICON-type oxides LATP, $Li_xLa_{2/3-x}TiO_3$, and $Li_{1.5}Al_xGe_{2-x}(PO_4)_3$ (LAGP) being thermodynamically stable up to ~4.2 V [21]. The garnet $Li_7La_3Zr_2O_{12}$ (cubic phase) shows the highest resistance to lithium reduction (theoretical reduction potential of 0.05 V vs. Li^+/Li), showing potential for future solid-state battery applications [12, 21–28]. The oxide electrolytes with outstanding electrochemical stability (e.g. NASICON-type $Li_{1.3}Al_xTi_{2-x}(PO_4)_3$ (LATP) and $Li_{1.5}Al_xGe_{2-x}(PO_4)_3$ (LAGP) with high oxidation stability; garnet-type LLZO) [21] present opportunities for pairing with high-voltage cathode materials [29], including $Li_2NiMn_3O_8$, $Li_2FeMn_3O_8$, and $LiCoMnO_4$, toward a high-energy-density battery beyond traditional graphite-based LIBs. Nonetheless, physiochemical (mechanical, chemical, thermal) stability issues accompanied by interfacial challenges at the cathode/electrolyte and anode/electrolyte interfaces require the implementation of interface engineering strategies, including compositional tuning [30, 31], coating and buffer layers [32–34], polymer interlayers [35, 36], and alloying [37, 38].

9.3 Cathode: Toward True Solid

Till date, interfacial impedance at the cathode side has been the main limiting factor in the overall performance of oxide solid state battery (SSBs) even under a moderate charging/discharging rate. Oxide (garnet)-type solid electrolytes face serious interfacial issues during contact formation and battery operation [39]. Oxide-based cathode composites such as those based on Li-garnets mainly suffer from chemical compatibility issues during cathode preparation. Regardless of the class of solid electrolyte, interfaces between the active material and solid electrolyte often delaminate and suffer from contact loss during battery cycling because of the volume change of the active material during the Li intercalation and deintercalation process [40, 41].

9.3.1 Origin of Interfacial Impedance and Current Pressing Issues at Cathode/Solid Electrolyte Interfaces

The interface between active materials and solid electrolyte can be defined as the bounding surface between and across the two components where a discontinuity in local structure, chemistry, and their properties arises. In general, an interface is the region through which material parameters, such as the concentration of elements/charge carriers, crystal structure, electrical conductivity, defect density, elastic modulus, and thermal expansion coefficient, change from one side to the other. One or more of these changes may be involved at any given interface and can also show a gradient in case of inter-diffusion at the phase boundary of the two constituent materials. The electrochemical performance of a composite cathode clearly depends on the combined property of the active materials, solid electrolyte, and their interfaces. The degree of discontinuity across the interface may be sharp or gradual depending on the interfacial mechanism [42]. An example of a sharp interface may be the result of the wettability that can assist or impede adhesion at the interfaces. However, by definition, the wettability is the ability of a liquid to spread on a solid surface or represents the extent of intimate contact between a liquid and a solid; it does not necessarily mean a strong bond to the surface. In a perspective on ASSLBs, the formation of a solid electrolyte/Li metal interface is an excellent example as one of the simplest methods to assemble them is by melting Li metal on the surface of the solid electrolyte. A high interfacial impedance at the Li/solid electrolyte interface was reported because of the poor wettability of Li metal on LLZO electrolyte [23]. Coating a thin metal or ceramic layer has been found to improve the wettability, lowering the interfacial resistance [43]. Another example is the role of the Li–B–O (Li_3BO_3, LBO) compound during the formation of oxide-based cathode composites for ASSLBs, for which a high-temperature (>1000 °C) sintering process is often needed. Here, the LBO is melted at ~700 °C inside a cathode composite and forms an ionically conducting liquid phase to wet its cathode constituent phases, providing intimate bonding between garnet solid electrolytes and cathode (e.g. LCO) particles, therefore lowering the interfacial resistance. Other types of interfacial joints for cathode composite constituents such as mechanical and physical bonding also create rather sharp interfaces for these. The degree of mechanical bonding is set by the contact area or length at the interfaces. For high mechanical bonding, one material must fill the hills and valleys on the surface of the other material. Surface roughness can contribute to mechanical bond strength depending on the void formation at the interface. The degree of mechanical bonding is reported to affect the interfacial formation. For instance, mechanical bonding may be considered the second important bonding mechanism at the Li/SE interface after wettability, and poor mechanical bonding can also increase interfacial resistance [44].

Another type of interfacial bonding involves atomic, molecule, or electron transport, by diffusion processes, and chemical reactions (e.g. interdiffusion or decomposition), resulting in gradual interfacial bonding. This directly creates new phase in the interfacial zone, affecting the interfacial adhesion and electrochemical properties. The interfacial zone consists of near-surface layers of active materials and solid electrolytes and any layer(s) of material (interphase) existing between these surfaces. Regardless of the thickness of the interfacial zone, any type of interface properties can be largely affected by external conditions such as the processing temperature, difference in chemical potential, and stress related to volume change of the active materials. (Figure 9.2). Thermal treatment for intimate bonding (interfacial neck growth), applied potential during battery operation, and volume change of active materials upon lithiation/de-lithiation may lead to different modes of interfacial failure. Current issues in the formation of composite cathodes for ASSLB require careful attention for the design of electrochemically, chemically, and mechanically stable interfaces within components as well as a large number of active reaction sites to ensure sufficient kinetics for redox reaction. A number of interfacial issues leading to high impedances in composite cathodes have been identified over the last decade. These issues are subdivided into the four major categories below depending on two important external conditions (fabrication ≈ processing temperature, operation ≈ chemical potential difference in Li) (Table 9.1):

A. Chemical reaction during cell fabrication
B. Electrochemical oxidation and chemical reaction during cycle
C. Chemical reaction during cycle
D. Chemo-mechanical degradation during cycle.

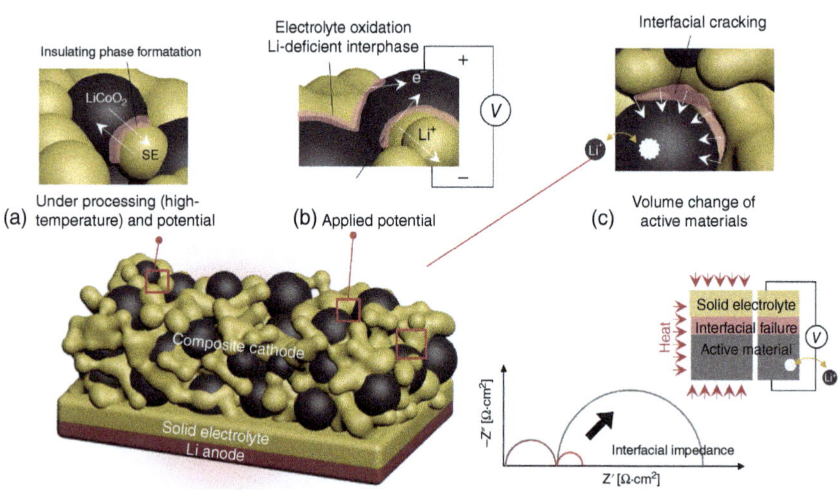

Figure 9.2 Common interfacial failure mechanism in all-solid-state Li metal batteries. Thermal treatment for intimate bonding (interfacial neck growth) (a), applied potential during battery operation (a, b), and volume change of active materials upon lithiation/de-lithiation (c) may lead to different modes of interfacial failures such as insulating phase formation due to interdiffusion (a), oxidation of the solid electrolyte (b), and interfacial cracking (c). Source: (b) Based on Xiao et al. [39], (c) Based on Tsai et al. [40], Koerver et al. [41, 45], Zhang et al. [46].

Table 9.1 List of reported interfacial degradation at garnet electrolyte and oxide cathode material interfaces.

Category	Solid electrolyte	Active material (other components)	Measurement	Degradation	References
A	Al-$Li_{6.6}La_3Zr_{1.6}Ta_{0.4}O_{12}$	$LiCoO_2$	XRD	Formation of $LaCoO_3$, $LiCoO_3$, Li_2CoZrO_4, and $Li_3Zr_{0.18}Ta_{0.82}O_4$ at 1100 °C	[47]
A	Al-$Li_7La_3Zr_2O_{12}$	$LiCoO_2$	XRD	Formation of t-LLZO at 700 °C	[48]
A	$Li_{6.25}Al_{0.25}La_3Zr_2O_{12}$	$LiCoO_2$	XRD	Stable up to 900 °C	[49]
A	$Li_{6.25}Al_{0.25}La_3Zr_2O_{12}$	$LiMn_2O_4$	XRD	Formation of Li_2MnO_3, La_2O_3, and $La_2Zr_2O_7$ at 600 °C	[49]
A	$Li_{6.25}Al_{0.25}La_3Zr_2O_{12}$	$LiFePO_4$	XRD	Formation of Li_3PO_4, $La_2Zr_2O_7$, and Fe at 400 °C	[49]
A	$Li_{6.6}La_3Zr_{1.6}Ta_{0.4}O_{12}$	$LiNi_{0.5}Mn_{1.5}O_4$	XRD	Formation of Li_2MnO_3, La_2O_3, and $La_2Zr_2O_7$ at 600 °C	[29]
A	$Li_{6.6}La_3Zr_{1.6}Ta_{0.4}O_{12}$	$LiCoMnO_4$	XRD	Formation of Li_2MnO_3 at 600 °C	[29]
A	$Li_{6.6}La_3Zr_{1.6}Ta_{0.4}O_{12}$	$Li_2FeMn_3O_8$	XRD	Formation of Li_2MnO_3 at 600 °C	[29]
A	$Li_{6.75}La_3Zr_{1.75}Ta_{0.25}O_{12}$	$LiCoO_2$	Raman XPS	Formation of $LaCoO_3$ at 700 °C	[50]
A	$Li_{6.75}La_3Zr_{1.75}Ta_{0.25}O_{12}$	$LiNi_{1/3}Co_{1/3}Mn_{1/3}O_2$	Raman XPS	Formation of $LaMO_3$(M = Ni, Mn, Co) at 800 °C	[50]
A	$Li_{6.75}La_3Zr_{1.75}Ta_{0.25}O_{12}$	$LiMn_2O_4$	XRD	Formation of $Li_2Mn_2O_3$ at 500 °C	[50]
A	$Li_{6.75}La_3Zr_{1.75}Ta_{0.25}O_{12}$	$LiFePO_4$	XRD	Formation of $La_xZr_{1-x}O_{2-2/x}$, $LaMO_3$, Li_3PO_4, and Fe at 500 °C	[50]
A	$Li_7La_3Zr_2O_{12}$	$LiCoO_2$	TEM	Formation of La_2CoO_4 at 664 °C	[51]
A	Al-$Li_{6.6}La_3Zr_{1.6}Ta_{0.4}O_{12}$	$LiCoO_2$	Raman	Co diffusion into LLZO at 1050 °C	[40]
B	$Li_{6.75}La_3Zr_{1.75}Ta_{0.25}O_{12}$ (carbon-coated)	No active material (carbon, binder, liquid electrolyte)	CV	Clear cathodic current rises around 4.0 V Release of O_2 when charged up to 4.5 V	[25]
B	$Li_{6.625}La_3Zr_{1.625}Ta_{0.375}O_{12}$	No active material (carbon)	CV	Clear cathodic current rises at 3.7 V	[52]
B	$Li_7La_3Zr_2O_{12}$	—	DFT	Decomposed to Li_2O_2, La_2O_3, $Li_6Zr_2O_7$ at 2.91 V	[21]
B	$Li_7La_3Zr_2O_{12}$	$LiCoO_2$	DFT	Decomposed to Zr, La_2O_3, Li_8ZrO_6 at c. 3.1 V	[53]
C	$Li_7La_3Zr_2O_{12}$	$LiCoO_2$	DFT	Formation of $La_2Zr_2O_7$, O_2, La_2O_3	[53]
C	$Li_7La_3Zr_2O_{12}$	$Li_{0.5}CoO_2$	DFT	Formation of La_2O_3, $La_2Zr_2O_7$, $Li_7Co_5O_{12}$, O_2	[26]
D	Al-$Li_{6.6}La_3Zr_{1.6}Ta_{0.4}O_{12}$	$LiCoO_2$	Raman	Trans- and inter- granular fracture after 100 cycles	[40]

For this, various types of intrinsic interfacial stabilities have been accessed in both computational [21, 26, 53] and experimental work [25, 29, 40, 47–52]. We begin to discuss the tendency of interfacial reaction and possible reaction products between the garnet electrolyte and commonly used active materials such as LCO, $LiNi_xCo_yMn_zO_2$ (NMC), $LiNi_xCo_yAl_zO_2$ (NCA), $LiMn_2O_4$ (LMO), and $LiFePO_4$ (LFPO) as well as high-voltage cathodes of $LiNi_{0.5}Mn_{1.5}O_4$, $LiCoMnO_4$, and $Li_2FeMn_3O_8$. Then, we discuss how the interfacial instability affects performance as well as suggested strategies to realize high-performance solid-state cathode formation.

9.3.1.1 Interfacial Reaction During Cell Fabrication

The high processing temperatures typically required to achieve intimate solid–solid contact within oxide-based composite cathodes however accelerate chemical reaction kinetics and diffusion processes. The calculation of the thermodynamic reaction energy and possible reaction products using density functional theory (DFT) has been critical to understanding experimental observations of the onset of reaction temperature, specified as T(rxt) [26, 29, 53, 54]. A cathode composite system with several components can rarely be considered in thermodynamic equilibrium. More often than not, there will be a driving force for interfacial reaction between the two components, leading to a new state of thermodynamic equilibrium for the composite system. The reaction kinetics, the diffusivities of one constituent in another, etc. provide key information on the rate at which the system would tend to attain the equilibrium state. Because of this complex balance between thermodynamic driving forces and kinetically accessible mechanisms at the reaction temperature, most of this information cannot currently be quantified computationally. Instead, the computational methods have focused on capturing the maximal possible chemical driving force at an interface and the possible reaction products [53]. The relevant energies as meV/atom can be calculated by DFT, and this quantification helps to understand experimental observation. In the absence of thermodynamic and kinetic data, experimental studies would be necessary to determine the compatibility of the components [29, 50]. In all-solid-state cathode preparation, interfacial interactions are more frequent, which can cause changes in the constituent properties and/or interface structure. Most of the interfacial reactions in the garnet/cathode interface can be found during fabrication (Table 9.1).

LCO and $Li(Ni_{1/3}Co_{1/3}Mn_{1/3})O_2$ with an LLZO interface have a lower driving force (1 meV/atom) for chemical reaction than with $LiMn_2O_4$ (63 meV/atom) or $LiFePO_4$ (94 meV/atom) interfaces [26, 55]. The reaction tendency between LLZO and half-lithiated active material (LCO, NMC, NMO, and LFP) is still the same but with increased reaction energy, indicating higher reactivity during charging. The reaction temperature T(rxt) based on a composite mixture (e.g. LLZO–LCO) formed to a pellet was experimentally investigated by annealing at different temperatures (400–900 °C) and investigating the interfacial reaction products using Raman spectroscopy and X-ray diffraction (XRD). Encouragingly, the predicted tendency from DFT calculations has been quite consistent with experimental observations, indicating a lower interfacial reactivity of LLZO with LCO (T(rxt): 700 °C) than with LMO or LFPO (both T(rxt): 500 °C) [50]. Similar work has been performed using only XRD, showing the same tendency with different $T_{rxt} > 900\,°C$ for the LCO/LLZO interface, $T_{rxt} \approx 600\,°C$ for the LMO/LLZO interface, and $T(rxt) \approx 400\,°C$ for the LFPO/LLZO interface [49]. In addition, the interfacial chemical stability showed $T(rxt) \approx 600\,°C$ between LLZO and high-voltage spinel-based active materials, including $Li_2NiMn_3O_8$, $LiCoMnO_4$, and $Li_2FeMn_3O_8$ [29]. We summarize experimental observation for chemical stability of LLZO electrolyte with oxide cathode materials as function of annealing temperature (Figure 9.3). Comparing the results detected by XRD, oxide cathode active materials with rock salt layered-based (LCO, NMC, LNO) [47, 50, 56] structure has apparently higher chemical stability against Li-garnet than ones with spinel-based ($LiNi_{0.5}Mn_{1.5}O_4$, $LiCo_{0.5}Mn_{1.5}O_4$, $LiCoMnO_4$, $LiFe_{0.5}Mn_{1.5}O_4$) [29, 56] and olivine-based (LFP) [49] structure. In addition, it is shown that studies on interface evolution by Raman spectroscopy, X-ray photoelectron spectroscopy (XPS), transmission electron microscopy (TEM), or nano-beam electron diffraction (NBD) are more powerful tools than XRD alone to precisely identify the evidence of the reaction at lower-temperature regime (rock-salt layered cathode against Li-garnet) [50, 51].

Thus far, LCO has been shown to form more stable interfacial stability with LLZO than with other active materials. This has motivated numerous experimental efforts to achieve a deeper understanding between the interfacial impedance and

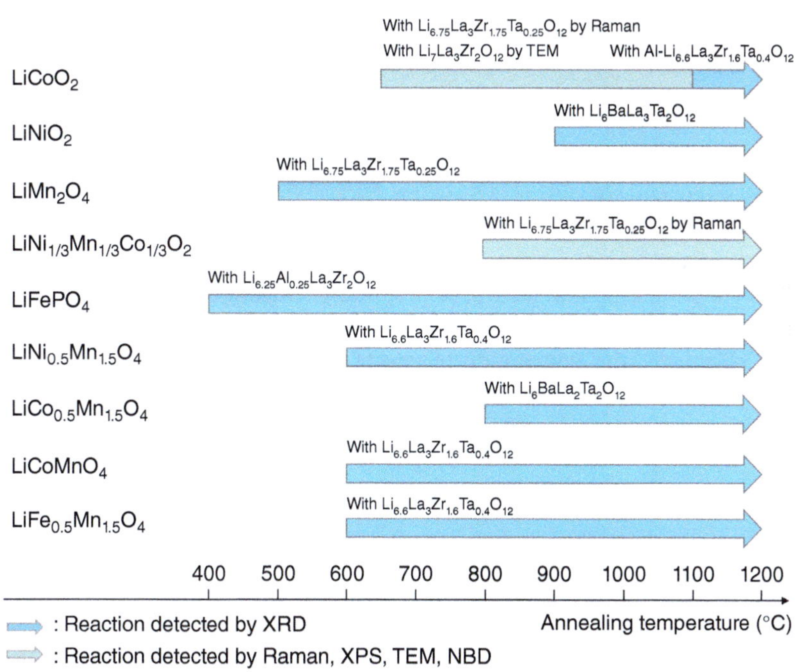

Figure 9.3 Chemical compatibility window of Li-garnet with oxide cathode-active materials. Corresponding data are available in Table 9.1, highlighting experimental results from Raman spectroscopy, X-ray diffraction (XRD), transmission electron microscopy (TEM), X-ray photoelectron spectroscopy (XPS), and nano-beam electron diffraction (NBD).

performance of a full cell with LLZO–LCO cathode composite, implying that co-assembly strategies of the composite cathode other than LCO require to be investigated in the future. For example, in a recent study, differential scanning calorimetry (DSC) was used to investigate the reactivity of an LCO–LLZTO composite pellet, showing negligible reaction up to their sintering temperature ≈ 1050 °C except a minor reaction event at 700 °C [47]. Subsequently, the same cathode preparation conditions were applied to ASSLB full-cell fabrication consisting of LCO–LLZTO as the cathode composite, LLZTO pellet as the solid electrolyte, and In–Li as the metallic anode, delivering a promising first discharge capacity of 113 mA h g^{-1} at a current density of 50 μA cm^{-2}at 50 °C [40]. To achieve intimate contact for the LCO/LLZO interface using a crystalline, micron-size LCO and LLZO powder, a sintering temperature over 1000 °C was required but with minor side reaction, which may be one of the reasons for the capacity being lower than the theoretical value.

To avoid potential reaction at high processing temperatures, LCO is often deposited at lower temperature (≲800 °C) either using a wet chemical method or physical vapor deposition. For instance, Li and Co acetate solution prepared by the sol–gel route was drop-coated on an LLZO pellet and then annealed at 800 °C, with no XRD evidence of interfacial reaction yet an interfacial resistance of several kΩ [57]. Nonetheless, the reaction product of $La_2Li_{0.5}Co_{0.5}O_4$ was observed for an LCO/LLZO bilayer thin film fabricated via the sol–gel route at 800 °C [58]. Likewise, an LCO thin film prepared on an LLZO pellet by pulsed laser deposition (PLD) and annealing at 664 °C resulted in an interfacial reaction product of La_2CoO_4, according to TEM–energy-dispersive X-ray spectroscopy (EDS) and nano-beam diffraction analyses [51]. Interestingly, another LCO thin film prepared by PLD at 600 °C on an LLZO pellet exhibited a relatively small interfacial resistance of 170 Ω cm^2 at the cathode interface, and the full cell delivered a reversible charge/discharge capacity at a current density of 3.5 μA cm^{-2} for 100 cycles [59]. Comparison of these low-temperature methods indicates that the LLZO/LCO interfaces formed at 800 °C [57] and 664 °C [51] produced more resistive interfaces than those formed at 600 °C [59], highlighting that a reduced processing temperature down to 600 °C is beneficial for lowered chemical reaction and cathode impedance for the LCO/LLZO system. Collectively, the preparation of one of the composite components via a wet chemical process and low temperature may avoid high-temperature-driven interfacial reactions; however, the issues of low loadings and poor utilization of the active material remain unresolved, making these approaches impractical for high-energy-density ASSLB fabrication.

9.3.1.2 Electrochemical Oxidation and Chemical Reaction During Cycle

Interfacial degradation may occur not only during cell fabrication but also during battery operation, typically at the cathode side upon charging. Cathode interfaces are subjected to applied voltage during cell operation; thus, solid electrolytes adjacent to active materials can be decomposed outside their electrochemical window or beyond their oxidation limit. Although the electrochemical stability window is a bulk property of a solid electrolytes rather than a property of the interface with active material interface, it is still important to the interface stability as the electrochemical decomposition of the solid electrolyte typically occurs at the interface with an electron source from active materials or carbon; thus, the solid electrolyte in a cathode composite directly experiences the applied potential difference (Figure 9.2b) [39]. DFT calculation predicts that the electrochemical window for most solid oxide electrolytes is below the redox potential of commonly used cathode active materials such as LCO (3.8 V), NMC (3.8 V), LMO (4.1 V), and LFP (3.5 V) [60].

Garnet-based cathode composites generally have higher thermodynamic stability against electrochemical decomposition and chemical reactions for room-temperature cycling. It is important to note that the difficulty of cell fabrication with garnet-based cathode composites, as discussed in Section 9.3.1.1, has resulted in limited reports on the performance of full cells (Table 9.1). Thus, most studies have either been based on computational work or cyclic voltammetry (CV) investigation using a model cathode (e.g. Li|LLZO|LLZO–carbon|Pt cell) and not a full-cell-level investigation. These reports suggest that the onset of oxidation of LLZO in a cathode composite starts at ~3.7 V [52] and ~4 V [25]. Interestingly, the values from the precise measurement setup were still higher than the DFT-predicted values (2.9 V [25] and 3.2 V [53] for LLZO), raising the question of why theoretical oxidation limit of LLZO is different than experimental observation. Similar effects are also known in other areas of solid-state batteries, if we think at the high cycle numbers of LIPON and stability vs. Li by formation of sub-oxides, despite the predicted higher cell voltage. In an analogy, we judge as critical for LLZO and cathodes obtaining precise information on the composition and thickness of the SEI as a function of voltage cut-off is critical to attain a better understanding of the stability of the SEI, which may assist future artificial SEI engineering. Regarding chemical reactions during cycling, only DFT calculation studies have reported interfacial products at the LLZO/LCO interface, suggesting products of La_2O_3, $La_2Zr_2O_7$, and Li_2CoO_4 at 3 V [54] and $La_2Zr_2O_7$ and $LaCoO_3$ above 4 V [26]. Unfortunately, all the reaction products above are likely to be poor ionic conductors, which may in turn increase interfacial impedance; however, no clear experimental evidence has been obtained yet.

9.3.1.3 Chemo-mechanical Degradation During Cycling

Volume expansion and contraction of the active materials occurs because of the compositional change during charge/discharge, continuously generating stress at the established interfaces and leading to the formation of micro-gaps between the active material and solid electrolyte and thus contact loss (Figure 9.2c) [40, 41, 45, 46]. Although maintaining the particle network in a cathode composite is important for Li-ion transport and the large contact is essential for charge transfer, redox reaction of the active materials leads to mechanical problems in the composite electrode. For example, NMC111 undergoes volume contraction of roughly 2%, whereas NMC622 and NMC811 undergo larger volume contraction up to 6% because they are more affected by the large change of the Ni-ion radius upon transition-metal oxidation [61]. Exceptionally, LCO exhibits the opposite volume change behavior [45]. Accordingly, the volume change severely affects the sustainability of the composite electrode geometry. Liquid electrolytes can maintain the Li-ion networks as well as contacts with the active material because of their fluidity, whereas solid electrolytes have difficulty maintaining the geometry because the mechanical stress from the active material causes interfacial delamination or crack formation in the composite cathode, increasing the charge-transport resistances. Once mechanical failure progresses, high interfacial impedance and capacity decay are likely to occur because of the loss of active reaction sites.

Recently, chemo-mechanical failure has been reported in the cathode composite of LLZO–LCO. The trans- and intergranular fractures within the cathode microstructure were hypothesized to be the degradation mechanism for the increased interfacial resistance (Figure 9.4a) [40]. A full-cell test of LCO–LLZO|LLZO|In–Li revealed that the first discharge capacity of 117 mAh g^{-1} decreased to 36 mAh g^{-1} after 100 cycles under a current density of 0.05 mA cm^{-1} and 50 °C (theoretical capacity of 140 mAh g^{-1}) [62]. The total resistance of the cell was increased from 1138 Ω cm^2 in the first cycle to 5807 Ω cm^2 after 100 cycles, where only the polarization resistance degraded with cycling (Figure 9.4b). Assuming that the LLZO-based cathode composite has higher stability against electrochemical oxidation and chemical reaction during room-temperature cycling than sulfide-based cathode composites, the observed micro-cracks in the LLZO–LCO composite cathode from the repetitive expansion and contraction of LCO are thought to be the main origin of the capacity degradation. To mitigate the interfacial degradation from the interfacial contact issues in the solid–solid interface system, hybrid-type cathodes on LLZO solid electrolytes have often been tested by adding small amounts of a liquid or polymer electrolyte in the composite electrode [63–65]. However, in this case, we lose the benefits of the solid electrolyte, e.g. safety and wide temperature ranges for operation.

9.3.2 Strategies and Approaches Toward Enhanced Stability and Performance

9.3.2.1 Cathode Coating

Given the conventional understanding, the engineering of composition and microstructure of cathode composites is recommended. By far, applying an artificial SEI layer, namely a cathode coating, on the active material surface has been the predominant approach for enhanced battery performance since the cathode coating has already been widely used in conventional liquid-electrolyte-based LIBs. However, the different criteria for liquid and solid systems and other fundamental considerations have recently been reviewed by [39, 66, 67]. Briefly, the implicit requirements for coating materials for solid-state cathodes can be summarized as follows:

a. High electrochemical decomposition voltage (>4.5 V vs. Li) with negligible electronic conductivity to avoid oxidation
b. Chemically stable with both active materials and solid electrolyte
c. Uniform coverage and/or morphology to maintain percolation
d. Mechanically "plastic" and deformable to accommodate volume change
e. Environmentally benign, non-hydroscopic, low cost, and offers ease of preparation
f. Diffusion barrier characteristics to block mutual interdiffusion during high-temperature processing.

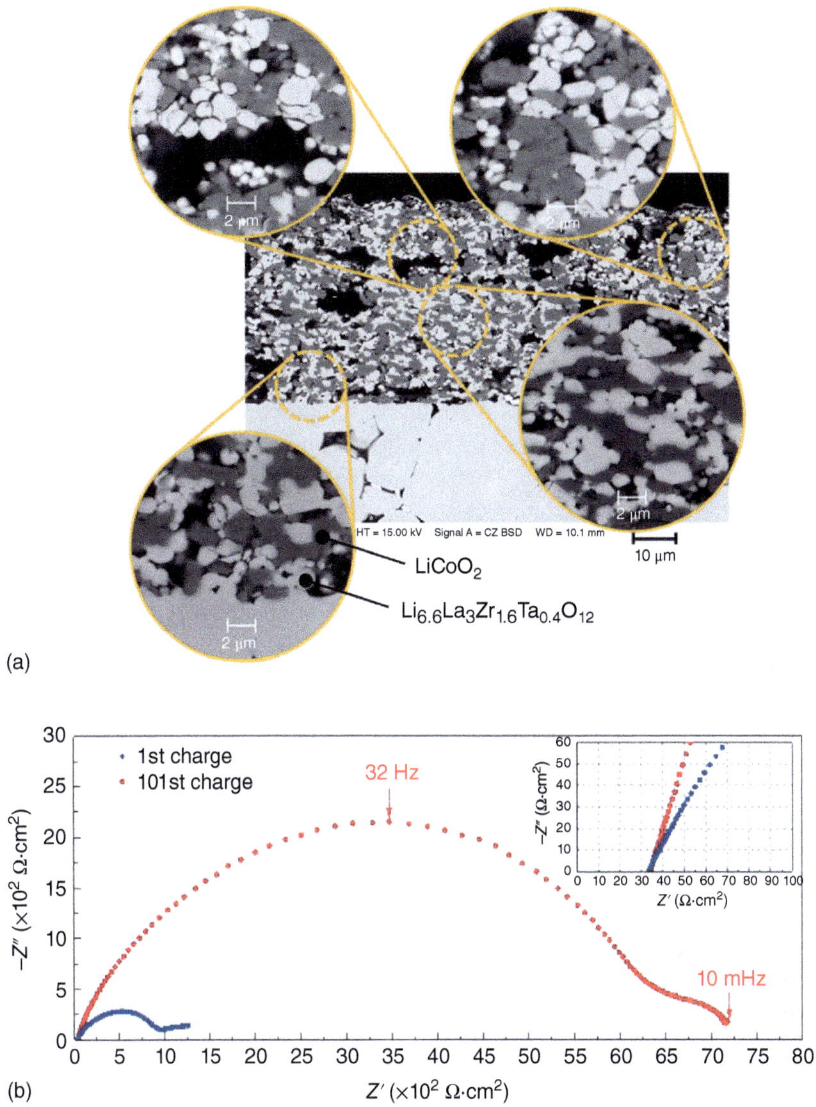

Figure 9.4 Mechanical degradation of garnet-based cathode composite in $Li_{6.6}La_3Zr_{1.6}Ta_{0.4}O_{12}$–$LiCoO_2$/$Li_{6.6}La_3Zr_{1.6}Ta_{0.4}O_{12}$/In-Li cell. (a) SEM cross-section images of a composite cathode of $Li_{6.6}La_3Zr_{1.6}Ta_{0.4}O_{12}$–$LiCoO_2$ that underwent 100 charge–discharge cycles at 50 °C. (b) Electrochemical impedance spectra before and after 100 cycling. Source: Tsai et al. [40]/with permission of Royal Society of Chemistry.

Garnet-based cathode composites must meet criteria *a–e* over the entire employed range of lithium activities and operation temperatures. For *f*, it is particularly necessary for the processing of oxide-based cathode composites. Based on the literature survey, most of the cathode coatings have been applied for sulfide-based cathode composites, with only one study on a garnet-based cathode composite [11]. More precisely, there are limited examples of electrochemical, chemical, and chemo-mechanical issues related to garnet-based cathode composites because of the fabrication difficulty.

The experimental validation of promising coating materials for garnet-based cathode composites has not yet been reported except for one model experiment in which a 10-nm-thick Nb coating was introduced by radio frequency (RF)-sputtering between an LLZO electrolyte pellet and thin-film LCO. Full cell of $LiCoO_2$|(Nb)|LLZO|Li was tested at a current density of 1 mA cm^{-2} with and without Nb film addition. The interface coated with the amorphous Li–Nb–O layer was found to effectively mitigate second-phase formation of La–Co–O, reduce the interfacial resistance from 2600 to 150 Ω cm^2 after charging at 4.0 V, and improve the overall performance [68]. Despite the improvement, the first discharge capacity achieved of 80 mAh g^{-1} is still below the expected discharge capacity (140 mAh g^{-1}), meaning that still there is the resistive interphase between LCO and LLZO, limiting the Li-ion motion at the interface thereby decreasing the cell's efficiency toward full utilization of LCO cathode. Further optimization or novel effective coating materials for garnet/cathode interface are required.

Based on first-principles calculation, the ternary oxides such as $LiNbO_3$, $Li_4Ti_5O_{12}$, Li_4SiO_4, Li_3PO_4, and Li_2ZrO_3 offer higher oxidation limits than garnet electrolytes but still lower than the cut-off voltage, for example, of LCO (4.2 V), implying that oxidation is still occurring [26, 53]. However, slow oxidation kinetics help to protect solid electrolytes from oxidation. Recently, poly-anionic materials such as borates were found to offer higher oxidation limits (>4.5 V) and excellent chemical compatibility compared with those of oxide coating materials because of the strong covalent bond (B—O) and sharing of the same ions with oxides [55]. In any case, understanding of the oxidation kinetics of SEs and the suppression of the interfacial degradation during room-temperature cycling and/or high-temperature co-firing is key to determining the lifetime of a coating layer; therefore, we need verify how these coating materials work through experiments. In particular, developing new coating materials and novel coating strategies (processing) for garnet-based cathode preparation is next urgent task. Probing the local chemistry and phase evolution over cycles at the grain boundaries will be important to optimize cathode coating toward ASSLB with high energy and long cycles.

9.3.2.2 Geometric Arrangement Concerns and Strategies Toward Maximizing Reaction Sites

When designing composites in general, one must consider their unique characteristics as follows. First, composite materials are inherently heterogeneous at the microstructural level, as they consist of two or three components with different strengths, different expansion properties upon bias, etc. Even after selecting the component characteristics and interface characteristics, one can obtain a range of electrochemical performances depending upon the geometric arrangement of the cathode composite. Physical properties such as the mechanical and electrical conductivity of the composite cathode can be controlled by the transport property of the active materials and solid electrolytes, their volume/size ratio, the number of interfaces (reaction sites), and characteristics of the interface region. In particular, the connection network and charge-transport properties of the ionic and electronic phases must be optimized for composite cathodes. It is expected that the particle size and ratio between the active material and solid electrolyte also significantly affect the charge-transport network and resistance in a cathode composite [69–73].

For oxide-based ASSLBs, there has been little success in applying similar research for microstructure optimization due possibly to the processing challenges. Rather, the more prominent issue is how to create chemically and mechanically stable interfaces between the active materials and solid electrolytes as well as between solid electrolytes. The garnet solid electrolytes require processing temperatures higher than 1000 °C to develop strong necks with another LLZO and LCO to form a dense cathode composite [74–77]. At this temperature, most of the active materials will be decomposed or reacted with LLZO [29, 50]. Lowering the co-firing temperature down to 700 °C is possible with promising performance but requires 12–34 wt% [9, 78] of inactive phases of sintering aids such as Li–B–O oxides, which currently requires sacrificing of a portion of the cathode loading. Additionally, it is not ideal to add such large quantities because of the poor ionic conductivity (10^{-6} S cm^{-1}) of Li–B–O oxides [79]. Another approach is to first prepare a porous-electrolyte network at high temperature and then introduce active materials by infiltration [80–85]. Although it has been challenging to achieve the theoretical capacity of active materials in many of these reports, mainly because of the high interfacial resistance, in very recent works, quite impressive advances for oxide-based lithium [85] and sodium metal batteries [84] were demonstrated. The strategy uses a pre-sintered porous electrolyte backbone as a reliable Li-ion transport network within the cathode composite, and the infiltrated precursor solution and subsequent annealing allows low-temperature (<740 °C) phase formation of desired active materials (Figure 9.5a). For example, oxide-based LCO–LLZO cathode composites were prepared using this manner (firing at 700 °C), showing the homogeneous distribution of LCO particles with a grain diameter of 0.29 ± 0.020 μm on the surface of porous LLZO (Figure 9.5b). The total interfacial resistance for the full cell (LCO–LLZO|LLZO|Li metal) is as low as 62 Ω cm^2 at initial cycle, and the cell delivers a discharge capacity of 118 mAh g^{-1}, which is near the theoretical capacity of LCO (115 mAh g^{-1}, 3–4.05 V) (Figure 9.5c,d). The improved capacity and interfacial resistance compared with those from co-sintering approach with and without Li–B–O sintering agent (Figure 9.5d) can be as a result of reinforcing ionic transport inside the cathode composite while avoiding detrimental interfacial reactions and maximizing active reaction sites [85]. The potential issues of this cathode architecture are limited loading capability and flexibility of material choice. Interestingly, recent work from oxide-based sodium metal batteries [84] reported high loading capability with $Na_3V_2P_3O_{12}$ (NVP) as active material by solution-assisted cathode composite formation method. The NVP active material of 1.7–3 mg cm^{-2} was introduced by chemical infiltration through a porous $Na_{3.4}Zr_2Si_{2.4}P_{0.6}O_{12}$ (NZSP) electrolyte scaffold and sintering at 740 °C (Figure 9.5e), with good specific capacity of the active positive electrode material (>95 mAh g^{-1}) and high coulombic efficiency (>99%) for 100 cycles at a current density of 35–127.6 μA cm^{-2}. Comparison of specific capacity with cycle number among the all-oxide Na and Li metal batteries

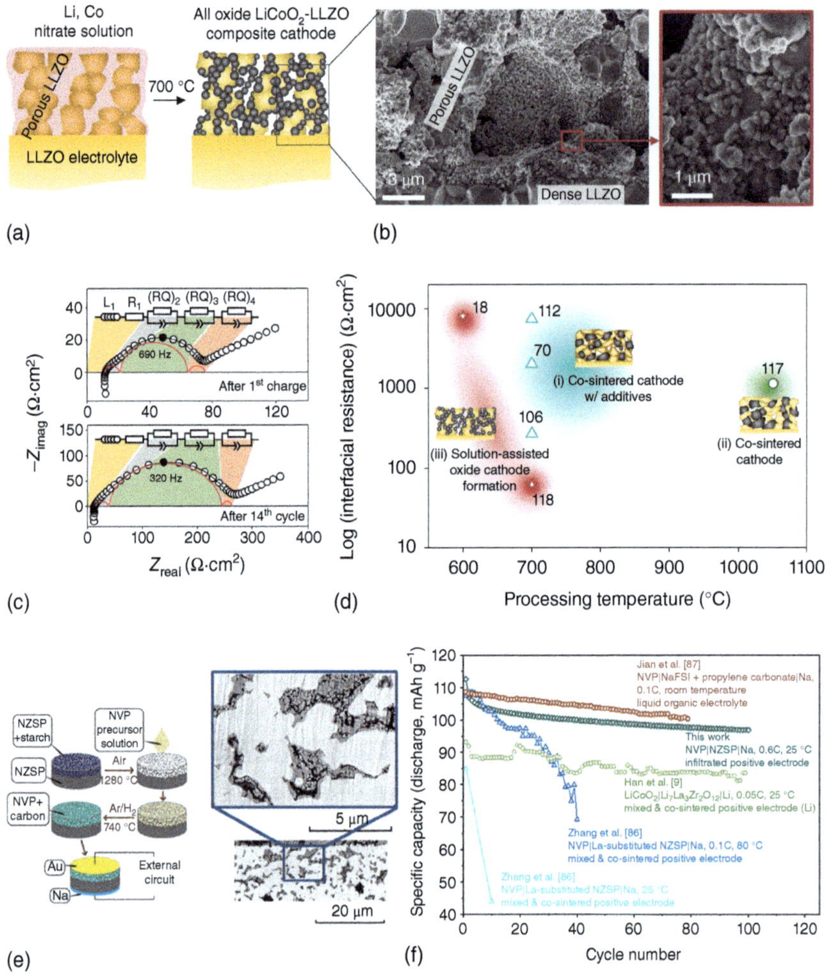

Figure 9.5 Electrochemical performance of oxide-based lithium and sodium metal batteries prepared by solution-assisted cathode composition formation. (a) Precursor solution infiltration into porous LLZO structure and low-temperature annealing at 700 °C, resulting in all oxide LCO–LLZO composite cathode. (b) Cross-sectional SEM images of LCO–LLZO composite. (c) Electrochemical impedance spectra of the full cell (LCO–LLZO|LLZO|Li metal) after 1st and 14th charging at 80 °C. (d) Comparison of the total interfacial resistance vs. processing temperature of oxide-based Li-garnet SSBs based on layered cathode composites. Three different categories for oxide-based cathode composite are shown. The numbers of 70–118 indicate the first discharge capacity as mAh g^{-1} and working temperature tested is 50–100 °C. (e) Design and fabrication of infiltrated all-solid-state sodium batteries. Manufacturing steps (left) for a $Na_3V_2P_3O_{12}$ (NVP)–$Na_{3.4}Zr_2Si_{2.4}P_{0.6}O_{12}$ (NZSP)–Na battery using the chemical infiltration method through porous NZSP scaffold for increased effective reaction area. SEM images (right) of a cross-sectional NVP (dark contrast)-infiltrated NZSP (bright contrast) pellet. (f) Discharge capacity degradation of different all-solid-state lithium and sodium battery with liquid electrolyte. Source: (a–d) Kim and Rupp [85]/with permission of Royal Society of Chemistry, (e, f) Lan et al. [84]/with permission of Elsevier.

indicates that infiltrated NVP–NZSP composite cathode outperforms state-of-art LBO–LLZO–LCO composite cathode (Figure 9.5f). For processing of oxide-based SEs, many techniques have been developed to attain good cathode interfaces with the low internal resistance. Based on the knowledge gained from prior art (additive approach, infiltration approach), the design of more practical solid-state cathodes through microstructure optimization, as shown in Ref. [85, 88, 89] is suggested.

9.3.2.3 Conductive Additives in Solid-State Cathode

The electronic conductivity in cathode composites is another important parameter to improve the utilization of the active material. Typically, ASSLB cathode composites consist of a mixture of the active material and SE without conductive additives. During oxide-based cathode composite preparation, carbon additives have been mostly excluded because the cathode-processing temperature is above the carbon-burnout temperature (~500 °C). Controlling sintering gas from the oxidizing to reducing atmosphere can be considered; however, the electrolyte and active materials must be stable without decomposition under such conditions [84]. Alternatively, inorganic electronic conductive materials, such as indium tin oxide ($In_{2(1-x)}Sn_{2x}O_3$) powders, can be used as oxide electronic conductive agents [90]. Nevertheless, an understanding of the rate-limiting step among the multiple transport steps inside the as-prepared cathode composite is critical. For example, the total conductivity and the partial electronic and ionic conductivity of the cathode composite can be characterized using the direct current (DC) polarization technique with an ion- or electronic-blocking electrode [72, 91]. These transport properties are strongly affected by the choice of active materials and solid electrolyte as well as their particle size/volume ratio and geometric arrangement, providing a concrete direction toward improved solid-state cathode performance.

9.4 Anode: Adopting Lithium Metal in the Solid

Replacing the conventional carbonaceous anode (LiC_6) in Li batteries with a Li metal anode can yield a 50% increase in the practical energy density (3860 vs. 372 mAh g^{-1}) and is considered one of the most promising approaches to realize high-energy-density batteries [92]. The US DOE targets a high per-cycle utilization (at least 80%) of Li metal present, a cumulative capacity plated before cell short circuiting > 10 mAh cm^{-2}, a plating current density > 3 mA cm^{-2}, and a high per-cycle areal capacity of >3 mAh cm^{-2} at the device level [92]. Ideally, a limited Li source (15–30 μm or 3–6 mAh cm^{-2}) and the complete stripping of Li metal during discharge are needed in Li-metal batteries to minimize "soft" shorts and support high-energy-density batteries [92]. Nonetheless, the preparation of ultrathin Li (<30 μm) by extrusion and subsequent calendaring ("rolling") is challenging because of the adhesiveness and reactivity of Li metal [93]. Spreading molten Li metal directly onto a metal current collector (or solid electrolyte) is a promising alternative strategy toward cost-effective and scalable ultrathin Li processing but is contingent upon tuning the surface energy ("wettability") of molten Li metal to facilitate its homogeneous spreading on various lithiophobic substrates, namely the metal current collector and ceramic solid electrolyte (e.g. Cu, Ni, LLZO, carbon) [94]. Another major predicament accompanying the use of a limited Li source is the 100% volume change of Li metal during discharge/charge (stripping/plating) that necessitates an excess lithium metal reservoir of 20–300% [95], which in turn reduces the volumetric energy density, induces stresses and shape change, destabilizes interfaces, and results in limited cycle life. Strategies to use layered reduced graphene sheets [96], a polyimide–ZnO core–shell structure [97], mixed ionic-electronic conductor tubular 3D host [98], or carbon spheres [99] as rigid and stable Li hosts to store Li metal to reduce volume change are still immature and prevent making full use of the alkali metal. Thus, the adoption of Li metal remains one of the most promising and challenging approaches to achieve safer high-energy-density batteries, especially in applications favoring ceramic or polymer solid electrolytes. There are other foreseen challenges associated with the morphological changes associated with the electrochemical deposition and dissolution of Li metal, especially during cycling at high current densities (>1 mA cm^{-2}), such as Li dendrite nucleation and propagation, interface delamination, and pore formation.

9.4.1 Li/Solid–Electrolyte Interface: Chemical, Electrochemical, and Mechanical Considerations, Including Mitigation Strategies

The coupling of Li metal with a solid electrolyte necessitates an interface that is chemically, electrochemically, and mechanically stable with sufficiently fast charge-transfer kinetics to support the fast Li stripping/plating during battery operation. Electrochemical compatibility (i.e. redox stability) between the Li metal and solid electrolyte is essential to prevent reduction of the electrolyte by the Li metal, meaning that the electrochemical potential of Li metal must be higher than the lowest occupied molecular orbital (LUMO) of the electrolyte, unless a stable passivation layer (i.e. an interphase layer) is formed. Broadly, Li solid electrolytes, except for Li binary compounds such as Li_3N, Li_2S, and Li_3P, are thermodynamically unstable against Li metal and decompose at low potentials, leading to the formation of an interphase layer (Table 9.2) [21, 26, 105–107].

The ideal interphase is a nanometrically thin good ionic conductor and an electronic insulator (Figure 9.6a). The electronically insulating character of the interphase reduces the Li chemical potential (sum of the electrochemical potential of the electronic and ionic carriers) to be within the electrochemical window of the solid electrolyte, inhibiting further electrode decomposition and stabilizing the Li/electrolyte interface. The high experimental electrochemical stability window, occasionally observed between thermodynamically unstable solid electrolytes when in contact with Li metal, originates from a kinetically stabilizing, self-limited, passivation interphase layer at the Li/electrolyte interface, facilitated by the electronic insulating and ion-conducting nature of the interphase [21], as observed for several solid electrolytes, including LLZO and LiPON [100, 110]. However, for solid electrolytes such as LAGP and LATP, which contain cations such as Ti^{4+}, Ge^{4+}, and Al^{3+}, the reduction of the cations at low potentials and the evolution

Table 9.2 List of stable and unstable passivation layers between solid oxide electrolyte and Li metal.

SEs		Reduction potential against Li (V)	Decomposition product	Detecting method	References
Stable passivation layer	LiPON $Li_{1.4}PO_{2.2}N_{0.7}$	0.69 [21, 26, 100]	Li_3PO_4, Li_3P, Li_3N, and Li_2O	XPS	[100]
			Li_3P, Li_3N, Li_2O	First-principles thermodynamic calculations	[26]
	LLZO	0.05 [21, 26]	Li_2O, La_2O_3, Zr_3O, and Zr metal at ~0 V	First-principles computations	[25, 101]
			Tetragonal LLZO	In situ scanning transmission electron microscope (STEM)	[102]
Unstable passivation layer	$Li_{0.33}La_{0.56}TiO_3$ LLTO	1.75 [21, 26]	Reduction of Ti^{4+} to Ti^{3+}, oxidized Li^+ inserting into La^{3+}/Li^+-site vacancies	SIMS, XPS	[103]
			Ti_6O, La_2O_3, and Li_2O at 0 V	First-principles thermodynamic calculations	[21, 26]
	$Li_{1.3}Al_{0.3}Ti_{1.7}(PO_4)_3$ (LATP)	2.17 [21, 26]	$Ti^{4+} \rightarrow Ti^{3+}$ Formation of a new mixed ionic–electronic thin conductive layer	XPS	[104]
			Ti_3P, TiAl, Li_3P, Li_2O at 0 V	First-principles thermodynamic calculations	[21, 26]
	$Li_{1.5}Al_{0.5}Ge_{1.5}(PO_4)_3$ (LAGP)	2.7 [21, 26]	$Ge^{4+} \rightarrow Ge^{x+}$; $Ti^{4+} \rightarrow Ti^{3+}$ Formation of a new mixed ionic–electronic thin conductive layer	XPS	[104]
			Li_9Al_4, $Li_{15}Ge_4$, Li_3P, Li_2O at 0 V	First-principles thermodynamic calculations	[21, 26]

Sources: Based on Zhu et al. [21, 26], Binninger et al. [105], Xu et al. [106], Chen and Amine [107].

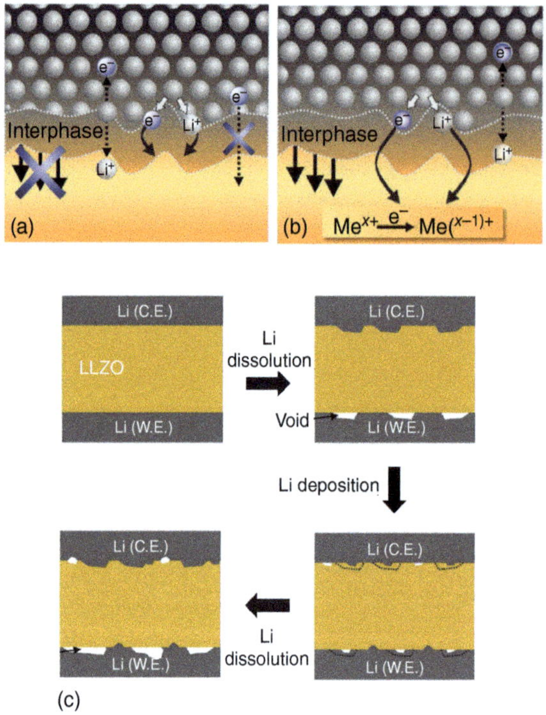

Figure 9.6 Formation mechanism of solid–electrolyte interphase (SEI) between solid electrolyte and Li metal anode. (a) Reactive and metastable solid–electrolyte interphase (SEI). (b) Reactive and mixed conducting interphase. (c) Schematic illustration of the dynamic changes at the Li/LLZO interface during repeated Li dissolution/deposition cycles. Source: (a, b) Wenzel et al. [108]/with permission of Elsevier, (c) Koshikawa et al. [109]/with permission of Elsevier.

of metallic states lead to high electronic conduction, continuous electrochemical reduction of the solid electrolyte, and interphase layer growth during cycling, which ultimately leads to the short circuit of the cell (Figure 9.6b) [104, 111]. The last factor typically negates their use as electrolytes in ASSLBs unless an electron-blocking and Li^+-ion-conducting interlayer is used [104, 112]. For example, NASICON-type LATP exhibits excellent resistance to oxidation with a high theoretical oxidation potential (based on Li grand potential phase diagram) of 4.21 V vs. Li^+/Li but has a relatively high theoretical reduction potential of 2.17 V vs. Li^+/Li because of the reduction of Ti from Ti^{4+} to Ti^{3+} (similar to LLTO and Ge in LAGP), leading to an ionic and electronic mixed conduction, which negatively affects the stability against Li metal [21]. XPS analysis has also confirmed that Li-conducting glass ceramics containing Ti^{4+}, Ta^{5+}, and Ge^{4+}, such as LATP and LAGP, readily react with Li metal with the formation of a new mixed ionic–electronic thin conductive layer that leads to a continuous increase in the electrolyte impedance with time [104]. A troublesome issue for Li oxides with high reduction potentials is the anode material being unable to react at potentials higher than its reduction potential, e.g. 1.7 V vs. Li^+/Li for perovskite-type $Li_{3x}La_{2/3-x}TiO_3$. Indeed, coupling $Li_{3x}La_{2/3-x}TiO_3$ and Li metal involves the insertion of Li-ions into La^{3+}/Li^+-site vacancies in $Li_{3x}La_{2/3-x}TiO_3$ accompanied by the reduction of Ti^{4+} to Ti^{3+} and an increase in electronic conductivity because of the existence of mixed Ti^{3+}/Ti^{4+} states [103]. The unstable interphase layer formed at the Li/electrolyte interface, where continuous "thickening" (growth) could effectively terminate Li dendrite propagation [113] but also inhibit further Li-ion transport, would in turn increase the cell impedance (manifesting in large polarization) and limit cycle life [104, 113]. When solid electrolytes are not stable in the presence of Li metal (i.e. they possess high reduction potentials) or have an unstable interphase layer, the need for an artificial protective film is sparked; Polyplus [114] is one example of a coating layer introduced to protect LATP against Li metal.

In contrast to LATP, LiPON and LLZO are Li-ion conductors with excellent compatibility against Li metal (reduction potential of 0.69 V vs. Li^+/Li for LATP and LiPON and 0.05 V for LLZO vs. Li^+/Li) because of the passivation mechanism [21]. The low thermodynamic driving force for LLZO reduction at 0 V (−0.02 eV/atom) and the electrochemical stability window experimentally determined by CV experiments [28, 115, 116] indicate that kinetic stabilization stemming from the formation of an interphase may result in an effectively wider stability window, suggesting its possible pairing with Li metal despite the thermodynamic instability. In a first-principles calculation study based on the evaluation of the intercalation voltages for various garnet-type materials $Li_xLa_3M_2O_{12}$ ($M = Zr^{4+}$, Ti^{4+}, Nb^{5+}, Ta^{5+}, Sb^{5+}, Bi^{5+}, etc.) and Li atoms, the redox potential of the lithium garnet-type structure was observed to be strongly related to the effective nuclear charge experienced by the valence electron of the cation M at octahedral sites [27]. Garnets with smaller effective nuclear charge cations such as Zr^{4+} or Ta^{5+} at the octahedral sites are thermodynamically nonreactive with Li metal because of the lower covalent-bonding character of the octahedral metal than Ta^{5+} in the perovskite $La_{1/3}TaO_3$ [27]. Nonetheless, computational analysis based on the grand canonical phase diagram revealed the instabilities of LLZO against Li metal having a low reduction potential of 0.05 V [25, 26], leading to possible cation reduction at low potentials with the subsequent evolution of metallic states at the Li/LLZO interface [25]. This last finding was also confirmed experimentally, with the coloration observed upon the immersion of LLZO in molten Li attributed to the reduction of Zr and/or Al in LLZO [25, 117] and Zr^{4+} reduction of Zr-3d by Li metal for a cycled Li/LLZO/Li symmetrical cell confirmed by XPS depth profile analysis [33]. LiPON is an excellent example of electrolyte that is expected to be unstable against Li metal due to the rather high theoretical reduction potentials of ~0.7 V according to first-principles computation based on the Li grand potential phase diagram [21, 53, 105]. However, a barrier layer with decomposition products of extended reduction stability toward Li metal forms at the interface, including Li_2O (0–3.1 V), Li_3P, Li_3PO_4 (0.7–4.2 V), $Li_4P_2O_7$, and Li_3N (0–0.6 V) for LiPON [21, 53, 100]. Importantly, the decomposition products at the interface such as Li_3N and Li_3P exhibit adequate Li ionic conductivity to the base solid-state electrolyte material, and are hypothesized to thereby contribute to high stability but also the excellent cyclability of thousands of cycles reported for LiPON-based thin-film solid-state batteries [4, 118].

The potentially high chemical reactivity between Li metal and solid electrolytes (especially for LATP and LAGP), poor solid–solid contact (especially for Li metal with LLZO), and significant volume changes during Li stripping and plating are predicted to cause continuous deterioration of the integrity of the Li/electrolyte interface. Even for a stable solid–electrolyte interphase (e.g. Li/LLZO, Li/LiPON), to realize an ASSLB with improved cycle life and longevity, the good chemical stability between the Li metal and solid electrolyte must be accompanied with excellent adhesion and "wettability" between the two components to ensure good interfacial contact. Nonetheless, maintaining good interfacial contact, improving the interface kinetics, and securing morphological stability between Li metal and the solid electrolyte under current load is considered a herculean task when considering the foreseen large volume changes and morphological instabilities (e.g. poor "wettability," pore formation, delamination, dendrites), and ultimately contact loss, associated with plating and stripping of the Li metal during cycling (Figure 9.6c) [24]. Volume changes may lead to pressure oscillations, where evolution of localized stresses at the Li/solid electrolyte interfaces may result in mechanical failure of the solid electrolyte (cracking, bending, etc.), contact loss, leading to low coulombic efficiency during cycling and limiting the safe practical applicability of high-energy-density ASSLBs [45]. Although the (electro)chemo-mechanics coupling at the Li/solid electrolyte has yet to be fully characterized and resolved, partially because of the buried nature of the interfaces, the morphological instabilities (namely pore/voids formation) at the Li metal/solid electrolyte interface have already been proven to cause contact loss during anodic loading [24]. A reference electrode was strategically placed in a three-electrode Li/LLZO/Li cell, and the dynamic changes in the charge-transfer resistance at the Li/LLZO interfaces during dissolution and deposition of Li metal were successfully decoupled, confirming that the deterioration of the interfacial resistance was in fact mainly due to Li dissolution (rather than deposition) and the formation of voids at the Li/LLZO interface (Figure 9.6c) [109]. In general, Li-metal dissolution (striping) at the Li/solid electrolyte interface is followed by the diffusion of Li ions across the interface to an available vacant site (or interstitial site) in the solid electrolyte, accompanied by the formation of a Li vacant site and an electron at the Li-metal surface [24]. The Li-stripping rate (i.e. the discharge rate) is thus kinetically limited by the diffusion rate of Li vacant sites created at the Li metal, which for a high discharge rate ($>0.2\,mA\,cm^{-2}$) may supersaturate and accumulate into voids near the interface, leading to morphological instabilities (pores) and loss of physical contact at the Li metal/electrolyte interface [119]. Insufficient contact (in terms of shape and size of the contact area) between Li metal and the solid electrolyte due to anodic dissolution (void formation) may lead to bending of the current lines at discrete contact spot near the interface (constriction/spreading resistance) [120] and to the high interfacial resistance typically observed at Li/electrolyte interfaces, such as for Li/LLZO [24, 121]. The mechanical properties and intrinsic limitation of the vacancy diffusion coefficient play a key role in the interface dynamics and electrode kinetics of the Li metal [24] and may require the use of Li-alloy interlayers, where higher diffusion coefficient compared with pure Li metal can be achieved, thereby alleviating the accumulation of vacancies, formation of pores, and loss of contact at high discharge rates ($>0.2\,mA\,cm^{-2}$). The aforementioned loss of contact between Li metal and the solid electrolyte is detrimental to the battery operation.

Dendrite formation and propagation have been highlighted as major failure mechanisms for chemically stable interfaces (Li/LLZO); in contrast, mechanically driven failure due to interphase growth may be the major deterioration mechanism for Li/electrolyte interfaces with poor chemical stability. Damaged, nonhomogeneous interfaces may instigate nonuniform Li-ion flux, preferable Li plating, and large mechanical stresses, further increasing the cell impedance and leading to inhomogeneous local potentials that may affect Li stripping/plating [122]. The nonuniform current density distributions can also lead to the formation and propagation of Li dendrites, possibly but not necessarily [123] accelerated through grain boundaries and voids, where Li-ion conduction is more prevalent, leading to the mechanical failure of the solid electrolyte, degradation of the interfaces, and deterioration of cell performance [38, 124, 125]. Namely, the contact between Li metal and the solid electrolyte and the formation of an unstable static interphase may instigate continuous growth of the interphase and massive volume changes (in a constrained interface) in addition to compositional, morphological, and structural change at the Li/solid electrolyte interface. The interphase cannot withstand mechanical deformation without mechanical degradation (e.g. cracks), which can further enhance the Li-ion flux and instigate Li dendrite growth and continuous growth of the interphase layer (and further consumption of the Li metal). Further interfacial deterioration may persist during the dynamic stripping and plating of Li metal upon battery cycling, where a nonuniform interface morphology may create stress concentrations that lead to the premature fracture of the solid electrolyte, as recently observed for Li/LAGP, and overall electrochemical degradation of the cell [126–128]. For an unstable, mixed-conducting interphase layer, as observed at the interphases of Li/LAGP, the continuous growth of the interphase and volume expansion exerts a tensile stress on the LAGP electrolyte, ultimately causing its fracture [126–128].

The intrinsic morphological instabilities of metal electrodes cannot be prevented but can be at least partially accounted for when applying external pressure by considering the cell assembly strategy in processing. Applying pressure during (i) cell fabrication to ensure good physical contact during Li/electrolyte interface formation and (ii) cell operation to ease loss of contact and lower interfacial resistance can have a significant effect on the interface kinetics and thus affect the overall cycling performance degradation associated with volume changes and "wettability" [24]. Moreover, it has been shown that maintaining low interfacial resistance (minimizing pore formation and growth by, for example, applying external load) during Li stripping is highly important to prevent short circuiting due to Li propagation through the solid electrolyte [24]. The magnitude of applied pressure will be dominated by the mechanical properties of the Li metal (plastic deformation, creep behavior, etc.). At high external pressure, the area of the contact spots between Li metal and the solid electrolyte will increase owing to plastic deformation of the soft Li metal [24]. When Li-metal stripping was conducted under an applied pressure of 35 MPa, no contact loss and/or interfacial resistance change were observed (Figure 9.7a) [24]. Applying a high external pressure of several hundreds of MPa to the Li/LLZO interface has proven effective in lowering the interfacial resistance to negligibly small values. Without external pressure, stripping experiment shows serious contact loss after 12 hours (1.2 mAh cm^{-2}) (Figure 9.7b).

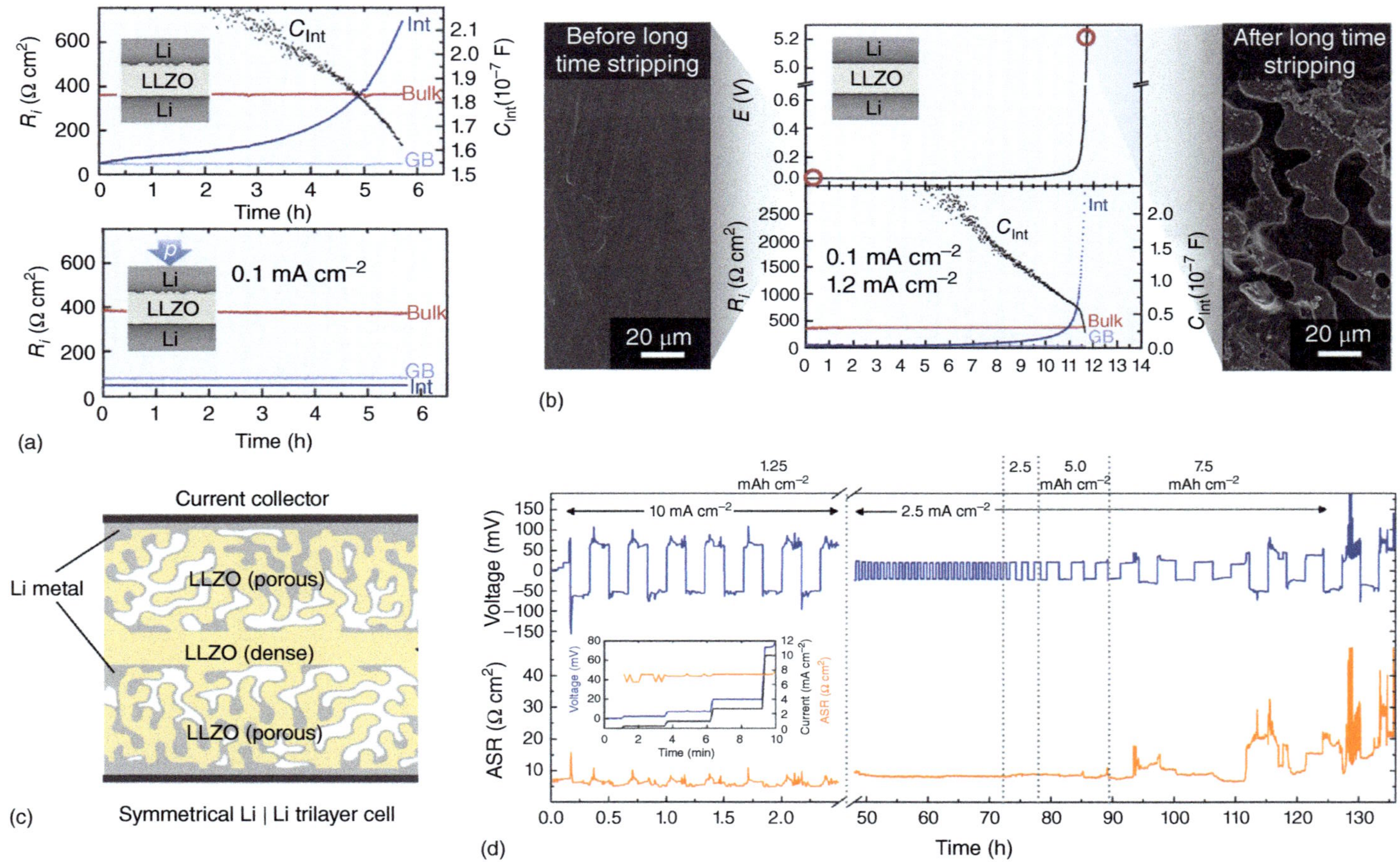

Figure 9.7 (a) Effect of pressure on interfacial resistance in Li|Li symmetrical LLZO cell. Results of long-time lithium-stripping experiments at 0.1 mA cm^{-2} without externally applied pressure (top) and ≈35 MPa external applied pressure perpendicular to the interface (down). By applying external pressure, pore formation that leads to contact loss is fully prevented. (b) Microstructure of Li metal contact with LLZO electrolyte before and after stripping at current density of 0.1 mA cm^{-2} for 12 hours (1.2 mAh cm^{-2}). (c, d) Three-dimensional solid electrolyte framework as a host for the plating and subsequent stripping of the Li metal; trilayer LLZO electrolyte with infiltrated Li metal anode shows cumulative areal capacity plated of 7.5 mAh cm^{-2} under two different current densities of 2.5 and 10 mA cm^{-2}. Source: (a, b) Krauskopf et al. [24]/with permission of American Chemical Society, (c, d) Hitz et al. [129]/with permission of Elsevier.

The charge-transfer resistance at the Li/electrolyte interface will be affected by both the applied external pressure and the mechanical properties of Li metal in addition to the ionic conductivity of the solid electrolyte (according to the constriction resistance theory) and the mechanical properties of the Li/solid electrolyte contact points [24]. Thus, compositional tuning (e.g. multiple elemental doping) of the solid electrolyte to improve its ionic conductivity can be considered as a valid strategy toward reducing the Li metal/electrolyte interfacial resistance, improving the Li electrode kinetics, and easing Li dendrite propagation [13, 130, 131]. The application of external pressure to improve physical contact and reduce the interfacial resistance to negligible small values has been reported for both LLZOs [24, 131, 132] but has limited applicability, as it may lead to mechanical failure because of the brittle nature and low fracture toughness of most thin solid electrolytes. Another strategy to mitigate the intrinsic challenges associated with utilizing a pure Li metal as the anode material, i.e. morphological interface instabilities, significant volume changes, and dendrite formation and propagation, is through the use of a three-dimensional solid–electrolyte framework as a host for the plating and subsequent stripping of the Li metal (Figure 9.7c) [129]. Generally speaking, current density plays a key role in the formation of Li dendrites. Thus, increasing the surface area of Li metal (for instance, by using Li powder or surface patterning of Li-metal foil [133] or by infiltrating Li metal into a porous framework) [134] and decreasing the effective current density may delay, to some extent, the formation of dendrites. When cycled for 7.5 mAh cm^{-2} at 10 and 2.5 mA cm^{-2} for 100 hours, Li metal grew in the pores of a garnet host (one side coated with Al_2O_3 to improve wettability with molten Li) with an interfacial resistance around 10 Ω cm^2 (Figure 9.7d) [129].

Substantial work is still needed to obtain a better understanding of the (electro)chemical and mechanical interrelation during the operation of a solid-state Li-metal battery at both stable and unstable interfaces (e.g. Li/LLZO and Li/LAGP, respectively) to better direct interface engineering efforts [135].

9.4.2 Li Dendrite Formation and Propagation in Solid Electrolytes: Challenges and Strategies

Inorganic ceramic solid electrolytes have often been regarded as a key component in achieving the goals of suppressing Li dendrite propagation, preventing short-circuit events, increasing the CCD (the maximum current before Li dendrite formation), and improving the overall battery cell safety. Although Li dendrites have been shown to grow in most types of considered battery electrolytes, including liquids [113] and polymers [136], solid-state inorganic ceramics were expected to be impervious to Li-dendrite propagation owing to their high Li transference number (close to unity) [136] and unique mechanical properties, including a high shear modulus on the order of tens to hundreds of GPa [137, 138] and low fracture toughness (K_{IC}) [11, 139]. Oxide solid electrolytes possess high shear moduli on the order of tens of GPa, up to 1 order of magnitude higher than that of Li metal [140, 141]. Nonetheless, it has been suggested that lithium dendrites penetrate preferably the solid electrolyte through or along grain boundaries [142], interconnected pores, and surface defects [13, 123], challenging the general consensus of the battery operation regime of solid electrolytes [142–146]. In particular, the growth of Li dendrites has been observed in polycrystalline [25, 38, 113, 123, 124, 139, 147–149] and single-crystal garnet-type LLZO [123], NASICON-type Li_2O–Al_2O_3–P_2O_5–TiO_2–GeO_2 (LATP) [113] as also in $Li_{1.5}A_{l0.5}Ge_{1.5}(PO_4)_3$ (LAGP) [150], to name a few. In dense LLZO (>99%) [38], Li dendrites were observed to propagate along grain boundaries at CCDs above ~0.6 mA cm^{-2}, respectively, setting the upper limit of operation and charging rates before failure [92, 124, 142, 151]. To put in context, stable Li metal stripping/plating of an LIB with a fluoroethylene-carbonate-based electrolyte solution was observed at 2 mA cm^{-2} with an areal capacity of 3.3 mAh cm^{-2} for more than 1100 cycles [152]. The nucleation and growth of the soft and ductile Li metal dendrites in various solid electrolytes under diverse cycling conditions, for instance in the hard and dense oxide garnet LLZO electrolyte [38] with a shear modulus of 58–60 GPa [140, 142], has proven that the purely mechanical criterion (shear modulus > 6.8 GPa) [137, 153] is unreliable for ceramic Li solid electrolytes because of volume changes of the electrode material, structure fragmentation, and deterioration of interfacial contact during battery operation. Needless to say, the shear modulus is not the sole parameter controlling dendritic growth, and parameters such as the electrolyte microstructure (e.g. pores, grain orientation, grain boundaries, dislocations) [23, 139, 154, 155], surface/interfacial chemistry [23], inhomogeneous Li/electrolyte contact [38], Li wettability [23], ionic conductivity at the grain boundaries [13], interfacial resistance [23, 124], and even proximity to current collectors [154], in addition to the mechanical considerations all contribute to different degrees to Li-metal propagation through the electrolyte and still remain under investigation.

The microstructure of the solid electrolyte, including pre-existing surface defects (e.g. pores, surface cracks, grains, gain boundaries, defects), can affect the local Li-ion concentration and transport properties, instigating dendrite formation, inducing crack opening, and affecting the Li/solid electrolyte interface mechanical integrity [123, 154, 156]. Generally speaking, the slower self-diffusion of Li metal compared with that of Li-ions in LLZO electrolyte was hypothesized to create a greater flux of Li-ions toward the interface relative to that of metallic lithium away from the interface, leading to Li metal build-up and "hot-spots" [124]. Similarly, as Li-ion migration cannot occur through voids and pores in the solid electrolyte, which in turn lowers the Li-ion flux, creating a nonuniform ionic charge distribution profile. A strong Li-ion flux gradient in the solid electrolyte may lead to high local concentration polarization and drive degradation mechanisms by creating "hot spots" and leading to Li deposition within those defected areas [156]. Once the flaws are filled with Li, mechanical stress is built within both Li metal and the solid electrolyte and is expected to continuously rise upon further Li plating/propagation, extending the surface defects and inducing crack opening, even in case of a low-shear-modulus metal such as Li [123]. Furthermore, the interfacial stress and electrical potential were combined to describe the local chemical potential of Li metal at the solid electrolyte interface and to further analyze the nucleation and growth of Li dendrites at grain boundaries [13]. Both the ionic conductivity and the mechanical properties (fracture stress, ability to resist fracture) of the solid electrolyte in addition to the interfacial stresses were observed to be critical parameters affecting the CCD leading to dendrite formation [13]. Low ionic conductivity at the grain boundaries, high electronic conductivity and/or physical irregularities in the shape of the Li/solid electrolyte interface may lead to a (electro)chemo-mechanical potential of Li that will provide a driving force for the nucleation of Li dendrites [13]. Moreover, as the specific grain boundary, resistance is up to 2 orders of magnitude larger with higher electronic conduction, compared to the grain lattice contribution, the transport of Li ion may be blocked at the grain boundary zones and thus becoming more susceptible for Li metal precipitation and propagation, which necessitates increasing the ionic conductivity or decreasing the electronic conductivity of the grain boundaries to suppress Li dendrite growth. In case of the low electronic conductivity LIPON (10^{-15}–10^{-12} S cm^{-1}) [157, 158], metallic Li formation was alleviated because of the lack of surface states that could trap excess electrons. The excess electrons would in turn migrated to the bulk of LIPON, which has a low Li nucleation tendency and thus effectively suppressing the formation of Li dendrites [155]. It was hypothesized that the reduction of LLZO at very high overpotentials and the formation of metallic states may facilitate Li-metal nucleation along grain boundaries [13], pore surfaces, and Li/LLZO interfaces for sufficient electronic conduction therein [25, 147, 155]. The ability of pore surfaces to trap excess electrons, localized around La^{+3}, thermodynamically favoring the reduction of Li^{+}, provides possible electron pathways for metallic lithium formation in LLZO [155]. Additionally, theoretical calculations [155] have also predicted that the formation of a stable 6-nm-thick interfacial tetragonal LLZO phase at the cubic LLZO/Li metal interface [102] will not prevent metallic Li formation due to trapping of excess electrons, essential to Li nucleation [155]. Thus, microstructural and grain boundary tuning are an important engineering tool toward achieving higher CCD for the stripping and plating of Li metal.

Although some efforts have been focused on eliminating grain boundaries (via single crystal, amorphous phases), the observation of lithium dendrites in single-crystal LLZO [154, 159] suggests that although lithium metal preferably propagates intergranularly in LLZO, grain boundaries are not a prerequisite for the evolution of Li filaments alone. Poor adhesion [38] and poor Li wettability [23], surface contamination (LiOH and Li_2CO_3) [23] may also have adverse effects on the Li/electrolyte interfacial resistance [23, 124], leading to localized high-current-density "hot-spots" and electrochemically driven mechanical stresses. As a general guideline, solid electrolytes should have a moderate elastic modulus (~tens of GPa) to be able to, on the one hand, suppress dendrite formation and, on the other hand, prevent stress evolution at the interface through elastic deformation [160]. Nevertheless, a high modulus is a fly in the ointment as, unlike liquid electrolytes, when Li metal recedes (several micrometers) during discharge, the adhesion between the Li metal and solid electrolyte cannot be maintained for electrolytes with moduli higher than a few MPa [138], leading to poor electrode–electrolyte adhesion, higher interfacial resistance, and non-uniform current densities, which may lead to dendrite growth [12, 124].

Strategies to suppress dendrite propagation while promoting fast charging rates with high lithium passage per cycle are needed [161]. One possible strategy involves reduction of the high Li/electrolyte interfacial resistance (by 1–2 orders of magnitude) by improving the physical contact between the two materials by applying pressure, mechanical polishing, or applying heat treatment in an inert environment [124, 125, 162] and by improving the Li metal wettability on the electrolyte by adding a thin coating or buffer layers of Al_2O_3 [163], ZnO [143], amorphous LPO [33], Au [38, 154], or Si [164]. Improved interfacial contact can be also achieved through organic–inorganic hybridization, where a polymer electrolyte (elastic modulus < 0.1 GPa) is sandwiched between the Li metal and ceramic electrolyte. Another strategy involves the addition of a Li-alloying interlayer [154] to reduce the energetic barrier for Li plating (overpotential), smooth the voltage drop between the electrolyte and electrode [12], and delay Li penetration and propagation. An ultra-low Li/LLZO interfacial resistance on the order of $2\,\Omega\,cm^2$ was achieved through a simple coating-free process of wet polishing followed by heat treatment (500 °C) in an inert atmosphere [23, 124], with a CCD of $1\,mA\,cm^{-2}$ [124]. Nevertheless, Li dendrite propagation along grain boundaries [154] makes the microstructure, electronic transport at the surface and bulk, and ionic diffusion kinetics at the grain boundaries important parameters [139]. A strategy involving the compositional tuning of the solid electrolyte (e.g. Li halides, doping in LLZO) [145, 150, 165–167] to tailor its electro–chemo–mechanics properties and alter the solid–electrolyte interphase composition and transport properties and thus manipulate the nature of the interface between the solid electrolyte and Li metal was largely explored [30]. Based on the understanding that the Li/electrolyte interface stability and nature (chemical, mechanical, microstructure, etc.) of the interphase layer largely affect Li dendrite formation and ultimately the battery rate and cycle performance, another strategy, similar to that used in LIBs where electrolyte fluorinated additives have been employed, has been suggested to tailor a stable and efficient interphase layer and suppress dendrite growth [152]. Another strategy involving the addition of a liquid additive to the solid electrolyte has also been explored as a tool to mitigate Li dendrite propagation by reacting (e.g. to form LiF) and consuming Li dendrites and thus suppressing their growth [167].

Pathways toward dendrite-free all-solid-state Li-based battery cells include but are not limited to suppressing Li dendrites and increasing the CCD by increasing the ionic conductivity or lowering the electronic conductivity at grain boundaries, introducing dendrite-free LIPON or Li_3PO_4 [166] thin layers between the electrolyte and Li metal, reducing the concentration of defects and flaws and the overall interfacial resistance, increasing solid electrolyte/Li reaction sites, and modifying the Li nucleation tendency through defect generation or doping (e.g. to affect the La atoms on the surface of LLZO). Other approaches include engineering of the current collector. A negative electrode Au- or Ni-coated porous, instead of a planar, current collector with micro-sized pores was utilized to instigate the preferable plating and stripping of Li metal in the pores and mitigate Li dendrite propagation through the solid electrolyte (Figure 9.8A) [168]. Another strategy involved the relief of compressive stresses during Li electroplating on Cu current, which is a major driving force for Li dendrite growth, through a surface-wrinkling-induced stress-relaxation mechanism enabled when the current collector was supported by a soft substrate (polydimethylsiloxane [PDMS]) (Figure 9.8B) [122].

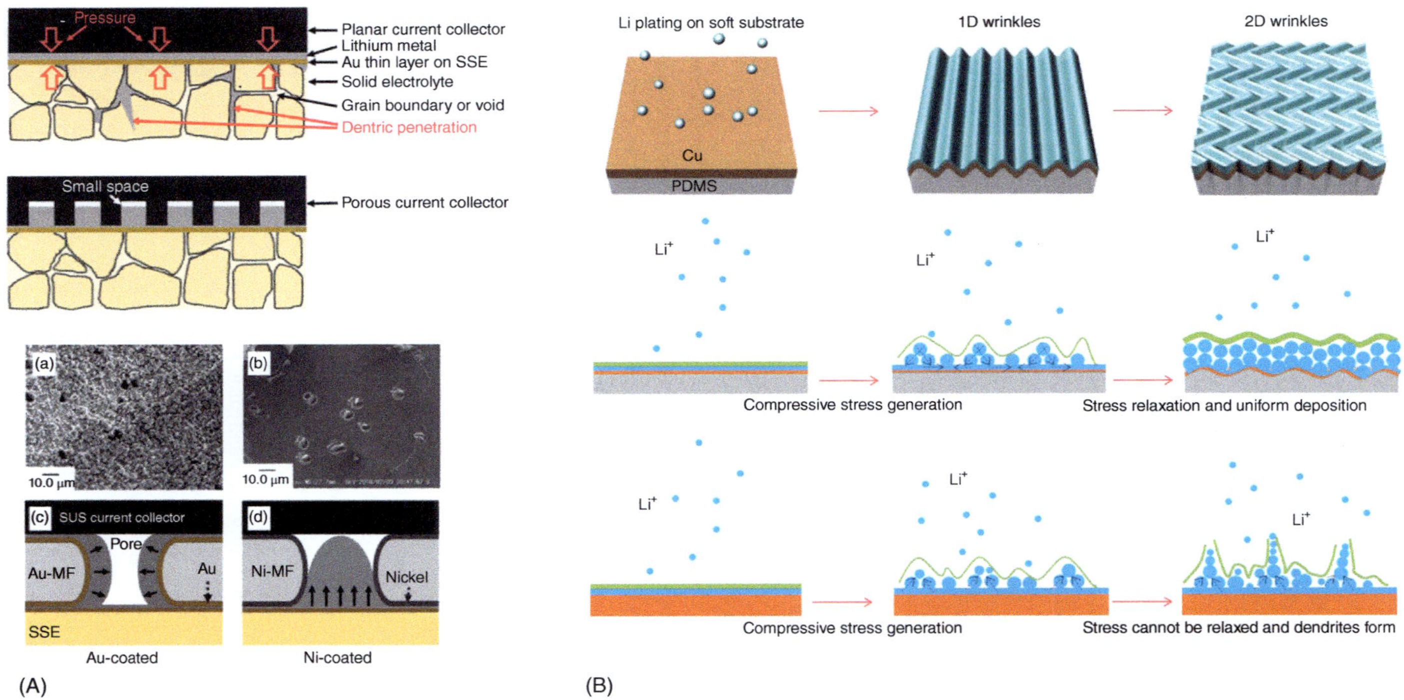

Figure 9.8 Strategies to mitigate Li dendrite formation by (A) porous current collector (Au and Ni), (B) soft PDMS substrate to accommodate compressive stress generated during Li plating. Source: (A) Shinzo et al. [168]/with permission of American Chemical Society, (B) Wang et al. [122]/with permission of Springer Nature.

9.5 Outlook and Perspective

Solid-state Li metal batteries employing inorganic oxide solid electrolytes are shedding light on a safer and more efficient next generation of rechargeable batteries. Although solid-state batteries can offer high gravimetric and volumetric energy of 250 Wh kg^{-1} and 700 Wh l^{-1}, respectively, the slow kinetics can impair the fast discharge and charge performance. This roadmap for the development of successful ASSLBs focuses on five parameters: energy density, power density, long-term stability, processing, and safety. Although solid electrolytes may offer the ultimate solution toward a safe high-energy-density battery, reduced at low potentials (by Li metal) and oxidized at intermediate potentials, the performance of ASSLBs still falls short of that of batteries using liquid electrolytes and requires attention. Safety will not automatically be enhanced in battery applications through the use of solid electrolytes, especially with the use of Li metal. Improving the safety of ASSLBs relies on mindful interface-tailored solutions and engineering strategies to secure good passivation layers and stabilized interfaces over a wide temperature range and under diverse battery operating conditions [169, 170]. Depending on the processing technique and preparation conditions, the electrode/electrolyte interface may become the bottleneck dictating the power density of the ASSLB. We highlight our perspective on the challenges and mitigation strategies associated with oxide solid electrolytes and their interfaces with Li-metal anodes and oxide-based cathodes toward the realization of high-energy-density ASSLBs.

On the anode side, the use of a Li-metal anode imposes challenges associated with Li dendrite growth and instabilities of the SEI at the Li/electrolyte interface in addition to possible "wettability" issues. The importance of adhesion with the Li-metal anode was illustrated by the improved mechanical strength and reduced resistance for Li-ion transport at the interface achieved by improving the "wettability" of Li metal on an LLZO electrolyte through simple surface treatment [163]. Transitioning from a liquid to a solid electrolyte toward an ASSLB introduces new interfacial challenges arising from (i) poor solid–solid contacts between the Li metal and the rigid oxide solid electrolyte or (ii) poor chemical stability between the Li metal and oxide solid electrolytes and thus the continuous growth of the SEI. Despite much progress in addressing this topic, cycling at high current densities (namely >3 mAh cm^{-2}) remains a major challenge that demands innovative approaches. Considering (i) adhesion to both Li and the solid electrolyte, (ii) chemical compatibility, and (iii) electrochemical compatibility, in addition to (iv) sufficient ionic and negligible electronic conductivity, more attention should be placed on Li binary compounds (Li_3N, LiF, Li_2O, etc.) as intermediate coating materials [21], in addition to coating layers with reduced electronic conductivities (e.g. LiPON, Li_3PO_4) at the Li/electrolyte interface to improve the coulombic efficiency and suppress dendrite growth. The appropriate pairing of the electrode material with a solid electrolyte to minimize mutual decomposition suggests the use of Li alloy. Increasing the interfacial area by incorporating Li metal into a highly porous electrolyte improves the critical current density for garnets. Nonetheless, cost and energy considerations are still needed to prove the superiority of such systems over typical LIBs. An approach where an anode-less cell with only a current collector is utilized and the cathode is the only source for Li metal plating, without any need for Li excess, may reduce the costs associated with the processing of ultrathin Li metal [171]. Currently, anode-less solid-state batteries cells with high energy density show great potential; however, cost-effective alternatives to precious metals are needed.

On the cathode side, when the electrolyte is placed in contact with a conductive additive such as carbon or the current collector, decomposition can occur, necessitating an expansion of coating strategies to additives and current collectors [172, 173]. Tailored interfaces with excellent adhesion between electrodes are mandatory to stabilize the electrode/electrolyte interfaces, resist delamination, and decrease interfacial resistance to achieve acceptable performance (thousands of cycles, $>1\ mA\ cm^{-2}$, >1C) [170]. Despite the excellent chemical, mechanical, and electrochemical properties of any type of solid electrolyte, low bulk, grain-boundary, and interfacial resistance between the electrode and electrolyte are prerequisites for their widespread commercial adoption. Interfacial failure processes originating from poor chemical stability, electrochemical oxidation, and chemo-mechanical degradation during cell fabrication or operation (cycling) are of prime concern and typically result in a continuous increase of the interfacial resistance between the active materials and solid electrolyte in a cathode composite. Considering the different major bonding mechanisms, oxide-based cathode composites are expected to exhibit stronger bonding driven by elevated-temperature sintering because both mechanical bonding and chemically induced bonding are involved. As soon as a processing method to produce more stable oxide interfaces is established, it could potentially provide opportunities to avoid interfacial cracking and the need for applied pressure by accommodating the stress level created during volume change of active materials [174, 175]. Overall, the formation of a mechanically and chemically stable SEI with high ionic conductivity but low electronic conductivity is a key condition for long-term battery operation. Regardless of the solid electrolyte selected, artificial protection on the cathode surface by introducing coating layers such as $Li_4Ti_5O_{12}$, $LiTaO_3$, $LiNbO_3$, Li_2SiO_3, or Li_3PO_4 has been shown to reduce the resistance at the cathode/electrolyte interface.

To conclude, the development of ASSLBs is very promising but remains in its infancy. Reports on the performance of full solid-state Li-based batteries involving garnet-type LLZO mainly with layered oxides (LCO, NMC) because of their excellent capacity, compatibility, and electronic conductivity, are growing; however, the full-cell performance requires further attention as it is currently representing only tens or hundreds of cycles and at low rates < 1C. SSBs employing oxides such as LLZO often exhibit strong charging polarization because of the cathode/electrolyte interface and interfaces within the composite cathode [176], leaving the state-of-the-art performance of oxide-based ASSLBs far behind in the race, limited to an irreversible capacity up to 70 [177] or 100 cycles [9]. Suitable coating strategy (composition, processing, thickness) combined with precise interfacial characterization of the buried interfaces should be further established and explored in a full SSB cell, preferably in a three-electrode cell configuration, which, although challenging in terms of fabrication, will better resolve the battery failure modes and clarify their potential feasibility if and when they are put into practical use.

Acknowledgments

This work was supported in parts by NGK Inc. and the National Science Foundation MRSEC Program (grant no. DMR-1419807). M.B. acknowledges financial support from the US–Israel Fulbright Program, the Zuckerman Israeli Postdoctoral Scholar Program, and the MIT-Technion Postdoctoral Fellowship. L.K. acknowledges financial support from Xi'an Jiaotong university and the office of the Chinese postdoctoral council. J.L.M.R. thanks the Thomas Lord Foundation for financial support.

Contributions

K.J.K and M.B contributed equally to this work. K.J.K and M.B initiated the concept of the paper, contributed equally to the research, figure preparation, discussion of data, writing, editing, and proofreading of the manuscript. M.W and L.K contributed to all figures and tables preparation. J.L.M.R. discussed, edited, and revised the manuscript in all parts.

Ethics Declarations

The authors declare no competing interests.

References

1 Judez, X., Eshetu, G.G., Li, C. et al. (2018). Opportunities for rechargeable solid-state batteries based on Li-intercalation cathodes. *Joule* 2 (11): 2208–2224.
2 Kato, Y., Hori, S., Saito, T. et al. (2016). High-power all-solid-state batteries using sulfide superionic conductors. *Nature Energy* 1 (4): 1–7.
3 Nagao, M., Hayashi, A., and Tatsumisago, M. (2012). High-capacity Li_2S-nanocarbon composite electrode for all-solid-state rechargeable lithium batteries. *Journal of Materials Chemistry* 22 (19): 10015–10020.
4 Li, J., Ma, C., Chi, M. et al. (2015). Solid electrolyte: the key for high-voltage lithium batteries. *Advanced Energy Materials* 5 (4): 1–6.
5 Balaish, M., Gonzalez-Rosillo, J.C., Kim, K.J. et al. (2020). A thin but solid perspective for oxide battery electrolytes. *Nature Energy* 6: 227–239.
6 Zhu, Y., Gonzalez-Rosillo, J.C., Balaish, M. et al. (2020). Lithium-film ceramics for solid-state lithionic devices. *Nature Reviews Materials* 6: 313–331.
7 Janek, J. and Zeier, W.G. (2016). A solid future for battery development. *Nature Energy* 1 (9): 1–4.
8 Randau, S., Weber, D.A., Kötz, O. et al. (2020). Benchmarking the performance of all-solid-state lithium batteries. *Nature Energy* 5 (March): 1–12.
9 Han, F., Yue, J., Chen, C. et al. (2018). Interphase engineering enabled all-ceramic lithium battery. *Joule* 2 (3): 497–508.

10 Gordon, I.A.J., Grugeon, S., Takenouti, H. et al. (2017). Electrochemical impedance spectroscopy response study of a commercial graphite-based negative electrode for Li-ion batteries as function of the cell state of charge and ageing. *Electrochimica Acta* 223: 63–73.

11 Kim, K.J., Balaish, M., Wadaguchi, M. et al. (2020). Solid-state Li-metal batteries: challenges and horizons of oxide and sulfide solid electrolytes and their interfaces. *Advanced Energy Materials* 11, 2002689.

12 Famprikis, T., Canepa, P., Dawson, J.A. et al. (2019). Fundamentals of inorganic solid-state electrolytes for batteries. *Nature Materials* 18 (12): 1278–1291.

13 Raj, R. and Wolfenstine, J. (2017). Current limit diagrams for dendrite formation in solid-state electrolytes for Li-ion batteries. *Journal of Power Sources* 343: 119–126.

14 Ishigaki, N., Kataoka, K., Morikawa, D. et al. (2020). Structural and Li-ion diffusion properties of lithium tantalum phosphate $LiTa_2PO_8$. *Solid State Ionics* 351 (April): 115314.

15 Kim, J., Kim, J., Avdeev, M. et al. (2018). $LiTa_2PO_8$: a fast lithium-ion conductor with new framework structure. *Journal of Materials Chemistry A* 6 (45): 22478–22482.

16 Pfenninger, R., Struzik, M., Garbayo, I. et al. (2019). A low ride on processing temperature for fast lithium conduction in garnet solid-state battery films. *Nature Energy* 4: 475–483.

17 Mahbub, R., Huang, K., Jensen, Z. et al. (2020). Text mining for processing conditions of solid-state battery electrolytes. *Electrochemistry Communications* 121: 106860.

18 Afyon, S., Kravchyk, K.V., Wang, S. et al. (2019). Building better all-solid-state batteries with Li-garnet solid electrolytes and metalloid anodes. *Journal of Materials Chemistry A* 7 (37): 21299–21308.

19 Hänsel, C., Afyon, S., and Rupp, J.L.M. (2016). Investigating the all-solid-state batteries based on lithium garnets and a high potential cathode-$LiMn_{1.5}Ni_{0.5}O_4$. *Nanoscale* 8 (43): 18412–18420.

20 Afyon, S., Krumeich, F., and Rupp, J.L.M. (2015). A shortcut to garnet-type fast Li-ion conductors for all-solid state batteries. *Journal of Materials Chemistry A* 3 (36): 18636–18648.

21 Zhu, Y., He, X., and Mo, Y. (2015). Origin of outstanding stability in the lithium solid electrolyte materials: insights from thermodynamic analyses based on first-principles calculations. *ACS Applied Materials & Interfaces* 7 (42): 23685–23693.

22 Gao, J., Guo, X.X., Li, Y. et al. (2019). The ab initio calculations on the areal specific resistance of Li-metal/$Li_7La_3Zr_2O_{12}$ interphase. *Advanced Theory Simulations* 2, 1900028.

23 Sharafi, A., Kazyak, E., Davis, A.L. et al. (2017). Surface chemistry mechanism of ultra-low interfacial resistance in the solid-state electrolyte $Li_7La_3Zr_2O_{12}$. *Chemistry of Materials* 29 (18): 7961–7968.

24 Krauskopf, T., Hartmann, H., Zeier, W.G., and Janek, J. (2019). Toward a fundamental understanding of the lithium metal anode in solid-state batteries – an electrochemo-mechanical study on the garnet-type solid electrolyte $Li_{6.25}Al_{0.25}La_3Zr_2O_{12}$. *ACS Applied Materials & Interfaces* 11 (15): 14463–14477.

25 Han, F., Zhu, Y., He, X. et al. (2016). Electrochemical stability of $Li_{10}GeP_2S_{12}$ and $Li_7La_3Zr_2O_{12}$ solid electrolytes. *Advanced Energy Materials* 6 (8): 1–9.

26 Zhu, Y., He, X., and Mo, Y. (2016). First principles study on electrochemical and chemical stability of solid electrolyte-electrode interfaces in all-solid-state Li-ion batteries. *Journal of Materials Chemistry A* 4 (9): 3253–3266.

27 Nakayama, M., Kotobuki, M., Munakata, H. et al. (2012). First-principles density functional calculation of electrochemical stability of fast Li ion conducting garnet-type oxides. *Physical Chemistry Chemical Physics* 14 (28): 10008–10014.

28 Ohta, S., Kobayashi, T., and Asaoka, T. (2011). High lithium ionic conductivity in the garnet-type oxide $Li_{7-X}La_3(Zr_{2-X}, Nb_X)O_{12}$ ($X = 0$–2). *Journal of Power Sources* 196 (6): 3342–3345.

29 Miara, L., Windmüller, A., Tsai, C.L. et al. (2016). About the compatibility between high voltage spinel cathode materials and solid oxide electrolytes as a function of temperature. *ACS Applied Materials & Interfaces* 8 (40): 26842–26850.

30 Zhang, Z., Zhang, L., Yan, X. et al. (2019). All-in-one improvement toward Li_6PS_5Br-based solid electrolytes triggered by compositional tune. *Journal of Power Sources* 410–411: 162–170.

31 Tao, Y., Chen, S., Liu, D. et al. (2016). Lithium superionic conducting oxysulfide solid electrolyte with excellent stability against lithium metal for all-solid-state cells. *Journal of the Electrochemical Society* 163 (2): A96–A101.

32 Kato, A., Hayashi, A., and Tatsumisago, M. (2016). Enhancing utilization of lithium metal electrodes in all-solid-state batteries by interface modification with gold thin films. *Journal of Power Sources* 309: 27–32.

33 Deng, T., Ji, X., Zhao, Y. et al. (2020). Tuning the anode – electrolyte Interface chemistry for garnet-based solid-state Li metal batteries. *Advanced Materials* 2000030 (23): 1–10.

34 Xu, B., Li, W., Duan, H. et al. (2017). Li_3PO_4-added garnet-type $Li_{6.5}La_3Zr_{1.5}Ta_{0.5}O_{12}$ for Li-dendrite suppression. *Journal of Power Sources* 354: 68–73.

35 Liu, B., Gong, Y., Fu, K. et al. (2017). Garnet solid electrolyte protected Li-metal batteries. *ACS Applied Materials & Interfaces* 9 (22): 18809–18815.

36 Zhou, W., Wang, S., Li, Y. et al. (2016). Plating a dendrite-free lithium anode with a polymer/ceramic/polymer sandwich electrolyte. *Journal of the American Chemical Society* 138 (30): 9385–9388.

37 Fu, K.K., Gong, Y., Liu, B. et al. (2017). Toward garnet electrolyte-based Li metal batteries: an ultrathin, highly effective, artificial solid-state electrolyte/metallic Li interface. *Science Advances* 3 (4): 1–12.

38 Tsai, C.L., Roddatis, V., Chandran, C.V. et al. (2016). $Li_7La_3Zr_2O_{12}$ interface modification for Li dendrite prevention. *ACS Applied Materials & Interfaces* 8 (16): 10617–10626.

39 Xiao, Y., Wang, Y., Bo, S.H. et al. (2020). Understanding interface stability in solid-state batteries. *Nature Reviews Materials* 5 (2): 105–126.

40 Tsai, C.L., Ma, Q., Dellen, C. et al. (2019). A garnet structure-based all-solid-state Li battery without interface modification: resolving incompatibility issues on positive electrodes. *Sustainable Energy & Fuels* 3 (1): 280–291.

41 Koerver, R., Aygün, I., Leichtweiß, T. et al. (2017). Capacity fade in solid-state batteries: interphase formation and chemomechanical processes in nickel-rich layered oxide cathodes and lithium thiophosphate solid electrolytes. *Chemistry of Materials* 29 (13): 5574–5582.

42 Chawla, K.K. (2003) *Composite Materials Science and Engineering*. Springer: New York.

43 Smith, B.J., Phillip, G.M., and Sweeney, M.E. (2019). A bird's-eye view of Li-stuffed garnet-type $Li_7La_3Zr_2O_{12}$ ceramic electrolytes for advanced all-solid-state Li batteries. *Environmental Sciences* 12: 1–502.

44 Kasemchainan, J., Zekoll, S., Spencer Jolly, D. et al. (2019). Critical stripping current leads to dendrite formation on plating in lithium anode solid electrolyte cells. *Nature Materials* 18 (10): 1105–1111.

45 Koerver, R., Zhang, W., De Biasi, L. et al. (2018). Chemo-mechanical expansion of lithium electrode materials-on the route to mechanically optimized all-solid-state batteries. *Energy & Environmental Science* 11 (8): 2142–2158.

46 Zhang, W., Schröder, D., Arlt, T. et al. (2017). (Electro)chemical expansion during cycling: monitoring the pressure changes in operating solid-state lithium batteries. *Journal of Materials Chemistry A* 5 (20): 9929–9936.

47 Gotzen, B., Dornseiffer, J., Tsai, C.-L.L. et al. (2017). Cathode-electrolyte material interactions during manufacturing of inorganic solid-state lithium batteries. *Journal of Electroceramics* 38 (2–4): 197–206.

48 Park, K., Yu, B.C., Jung, J.W. et al. (2016). Electrochemical nature of the cathode interface for a solid-state lithium-ion battery: interface between $LiCoO_2$ and garnet-$Li_7La_3Zr_2O_{12}$. *Chemistry of Materials* 28 (21): 8051–8059.

49 Wakasugi, J., Munakata, H., and Kanamura, K. (2017). Thermal stability of various cathode materials against $Li_{6.25}Al_{0.25}La_3Zr_2O_{12}$ electrolyte. *Electrochemistry* 85 (2): 77–81.

50 Ren, Y., Liu, T., Shen, Y. et al. (2016). Chemical compatibility between garnet-like solid state electrolyte $Li_{6.75}La_3Zr_{1.75}Ta_{0.25}O_{12}$ and major commercial lithium battery cathode materials. *Journal of Materials* 2 (3): 256–264.

51 Kim, K.H., Iriyama, Y., Yamamoto, K. et al. (2011). Characterization of the interface between $LiCoO_2$ and $Li_7La_3Zr_2O_{12}$ in an all-solid-state rechargeable lithium battery. *Journal of Power Sources* 196 (2): 764–767.

52 Jalem, R., Morishita, Y., Okajima, T. et al. (2016). Experimental and first-principles DFT study on the electrochemical reactivity of garnet-type solid electrolytes with carbon. *Journal of Materials Chemistry A* 4 (37): 14371–14379.

53 Richards, W.D., Miara, L.J., Wang, Y. et al. (2016). Interface stability in solid-state batteries. *Chemistry of Materials* 28 (1): 266–273.

54 Miara, L.J., Richards, W.D., Wang, Y.E., and Ceder, G. (2015). First-principles studies on cation dopants and electrolyte|cathode interphases for lithium garnets. *Chemistry of Materials* 27: 4040–4047.

55 Xiao, Y., Miara, L.J., Wang, Y., and Ceder, G. (2019). Computational screening of cathode coatings for solid-state batteries. *Joule* 3 (5): 1252–1275.

56 Thangadurai, V. and Weppner, W. (2005). Investigations on electrical conductivity and chemical compatibility between fast lithium ion conducting garnet-like $Li_6BaLa_2Ta_2O_{12}$ and lithium battery cathodes. *Journal of Power Sources* 142 (1–2): 339–344.

57 Kotobuki, M., Munakata, H., Kanamura, K. et al. (2010). Compatibility of $Li_7La_3Zr_2O_{12}$ solid electrolyte to all-solid-state battery using Li metal anode. *Journal of the Electrochemical Society* 157 (10): A1076.

58 Bitzer, M., Van Gestel, T., and Uhlenbruck, S. (2016). Sol-gel synthesis of thin solid $Li_7La_3Zr_2O_{12}$ electrolyte films for Li-ion batteries. *Thin Solid Films* 615: 128–134.

59 Ohta, S., Kobayashi, T., Seki, J., and Asaoka, T. (2012). Electrochemical performance of an all-solid-state lithium ion battery with garnet-type oxide electrolyte. *Journal of Power Sources* 202: 332–335.

60 Ellis, B.L., Lee, K.T., and Nazar, L.F. (2010). Positive electrode materials for Li-ion and Li-batteries. *Chemistry of Materials* 22 (3): 691–714.

61 Ishidzu, K., Oka, Y., and Nakamura, T. (2016). Lattice volume change during charge/discharge reaction and cycle performance of $Li[Ni_xCo_yMn_z]O_2$. *Solid State Ionics* 288: 176–179.

62 Manthiram, A. (2020). A reflection on lithium-ion battery cathode chemistry. *Nature Communications* 11 (1): 1–9.

63 Shao, Y., Wang, H., Gong, Z. et al. (2018). Drawing a soft interface: an effective interfacial modification strategy for garnet-type solid-state Li batteries. *ACS Energy Letters* 3 (6): 1212–1218.

64 Fu, K., Gong, Y., Hitz, G.T. et al. (2017). Three-dimensional bilayer garnet solid electrolyte based high energy density lithium metal-sulfur batteries. *Energy & Environmental Science* 10 (7): 1568–1575.

65 Tao, X., Liu, Y., Liu, W. et al. (2017). Solid-state lithium-sulfur batteries operated at 37 °C with composites of nanostructured $Li_7La_3Zr_2O_{12}$/carbon foam and polymer. *Nano Letters* 17 (5): 2967–2972.

66 Culver, S.P., Koerver, R., Zeier, W.G., and Janek, J. (2019). On the functionality of coatings for cathode active materials in thiophosphate-based all-solid-state batteries. *Advanced Energy Materials* 9 (24): 1–14.

67 Nolan, A.M., Zhu, Y., He, X. et al. (2018). Computation-accelerated design of materials and interfaces for all-solid-state lithium-ion batteries. *Joule* 2 (10): 2016–2046.

68 Kato, T., Hamanaka, T., Yamamoto, K. et al. (2014). In-situ $Li_7La_3Zr_2O_{12}$/$LiCoO_2$ interface modification for advanced all-solid-state battery. *Journal of Power Sources* 260 (50921061): 292–298.

69 Sakuda, A., Takeuchi, T., and Kobayashi, H. (2016). Electrode morphology in all-solid-state lithium secondary batteries consisting of $LiNi_1/3Co_1/3Mn_1/3O_2$ and Li_2S-P_2S_5 solid electrolytes. *Solid State Ionics* 285: 112–117.

70 Calpa, M., Rosero-Navarro, N.C., Miura, A., and Tadanaga, K. (2019). Electrochemical performance of bulk-type all-solid-state batteries using small-sized

$Li_7P_3S_{11}$ solid electrolyte prepared by liquid phase as the ionic conductor in the composite cathode. *Electrochimica Acta* 296: 473–480.

71 Wan, H., Mwizerwa, J.P., Qi, X. et al. (2018). Nanoscaled Na_3PS_4 solid electrolyte for all-solid-state FeS_2/Na batteries with ultrahigh initial coulombic efficiency of 95% and excellent cyclic performances. *ACS Applied Materials & Interfaces* 10 (15): 12300–12304.

72 Strauss, F., Bartsch, T., De Biasi, L. et al. (2018). Impact of cathode material particle size on the capacity of bulk-type all-solid-state batteries. *ACS Energy Letters* 3 (4): 992–996.

73 Bielefeld, A., Weber, D.A., and Janek, J. (2019). Microstructural modeling of composite cathodes for all-solid-state batteries. *Journal of Physical Chemistry C* 123 (3): 1626–1634.

74 Murugan, R., Thangadurai, V., and Weppner, W. (2007). Fast lithium ion conduction in garnet-type $Li_7La_3Zr_2O_{12}$. *Angewandte Chemie, International Edition* 46 (41): 7778–7781.

75 Kotobuki, M., Kanamura, K., Sato, Y., and Yoshida, T. (2011). Fabrication of all-solid-state lithium battery with lithium metal anode using Al_2O_3-added $Li_7La_3Zr_2O_{12}$ solid electrolyte. *Journal of Power Sources* 196 (18): 7750–7754.

76 Li, Y., Han, J.T., Wang, C.A. et al. (2012). Optimizing Li+ conductivity in a garnet framework. *Journal of Materials Chemistry* 22 (30): 15357–15361.

77 El Shinawi, H. and Janek, J. (2013). Stabilization of cubic lithium-stuffed garnets of the type "$Li_7La_3Zr_2O_{12}$" by addition of gallium. *Journal of Power Sources* 225: 13–19.

78 Ohta, S., Seki, J., Yagi, Y. et al. (2014). Co-sinterable lithium garnet-type oxide electrolyte with cathode for all-solid-state lithium ion battery. *Journal of Power Sources* 265: 40–44.

79 Tatsumisago, M., Machida, N., and Minami, T. (1987). Mixed anion effect in conductivity of rapidly quenched Li_4SiO_4–Li_3Bo_3 glasses. *Yogyo Kyokai Shi/ Journal of the Ceramic Society of Japan* 95 (2 Pt 2): 197–201.

80 Kotobuki, M., Munakata, H., and Kanamura, K. (2011). Fabrication of all-solid-state rechargeable lithium-ion battery using mille-feuille structure of $Li_{0.35}La_{0.55}TiO_3$. *Journal of Power Sources* 196 (16): 6947–6950.

81 Kotobuki, M., Suzuki, Y., Kanamura, K. et al. (2011). A novel structure of ceramics electrolyte for future lithium battery. *Journal of Power Sources* 196 (22): 9815–9819.

82 van den Broek, J., Afyon, S., and Rupp, J.L.M. (2016). Interface-engineered all-solid-state Li-ion batteries based on garnet-type fast Li+ conductors. *Advanced Energy Materials* 6 (19), 1600736.

83 Ren, Y., Liu, T., Shen, Y. et al. (2017). Garnet-type oxide electrolyte with novel porous-dense bilayer configuration for rechargeable all-solid-state lithium batteries. *Ionics (Kiel)* 23 (9): 2521–2527.

84 Lan, T., Tsai, C.L., Tietz, F. et al. (2019). Room-temperature all-solid-state sodium batteries with robust ceramic interface between rigid electrolyte and electrode materials. *Nano Energy* 65 (July): 104040.

85 Kim, K.J. and Rupp, J.L.M. (2020). All ceramic cathode composite design and manufacturing towards low interfacial resistance for garnet-based solid-state lithium batteries. *Energy & Environmental Science* 12: 4930–4945.

86 Zhang, Z., Zhang, Q., Shi, J., et al. (2017). A self-forming composite electrolyte for solid-state sodium battery with ultralong cycle life. *Adv. Energy Mater.* 7, 1601196–n/a.

87 Jian, Z.L., Han, W.Z., Lu, X., et al. (2013). Superior electrochemical performance and storage mechanism of Na_3V_2 $(PO_4)_3$ cathode for room-temperature sodium-ion batteries. *Adv. Energy Mater.* 13: 156–160.

88 Shi, T., Tu, Q., Tian, Y. et al. (2019). High active material loading in all-solid-state battery electrode via particle size optimization. *Advanced Energy Materials* 1902881 (1), 1902881.

89 Zhang, W., Weber, D.A., Weigand, H. et al. (2017). Interfacial processes and influence of composite cathode microstructure controlling the performance of all-solid-state lithium batteries. *ACS Applied Materials & Interfaces* 9 (21): 17835–17845.

90 Liu, T., Zhang, Y., Zhang, X. et al. (2018). Enhanced electrochemical performance of bulk type oxide ceramic lithium batteries enabled by interface modification. *Journal of Materials Chemistry A* 6 (11): 4649–4657.

91 Neudecker, B.J. (1996). Li_9SiAlO_8: a lithium ion electrolyte for voltages above 5.4 V. *Journal of the Electrochemical Society* 143 (7): 2198.

92 Albertus, P., Babinec, S., Litzelman, S., and Newman, A. (2018). Status and challenges in enabling the lithium metal electrode for high-energy and low-cost rechargeable batteries. *Nature Energy* 3 (1): 16–21.

93 Schnell, J., Knörzer, H., Imbsweiler, A.J., and Reinhart, G. (2020). Solid versus liquid – a bottom-up calculation model to analyze the manufacturing cost of future high-energy batteries. *Energy Technology* 8 (3), 1901237.

94 Wang, S.H., Yue, J., Dong, W. et al. (2019). Tuning wettability of molten lithium via a chemical strategy fora lithium metal anodes. *Nature Communications* 10 (1): 4930.

95 Schmuch, R., Wagner, R., Hörpel, G. et al. (2018). Performance and cost of materials for lithium-based rechargeable automotive batteries. *Nature Energy* 3: 267–278.

96 Lin, D., Liu, Y., Liang, Z. et al. (2016). Layered reduced graphene oxide with nanoscale interlayer gaps as a stable host for lithium metal anodes. *Nature Nanotechnology* 11 (7): 626–632.

97 Liu, Y., Lin, D., Liang, Z. et al. (2016). Lithium-coated polymeric matrix as a minimum volume-change and dendrite-free lithium metal anode. *Nature Communications* 7: 1–9.

98 Chen, Y., Wang, Z., Li, X. et al. (2020). Li metal deposition and stripping in a solid-state battery via Coble creep. *Nature* 578 (7794): 251–255.

99 Yan, K., Lu, Z., Lee, H.W. et al. (2016). Selective deposition and stable encapsulation of lithium through heterogeneous seeded growth. *Nature Energy* 1 (3): 16010.

100 Schwöbel, A., Hausbrand, R., and Jaegermann, W. (2015). Interface reactions between LiPON and lithium studied by in-situ X-ray photoemission. *Solid State Ionics* 273: 51–54.

101 Schwietert, T.K., Arszelewska, V.A., Wang, C. et al. (2020). Clarifying the relationship between redox activity and electrochemical stability in solid electrolytes. *Nature Materials* 19 (April): 428–435.

102 Ma, C., Cheng, Y., Yin, K. et al. (2016). Interfacial stability of Li metal-solid electrolyte elucidated via in situ electron microscopy. *Nano Letters* 16 (11): 7030–7036.

103 Yang, K.Y., Leu, I.C., Fung, K.Z. et al. (2008). Mechanism of the interfacial reaction between cation-deficient $La_{0.56}Li_{0.33}TiO_3$ and metallic lithium at room temperature. *Journal of Materials Research* 23 (7): 1813–1825.

104 Hartmann, P., Leichtweiss, T., Busche, M.R. et al. (2013). Degradation of NASICON-type materials in contact with lithium metal: formation of mixed conducting interphases (MCI) on solid electrolytes. *Journal of Physical Chemistry C* 117 (41): 21064–21074.

105 Binninger, T., Marcolongo, A., Mottet, M. et al. (2020). Comparison of computational methods for the electrochemical stability window of solid-state electrolyte materials. *Journal of Materials Chemistry A* 8 (3): 1347–1359.

106 Xu, X., Wen, Z., Yang, X. et al. (2006). High lithium ion conductivity glass-ceramics in Li_2O-Al_2O_3-TiO_2-P_2O_5 from nanoscaled glassy powders by mechanical milling. *Solid State Ionics* 177 (26–32 Special Issue): 2611–2615.

107 Chen, C.H. and Amine, K. (2001). Ionic conductivity, lithium insertion and extraction of lanthanum lithium titanate. *Solid State Ionics* 144 (1–2): 51–57.

108 Wenzel, S., Leichtweiss, T., Krüger, D. et al. (2015). Interphase formation on lithium solid electrolytes – an in situ approach to study interfacial reactions by photoelectron spectroscopy. *Solid State Ionics* 278: 98–105.

109 Koshikawa, H., Matsuda, S., Kamiya, K. et al. (2018). Dynamic changes in charge-transfer resistance at Li metal/$Li_7La_3Zr_2O_{12}$ interfaces during electrochemical Li dissolution/deposition cycles. *Journal of Power Sources* 376 (November 2017): 147–151.

110 Wang, Z., Santhanagopalan, D., Zhang, W. et al. (2016). In situ STEM–EELS observation of nanoscale interfacial phenomena in all-solid-state batteries. *Nano Letters* 16 (6): 3760–3767.

111 Wenzel, S., Weber, D.A., Leichtweiss, T. et al. (2016). Interphase formation and degradation of charge transfer kinetics between a lithium metal anode and highly crystalline $Li_7P_3S_{11}$ solid electrolyte. *Solid State Ionics* 286: 24–33.

112 Direct observation of the interfacial instability of the fast ionic conductor $Li_{10}GeP_2S_{12}$ at the lithium metal anode. *ACS Publications* 286: 24–33.

113 Wu, B., Wang, S., Lochala, J. et al. (2018). The role of the solid electrolyte interphase layer in preventing Li dendrite growth in solid-state batteries. *Energy & Environmental Science* 11 (7): 1803–1810.

114 Visco, S.J., Nimon, V.Y., Petrov, A. et al. (2014). Aqueous and nonaqueous lithium-air batteries enabled by water-stable lithium metal electrodes. *Journal of Solid State Electrochemistry* 18 (5): 1443–1456.

115 Il, E.A., Druzhinin, K.V., Saetova, N.S. et al. (2018). Electrochimica acta and $Li_7La_3Zr_2O_{12}$ solid electrolyte. *Electrochimica Acta* 285: 326–335.

116 Thompson, T., Yu, S., Williams, L. et al. (2017). Electrochemical window of the Li-ion solid electrolyte $Li_7La_3Zr_2O_{12}$. *ACS Energy Letters* 2 (2): 462–468.

117 Wolfenstine, J., Allen, J.L., Read, J., and Sakamoto, J. (2013). Chemical stability of cubic $Li_7La_3Zr_2O_{12}$ with molten lithium at elevated temperature. *Journal of Materials Science* 48 (17): 5846–5851.

118 Bates, J.B., Dudney, N.J., Neudecker, B. et al. (2000). Thin-film lithium and lithium-ion batteries. *Solid State Ionics* 135 (1–4): 33–45.

119 Jow, T.R. (1983). Interface between solid electrode and solid electrolyte – a study of the Li/LiI(Al_2O_3) solid-electrolyte system. *Journal of the Electrochemical Society* 130 (4): 737.

120 Meyer, M., Rickert, H., and Schwaitzer, U. (1983). Investigations on the kinetics of the anodic dissolution of lithium at the interface Li/Li3N. *Solid State Ionics* 9–10 (Part 1): 689–693.

121 Slade, P.G. (ed.) (2014). *Electrical Contacts: Principles and Applications*. CRC Press/Taylor and Francis.

122 Wang, X., Zeng, W., Hong, L. et al. (2018). Stress-driven lithium dendrite growth mechanism and dendrite mitigation by electroplating on soft substrates. *Nature Energy* 3 (3): 227–235.

123 Porz, L., Swamy, T., Sheldon, B.W. et al. (2017). Mechanism of lithium metal penetration through inorganic solid electrolytes. *Advanced Energy Materials* 7 (20): 1–12.

124 Wang, M., Wolfenstine, J.B., and Sakamoto, J. (2019). Temperature dependent flux balance of the Li/$Li_7La_3Zr_2O_{12}$ interface. *Electrochimica Acta* 296: 842–847.

125 Sudo, R., Nakata, Y., Ishiguro, K. et al. (2014). Interface behavior between garnet-type lithium-conducting solid electrolyte and lithium metal. *Solid State Ionics* 262: 151–154.

126 Lewis, J.A., Cortes, F.J.Q., Boebinger, M.G. et al. (2019). Interphase morphology between a solid-state electrolyte and lithium controls cell failure. *ACS Energy Letters* 4 (2): 591–599.

127 Tippens, J., Miers, J.C., Afshar, A. et al. (2019). Visualizing chemomechanical degradation of a solid-state battery electrolyte. *ACS Energy Letters* 4 (6): 1475–1483.

128 Chung, H. and Kang, B. (2017). Mechanical and thermal failure induced by contact between a $Li_{1.5}Al_{0.5}Ge_{1.5}(PO_4)_3$ solid electrolyte and Li metal in an all solid-state Li cell. *Chemistry of Materials* 29 (20): 8611–8619.

129 Hitz, G.T., McOwen, D.W., Zhang, L. et al. (2019). High-rate lithium cycling in a scalable trilayer Li-garnet-electrolyte architecture. *Materials Today* 22 (February): 50–57.

130 Buschmann, H., Berendts, S., Mogwitz, B., and Janek, J. (2012). Lithium metal electrode kinetics and ionic conductivity of the solid lithium ion conductors "'$Li_7La_3Zr_2O_{12}$'" and $Li_{7-x}La_3Zr_{2-x}Ta_xO_{12}$ with garnet-type structure. *Journal of Power Sources* 206: 236–244.

131 Ishiguro, K., Nakata, Y., Matsui, M. et al. (2013). Stability of Nb-doped cubic $Li_7La_3Zr_2O_{12}$ with lithium metal. *Journal of the Electrochemical Society* 160 (10): A1690–A1693.

132 Ishiguro, K., Nemori, H., Sunahiro, S. et al. (2014). Ta-doped $Li_7La_3Zr_2O_{12}$ for water-stable lithium electrode of lithium-air batteries. *Journal of the Electrochemical Society* 161 (5): A668–A674.

133 Ryou, M.H., Lee, Y.M., Lee, Y. et al. (2015). Mechanical surface modification of lithium metal: towards improved Li metal anode performance by directed Li plating. *Advanced Functional Materials* 25 (6): 834–841.

134 Yang, C., Zhang, L., Liu, B. et al. (2018). Continuous plating/stripping behavior of solid-state lithium metal anode in a 3D ion-conductive framework. *Proceedings of the National Academy of Sciences of the United States of America* 115 (15): 3770–3775.

135 Lewis, J.A., Tippens, J., Javier, F. et al. (2019). Chemo-mechanical challenges in solid-state batteries. *Trends in Chemistry* 1: 1–14.

136 Métivier, Y. and Richomme, G. (1994). On the star operation and the finite power property in free partially commutative monoids. In: *Lecture Notes in Computer Science*, vol. 775 (eds. P. Enjalbert, E.W. Mayr and K.W. Wagner), 341–352. Berlin, Heidelberg: Springer https://doi.org/10.1007/3-540-57785-8_153.

137 Monroe, C. and Newman, J. (2005). The impact of elastic deformation on deposition kinetics at lithium/polymer interfaces. *Journal of the Electrochemical Society* 152 (2): A396.

138 Stone, G.M., Mullin, S.A., Teran, A.A. et al. (2012). Resolution of the modulus versus adhesion dilemma in solid polymer electrolytes for rechargeable lithium metal batteries. *Journal of the Electrochemical Society* 159 (3): A222–A227.

139 Sharafi, A., Haslam, C.G., Kerns, R.D. et al. (2017). Controlling and correlating the effect of grain size with the mechanical and electrochemical properties of $Li_7La_3Zr_2O_{12}$ solid-state electrolyte. *Journal of Materials Chemistry A* 5 (40): 21491–21504.

140 Wolfenstine, J., Allen, J.L., Sakamoto, J. et al. (2018). Mechanical behavior of Li-ion-conducting crystalline oxide-based solid electrolytes: a brief review. *Ionics (Kiel)* 24 (5): 1271–1276.

141 Ni, J.E., Case, E.D., Sakamoto, J.S. et al. (2012). Room temperature elastic moduli and Vickers hardness of hot-pressed LLZO cubic garnet. *Journal of Materials Science* 47 (23): 7978–7985.

142 Nagao, M., Hayashi, A., Tatsumisago, M. et al. (2013). In situ SEM study of a lithium deposition and dissolution mechanism in a bulk-type solid-state cell with a Li_2S-P_2S_5 solid electrolyte. *Physical Chemistry Chemical Physics* 15 (42): 18600–18606.

143 Wang, C., Gong, Y., Liu, B. et al. (2017). Conformal, nanoscale ZnO surface modification of garnet-based solid-state electrolyte for lithium metal anodes. *Nano Letters* 17 (1): 565–571.

144 Yamada, T. (2015). All solid-state lithium–sulfur battery using a glass-type P_2S_5–Li_2S electrolyte: benefits on anode kinetics. *Journal of the Electrochemical Society* 162: A646–A651.

145 Han, F., Yue, J., Zhu, X., and Wang, C. (2018). Suppressing Li dendrite formation in Li_2S-P_2S_5 solid electrolyte by LiI incorporation. *Advanced Energy Materials* 8 (18): 6–11.

146 Nagao, M., Hayashi, A., and Tatsumisago, M. (2012). Fabrication of favorable interface between sulfide solid electrolyte and Li metal electrode for bulk-type solid-state Li/S battery. *Electrochemistry Communications* 22 (1): 177–180.

147 Han, F., Westover, A.S., Yue, J. et al. (2019). High electronic conductivity as the origin of lithium dendrite formation within solid electrolytes. *Nature Energy* 4: 187–196.

148 Manalastas, W., Rikarte, J., Chater, R.J. et al. (2019). Mechanical failure of garnet electrolytes during Li electrodeposition observed by in-operando microscopy. *Journal of Power Sources* 412 (August 2018): 287–293.

149 Cheng, E.J., Sharafi, A., and Sakamoto, J. (2017). Intergranular Li metal propagation through polycrystalline $Li_{6.25}Al_{0.25}La_3Zr_2O_{12}$ ceramic electrolyte. *Electrochimica Acta* 223: 85–91.

150 Hou, G., Ma, X., Sun, Q. et al. (2018). Lithium dendrite suppression and enhanced interfacial compatibility enabled by an ex situ SEI on Li anode for LAGP-based all-solid-state batteries. *ACS Applied Materials & Interfaces* 10 (22): 18610–18618.

151 Garcia-Mendez, R., Mizuno, F., Zhang, R. et al. (2017). Effect of processing conditions of $75Li_2S$-$25P_2S_5$ solid electrolyte on its DC electrochemical behavior. *Electrochimica Acta* 237: 144–151.

152 Markevich, E., Salitra, G., Chesneau, F. et al. (2017). Very stable lithium metal stripping-plating at a high rate and high areal capacity in fluoroethylene carbonate-based organic electrolyte solution. *ACS Energy Letters* 2 (6): 1321–1326.

153 Masias, A., Felten, N., Garcia-Mendez, R. et al. (2019). Elastic, plastic, and creep mechanical properties of lithium metal. *Journal of Materials Science* 54 (3): 2585–2600.

154 Krauskopf, T., Dippel, R., Hartmann, H. et al. (2019). Lithium-metal growth kinetics on LLZO garnet-type solid electrolytes. *Joule* 3 (8): 2030–2049.

155 Tian, H.K., Xu, B., and Qi, Y. (2018). Computational study of lithium nucleation tendency in $Li_7La_3Zr_2O_{12}$ (LLZO) and rational design of interlayer materials to prevent lithium dendrites. *Journal of Power Sources* 392 (April): 79–86.

156 Dixit, M.B., Regala, M., Shen, F. et al. (2019). Tortuosity effects in garnet-type $Li_7La_3Zr_2O_{12}$ solid electrolytes. *ACS Applied Materials & Interfaces* 11 (2): 2022–2030.

157 Li, J., Dudney, N.J., Nanda, J., and Liang, C. (2014). Artificial solid electrolyte interphase to address the electrochemical degradation of silicon electrodes. *ACS Applied Materials & Interfaces* 6 (13): 10083–10088.

158 Su, Y., Falgenhauer, J., Polity, A. et al. (2015). LiPON thin films with high nitrogen content for application in lithium batteries and electrochromic devices prepared by RF magnetron sputtering. *Solid State Ionics* 282: 63–69.

159 Swamy, T. and Chiang, Y.-M. (2011). Lithium metal penetration induced by electrodeposition through solid electrolytes: example in single-crystal $Li_6La_3ZrTaO_{12}$ garnet. 9 (6): 2–7.

160 Sakuda, A., Hayashi, A., and Tatsumisago, M. (2013). Sulfide solid electrolyte with favorable mechanical property for all-solid-state lithium battery. *Scientific Reports* 3: 2–6.

161 Cao, Y., Li, M., Lu, J. et al. (2019). Bridging the academic and industrial metrics for next-generation practical batteries. *Nature Nanotechnology* 14 (3): 200–207.

162 Cheng, L., Chen, W., Kunz, M. et al. (2015). Effect of surface microstructure on electrochemical performance of garnet solid electrolytes. *ACS Applied Materials & Interfaces* 7 (3): 2073–2081.

163 Han, X., Gong, Y., Fu, K. et al. (2017). Negating interfacial impedance in garnet-based solid-state Li metal batteries. *Nature Materials* 16 (5): 572–579.

164 Luo, W., Gong, Y., Zhu, Y. et al. (2016). Transition from superlithiophobicity to superlithiophilicity of garnet solid-state electrolyte. *Journal of the American Chemical Society* 138 (37): 12258–12262.

165 Zhang, H., Li, X., Hao, S. et al. (2019). Inducing interfacial progress based on a new sulfide-based composite electrolyte for all-solid-state lithium batteries. *Electrochimica Acta* 325: 134943.

166 Li, N.W., Yin, Y.X., Yang, C.P., and Guo, Y.G. (2016). An artificial solid electrolyte interphase layer for stable lithium metal anodes. *Advanced Materials* 28 (9): 1853–1858.

167 Xu, R., Han, F., Ji, X. et al. (2018). Interface engineering of sulfide electrolytes for all-solid-state lithium batteries. *Nano Energy* 53: 958–966.

168 Shinzo, S., Higuchi, E., Chiku, M. et al. (2020). Control of dendritic growth of lithium metal in all-solid-state lithium metal batteries: effect of current collector with micro-sized pores. *ACS Applied Materials & Interfaces.*

169 Aurbach, D., Talyosef, Y., Markovsky, B. et al. (2004). Design of electrolyte solutions for Li and Li-ion batteries: a review. *Electrochimica Acta* 50 (2–3 Special Issue): 247–254.

170 Luntz, A.C., Voss, J., and Reuter, K. (2015). Interfacial challenges in solid-state Li ion batteries. *Journal of Physical Chemistry Letters* 6 (22): 4599–4604.

171 Lee, Y.G., Fujiki, S., Jung, C. et al. (2020). High-energy long-cycling all-solid-state lithium metal batteries enabled by silver–carbon composite anodes. *Nature Energy* 5 (April): 299–308.

172 Tian, Y., Shi, T., Richards, W.D. et al. (2017). Compatibility issues between electrodes and electrolytes in solid-state batteries. *Energy & Environmental Science* 10 (5): 1150–1166.

173 Koerver, R., Walther, F., Aygün, I. et al. (2017). Redox-active cathode interphases in solid-state batteries. *Journal of Materials Chemistry A* 5 (43): 22750–22760.

174 Bucci, G., Swamy, T., Chiang, Y.M., and Carter, W.C. (2017). Modeling of internal mechanical failure of all-solid-state batteries during electrochemical cycling, and implications for battery design. *Journal of Materials Chemistry A* 5 (36): 19422–19430.

175 Bucci, G., Talamini, B., Renuka Balakrishna, A. et al. (2018). Mechanical instability of electrode-electrolyte interfaces in solid-state batteries. *Physical Review Materials* 2 (10): 1–11.

176 Liu, T., Ren, Y., Shen, Y. et al. (2016). Achieving high capacity in bulk-type solid-state lithium ion battery based on $Li_{6.75}La_3Zr_{1.75}Ta_{0.25}O_{12}$ electrolyte: interfacial resistance. *Journal of Power Sources* 324: 349–357.

177 Finsterbusch, M. (2018). High capacity garnet-based all-solid-state lithium batteries: fabrication and 3D-microstructure resolved modeling. *ACS Applied Materials & Interfaces* 10.

10

Degradation and Life Performance of Transition Metal Oxide Cathodes used in Lithium-Ion Batteries

Satish B. Chikkannanavar[1], Jong H. Kim[2], and Wangmo Jung[3]

[1] *Cell Testing and Integration, Electrified Powertrain Engineering, Ford Motor Company, Dearborn, MI 48124, USA*
[2] *L G Chem, Development Department, ul. LG 1A, 55-040 Kobierzyce, Poland*
[3] *L G Chem, Battery R&D Department, 188, Moonji-ro, Yuseong-gu, 305-380, Daejeon, South Korea*

10.1 Introduction

Li ion batteries are perhaps the most ubiquitous energy storage systems of versatile usage in daily life [1, 2]. The energy density is an important attribute for the batteries, which in turn drives their application and adaptation for propulsion in the electrified vehicles (xEVs) [3]. The need for higher EV range and higher fuel economy are the main drivers for exploration in this field. Also, energy density and other aspects of batteries are mostly decided by the combination of cathode and anode materials paired to form the battery. Due to versatility of anode being based mostly on graphite, and for reasons such as graphite anode chemistry and performance being understood so extensively, this leaves us with cathode as the main rider for deciding the main battery attributes. For more than two decades, cathodes based on transition metal oxides (TMOs) have dominated the field [4–6]. Most of the initial advances are centered around how to address various degradation processes affecting the cathode performance. In this chapter, we survey advances made in the understanding of the degradation processes, which affect the performance of various TMO cathodes, namely $LiMn_2O_4$ (or spinel), $LiNi_{0.33}Co_{0.33}Mn_{0.33}O_2$ (NCM), Ni-rich NCM ($LiNi_xCo_yMn_zO_2, x \geq 0.5, x+y+z=1$), $LiNi_{1-x-y}Co_xAl_yO_2$ (NCA), and $LiNi_{1-x-y-z}Co_xMn_yAl_zO_2$ (NCMA). We shall also discuss the battery usage–related degradation aspects and performance-related failure in the last section. We shall also discuss the progress made over the years at improving the performance of various TMO cathodes. In the last section, we cover the degradation processes in the batteries over the usage life and ways to quantify and analyze them.

10.2 Degradation Trends

In general terms, the degradation processes in batteries are associated with battery capacity loss over the usage life. The usage life is a term used to define the useful

Transition Metal Oxides for Electrochemical Energy Storage, First Edition.
Edited by Jagjit Nanda and Veronica Augustyn.

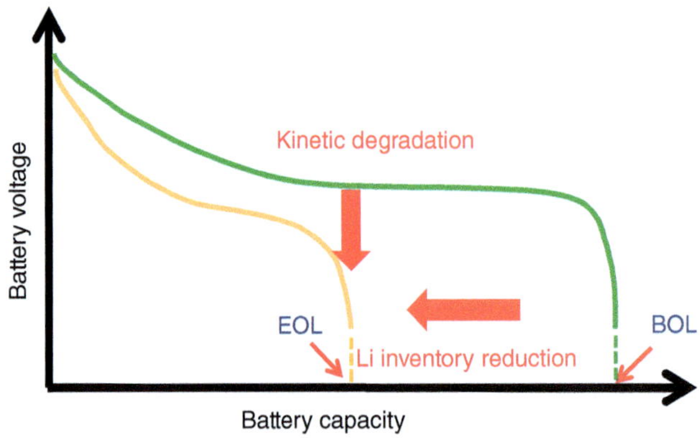

Figure 10.1 Illustration of capacity or life degradation caused by electrochemical reactions, gas generation, changes at surface layer, change in structure, volume of electrodes, etc.

life time for the battery, during which the battery delivers predefined charge and discharge power per demand in a given xEV. Here the term, "xEVs" refers to various types of vehicle electrification covering hybrid, plug-in hybrid, and electric vehicles (FHEVs, PHEVs, and EVs). Beyond the usage life time, the battery power capability during charge and discharge is supposed to be declining beyond a certain threshold. We have to note that the degradation is a function of battery usage history. For a battery with capacity at beginning of life (BOL), C_{BOL} (t, I) is both dependent on the usage time, t, and the usage load, I (here I refers to battery current). So, battery capacity at later stage of life (LSOL) is given as,

$$C_{LSOL}(t,I) = \alpha C_{BOL}(t,I) = f(t,I,\mathrm{SOC},T) \tag{10.1}$$

where α is a factor, which represents the sum of degradation contributions from various sources. Note that the capacity or life degradation is a function of time (t), current load (during discharge and charge events), state of charge (SOC), and temperature (T). There are other attributes such as depth of discharge (DOD), rest time between cycles, time duration of cycle, and calendar aging. Also, as a way to identify the useful life for the battery, we define the battery end of life (EOL) as (see Figure 10.1),

$$C_{EOL}(t,I) = 0.7C_{BOL}(t,I) \tag{10.2}$$

Note that the factor 0.7 relates to 70% of the useful life from start is a reference marker for battery life measurement. The sources of degradation are mainly from the physical and chemical changes in cell components, namely cathode, anode, electrolyte, binder, current collectors, etc. Most of the changes are chemically driven due to the electrochemical aspects of battery operation. Also, chemical reactions lead to physical changes, such as electrode thickness change, electrode volume change, particle cracking, and separator shrinkage. These changes are taking place during the battery usage, whether it is under storage (or idle status) and during cycling usage under various temperatures. Another way to substantiate capacity degradation is to treat it as the sum of Li inventory loss and kinetic degradation (see Table 10.1),

Table 10.1 Degradation root causes based on origin, which account for Li loss during usage life for Li ion batteries.

	Electrochemical root causes	**Physical root causes**
Reduction of Li inventory	Loss of lithium by electrochemical or chemical reactions at the interfaces between electrodes and electrolyte	Physical isolation and deactivation of active materials (loss of active sites) cell swelling, irreversible volume change (e.g. Si)
Kinetic degradation	Proliferation of passive layer and increase of concentration over potential at high C-rate operation conditions	Partial delamination of current collector or clogging of pores in electrodes and separators

with each having factors responsible for Li inventory loss. Figure 10.1 illustrates the change in battery useful life from BOL to EOL over time, attributed to the degradation effects. The general discussion of degradation for Li ion batteries would relate to various degradation mechanisms at material level for cathode, anode, electrolyte reaction processes, parasitic reactions, surface reactions, etc. [7, 8]. For surface reaction schemes occurring during battery operation and usage, and the changes at the surface of graphite anodes, readers are advised to the research articles [9, 10] and references therein. We will survey the degradation trends for cathodes, in particular those made of TMOs in the following sections.

To achieve high-energy density, the focus for the research community and for material suppliers is to design new cathodes and, in this pursuit, increasing Ni content in NCM cathode is the most successful effort [11–15]. Doping Ni in the parent-layered cathode, $LiNi_{0.33}Co_{0.33}Mn_{0.33}O_2$, gives rise to Ni-rich NCM cathodes, with composition, $LiNi_xCo_yMn_zO_2$ ($x \geq 0.5, x+y+z=1$), which exhibit a reversible capacity of 200–230 mAh g^{-1} under optimal conditions. Figure 10.2 illustrates the composition diagram of the NCM system of TMO cathodes along with capacity in mAh g^{-1} for various compositions, namely NCM (111), NCM (532), NCM (622), and

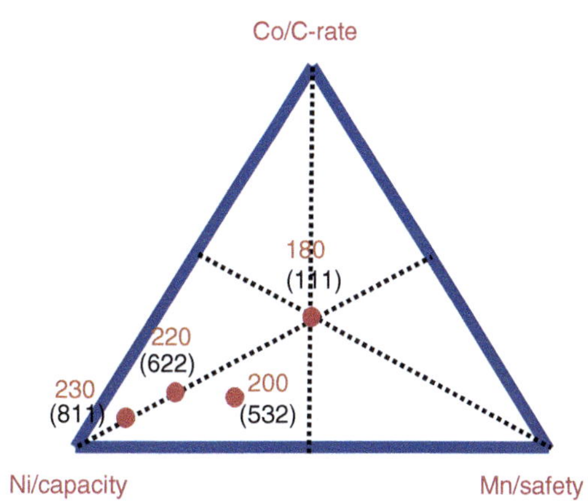

Figure 10.2 Composition diagram for the NCM system of cathodes. Ni-rich compositions are shown along with capacity in mAh g^{-1}. Increasing the amount of one particular element in the ternary system is expected to optimize the main attributes such as rate performance, capacity, and safety.

NCM (811). Note that the capacity numbers listed relate to the values for various NCM compositions when cycled up to an upper voltage limit of 4.6 V. However, electrolyte stability (for electrolytes EC, DMC, etc.) and cathode degradation concerns make it challenging to tap into voltage limits beyond 4.20 V for xEV applications [4]. In the further sections, we shall discuss degradation schemes on spinel cathodes and Ni-rich NCM cathodes.

10.3 Transition Metal Oxide Cathodes

From the early stage of research work on rechargeable batteries, cathodes made of transition metal sulfides, selenides, and oxide systems have been the focus of research [2]. These were basically layered materials, enabling Li ion to intercalate in the spaces between ion layers of transition metal and sulfur/oxygen. For the $LiTiS_2$ cathode, operation voltage of 2 V was lower [16]. It was succeeded by an oxide cathode, namely $LiCoO_2$ (LCO) [17], due to higher electrode potential extending up to 4.2 V. The initial Li-based anodes in the batteries were not uniform for Li intercalation, showed limited cycle life along with lower columbic efficiency [18], and also posed challenges with Li dendrite growth, causing battery internal short, failure, and also posed safety concerns. Later, research was focused on using alternate anode candidates, and graphite proved out to be a reliable candidate [19]. $LiCoO_2$ was the prominent cathode in the batteries for over a decade, due to superior life performance and that has enabled their use in consumer electronic devices and laptop computers. Later, in efforts to further improve safety, doping LCO with other transition metals led to the NCM cathode, which we shall discuss in a later section. Besides layered structures, another prominent cathode is $LiMn_2O_4$, which is also called as spinel and is a three-dimensional structure.

10.3.1 Spinel Cathodes

Spinel or $LiMn_2O_4$ (LMO) is an environmentally benign cathode, possesses a higher operating voltage, higher rate capability [20, 21], and better thermal stability than NCM cathodes [22, 23]. Spinel has a lower coulombic capacity [20] and disadvantages such as crystal-structure transformations during cycling at high temperatures, oxygen deficiency, and Mn dissolution [24–31], all of which can lead to higher capacity fade than observed in the layered oxides (NCM). Among the cathode materials, LMO offers us a classic example where combination of structure change and chemical factors engender battery life degradation. LMO undergoes what is called "Jahn–Teller distortion" during late stage of lithiation, where the structure change from orthorhombic to tetragonal is accompanied by the volume change in the structure [24–31]. This structure change has been studied very well through *in situ* X-ray diffraction studies [25]. The associated volume change is substantial (c.a. 3%), which manifests in LMO particles cracking around the edges and corners. The detached LMO nanoparticles dissolve in the electrolyte following the cracking, during the cycle aging at high temperatures. Figure 10.3a

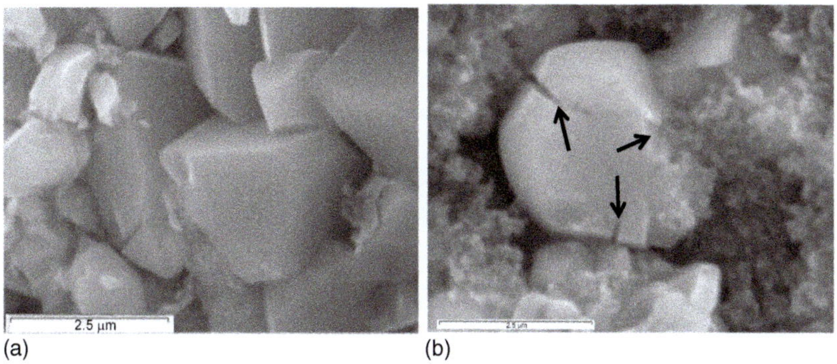

Figure 10.3 SEM micrographs of cathodes with spinel; (a) fresh electrode and (b) cycle aged electrode, where spinel particles show cracks due to structure change.

shows an example of fresh spinel cathode microstructure, where LMO crystalline particles (0.5–3 μm) exhibit sharp edges. However, LMO particles undergo change in morphology, where particle edges get eroded and cracks appear (see Figure 10.3b) following cycle aging at 45 °C. The dissolved Mn ions migrate toward anode during the battery usage cycles and deposit on the anode surface. Eventually, MnO_2 undergo reduction on the graphite surface to form a passive Mn layer on the anode surface, which inhibits Li ion access to the active sites on the anode. This is referred to as Mn dissolution and is the major reason why LMO suffers from severe degradation. Figure 10.4a,b shows the XPS spectrum showing Mn 2p features from cycled cathode and anode, respectively. We notice that anode surface shows far intense elemental feature showing Mn ions deposited on the anode surface. Mn dissolution and related degradation of cathode and anode is a major challenge for using LMO as a cathode. Table 10.2 lists the binding energies of Mn species on cathode and anode, and Mn deposition on anode in particular deprives access to Li for anode. Recently, Kim et al. [32] have shown a major improvement to the spinel by means of

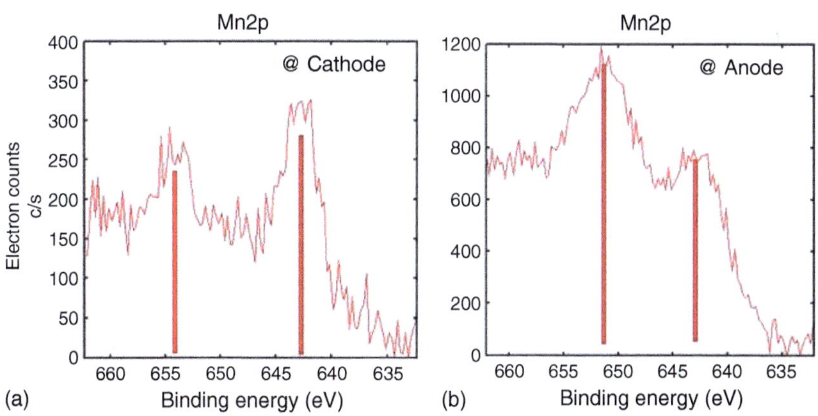

Figure 10.4 XPS spectrum of cycle aged electrodes showing Mn presence on the cathode and anode surface. Mn deposition passivates and deprives Li access to anode active sites@ cathode and @ anode for (a) and (b).

Table 10.2 XPS binding energies for Mn species detected on cathode and anode.

Sample	Mn $2p_{1/2}$ (eV)	Mn $2p_{3/2}$ (eV)	Possible Mn chemical states
Aged cathode	654	643	MnO, MnO_2, MnF_2
Aged anode	651	643	MnO_x/Mn

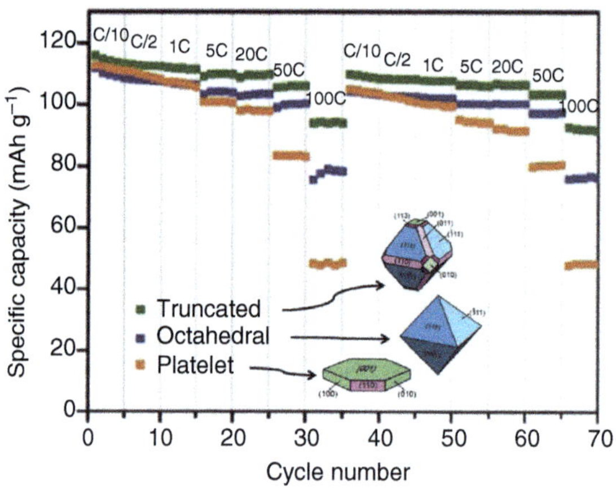

Figure 10.5 Rate performance of $LiMn_2O_4$ electrode particles of truncated, octahedral, and platelet shape during cycling at 45 °C. Source: Kim et al. [32]/with permission of American Chemical Society.

shape-selective synthesis strategy. The authors followed a hydrothermal synthesis approach for shape-selective synthesis and control of surface structure, to reduce volume change and to minimize the metal dissolution. Such improved surface and size scaling yielded far improved cycle life performance for the spinel cathode at high temperatures (see Figure 10.5). Note that the truncated octahedral shape for spinel particles offers the best cycle performance. Thus, concerns such as structure change and resulting Mn dissolution have been overcome, giving us an opportunity to use spinel cathodes into high voltage-operating window.

10.3.2 NCM System of Cathodes

From early synthesis efforts and studies on $LiCoO_2$ (LCO) cathode in 1980 [17, 33], followed by synthesis of $LiNiO_2$ (LNO) cathodes in 1990s [34, 35] and LMO cathode [23, 24], has led to the synthesis of the ternary system, namely NCM cathode with composition, $LiNi_{1-x-y}Co_xMn_yO_2$ [36]. We discussed in an earlier section about spinel cathode particles undergoing cracking due to structure and volume change. For Ni-rich NCM cathodes, with Ni content increase, we expect to see various challenges such as reactive species generation followed by gas generation, inactive phase formation on surfaces, crystal structure strain with associated volume change, and material instability at high temperature leading to thermal events. All of these would call for new approaches for improving surface structure integrity, protection against gas generation and volume change, protection against reactivity at high temperatures, and improving the life performance for high-temperature storage and cycling.

Ni-rich cathodes pose unique challenges in spite of the high capacity or energy density they can offer. Operating in high-voltage range extending into 4.3 V degrades capacity and life. With Ni content increase, there arises a problem due to similarity in ionic radius for Ni^{2+} and Li^{+} ions. This means Ni ions may occupy Li sites, thus referred to as "ion mixing," causing phase change and volume change, which in turn leads to particle cracking [37, 38]. In the highly delithiated state, Ni-rich cathodes are highly reactive and can oxidize the electrolyte, leading to gas generation [39–43]. Also, active material loss at the cathode can lead to formation of a thicker surface film, referred to as cathode–electrolyte interphase (CEI) [44, 45], which causes resistance increase [37, 46]. A more challenging aspect for high Ni cathode materials is the poor thermal stability or high reactivity in highly delithiated state [47]. This is due to activated state of Ni oxide, which triggers thermal events at lower threshold temperature than expected for NCM (111) cathode [11–15].

An illustration of particle cracking effects on cycle life, we show results of cycling at high temperature for NCM (622). During 1C/1C cycling at 45 °C, standard-grade NCM (622) cathode is known to exhibit cracking at the particle level due to volume change, as illustrated in Figure 10.6a. Various approaches, such as doping and surface coating, are employed to address the cracking concern. In one such example, NCM (622) cathode, which has been doped and has surface oxide coating, shows improved performance in cycle life. Figure 10.6b shows the microstructure of particles from an improved grade cathode after cycling for 1800 cycles.

It is obvious that doping and surface coating help the cathode particles to withstand structure change and volume change at 45 °C. The cathode particles suffer strain during lithiation and delithiation cycles due to the structure change and associated volume change. X-ray diffraction studies indicate that for various compositions, Ni-rich cathodes show susceptibility to strain with Ni concentration increases from Ni-40 to Ni-80 (811), as shown in Figure 10.7. The Ni-rich compositions for NCM exhibit a volume change in the range 1–5.5% as Ni composition is increased from 50% all the way to 80%.

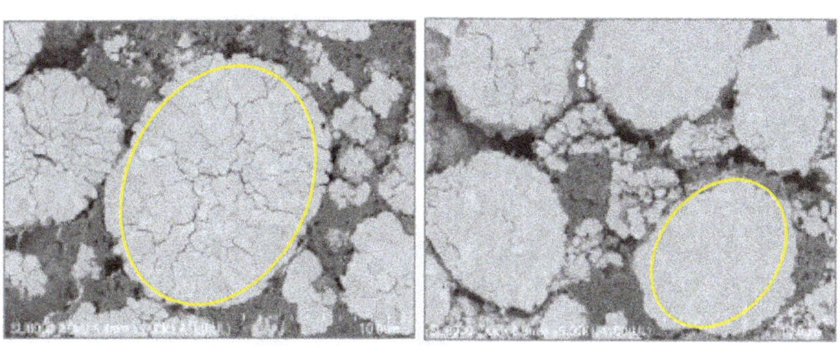

(a) (b)

Figure 10.6 NCM (622) cathode particles exhibiting cracks in particles due to structure and volume change during cycling at 45 °C. (a) Standard-grade cathode shows cracks in the particles after 600 cycles and (b) special-grade cathode particles show less cracking after 1800 cycles (LG Chem provided material).

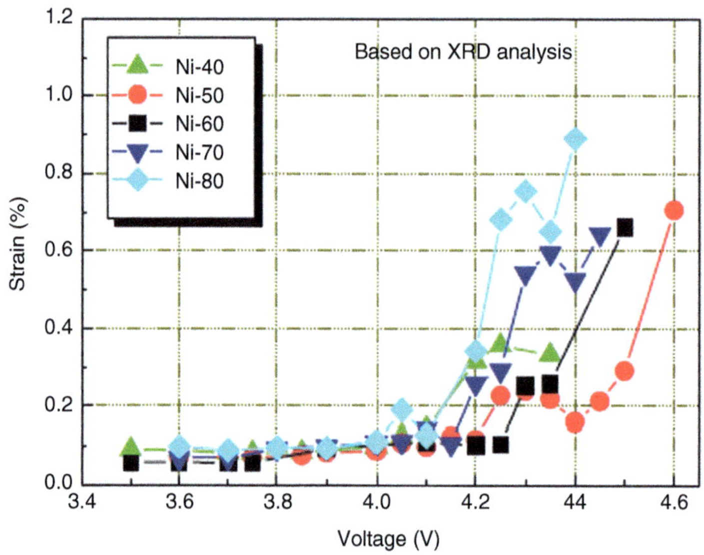

Figure 10.7 Effect of strain due to structure change in various Ni-rich compositions for NCM cathodes. Such strain affects the life performance (LG Chem provided material).

On the gassing behavior of TMO cathodes, in particular Ni-rich NCM cathodes, recent advances have been made at *in situ* analysis of gas generation characteristics [48–50]. Berkes et al. investigated NCM (523) cathode by assembling a full cell and investigated gassing behavior during low rate charge–discharge cycle, formation cycle, and regular cycling [48]. The authors used differential electrochemical mass spectroscopy (DEMS) and differential electrochemical infrared mass spectroscopy (DEIRS). Figure 10.8 shows the example of gas pressure build up during the low rate charge–discharge of the cell, followed by cycle tests. The type of gases generated relates to SEI formation, side reactions, etc., which contribute to swelling and pressure effects over the usage life of the batteries. Such effects call for appropriate case design for the batteries, which come in various formats, namely pouch, cylindrical, and prismatic can. For NCM cathodes, the gassing behavior is expected to get worse with increase in the Ni content. So, coating approaches have to be followed to reduce the gas generation and related degradation on battery performance and life [40]. For TMO cathodes, the degradation scheme is such that active particles develop surface layer called CEI layer and over the usage time, a new layer deposit is observed [15, 45], which is inactive layer. This is mainly due to disordered phase, and, is the result of particle cracking and reaction product depositions (see Figure 10.9). This is accompanied by volume change, gas evolution, etc., which would cause cracking as shown in Figure 10.9. Note that all of these degradation processes become aggravated with the battery operation in the voltage window higher than 4.15 V. Here again, surface modification and doping approaches should help address such a limitation; however, research into this area is still open to exploration.

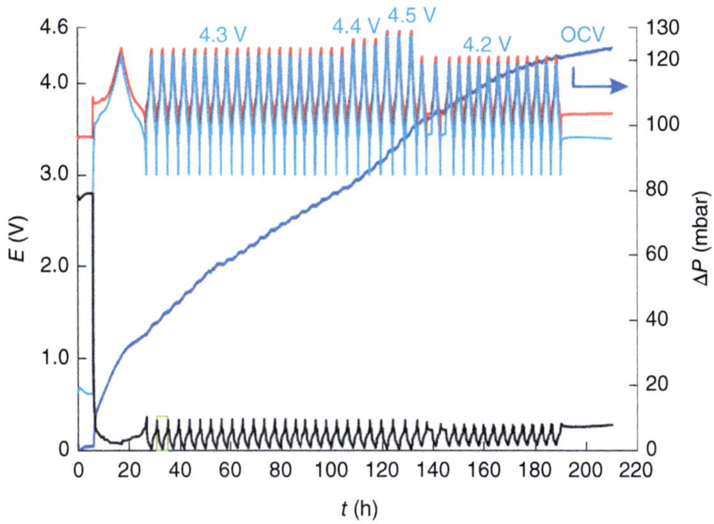

Figure 10.8 Voltage traces of cathode (blue), anode (black), and full cell (red) during C/10 cycle followed by cycling tests. Gas pressure curve (bright blue) indicates the pressure generated within NCM (523). Source: Berkes et al. [48]/Springer Nature.

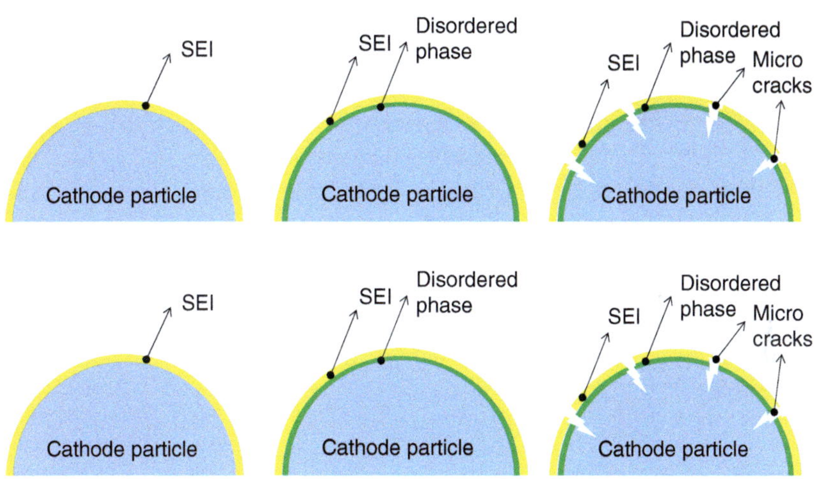

Figure 10.9 Effect of aging and surface layer growth for various Ni-rich compositions for NCM cathodes. Particle volume change and strain cause cracking, in turn affecting the life performance. Source: Modified from Schipper et al. [15].

10.3.3 NCMA Cathodes

With the proliferation of EVs, one could expect that demand for higher range is always expected by the EV customers and also by the auto manufacturers. This further calls for enhancing of energy density for the cathodes and other advances in the battery technology. Over the past few years, there is an enhanced interest to improve the Ni-rich NCM cathodes (Ni content >80%) for better cycle

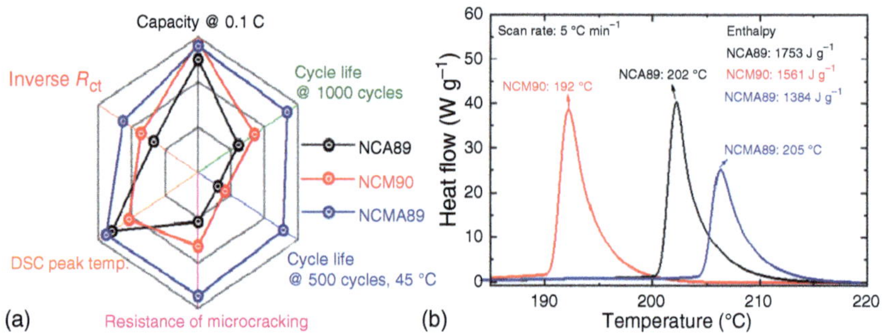

Figure 10.10 (a) The spider chart illustrates the comparison for various performance attributes of cathodes with 89/90% Ni rich composition for NCA, NCM, and NCMA. (b) Shows the thermal activation temperatures for various Ni-rich NCM, NCA, and NCMA. The advantage of Al doping in NCM90 cathode is shown to lower the thermal activation temperature. Source: Kim et al. [51]/with permission of American Chemical Society.

life and thermal stability by means of doping with Al to achieve a stable and high performance NCMA cathode. In other words, doping of Ni-rich NCM (Ni ~90 atom%) with low levels of Al imparts advantages, such as reduced particle cracking, improved cycle life at high temperature, and improved thermal stability [51, 52]. NCA is a high-energy density cathode; however, thermal stability is a big concern [53]. The primary degradation route is that volume change causes particle cracking, which in turn exposes the cathode active surface to electrolyte. Also, in high-voltage operation state, reaction of Ni^{4+} causes impurity phase formation, surface layer growth, and resistance increase [53–55]. In Figure 10.10a, a comparison for NCM90, NCA89, and NCMA89 cathodes is shown in a spider chart, with battery attributes such as cathode capacity, cycle life at 1000 cycles, cycle life at 500 cycles at 45 °C, resistance to microcracking, and DSC peak temperature. Low-level Al doping drastically improves all the performance attributes for NCMA89. We also notice that Al doping in NCM90 dramatically improves the onset temperature for thermal activation (see Figure 10.10b). These studies bode well toward realization of a cathode with further enhanced energy density and performance characteristics.

10.4 Degradation Mechanism

In the above sections, we looked into the degradation pathways, which affect the battery performance and life for cathode materials based on TMOs. The various degradation pathways for cathodes or for the batteries in general and how such pathways affect the observed material structural and chemical properties, electrical properties, etc. are depicted schematically in Figure 10.11. These are considered from the perspective of battery performance and usage life over usage time. In the automotive world, for xEVs (includes FHEVs, PHEVs, and EVs) in general, we define usage life for a battery as 10 years of use or lasting for 150 000 mi driven by the customer. However, auto manufacturers provide warranty on xEV batteries for 100 000 mi or

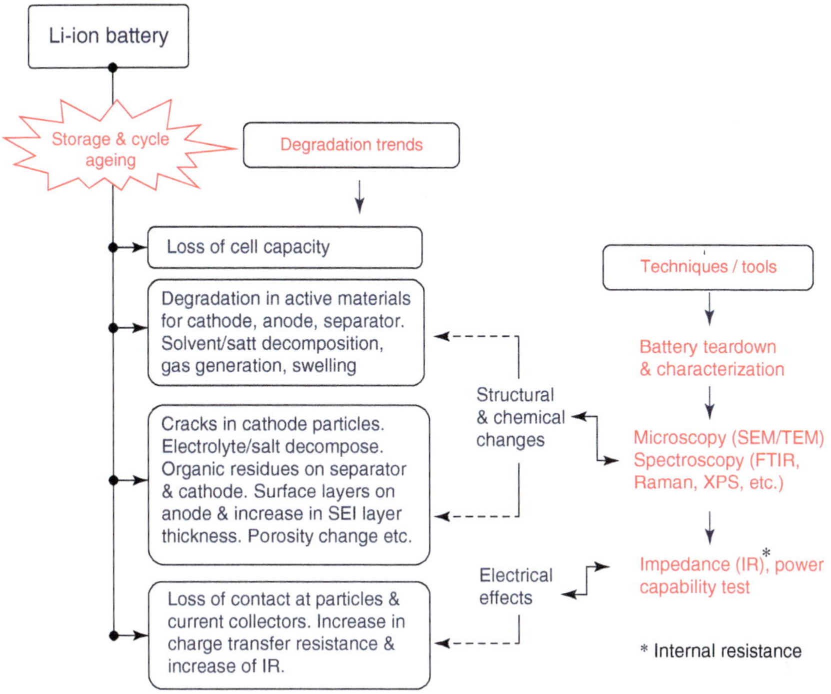

Figure 10.11 The flow chart describes the degradation process for Li ion batteries during usage. Various degradation processes at component level can be characterized and quantified by a host of characterization techniques.

usage life of 8 years. Over many years, many advances have yielded positive results to limit degradation pathways and improve the life performance for the TMO batteries. Also, advances in physical characterization technique and tools have yielded new insights at understanding limitations and ways to address them. Figure 10.11 also depicts the techniques and tools necessary for studying the degradation processes and pathways. With the extensive usage of Li ion batteries in xEVs, one could expect a degradation pathway, wherein, Li plating can cause accelerated degradation of battery capacity and life. The concern for Li plating is particularly relevant to battery usage during cold-temperature charging. This has particularly garnered more attention during recent years, and readers can refer to the recent review and references therein [56]. The plated Li on the anode during operation in high SOC window at cold temperatures is accompanied by resistance increase, accelerated side reactions, and decrease of ionic conductivity [57], all of which affect the battery life. With increase of energy density for batteries, Li plating becomes even more concerning, due to slow solid-state diffusion kinetics with large anode particles, thus leading to increased propensity for Li plating. This calls for further improvement in cell design by means of avoiding plating using better design of anode, use of electrolyte additives, etc. The strategies such as operation of batteries in warmer conditions and avoiding high charging rates are shown to help improve the performance and avoid accelerated degradation [58].

One highly effective means of understanding the degradation pathways for batteries under various aging conditions is to carry out the battery teardown analysis. The teardown process goal is to look into physical and chemical changes in the battery components (electrodes, separator, electrolyte, current collector, battery swelling or venting, safety devices or circuits, etc.). The battery usage is what dictates the degradation process and one can correlate usage history to the degradation processes. Thus, battery teardown analysis is highly emphasized to assess and understand the cause of degradation or performance failure over the usage life (Readers are advised to the listed articles in the links provided: http://nexceris.com/learning-from-4-damaging-lithium-ion-battery-failures/; https://www.ntsb.gov/investigations/pages/boeing_787.aspx; https://news.samsung.com/us/Samsung-Electronics-Announces-Cause-of-Galaxy-Note7-Incidents-in-Press-Conference; https://arstechnica.com/gadgets/2017/01/galaxy-note-7-investigation-blames-small-battery-cases-poor-welding/). Generally, this is expected to shed deep insight into degradation processes and pathways and also failure-related root causes. Thus, the electrical characterization of aging process during the course of battery usage time can be compared for correlation to the battery life performance.

10.5 Concluding Remarks

In summary, we have surveyed the degradation processes and pathways for Li ion battery cathodes made with transition metal oxides. The overall goal of research into these cathodes is not only to achieve higher energy density, but also to address the degradation processes and to improve the battery life performance. The demand for life expectation of batteries for propulsion in the xEVs is pretty demanding and that in turn calls for advances in chemistry update, doping, surface treatment, electrolyte formulation, electrolyte additive formulation, etc. All in all, various advances in the past decade have indeed improved the performance characteristics and life expectation for the Li ion batteries. At present, Ni-rich cathodes are being formulated in high-energy density batteries for use in xEVs, in particular for applications involving PHEVs and EVs. This has led to current state-of-art battery chemistries based on Ni-rich NCM cathodes (with Ni content $\geq$90%) capable of delivering EVs with electric range over 200 mi. This in itself is a great testimonial for the advances and efforts made by the scientific research community and the battery suppliers alike. All such advances indeed portend a great potential for TMO cathodes for energy storage applications involving xEVs and also for grid storage. In the near future, emphasis would be at developing Ni-rich and Co-free cathodes to further reduce cost. Another frontier in exploratory stage in this field is developing the solid-state batteries for commercial and EV applications. They are of great interest due to inherent attributes such as high energy density and safety.

References

1 Julien, C., Mauger, A., Vijh, A., and Zaghib, K. (eds.) (2016). *Lithium Batteries: Science and Technology*. Springer.

2 Whittingham, M.S. (2004). *Lithium Batteries and Cathode Materials. Chemical Reviews* 104: 4271–4302.

3 Thaler, A. and Watzenig, D. (eds.) (2014). *Automotive Battery Technology*. Springer.

4 Etacheri, V., Marom, R., Elazari, R. et al. (2011). Challenges in the development of advanced Li-ion batteries: a review. *Energy & Environmental Science* 4: 3243–3262.

5 Xu, B., Qian, D., Wang, Z., and Meng, Y.S. (2012). *Recent progress in cathode materials research for advanced lithium ion batteries. Materials Science and Engineering R* 73: 51–65.

6 Zhang, X.-Q., Zhao, C.-Z., Huang, J.-Q., and Zhang, Q. (2018). *Recent advances in energy chemical engineering of next-generation lithium batteries. Engineering* 4: 831–847.

7 Vetter, J., Novak, P., Wagner, M.R. et al. (2005). Ageing mechanisms in lithium-ion batteries. *Journal of Power Sources* 147: 269–281.

8 Ecker, M., Nieto, N., Kabitz, S. et al. (2014). Calendar and cycle life study of $Li(NiMnCo)O_2$-based 18650 lithium ion batteries. *Journal of Power Sources* 248: 839–851.

9 Aurbach, D. (2000). Review of selected electrode–solution interactions which determine the performance of Li and Li ion batteries. *Journal of Power Sources* 89: 206–218.

10 Agubra, V. and Fergus, J. (2013). Lithium ion battery anode aging mechanisms. *Materials* 6: 1310–1325.

11 Kim, J., Lee, H., Cha, H. et al. (2018). *Nickel-rich cathodes: prospect and reality of Ni-rich cathode for commercialization. Advanced Energy Materials* 8: 1702028.

12 Myung, S.-T., Maglia, F., Park, K.-J. et al. (2017). *Nickel-rich layered cathode materials for automotive lithium-ion batteries: achievements and perspectives. ACS Energy Letters* 2: 196–223.

13 Xu, J., Lin, F., Doeff, M.M., and Tong, W. (2017). *A review of Ni-based layered oxides for rechargeable Li-ion batteries. Journal of Materials Chemistry A* 5: 874–901.

14 Roziera, P. and Tarascon, J.M. (2015). *Review – Li-rich layered oxide cathodes for next-generation Li-ion batteries: chances and challenges. Journal of the Electrochemical Society* 162: A2490–A2499.

15 Schipper, F., Erickson, E.M., Erk, C. et al. (2017). Review – recent advances and remaining challenges for lithium ion battery cathodes: I. nickel-rich, $LiNi_xCo_yMn_zO_2$. *Journal of the Electrochemical Society* 164: A6220–A6228.

16 Whittingham, M.S. (1976). Electrical energy storage and intercalation chemistry. *Science* 92: 1126–1127.

17 Mizushima, K., Jones, P.C., Wiseman, P.J., and Goodenough, J.B. (1980). $Li_xCoO_2(0<x<-1)$: A new cathode material for batteries of high energy density. *Materials Research Bulletin* 15: 783–789.

18 Rao, B., Eustace, D., and Shropshire, J. (1980). The Li/TiS_2 cell with LiSCN electrolyte. *Journal of Applied Electrochemistry* 10: 757–763.

19 Fong, R., Vonsacken, U., and Dahn, J.R. (1990). Studies of lithium intercalation into carbons using nonaqueous electrochemical cells. *Journal of the Electrochemical Society* 137: 2009–2013.

20 Ohzuku, T., Kitagawa, M., and Hirai, T. (1990). *Electrochemistry of manganese dioxide in lithium nonaqueous cell. Journal of the Electrochemical Society* 137: 769–775.

21 Deng, B., Nakamura, H., and Yoshio, M. (2005). Superior capacity retention of oxygen stoichiometric spinel $Li_{1+x}Mn_{2-x}O_{4+\delta}$ at elevated temperature. *Electrochemical and Solid-State Letters* 8: A171–A174.

22 Maleki, H., Deng, G., Anani, A., and Howard, J. (1999). *Thermal stability studies of Li-ion cells and components. Journal of the Electrochemical Society* 146: 3224–3229.

23 Zhang, Z., Fouchard, D., and Rea, J.R. (1998). *Differential scanning calorimetry material studies: implications for the safety of lithium-ion cells. Journal of Power Sources* 70: 16–20.

24 Xia, Y., Sakai, T., Fujieda, T. et al. (2001). Correlating capacity fading and structural changes in $Li_{1+y}Mn_{2-y}O_{4-\delta}$ spinel cathode materials – a systematic study on the effects of Li/Mn ratio and oxygen deficiency. *Journal of the Electrochemical Society* 148: A723–A729.

25 Nam, K.-W., Yoon, W.-S., Shin, H. et al. (2009). In situ X-ray diffraction studies of mixed $LiMn_2O_4$–$LiNi_{1/3}Co_{1/3}Mn_{1/3}O_2$ composite cathode in Li-ion cells during charge–discharge cycling. *Journal of Power Sources* 192: 652–659.

26 Blyr, A., Sigala, C., Amatucci, G. et al. (1998). Self-discharge of $LiMn_2O_4$/C Li-ion cells in their discharged state understanding by means of three-electrode measurements. *Journal of the Electrochemical Society* 145: 194–209.

27 Thackeray, M.M., Johnson, C.S., Kahaian, A.J. et al. (1999). Stabilization of insertion electrodes for lithium batteries. *Journal of Power Sources* 81–82: 60–66.

28 Shao-Horn, Y., Hackney, S.A., Kahaian, A.J. et al. (1999). Structural fatigue in spinel electrodes in $Li/Li_x[Mn_2]O_4$ cells. *Journal of Power Sources* 81–82: 496–499.

29 Chung, K.Y. and Kim, K.-B. (2002). Investigation of structural fatigue in spinel electrodes using in situ laser probe beam deflection technique. *Journal of the Electrochemical Society* 149: A79–A85.

30 Chung, K.Y., Yoon, W.-S., Kim, K.-B. et al. (2004). *Suppression of structural fatigue by doping in spinel electrode probed by in situ bending beam method. Journal of the Electrochemical Society* 151: A484–A490.

31 Chung, K.Y. and Kim, K.-B. (2004). Investigations into capacity fading as a result of a Jahn–Teller distortion in 4 V $LiMn_2O_4$ thin film electrodes. *Electrochimica Acta* 49: 3327–3337.

32 Kim, J.-S., Kim, K., Cho, W. et al. (2012). A truncated manganese spinel cathode for excellent power and lifetime in lithium-ion batteries. *Nano Letters* 12: 6358–6365.

33 Amatucci, G.G., Tarascon, J.M., and Klein, L.C. (1996). CoO_2, the end member of the Li_xCoO_2 solid solution. *Journal of the Electrochemical Society* 143: 1114–1123.

34 Dahn, J.R., von Sacken, U., Juzkow, M.W., and Al-Janaby, H. (1991). Rechargeable $LiNiO_2$/carbon cells. *Journal of the Electrochemical Society* 138: 2207–2211.

35 Dahn, J., von Sacken, U., and Michal, C. (1990). Structure and electrochemistry of $Li_{1\pm y}NiO_2$ and a new Li_2NiO_2 phase with the $Ni(OH)_2$ structure. *Solid State Ionics* 44: 87–97.

36 Liu, Z., Yu, A., and Lee, J.Y. (1999). Synthesis and characterization of $LiNi_{1-x-y}Co_xMn_yO_2$ as the cathode materials of secondary lithium batteries. *Journal of Power Sources* 81–82: 416–419.

37 Liu, W., Oh, P., Liu, X. et al. (2015). Nickel-rich layered lithium transition-metal oxide for high-energy lithium-ion batteries. *Angewandte Chemie International Edition* 54: 4440–4457.

38 Abdellahi, A., Urban, A., Dacek, S., and Ceder, G. (2016). Understanding the effect of cation disorder on the voltage profile of lithium transition-metal oxides. *Chemistry of Materials* 28: 5373–5383.

39 Kim, Y. (2013). Encapsulation of $LiNi_{0.5}Co_{0.2}Mn_{0.3}O_2$ with a thin inorganic electrolyte film to reduce gas evolution in the application of lithium ion batteries. *Physical Chemistry Chemical Physics* 15: 6400–6405.

40 Kim, Y. and Cho, J. (2007). Lithium-reactive $Co_3(PO_4)_2$ nanoparticle coating on high-capacity $LiNi_{0.8}Co_{0.16}Al_{0.04}O_2$ cathode material for lithium rechargeable batteries. *Journal of the Electrochemical Society* 154: A495–A499.

41 Sun, F., Markötter, H., Manke, I. et al. (2016). Three-dimensional visualization of gas evolution and channel formation inside a lithium-ion battery. *ACS Applied Materials & Interfaces* 8: 7156–7164.

42 Zhang, S.S. (2014). Insight into the gassing problem of Li-ion battery. *Frontiers in Energy Research* 2 (8694): 1–4.

43 Wagner, M.R., Raimann, P.R., Trifonova, A. et al. (2004). Dilatometric and mass spectrometric investigations on lithium ion battery anode materials. *Analytical and Bioanalytical Chemistry* 379: 272–276.

44 Qian, Y., Niehoff, P., Börner, M. et al. (2016). Influence of electrolyte additives on the cathode electrolyte interphase (CEI) formation on $LiNi_{1/3}Mn_{1/3}Co_{1/3}O_2$ in half cells with Li metal counter electrode. *Journal of Power Sources* 329: 31–40.

45 Becker, D., Börner, M., Nölle, R. et al. (2019). Surface modification of Ni-rich $LiNi_{0.8}Co_{0.1}Mn_{0.1}O_2$ cathode material by tungsten oxide coating for improved electrochemical performance in lithium-ion batteries. *ACS Applied Materials & Interfaces* 11: 18404–18414.

46 Xiong, X., Wang, Z., Yue, P. et al. (2013). Washing effects on electrochemical performance and storage characteristics of $LiNi_{0.8}Co_{0.1}Mn_{0.1}O_2$ as cathode material for lithium-ion batteries. *Journal of Power Sources* 222: 318–325.

47 Bak, S.-M., Hu, E., Zhou, Y. et al. (2014). Structural changes and thermal stability of charged $LiNi_xMn_yCo_zO_2$ cathode materials studied by combined in situ time-resolved XRD and mass spectroscopy. *ACS Applied Materials & Interfaces* 6: 22594–22601.

48 Berkes, B.B., Schiele, A., Sommer, H. et al. (2016). On the gassing behavior of lithium-ion batteries with NCM523 cathodes. *Journal of Solid State Electrochemistry* 20: 2961–2967.

49 Berkes, B.B., Jozwiuk, A., Sommer, H. et al. (2015). Simultaneous acquisition of differential electrochemical mass spectrometry and infrared spectroscopy data for in situ characterization of gas evolution reactions in lithium-ion batteries. *Electrochemistry Communications* 60: 64–69.

50 Berkes, B.B., Jozwiuk, A., Vračar, M. et al. (2015). On-line continuous flow differential electrochemical mass spectrometry with a realistic battery setup for high-precision, longterm cycling tests. *Analytical Chemistry* 87: 5878–5883.

51 Kim, U.-H., Kuo, L.-Y., Kaghazchi, P. et al. (2019). Quaternary layered Ni-rich NCMA cathode for lithium-ion batteries. *ACS Energy Letters* 4: 576–582.

52 Kim, U.-H., Kim, J.-H., Hwang, J.Y. et al. (2019). Compositionally and structurally redesigned high-energy Ni-rich layered cathode for next-generation lithium batteries. *Materials Today* 23: 26–36.

53 Park, K.-J., Hwang, J.Y., Ryu, H.H. et al. (2019). Degradation mechanism of Ni-enriched NCA cathode for lithium batteries: are microcracks really critical? *ACS Energy Letters* 4: 1394–1400.

54 Huang, S., Kim, S.M., Bak, S.M. et al. (2014). Investigating local degradation and thermal stability of charged nickel-based cathode materials through real-time electron microscopy. *ACS Applied Materials & Interfaces* 6: 15140–15147.

55 Huang, Y., Lin, Y.C., Jenkins, D.M. et al. (2016). Thermal stability and reactivity of cathode materials for Li-ion batteries. *ACS Applied Materials & Interfaces* 8: 7013–7021.

56 Waldmann, T., Hogg, B.-I., and Wohlfahrt-Mehrens, M. (2018). Li plating as unwanted side reaction in commercial Li-ion cells – a review. *Journal of Power Sources* 284: 107–124.

57 Petzl, M., Kasper, M., and Danzer, M.A. (2015). Lithium plating in a commercial lithium-ion battery – a low temperature aging study. *Journal of Power Sources* 275: 799–807.

58 Yang, X.-G. and Wang, C.-Y. (2018). Understanding the trilemma of fast charging, energy density and cycle life of lithium-ion batteries. *Journal of Power Sources* 402: 489–498.

11

Mechanical Behavior of Transition Metal Oxide-Based Battery Materials

Truong Cai, Jung Hwi Cho, and Brian W. Sheldon

Brown University, School of Engineering, 184 Hope St., Providence, RI, 02912, USA

11.1 Introduction

From a mechanical properties' perspective, the transition metal (TM) oxides that are employed in energy-storage systems are typical ceramics. Because the relevant applications occur at low-to-moderate temperatures, plasticity and creep are minimal and thus deformation is largely elastic. The relatively high modulus of these materials means that they can support relatively high stresses. In configurations where large stresses are possible, brittle fracture is then an important concern.

In Li-based batteries, TM oxides are widely used as cathodes. The compositional changes that occur repeatedly during the use of these materials leads directly to volume changes. The corresponding strains are significantly smaller than those that occur in high-capacity alloys like silicon that are currently candidates as next-generation anode materials. However, the relatively large modulus of oxides means that strains of several percentages can still lead to GPa level stresses. The resulting deformation and cracking in individual cathode particles are essentially mechanical responses to chemically induced changes as shown in Figure 11.1a, with Ni-rich $Li[Ni_xCo_yMn_{1-x-y}]O_2$ (NMC) cathode as an example. These are addressed in Section 11.2 of this chapter, where criteria for brittle fracture are reviewed. Because the stresses in TM oxides can be relatively large, it is also possible that the corresponding elastic energies will alter chemical thermodynamics. This formalism is described in Section 11.3, where it is shown that it is possible for large stresses to alter lithium concentrations, but that the degree to which this occurs depends strongly on the solution thermodynamics of individual compounds.

Chemo-mechanical phenomena in TM oxide solid electrolytes (SEs) introduce additional issues, which are described in Section 11.4. This includes general phenomena at the cell level such as interfacial contact and fracture. The possible relationship between mechanical stress and electrochemical failure of solid electrolyte is also discussed, including the possible role that this plays in dendrite-like lithium metal penetration through the solid electrolyte, as illustrated in Figure 11.1b.

Transition Metal Oxides for Electrochemical Energy Storage, First Edition.
Edited by Jagjit Nanda and Veronica Augustyn.

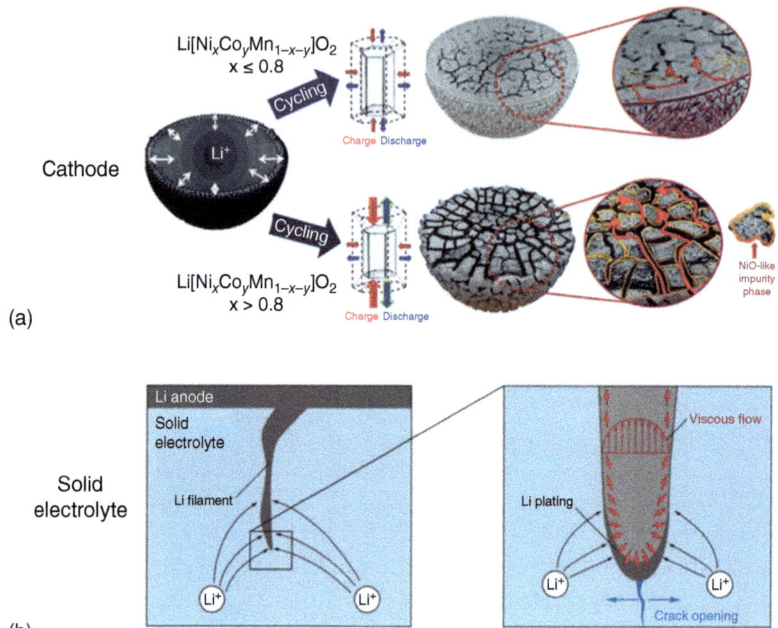

Figure 11.1 Illustration of the effect of mechanical stress in transition metal (TM) oxides used in energy-storage systems. (a) Schematic figure showing capacity-fading mechanism of Ni-rich Li[Ni$_x$Co$_y$Mn$_{1-x-y}$]O$_2$ cathodes with repeated volume changes during cycling. Source: Xu et al. [1]/with permission of Royal Society of Chemistry. (b) Schematic of stress fields accumulated due to lithium filament penetration in ceramic solid electrolytes. Source: LePage et al. [2]/IOP Publishing/CC BY 4.0.

11.2 Mechanical Responses to Compositional Changes

11.2.1 Volume Changes and Deformation in Electrode Particles

Consider first the volume expansion induced by adding Li to a free-standing host material, $V' = N_o V_o + \int_0^{N_{\mathrm{Li}}} \overline{V_{\mathrm{Li}}}\, \mathrm{dn}$, where V' is the total volume, N_o and N_{Li} are the moles of host material and Li, respectively, V_o is the molar volume of the pure host material, and $\overline{V_{\mathrm{Li}}}$ is the partial molar volume of Li. This total volume change can also be expressed in terms of the free strains due to expansion in the three principal directions, f_x, f_y, and f_z:

$$\frac{V'(c)}{N_o} = V_o + \int_{c_C}^{c_D} \overline{V_{\mathrm{Li}}}\, dc = V_o\,(1 + f_x)(1 + f_y)(1 + f_z) \tag{11.1}$$

where $c = N_{\mathrm{Li}}/N_o$. During discharge, the Li concentration then increases from c_C to c_D.

While $\overline{V_{\mathrm{Li}}}$ typically varies with both composition and crystallographic orientation, it is common to use a simplified description where $\overline{V_{\mathrm{Li}}}$ is treated as a constant. These values for some common transition metal oxide electrode materials are listed

Table 11.1 Mechanical properties of common transition metal oxide cathode materials.

Materials	V_o (cm^3 mol^{-1})	$\overline{V_{Li}}$ (cm^3 mol^{-1}) [3]	f [4]	$E^{a)}$ (GPa) [4]	ν [4]
$LiCoO_2$	20	~0.1	0.01	264	0.32
$LiMn_2O_4$	43	~1	0.02	194	0.26
$LiFePO_4$	44	~4	0.02	124	0.26
$LiNi_{0.33}Mn_{0.33}Co_{0.33}O_2$	42	0.25	0.01	199	0.25

a) Young's modulus is computed from stiffness matrix using Voigt–Reuss–Hill homogenization.

in Table 11.1. With the relatively small volume changes that occur in TM oxides, Eq. (11.1) leads to the following simplified description of the strain for the free expansion of the electrode material:

$$f = \left[1 + \frac{1}{V_o}\int_{c_C}^{c_D} \overline{V_{\mathrm{Li}}}\, dc\right]^{1/3} - 1 \cong \frac{\overline{V_{\mathrm{Li}}}}{3\,V_o}(c_D - c_C) \tag{11.2}$$

where $f = f_x = f_y = f_z$ corresponds to isotropic behavior. For this case, it is relatively straight forward to analyze chemical expansion effects in spherical electrode particles. An isolated, unconstrained particle will expand uniformly in all directions. Composition gradients in this type of isolated particle will lead to internal stresses, when charge or discharge rates are fast enough to induce non-uniform Li profiles. As seen in Table 11.1, the strains that occur in most transition metal oxides are relatively modest. Thus, analytical solutions provide reasonable approximations. For example, for a spherical cathode particle, full discharge under galvanostatic conditions can be assessed with the following initial and boundary conditions:

$$c\,(r,0) = 0 \tag{11.3a}$$

$$D\left.\frac{\partial c}{\partial r}\right|_{r=0} = 0 \tag{11.3b}$$

$$D\left.\frac{\partial c}{\partial r}\right|_{r=R} = \frac{I}{\mathcal{F}} \tag{11.3c}$$

where D is the Li diffusivity in the particle, I is the current density at the particle surface, and $\mathcal{F}$ is Faraday's constant. Solving the corresponding diffusion problem (i.e. Fick's second law for constant D) and applying Hooke's law (with constant elastic constants) then give the following expressions for the radial and tangential stresses in a spherical particle [4]:

$$\sigma_r(r,\tau) = \rho\left[\frac{(1-x^2)}{5} + \frac{4}{x^3}\sum_{n=1}^{\infty}\left(\frac{\sin(\lambda_n x) - (\lambda_n x)\cos(\lambda_n x)}{\lambda_n^4\,\sin(\lambda_n x)}\right)\exp(-\lambda_n^4\,\tau)\right] \tag{11.4}$$

$$\sigma_\theta(r,\tau) = \rho \left[\frac{(1-2x^2)}{5} + 2\sum_{n=1}^{\infty} \frac{\exp(-\lambda_n^2\,\tau)}{\lambda_n \sin(\lambda_n x)} \times \left(\frac{\sin(\lambda_n x)}{\lambda_n x} - \frac{\sin(\lambda_n x) - (\lambda_n x)\cos(\lambda_n x)}{\lambda_n^3\, x^3} \right) \right] \tag{11.5}$$

where R is the particle radius, r is the radial coordinate, $x = r/R$, and λ_n ($n = 1, 2, 3, \ldots$) are the positive roots of $\tan(\lambda_n) = \lambda_n$. The dimensionless time is $\tau = Dt/R^2$, and the stresses here are scaled by the quantity:

$$\rho = \frac{E\overline{V_{\text{Li}}}}{3(1-\nu)} \frac{RI}{\mathcal{F}D} \tag{11.6}$$

where E is Young's modulus and ν is the Poisson ratio. Similar results are also available for basic potentiostatic conditions (i.e. obtained for a fixed surface concentration, instead of the constant flux boundary condition in Eqs. (11.6) and (11.7)). Predicted stress profiles based on these results are shown in Figure 11.2. These results show that diffusion-induced stress (DIS) is primarily determined by the particle size and lithiation/delithiation rate.

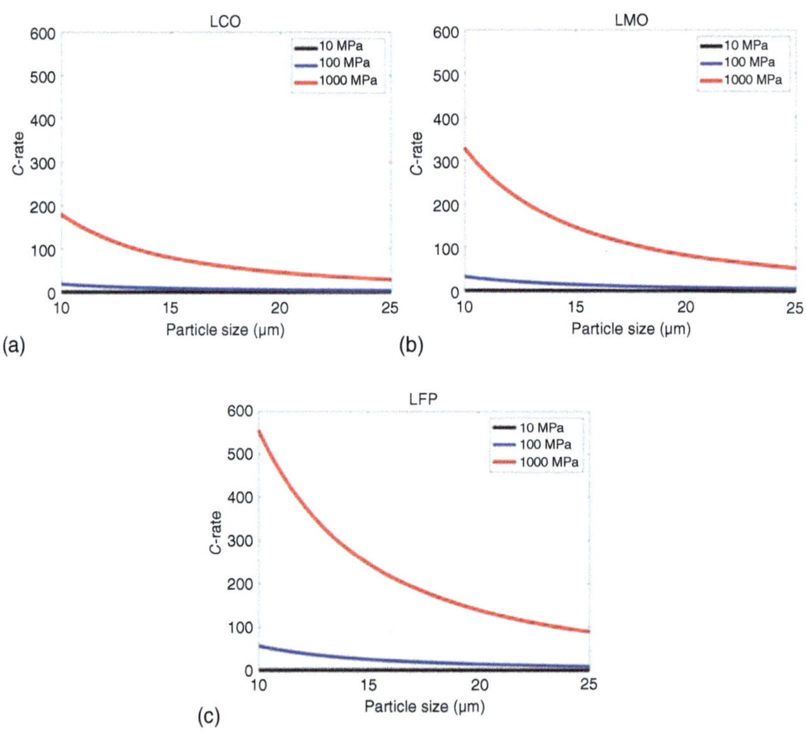

Figure 11.2 Maximum stress ($\sigma^{\max}$) DIS estimates during lithiation of spherical particles, during galvanostatic cyclic. The C rates are calculated, where $C_{\max}$ is the full capacity of the material. The values plotted here are based on Eq. (11.7) for: (a) LCO; (b) LMO; (c) LFP. Source: Based on Woodford et al. [5].

The solution for the galvanostatic case above gives maximum stresses [6]:

$$\sigma^{\max} = \sigma_r^{\max}(0) = -\sigma_\theta^{\max}(R) \cong \frac{\rho}{15} \tag{11.7}$$

As shown here, this is the magnitude of both the maximum radial tensile stress at $r = 0$ and the maximum tangential compressive stress at $r = R$ [4]. For a given cathode material, $\sigma^{\max}$ is largely dictated by R and I. With this in mind, the $\sigma^{\max}$ contour lines in Figure 11.2 provide guidelines for understanding conditions that will lead to significant DIS in several common transition metal oxide cathode materials.

The treatment outlined above does not explicitly include the actual volume change in the particle (i.e. the size is held constant). However, for the relatively small strains expected in most cathode particles, the analytical result gives a reasonable estimate of the stress gradients that occur across the particle. Other treatments for spherical particles that account for the volume change and composition-dependent elastic constants have also been presented [7–10]. Real battery particles will also deviate from the assumptions employed here for a variety of reasons. Examples that are relevant in TM oxides include anisotropic volume changes due to crystallographic orientation, anisotropic elastic constants, grain boundaries (GBs), phase boundaries, and internal porosity. Descriptions of chemical expansion in these more complex structures typically require finite element analysis (FEA) or other numerical models.

11.2.2 Particle Fracture

Cracking in electrode particles has been examined with electron microscopy. The complex, multiphase structure of battery electrodes makes it difficult to directly observe particle fracture during electrochemical cycling, and thus some researchers have questioned whether most of the observed particle cracking occurs during the initial synthesis of the material. However, several careful studies have provided direct evidence of particle fracture during cycling, Figure 11.3 [11] and Figure 11.4 [12].

Figure 11.3 Cross-sectional SEM images of (a) pristine NCA95 particles and (b) the same region after 50 cycles at a rate of 0.5C. Source: Park et al. [11].

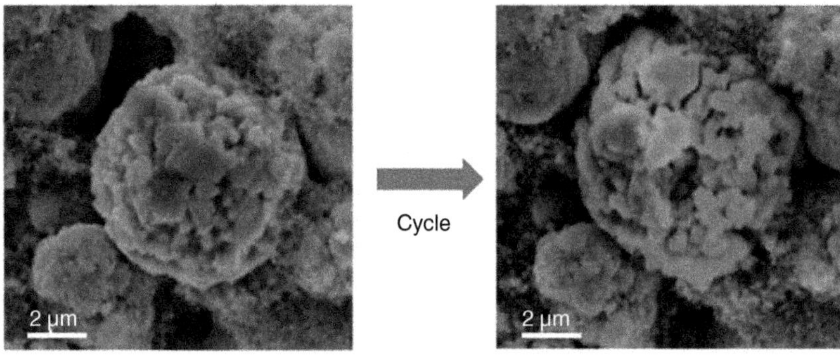

Figure 11.4 SEM images of pristine NMC particles and the same region after 100 cycles at a rate of 0.5C. Source: Xu and Zhao [12].

There are several general properties of TM oxide materials that have an important impact on particle fracture:

- Elastic modulus is relatively high (>100 GPa). This means that the relatively small compositional strains that were noted above can still lead to large stresses.
- Plastic deformation is minimal (and possibly irrelevant).
- Fracture toughness values are relatively low (<2 MPa $m^{1/2}$).

Based on these factors, TM oxide electrode particles can undergo brittle fracture when stresses are sufficiently high. With isotropic behavior, the results in Figure 11.2 show that large stresses will only occur at high C rates. However, anisotropy in the particle structure can lead to fracture at lower C rates. Several examples of this have been reported in the literature. Two examples are shown in Figures 11.4 and 11.5 [12, 13].

The basic analysis of diffusion-induced stress outlined in Section 11.2.1 provides necessary information, but this is not sufficient to predict particle fracture. Because dislocation-controlled plastic deformation is limited, the yield stress and von Mises equivalent failure criteria are not dominant effects and linear elastic fracture mechanics can be employed. The standard treatments lead to stress intensity factors, which are a function of the stress field and the initial, strength-limiting flaw size.

To develop and quantify a fracture mechanics failure criterion for electrochemical influx of lithium to TM oxide, Carter and coworkers [5] expanded on earlier analyses

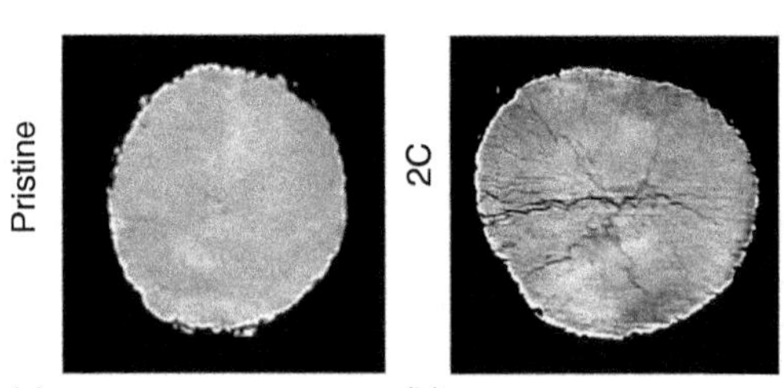

Figure 11.5 Transmission X-ray microscopy images of: (a) a pristine layered NMC particle and (b) after 50 cycles at 2C. Source: Xia et al. [13].

[14, 15], using individual spherical particles of $Li_XMn_2O_4$ as a model system to obtain a map of fracture dependence on particle size and C-rate. This is based on finite-sized semi-elliptical surface cracks of length a, in a particle of radius r_{max}. The crack faces are assumed to be electrochemically inert so that only the spherical outer surface of the particle is reactive, and the $Li_XMn_2O_4$ is treated as a continuous solid solution $0<x<1$ with uniform elastic and fracture resistance (independent of lithium composition). To develop an expression for the stress intensity factor, the lithium distribution and the corresponding stress profiles for different galvanostatic charge rates are first computed numerically. The diffusion-induced stress profiles at the end of charging are then used to calculate the stress intensity factor as a function of flaw size. This provides the basis for maps of particle fracture regimes based on C-rates and particle sizes, as seen in Figure 11.2.

To obtain diffusion-induced stress profiles, Carter et al. solved the diffusion equation with a linearized flux and obtained an expression for the stress profiles in the same manner presented in Section 11.2.1. The stress intensity factor is then obtained with the approach of Mattheck et al. [16] via the integral equation:

$$K_I = \frac{E}{K_r(1-n^2)} \int_0^a s_q(x)\, m(x,a)\, dx \tag{11.8}$$

where K_r is the reference stress intensity factor, $s_q(x)$ is the tangential stress in Eq. (11.5) at the end of charge, and $m(x, a)$ is a weighting function. The reference here is uniform tensile loading in the z-direction of the semi-elliptical surface cracked plate [17]. Integration over the crack length accounts for spatial variations in the stress field. The impact of this can be seen by noting that a moderate stress with a small gradient can produce a stress intensity factor greater than a sharply peaked but rapidly decaying stress distribution. The analysis in Woodford et al. is based on semi-circular cracks, where the results in Figure 11.2 show a decreasing critical C-rate with increasing particle size [5]. This is consistent with expectations based on DIS results (see Section 11.2.1), where higher rates and larger particles lead to larger stresses and should thus induce more damage. This type of fracture mechanics analysis provides quantitative estimates of particle size and charging procedures that can prevent diffusion-mediated fracture in TM oxide particles. For example, a controlled variable current charge that starts at low rate and increases should make it possible to charge in shorter times without inducing fracture. Here, the low initial leads to a more uniform concentration in the core, which then limits the internal composition gradients and lowers the stress intensity factor when higher rates are employed later in the cycle [5].

The accuracy of electrochemical shock predictions is currently limited by the accuracy of lithium diffusivity data. For example, reported values for $Li_XMn_2O_4$ span at least three orders of magnitude. Figure 11.6, using values in the middle of this range, shows that crack formation is not expected to occur for a particle size of 2.5 μm an 1C. However, Figure 11.5 shows cracking after 50 cycles for these conditions. A lower diffusivity is one possible explanation for this discrepancy, since this will shift the boundary in Figure 11.6 and lead to fracture under the conditions employed in Figure 11.7 [18].

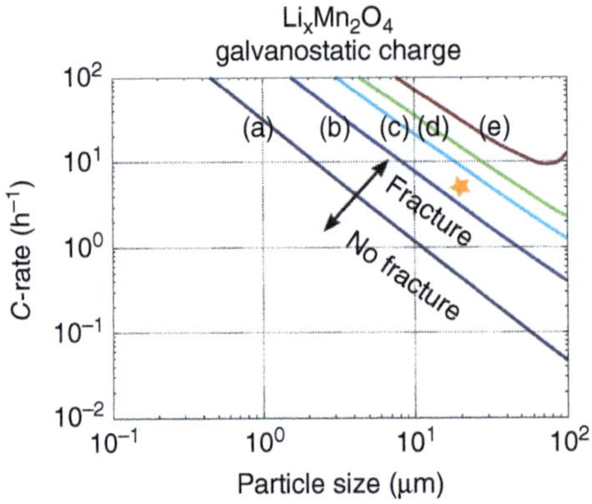

Figure 11.6 Electrochemical fracture maps for galvanostatic charging of LMO. The curves represent the onset of fracture for K_{Ic} values of: (a) 0.1, (b), 1, (c) 3, (d) 5, and (e) 10 MPa $m^{1/2}$. The orange star is a specific example for 21 mm particles at 5C. Source: Woodford et al. [5]/with permission of IOP Publishing.

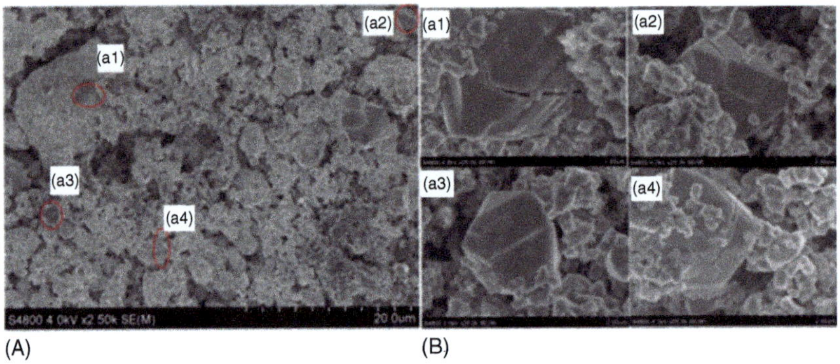

Figure 11.7 SEM image of LMO after 350 cycles. (a1)–(a4) are the enlarged views of specific particles [18]. (A) SEM images of LMO after 350 cycles (B) Enlarged view of particles in (A).

11.3 Impact of Strain Energy on Chemical Phenomena

11.3.1 Thermodynamics

Chemical and electrochemical phenomena in battery materials are fundamentally linked to mechanical deformation through thermodynamics. As noted in Section 11.2.2, TM oxides can generally be described as elastic solids. Here, it is convenient to describe the total Gibbs Free energy, $\widehat{G}$, with:

$$\widehat{G} = \widehat{G}_{\text{ref}} + \widehat{E}_{\text{elastic}} \tag{11.9}$$

where $\widehat{G}_{\text{ref}}$ corresponds to an undeformed reference state and $\widehat{E}_{\text{elastic}}$ is the excess elastic strain energy for an arbitrary deformed configuration. An equivalent molar expression can also be used: $G = G_{\text{ref}} + E_{\text{elastic}}$.

The differential form of the Gibbs free energy is given by:

$$d\widehat{G} = \widehat{S}\, dT + \widehat{V}\, \varepsilon_{jk}\, d\sigma_{jk} + \sum_i \Gamma_i\, dn_i = 0 \tag{11.10}$$

where σ_{jk} and ε_{jk} are the stress and strain tensors, and $\widehat{V}$ is the total volume. As noted in Section 11.2, the treatment of TM oxide electrodes can generally be simplified because plasticity is negligible and small strain approximations are reasonably accurate. The chemical potential is defined by:

$$\mu_i \equiv \left(\frac{\partial \hat{G}_{\text{ref}}}{\partial N_i}\right)_{\text{T,P},N_j \neq N_i} = \mu_i^o + RT \ln\left[\frac{\gamma_i N_i}{\sum N_i}\right] \tag{11.11}$$

where μ_i^o and γ_i are the reference chemical potential and activity coefficient for i. A modified chemical potential, Γ_i, that includes elastic strain energy contributions is then given by:

$$\Gamma_i \equiv \left(\frac{\partial \hat{G}}{\partial N_i}\right)_{\text{T,P},N_j \neq N_i} = \mu_i + \left(\frac{\partial \hat{E}_{\text{elastic}}}{\partial N_i}\right)_{\text{T,P},N_j \neq N_i} \tag{11.12}$$

This quantity is also sometimes referred to as a diffusion potential [19].

Oxides used as battery electrodes are necessarily multicomponent materials, where Gibbs–Duhem relationships are useful for understanding solution thermodynamics for different components. These can be obtained by invoking the 1st and 2nd laws of thermodynamics to obtain (molar form):

$$S\,dT + V_m\,\varepsilon_{jk}\,d\sigma_{jk} + \chi_i\,d\Gamma_i = 0 \tag{11.13}$$

where χ_i is the mole fraction of component i.

In principle, large stresses in battery electrodes can alter the Li solubility and open-circuit voltage. The variation in the Li concentration depends on both the stress level and the solution thermodynamics for a particular material. Basic thermodynamics describes the relationship between the stress state and the open-circuit voltage of a solid. The modified chemical potential in Eq. (11.13) can be extended to include electrical contributions [20]:

$$\Gamma_i \equiv \left(\frac{\partial \hat{G}}{\partial N_i}\right)_{\text{T,P}} = \mu_{\text{Li}}^o + RT \ln\left[\frac{\gamma_i N_i}{\sum N_i}\right] + z_i\,\mathcal{F}\,\phi + \hat{V}\,\sigma_{jj}\left(\frac{\partial \varepsilon_{jj}}{\partial N_i}\right)_{\text{T,P}} \tag{11.14}$$

where ϕ is the electric potential and z_i is the charge of i (+1 for Li ions, etc.). When the stiffness tensor varies with composition, an additional elastic contribution associated with this gradient should be added to Eq. (11.14); however, this term is neglected here because the elastic term in Eq. (11.14) is usually much larger. The description in Eq. (11.14) leads directly to a modified Nernst equation [20].

$$\mathcal{F}\,\Delta\phi_{\text{o}} \cong -RT \ln\left[\frac{\gamma_{Li}\,\text{c}}{(1+\text{c})}\right] + \overline{V_{\text{Li}}}\,\frac{\sigma_{jj}}{3} \tag{11.15}$$

where $c = \chi_{\text{Li}}/(1-\chi_{\text{Li}})$ and $\Delta\phi_{\text{o}}$ is the open-circuit potential relative to Li metal. This form again assumes that $\overline{V_{\text{Li}}}$ is isotropic, but can be modified to treat crystallographic anisotropy. Also, while $\overline{V_{\text{Li}}}$ is often treated as a constant, variations in $\overline{V_{\text{Li}}}$ with composition may be significant in some materials.

The implications of Eq. (11.15) will depend on the configuration of interest, and also on the solution thermodynamics of a given material. It is convenient to treat the TM oxide as a pseudo-binary that is described with a regular solution model, such that the activity coefficient is described by:

$$\gamma_{\text{Li}} = \exp\left[\frac{\Delta\overline{H_{\text{Li}}}}{RT}\right] \tag{11.16}$$

where the partial molar enthalpy of Li, $\Delta\overline{H_{\text{Li}}}$, usually varies with composition. To assess the impact of chemo-mechanical coupling, Li insertion in a spherical particle is used as a basic example (see Section 11.2.1). Under potentiostatic conditions, the initial tangential stress at the particle surface when Li is first inserted can be described by [21]:

$$\sigma_\theta^{\max}(R) \cong -\frac{E}{3(1-\nu)}\frac{\overline{V_{\mathrm{Li}}}}{3V_o}[c_R - c_o] \tag{11.17}$$

where c_o is the initial uniform Li concentration in the particle and $c_R = c(R)$ is the equilibrium concentration at the surface (i.e. set by the potentiostatic boundary condition). Note that this expression employs the small strain approximation that is used in Eq. (11.2). To illustrate how variations in the solution behavior are related to impact of stress, consider a standard one parameter regular solution model where:

$$\Delta\overline{H_{\mathrm{Li}}} = \Omega(1-\chi_{\mathrm{Li}})^2 = \frac{\Omega}{(1+c)^2} \tag{11.18}$$

At a given value of $\Delta\phi_o$ the equilibrium concentration at the particle surface, c_R, is then described by Eqs. (11.15)–(11.18), which can be combined to give:

$$\frac{\mathcal{F}\,\Delta\phi_o}{RT} \cong -\ln\left[\frac{c_R}{1+c_R}\right] - \frac{\alpha}{(1+c_R)^2} - \beta[c_R - c_o] \tag{11.19a}$$

$$\alpha = \frac{\Omega}{RT};\ \beta = \frac{2E(\overline{V_{\mathrm{Li}}})^2}{9(1-\nu)V_o RT} \tag{11.19b}$$

For an electrode particle that is initially in a charged state (i.e. $c_o = c_C$), the stress in Eq. (11.18) is maximized at a potential ($\Delta\phi_D$) that corresponds to the full capacity of the material, $\Delta c_{\max} = c_D - c_C$, which is defined in the absence of any stress effects via:

$$\frac{\mathcal{F}(\phi_C - \phi_D)}{RT} \cong \ln\left[\frac{(1+c_C)(c_C+\Delta c_{\max})}{c_C(1+c_C+\Delta c_{\max})}\right] + \frac{\alpha}{(1+c_C)^2}\left[\frac{2c_C + \Delta c_{\max}(2+\Delta c_{\max})}{(1+c_C+\Delta c_{\max})^2}\right] \tag{11.20}$$

The analysis presented here considers a hypothetical cathode material, operating over a range where $\Delta c_{\max} = 0.3$ mol Li/moles oxide corresponds to $(\phi_C - \phi_D) = 0.1$ V. For this case, variations in α can be compared by varying c_C accordingly (i.e. based on Eq. (11.20)). The impact of chemo-mechanical coupling is then evaluated by determining the extent to which the stress, $\sigma_\theta^{\max}(R)$ at $\Delta\phi_D$, alters the surface concentration:

$$c_R = c_D^* = c_C + \Delta c_{\max}^* \tag{11.21}$$

where the elastic energy contribution at the particle surface causes $\Delta c_{\max}^*$ to differ from $\Delta c_{\max}$. For example, with a positive $\overline{V_{\mathrm{Li}}}$ value, compressive stress will lead to $c_D^* < c_D$. Assessing this variation at $\Delta\phi_D$ provides an upper bound estimate for the impact of chemo-mechanical coupling. This can be accomplished by applying Eq. (11.19a,b) with and without the stress contribution, to give:

$$\ln\left[\frac{(c_C+\Delta c_{\max}^*)(1+c_C+\Delta c_{\max})}{(c_C+\Delta c_{\max})(1+c_C+\Delta c_{\max}^*)}\right] + \alpha\left[\frac{1}{(1+c_C+\Delta c_{\max}^*)^2} - \frac{1}{(1+c_C+\Delta c_{\max})^2}\right] = -\beta\,\Delta c_{\max}^* \tag{11.22}$$

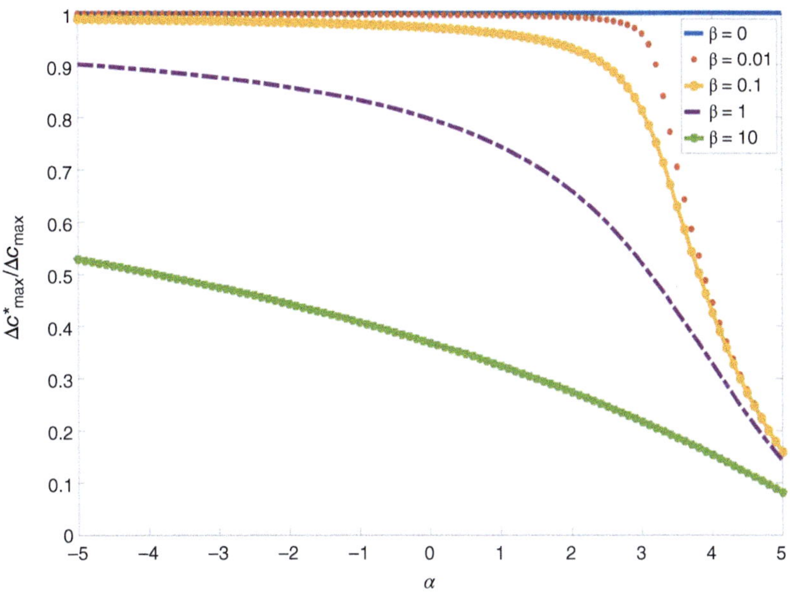

Figure 11.8 Computed values of $\Delta c^*_{max}/\Delta c_{max}$ vs. α, based on Eq. (11.22) for different values of β.

Values of Δc^*_{max} obtained with Eq. (11.22) as a function of α are plotted in Figure 11.8 for different values of β. The strain energy increases with β, and thus the decrease in Δc^*_{max} (i.e. lower c_R) is more pronounced with larger β values. For most of the common positive electrodes, β<1. The variation with α in Figure 11.8 demonstrates that the impact of stress on the composition at a fixed potential will depend strongly on $\Delta\overline{H_{Li}}$ (which is reflected in the value of α). With small or positive values of α, the results show that compressive stress in a material with positive $\overline{V_{Li}}$ can lead to a significant decrease in the Li concentration for a given $\Delta\phi_o$ value, when β is sufficiently large. However, this effect is much smaller for $\alpha < 0$ (i.e. negative $\Delta\overline{H_{Li}}$ values), where chemical bonding effects strongly promote mixing, such that the elastic energy effects lead to a relatively small decrease in Δc^*_{max}. The general conclusion from Figure 11.8 is that negative $\Delta\overline{H_{Li}}$ values will counteract demixing due to elastic strain energy effects. While the simple regular solution model in Eq. (11.18) is not expected to be valid for Li in most TM oxide materials (except for narrow composition ranges), more complex solution models can be employed to accurately describe these materials. For example, more complex multi-parameter regular solution models have been used to describe $\Delta\phi_o(c)$ for silicon-negative electrodes. In these materials, negative $\Delta\overline{H_{Li}}$ values lead to only modest stress-induced changes in the lithium content, in spite of the large volume changes that occur [20].

11.3.2 Two-Phase Equilibrium

The basic relationships above (Eqs. (11.16)–(11.20)) can also be applied to TM oxides where multiple phases are present. However, with multiphase equilibrium, the interplay between solution thermodynamics and strain energy contributions can lead to more complex behavior. Cycling conditions are generally designed to avoid phase transformations that produce large volume changes, since these typically cause excessive degradation of the electrode particles (e.g. Jahn–Teller distortion in LMO [22, 23]). However, in some technologically important cathode materials, phase transformations during normal cycling can occur without leading to this type of mechanical degradation. In NMC and $LiFePO_4$ materials, there is still some controversy about the full thermodynamic nature of the relevant phase transformations. Thus, there is also considerable uncertainty about the elastic contributions to these transformations. Models that include detailed descriptions of the elastic energies have still been developed in some cases. For example, Bazant generalized both Marcus and Butler–Volmer kinetics for condensed phases and unified Cahn–Hilliard and Allen–Cahn equations to describe the dynamics resulting from such kinetics [24]. He and Cogswell also incorporated elastic strain effect into this framework using LFP as a specific example [7]. More recently, Kejie Zhao has been developing continuum models that coupled chemical dynamics and mechanics to describe evolution of lithium concentration and stress for NMC [25, 26].

The formation of surface layers on TM oxides also introduces multiphase equilibrium considerations. For example, cathode materials often react with the electrolyte to form a cathode electrolyte interphase (CEI) that must permit fast Li transport and also provide passivation against the deterioration of the active material. These are generally less well understood than the solid electrolyte interphase (SEI) that forms on battery anodes. In some NMC materials, irreversible structural changes that produce surface damage layers have been carefully documented. The corresponding volume changes are expected to produce significant stresses that will lead to the type of chemo-mechanical coupling that is described in section 3.1. Evaluation of these effects will ultimately require detailed investigations.

11.4 Solid Electrolytes

11.4.1 Electrode/Electrolyte Interfaces

Transition metal oxide ceramic solid electrolytes (SE) such as LLZO (nominal composition $Li_7La_3Zr_2O_{12}$) exhibit very high ionic conductivities (0.1–1.0 mS cm^{-1}); however, maintaining low impedance at electrode/SE interfaces in these materials remains a key challenge [27–30]. The combined effects of chemical and mechanical effects are particularly critical in all solid-state batteries (Figure 11.9). Strains due to chemically induced volume changes on one or both sides of the interface can lead directly to significant stresses that can either facilitate or impede transport across the interface [29, 30]. This chemical–mechanical coupling is further complicated by surface films that form due to electrode–electrolyte reactions [32, 33], or that are applied intentionally to improve contact and/or conductivity [34–36]. In either case, the combined chemical and mechanical properties of these films will dictate interfacial impedance. Researchers have explored chemical effects at electrode–solid

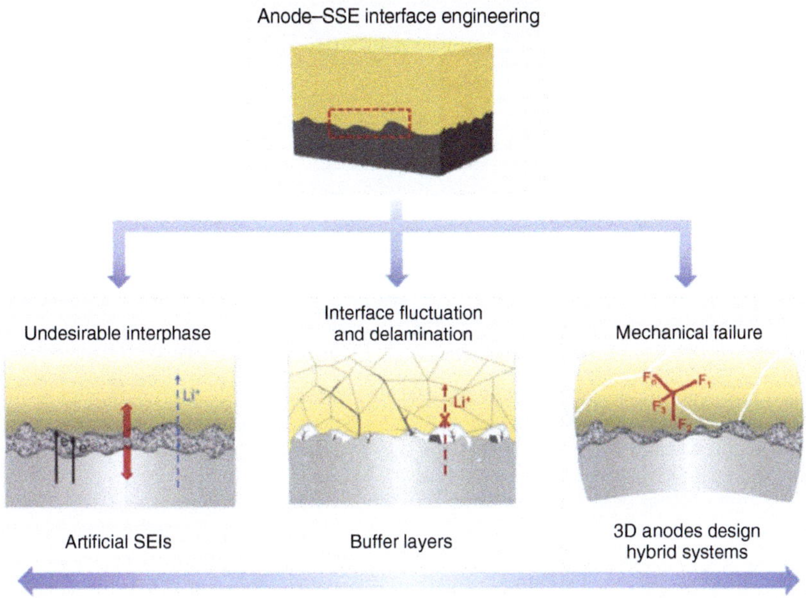

Figure 11.9 Overview of main challenges and strategies for the anode–SE interface. Source: Shen et al. [31]/with permission of John Wiley & Sons.

electrolyte interfaces (e.g. phase stability, interfacial coating, dopant, etc. [32–37]). However, chemo-mechanical coupling effects have only recently been addressed [2, 38–43].

Thermodynamic calculations and experiments reveal that most solid-electrolytes are not intrinsically stable against Li metal or cathode materials, such as $LiCoO_2$ [44, 45]. For example, LLZO can be lithiated at a voltage slightly above zero and oxidized at 2.9 V, leading to reaction layer formation at both the anode and cathode interfaces. Some of these decomposed products have been identified in experiments as shown in Figure 11.10 [33, 45, 46]. Furthermore, the loss of contact during cycling at the interface due to contractions of the active particle can result in serious ion-transport tortuosity, and hence increases interfacial resistance and capacity fade [47]. The corresponding volume changes of these interphase transformation have not been proved directly. However, reaction layers that form at these interfaces are likely to induce large stresses. Volume expansion here is likely to create compressive stress in the interlayer. Compression typically reduces ion diffusivities, but at the same time large stresses will reduce the driving force for the formation of surface damage layers (i.e. similar to some of the effects discussed for NMC cathodes in Section 11.2). While recent work reports that high interfacial impedance is a major issue in all-solid-state batteries, the impact of stress on this is just beginning to receive attention via in situ and operando experiments along with modeling studies [38–40, 42, 43, 48].

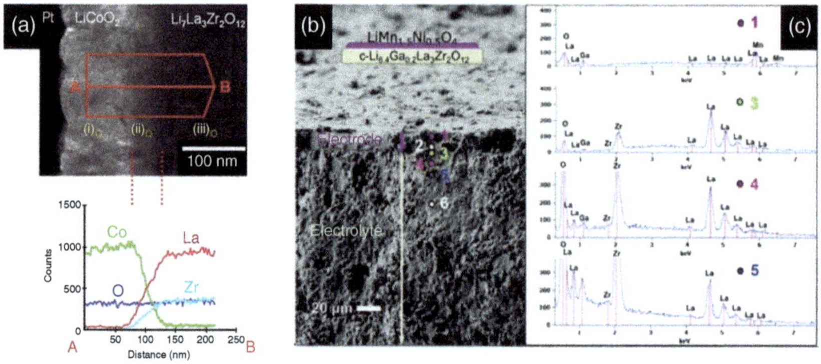

Figure 11.10 (a) Cross-sectional transmission electron microscopy (TEM) image of an LLZO/$LiCoO_2$ thin film interface and the energy-dispersive X-ray spectroscopy (EDS) line profile obtained from the region indicated by the red arrow in the direction A–B. The broken red lines indicate the reaction layer at the LLZ/$LiCoO_2$ interface. Points (i), (ii), and (iii) indicate locations used for nano-beam diffraction (NBD) analysis (for more details about the color in this figure legend, the reader is *referred* to the online publication). Source: Kim et al. [44]/with permission of Elsevier. (b) The cross section scanning electron microscopy (SEM) image of $LiMn_{1.4}Ni_{0.5}O_4$ cathode and c-$Li_{6.4}Ga_{0.2}La_3Zr_2O_{12}$ SE after cycling. (c) Energy-dispersive X-ray spectroscopy (EDXS) data from the parked regions in (a). Source: (c) Hänsel et al. [45]/with permission of Royal Society of Chemistry.

Another related issue is that stress and interfacial morphology, chemistry, and kinetics can cause fracture through the bulk SE and loss of contact at electrode/electrolyte interfaces [38–41, 43]. One recent electrochemo-mechanical finite element method (FEM) simulation predicts that fracture in the SE can be prevented when the fracture energy is higher than $4\,J\,m^{-2}$, for electrode particle expansion of 7.5% [39].

On the anode side, the creep-dominant deformation of soft Li metal is expected to form a more conformal interface with solid electrolytes, as interfacial contact between SE and lithium metal has been shown to significantly improve from applying pressure and deforming lithium metal onto the SE close to its melting temperature. Under cycling conditions, however, the lithium metal near the interface is expected to go under inhomogeneous volume change, resulting in pores and poor interfacial contact. Nonuniform stress at the interface will lead to internal cracks and bending near the interface, resulting in further capacity drop and increased interfacial resistance [3, 29, 49]. Therefore, maintaining a good contact and mechanical stability in the interface during cycling (under large volume change) is vital. Much of the work to date has focused on improving the "wettability" of lithium metal in the interface by using surface coatings [36, 50–52], and removing surface contaminants such as carbonate and hydroxide species from atmospheric exposure (Figure 11.11) [37]. Probing the dynamic stress evolution within the interface region is an important challenge that has not yet been addressed.

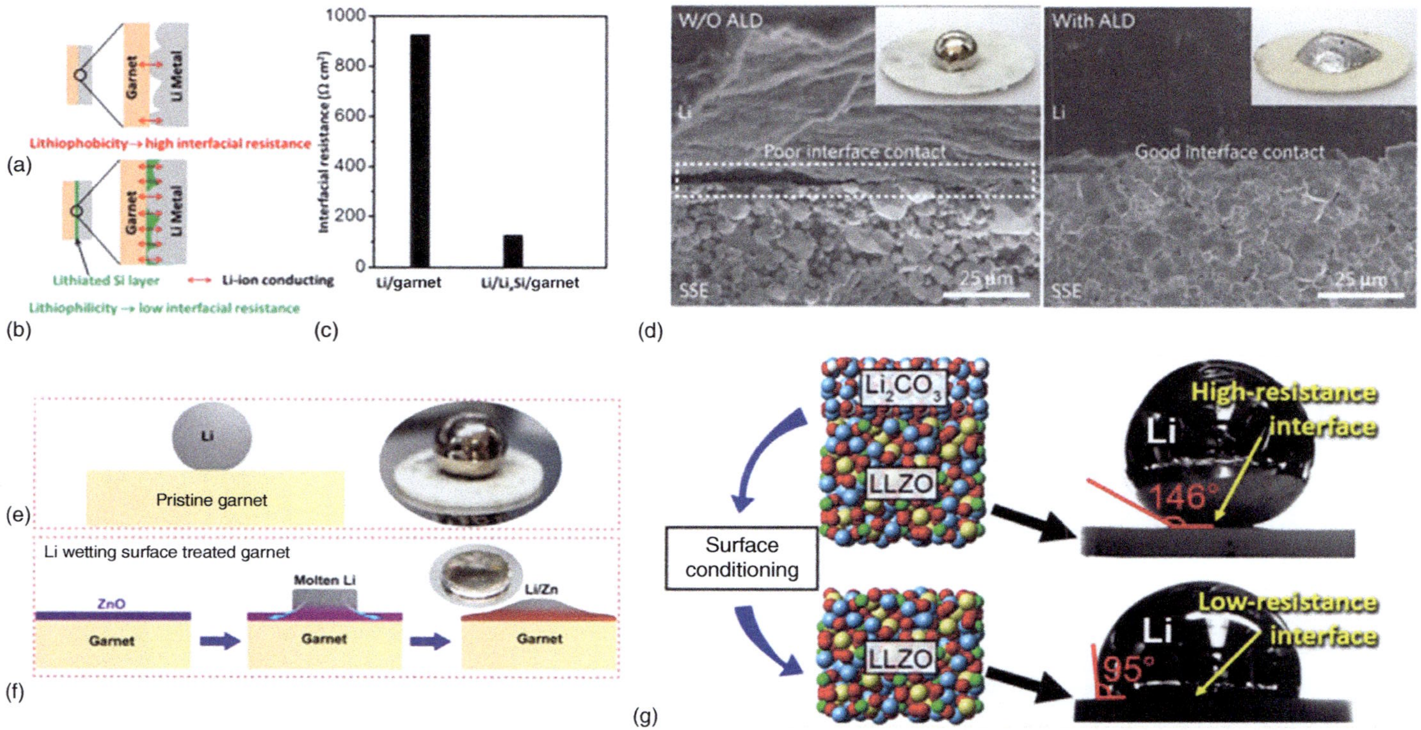

Figure 11.11 (a) Lithiophobic nature of garnet SE is shown (b) Improved wettability from reacted Li–Si alloy at the interface. (c) Better interfacial contact from the Li–Si phase leads to decreased interfacial resistance [51]. (d) SEM images showing improved interface contact between lithium metal and garnet SE interface via Al_2O_3 atomic layer deposition (ALD) coating [36]. (e–f) schematic of Li wetting on ZnO surface-treated garnet SE. Li metal reacts with ZnO-coated surface of the SE, where Li diffuses into the ZnO layer to form Li–Zn alloy [52]. (g) Wettability of Li can also be improved by removing the native Li_2CO_3 layer via heat-treatment and various polishing procedures [37]. Source: (a–c) Luo et al. [51], (d) Han et al. [36]/with permission of American Chemical Society, (e, f) Wang et al. [52]/with permission of American Chemical Society, (g) Sharafi et al. [37]/with permission of American Chemical Society.

11.4.2 Electrolyte Fracture

Because of the low inherent fracture toughness of ceramic electrolytes, these materials are susceptible to cracking during handling and cell assembly. As noted above, the stresses that can occur during electrochemical cycling introduce additional complexity. In general, a variety of stress states are expected to occur in solid electrolytes, based on different battery architectures, material properties, and external mechanical constraints. Reported K_{Ic} values for some LLZO-based SE materials are listed in Table. 11.2 showing with all K_{Ic} values at ~1 MPa $m^{1/2}$. These low values, typical for many oxide ceramics, imply that solid electrolytes are extremely brittle and are prone to cracking due to tensile stress (e.g. during bending, etc.).

Strategies for increasing K_{Ic} have been widely studied in other ceramic materials for many years. Engineering the relative density, grain structure, or grain boundary chemistry will lead to some improvements in the fracture toughness of single-phase solid electrolytes as shown in Figure 11.12 [53, 55]. However, these established methods for increasing K_{Ic} in single-phase ceramics are somewhat limited, as trade-off between fracture toughness and ionic conductivity is expected due to increased grain boundary resistance [53]. Furthermore, relatively thin solid electrolyte layers are desirable for faster ion transport, and these length scales will restrict the approaches that can be used to engineer higher fracture resistance.

A broader range of toughening strategies are possible with composite materials. Attempts to achieve this with polymer–ceramic composite solid electrolytes show promising results for suppressing dendritic growth and improving cycling behavior, but these have encountered difficulties with high interface impedance [57–59]. Another approach recently demonstrated in our lab shows that adding only 1% reduced graphene oxide leads to significant fracture toughness improvements in LATP, without substantially altering the ionic and electronic behavior of the material [60].

Dendrite-like lithium filaments that propagate through ceramic electrolytes have recently been identified as a serious problem. It is well known that elastic energy in a solid can induce surface-roughening phenomena such as wrinkling

Table 11.2 Grain size, relative density, and fracture toughness of LLZO-based solid electrolytes.

References	Material	Grain size (μm)	Relative density (%)	Technique	K_{Ic} (MPa $m^{1/2}$)
Kim et al. [53]	$Li_{6.19}Al_{0.27}La_3Zr_2O_{12}$	4	98	Indentation	0.97 ± 0.11
Wolfenstine et al. [54]	$Li_{6.28}Zr_2La_3Al_{0.24}O_{12}$	5	97	Indentation	1.25 ± 0.32
Sharafi et al. [55]	$Li_{6.25}Al_{0.25}La_3Zr_2O_{12}$	6	96	Indentation	0.82 ± 0.07
Yan et al. [42]	$Li_7La_3Zr_2O_{12}$ w/2 mol% Al and 40 mol% Ta doping	30	93	Indentation	0.94 ± 0.19
Wang et al. [56]	$Li_{6.91}Zr_{1.98}La_3Al_{0.13}O_{12}$	100	93	Indentation Micro-pillar indentation	1.19 ± 0.18 0.99 ± 0.05

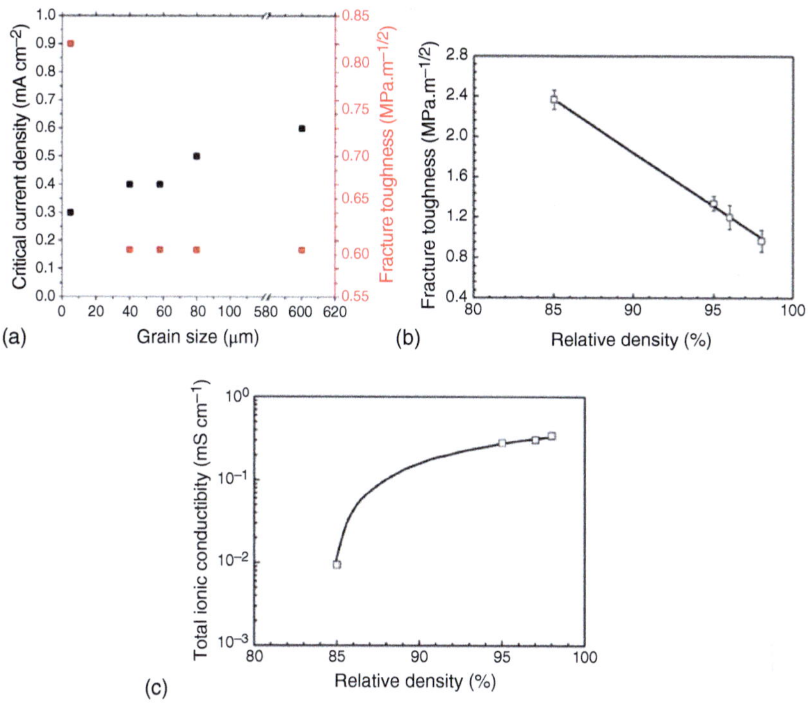

Figure 11.12 (a) Critical current density and fracture toughness as a function of grain size. Source: Sharafi et al. [55]/with permission of Royal Society of Chemistry. (b) Fracture toughness as a function of relative density of SE. (c) Total ionic conductivity as a function of relative density of SE. Source: (b, c) Kim et al. [53]/with permission of John Wiley & Sons.

or whisker formation. A basic linear stability analysis for solid electrolytes was first developed by Monroe and Newman, who showed that an SE with a shear modulus that is approximately twice that of lithium metal can effectively suppress surface roughening. However, numerous experimental observations of lithium filament propagation through both single and poly-crystalline ceramic electrolytes with high shear modulus values indicate the need for better understanding of electrochemo-mechanical phenomena in the SE. Recent studies suggest that much more complex behavior occurs at the electrode–SE interface than previously thought, and that the existing models must be modified to provide accurate design guide. Current research on lithium metal penetration through the SE has not resolved the role of mechanical driving forces. Many reports show that lithium metal penetration occurs preferentially along grain boundaries (GB) as shown in Figure 11.13A–C [61, 63], but different mechanisms have been proposed for the origin. Some argue that the filament growth and penetration is largely driven by the interfacial phenomena such as poor contact, surface flaws, and defects on the SE [35, 38]. Others have proposed that different transport properties between the bulk and GBs of ceramic SE result in more favorable growth along the GBs [63]. Relative ionic conductivity differences between bulk and GB also remain unresolved as shown by simulation results in Figure. 11.13D. [62, 64]. Recent simulation results

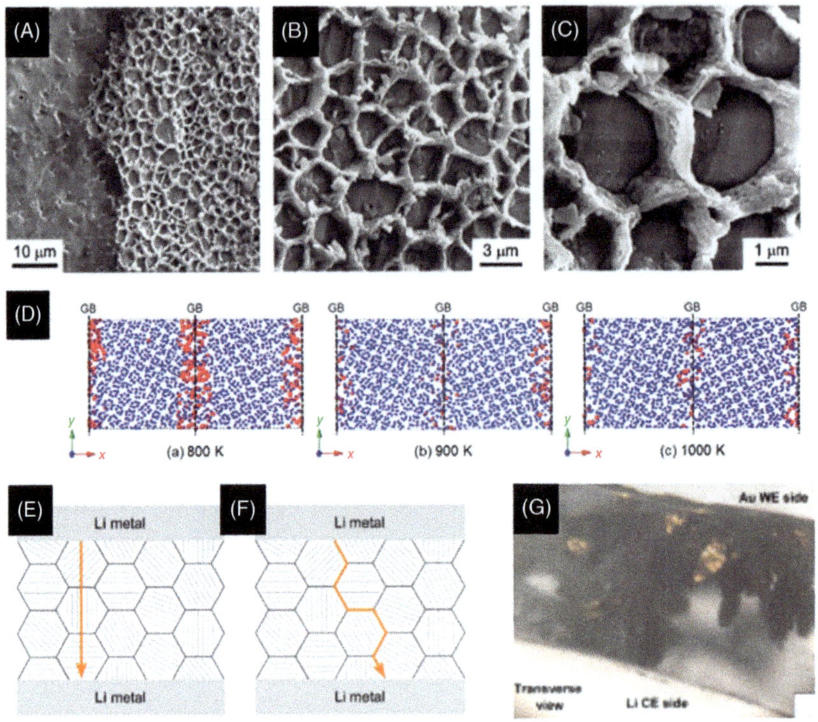

Figure 11.13 (A–C) SEM micrographs of deposited Li metal along the grain boundaries of cycled LLZO [61]. (D) Mobile (blue) and immobile (red) Li ions in the GB simulation cell during 5 ns of molecular dynamics (MD) at (a) 800; (b) 900; (c) 1000 K [62]. (E, F) Illustration of (E) transgranular and (F) intergranular Li metal plating through polycrystalline LLZO SE [61]. (G) Leaf-like morphology of Li metal penetration through single-crystalline LLZO SE [40]. Source: (D) Yu and Siegel [62]/with permission of American Chemical Society, (E, F) Cheng et al. [61], (G) Swamy et al. [40]/with permission of Elsevier.

from Barai et al. show that stress effects can lead to "current focusing" and can still persist at the GB even when the conductivity of GB is considered much smaller than that of the bulk [65]. This study highlights one way that the elastic energies can interact with electrochemical phenomena such as the current distribution. Also, as lithium penetration occurs along the GB, it will induce substantial strain energy in the stiff electrolyte, which can result in crack propagation. Here, reducing the grain size is suggested to minimize the stress-induced current focusing. Also, it has been shown that lithium can also penetrate single-crystal LLZO (Figure 11.13G) [38, 40]; thus it is clear that controlling only the grain boundary properties of the SE is not sufficient for dendrite suppression. Two main modes of lithium metal penetration, both intergranular and trans-granular penetration, have been identified in recent years and are highlighted in Figure 11.13E,F. Other non-mechanical effects have also been proposed. For example, the presence of mobile electrons in the ceramic may also drive lithium nucleation and growth in the SE [66]. Clearly more thorough work is needed to accurately evaluate the relevant electrochemo-mechanical phenomena in all solid-state systems.

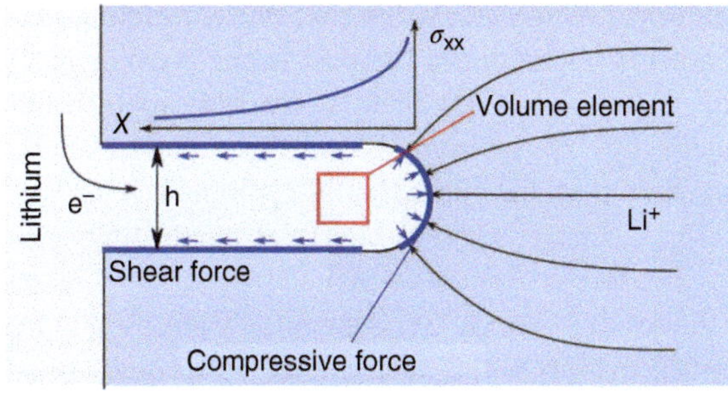

Figure 11.14 Simplified schematic of a Li filament in a solid electrolyte matrix. The arrows at the rounded end of the filament depict the applied pressure from the Li metal (and are also present along the length of the filament), and the arrows along the side show shear tractions due to friction along this interface. The model predicts a maximum stress at the filament tip that decays along the length from the tip backward. Note that the actual filaments observed have a much higher aspect ratio than what is shown here, and that the crack extension model is based on an atomically sharp tip. Source: Porz et al. [38]/with permission of John Wiley & Sons.

Although lithium metal penetration in ceramic electrolytes appears to involve a variety of phenomena, fracture that occurs during electrochemical cycling is a likely explanation in at least some of these materials. Several studies [2, 38, 40, 43] have proposed that Li-filled flaws can create relatively high stress inside of the solid electrolyte, as shown in Figure 11.14. This type of surface flaw should first fill with lithium during plating. Current focusing at the tip will then lead to a rapid stress build up inside of the flaw. If the flaw does not grow into the electrolyte, the tip is sufficiently far away from the surface, and lithium atoms are immobile, then the stress inside of the flaw should reach a limiting hydrostatic stress, $\hat{\sigma}_P^{\max}$. This is the value that exactly counters the local overpotential, $\Delta\phi_P$, such that $\Delta\Gamma_{\text{Li}} = 0 = \mathcal{F}\Delta\phi_P - V_m^{\text{Li}}\,\hat{\sigma}_P$, which leads to:

$$\hat{\sigma}_P^{\max} = \frac{\mathcal{F}\,\Delta\phi_P}{V_m^{\text{Li}}} \tag{11.23}$$

If mechanisms that can relax this internal stress are neglected, then the stress in the flaw reaches $\hat{\sigma}_P^{\max}$ relatively quickly and the lithium flux into the flaw ceases.

The propagation of a lithium-filled flaw due to internal pressure is readily evaluated with linear elastic fracture mechanics. The mode I stress intensity factor at the tip of a narrow, straight flaw is given by [67]:

$$K_I = 2\sqrt{\frac{c}{\pi}} \int_0^a \frac{\sigma_{\text{yy}}^{\text{flaw}}(x)}{\sqrt{c^2 - x^2}}\,dx \tag{11.24}$$

where $\sigma_{\text{yy}}^{\text{flaw}}(x)$ is the normal force acting on the flaw faces. When this value exceeds the fracture toughness of the solid electrolyte (i.e. $K_{\text{Ic}} > K_I$), the lithium-filled flaw will grow. To better understand this behavior, it is convenient to set $\sigma_{\text{yy}}^{\text{flaw}} = \hat{\sigma}_P^{\max}$ (constant value), to give the following upper bound result:

$$K_I = \sqrt{\pi\,c_o}\,\hat{\sigma}_P^{\max} \tag{11.25}$$

As noted above, $\hat{\sigma}_P^{max}$ here is an upper bound on the internal stress, which neglects relaxation processes. Extrusion of the metal out of the flaw is one possible mechanism that can reduce the stress. While the low yield stress of Li metal would appear to promote this behavior, one recent analysis by Klinsmann et al. [43] uses indentation mechanics to show that this effect should be limited. Also, work showing that the yield stress of Li metal is much higher at small dimensions suggests that relaxation due to extrusion is likely to be limited [68].

It is interesting to note that the result in Eq. (11.25) is essentially identical to that obtained for standard mode I brittle fracture with a fixed external load equal to $\hat{\sigma}_P^{max}$. However, with an internal stress due to lithium metal, there are several important differences. In both cases, the strain energy release rate provides the driving force for crack extension. This quantity is given here by:

$$G_E = -\frac{dU_E}{dc} = \frac{K_I{}^2}{E'_E} \tag{11.26}$$

where U_E is the stored elastic strain energy in the solid electrolyte. The "resistance" for lithium penetration is then given by:

$$R = \frac{dU_S}{dc} + \frac{dU_P}{dc} \tag{11.27}$$

where the first term on the right describes the creation of new interfaces (i.e. the standard Griffith definition), and U_P is the stored elastic strain energy in the lithium metal. The criterion for lithium penetration is then:

$$G_E \geq R \tag{11.28}$$

It is convenient to apply this condition to the situation where $\Delta\phi_P$ is gradually increased to the point where $G_E < R$. When this condition is first reached, filament extension will then occur at this critical plating potential [38]:

$$\Delta\phi_P^c \cong \frac{V_m^{Li}}{\mathcal{F}}\sqrt{\frac{R\,E'}{\pi\,c}} \tag{11.29}$$

Note that this is a lower bound value since any stress relaxation mechanisms inside of the filament will reduce the overall driving force (i.e. G_E).

For the assumptions considered above, the static case that corresponds to Eq. (11.23) (zero Li flux into the flaw) is no longer operative when $\Delta\phi_P^c$ is reached. At this point, crack extension occurs, and with a fixed quantity of lithium in the flaw, the following condition is expected to hold:

$$\frac{dG_E}{dc} < \frac{dR}{dc} \tag{11.30}$$

The change in G_E (left-hand side) will be negative because the incremental crack growth will increase the crack opening and relax the stress in the Li, such that the internal pressure decreases, thus reducing the strain energy-release rate (i.e. the driving force for extension). The right-hand side of Eq. (11.30) is zero for the standard Griffith result (R = constant). Note that Eq. (11.30) is the standard condition for stable crack growth, which indicates that as c increases the driving force for further propagation (G_E) is now less than the fracture resistance.

Continued stress-driven extension of the lithium-filled flaws will then be linked to the net flux of lithium into the flaw, which will increase the internal pressure (i.e. counteracting the stress-drop associated with the increase in c). With this in mind, the extension of a lithium-filled filament that occurs when the condition in Eq. (11.29) is reached can be described with the following steady-state condition:

$$\frac{dG_E}{dc} = \frac{dR}{dc} \tag{11.31}$$

This condition means that the filament extension criterion ($G_E = R$) is maintained as c increases.

The basic model outlined above implies that Li-filled flaws will move through the electrolyte at an observable rate dictated by Eq. (11.31) (in contrast to unstable crack propagation, which generally occurs too quickly to be directly observed). This assessment is consistent with experimental observations, particularly in situ experiments which show that Li-filled flaws can be observed as they move through the electrolyte [38, 40, 69].

In conventional fracture analysis, R is also equated to the fracture toughness via:

$$R = \frac{{K_{\text{Ic}}}^2}{E'_E} \tag{11.32}$$

A basic Griffith-style interpretation corresponds to $R \cong \Delta\gamma$, where $\Delta\gamma$ is the total energy change associated with creating new electrolyte interfacial area during crack extension. Equating this description with Eq. (11.32) then implies that $\Delta\gamma = 2\,\gamma_o$, where γ_o is the decohesion energy of the fully separated crack faces in a vacuum (or in the atmosphere where the K_{Ic} measurement was conducted). This basic interpretation assumes that crack extension from the tip initially occurs without filling the new incremental space with Li. This might be expected to occur under conditions where Li does not readily wet the solid electrolyte. However, the presence of Li metal in the crack alters this interpretation, such that the fracture resistance associated with the measured K_{Ic} value might be quantitatively different than the fracture resistance associated with Li metal penetration. It is also possible, and perhaps likely that chemistry at the crack tip alters the fracture resistance (i.e. analogous to stress-corrosion effects). Thus, while measured K_{Ic} values for different electrolytes can provide relevant guidance, these considerations suggest that the value of this property may not be quantitatively accurate for full interpretation of lithium filament propagation.

11.5 Summary

In transition metal oxides that are employed in lithium-ion batteries (LIBs), volume changes that occur during electrochemical cycling can lead to significant mechanical stresses when the expansion or contraction of the material is constrained in some way. As noted in Section 11.2, a relevant example of this is a single-cathode particle where charging or discharging rates are fast enough to produce a significant composition gradient. A variety of stress states are also likely to occur in solid electrolytes, based on different battery architectures, material properties, and external mechanical constraints. Because of the relatively high modulus and absence of plastic deformation in most oxides, small-to-moderate chemically induced strains can result in large elastic strain energies. As noted in Section 11.3, these can be large enough to noticeably alter thermodynamic equilibrium in some situations. High stresses can ultimately lead to fracture in both oxide electrodes and electrolytes. Thus, the fracture toughness in these materials is an important material property. In general, the full range of chemo-mechanical phenomena that are relevant in energy-storage materials have not been studied in detail. However, recent work, some of which is highlighted in this chapter, has begun to address many of the important issues.

References

1 Xu, Z., Rahman, M.M., Mu, L. et al. (2018). Chemomechanical behaviors of layered cathode materials in alkali metal ion batteries. *Journal of Materials Chemistry A* 6 (44): 21859–21884.

2 LePage, W.S., Chen, Y., Kazyak, E. et al. (2019). Lithium mechanics: roles of strain rate and temperature and implications for lithium metal batteries. *Journal of the Electrochemical Society* 166 (2): A89–A97.

3 Koerver, R., Zhang, W., Biasi, L. et al. (2018). Chemo-mechanical expansion of lithium electrode materials – on the route to mechanically optimized all-solid-state batteries. *Energy & Environmental Science* 11 (8): 2142–2158.

4 Qi, Y., Hector, L.G., James, C. et al. (2014). Lithium concentration dependent elastic properties of battery electrode materials from first principles calculations. *Journal of the Electrochemical Society* 161 (11): F3010–F3018.

5 Woodford, W.H., Chiang, Y.-M., and Carter, W.C. (2010). "Electrochemical shock" of intercalation electrodes: a fracture mechanics analysis. *Journal of the Electrochemical Society* 157 (10): A1052–A1059.

6 Verbrugge, M. and Cheng, Y.-T. (2008). Stress distribution within spherical particles undergoing electrochemical insertion and extraction. *ECS Transactions* 16 (13): 127–139.

7 Cogswell, D.A. and Bazant, M.Z. (2012). Coherency strain and the kinetics of phase separation in $LiFePO_4$ nanoparticles. *ACS Nano* 6 (3): 2215–2225.

8 Bower, A.F., Chason, E., Guduru, P. et al. (2015). A continuum model of deformation, transport and irreversible changes in atomic structure in amorphous lithium–silicon electrodes. *Acta Materialia* 98: 229–241.

9 Bower, A.F., Guduru, P.R., and Chason, E. (2015). Analytical solutions for composition and stress in spherical elastic–plastic lithium-ion electrode particles containing a propagating phase boundary. *International Journal of Solids and Structures* 69-70: 328–342.

10 Baker, D.R., Verbrugge, M.W., and Bower, A.F. (2016). Swelling and elastic deformation of lithium-silicon electrode materials. *Journal of the Electrochemical Society* 163 (5): A624–A631.

11 Park, K.-J., Hwang, J.-Y., Ryu, H.-H. et al. (2019). Degradation mechanism of Ni-enriched NCA cathode for lithium batteries: are microcracks really critical? *ACS Energy Letters* 4 (6): 1394–1400.

12 Xu, R. and Zhao, K. (2018). Corrosive fracture of electrodes in Li-ion batteries. *Journal of the Mechanics and Physics of Solids* 121: 258–280.

13 Xia, S., Mu, L., Xu, Z. et al. (2018). Chemomechanical interplay of layered cathode materials undergoing fast charging in lithium batteries. *Nano Energy* 53: 753–762.

14 Aifantis, K.E. and Dempsey, J.P. (2005). Stable crack growth in nanostructured Li-batteries. *Journal of Power Sources* 143 (1): 203–211.

15 Zhang, X., Shyy, W., and Marie Sastry, A. (2007). Numerical simulation of intercalation-induced stress in Li-ion battery electrode particles. *Journal of the Electrochemical Society* 154 (10): A910–A916.

16 Mattheck, C., Munz, D., and Stamm, H. (1983). Stress intensity factor for semi-elliptical surface cracks loaded by stress gradients. *Engineering Fracture Mechanics* 18 (3): 633–641.

17 Newman, J.C. and Raju, I.S. (1981). An empirical stress-intensity factor equation for the surface crack. *Engineering Fracture Mechanics* 15 (1): 185–192.

18 Zhang, Z., Chen, Z., Wang, G. et al. (2016). Dual-doping to suppress cracking in spinel $LiMn_2O_4$: a joint theoretical and experimental study. *Physical Chemistry Chemical Physics* 18 (9): 6893–6900.

19 Balluffi, R.W., Allen, S.M., Carter, W.C. et al. (2005). *Kinetics of Materials*. Hoboken, N.J.: Wiley.

20 Sheldon, B.W., Soni, S.K., Xiao, X. et al. (2011). Stress contributions to solution thermodynamics in Li-Si alloys. *Electrochemical and Solid-State Letters* 15 (1): A9–A11.

21 Cheng, Y.-T. and Verbrugge, M.W. (2009). Evolution of stress within a spherical insertion electrode particle under potentiostatic and galvanostatic operation. *Journal of Power Sources* 190 (2): 453–460.

22 Yamada, A. (1996). Lattice instability in $Li(Li_xMn_{2-x})O_4$. *Journal of Solid State Chemistry* 122 (1): 160–165.

23 Chung, K.Y., Ryu, C.-W., and Kim, K.-B. (2005). Onset mechanism of Jahn-Teller distortion in 4 V $LiMn_2O_4$ and its suppression by $LiM_{0.05}Mn_{1.95}O_4$ (M = Co, Ni) coating. *Journal of the Electrochemical Society* 152 (4): A791–A795.

24 Bazant, M.Z. (2013). Theory of chemical kinetics and charge transfer based on nonequilibrium thermodynamics. *Accounts of Chemical Research* 46 (5): 1144–1160.

25 Xu, R., de Vasconcelos, L.S., Shi, J. et al. (2018). Disintegration of meatball electrodes for $LiNi_xMn_yCo_zO_2$ cathode materials. *Experimental Mechanics* 58 (4): 549–559.

26 Xu, R., Scalco de Vasconcelos, L., and Zhao, K. (2016). Computational analysis of chemomechanical behaviors of composite electrodes in Li-ion batteries. *Journal of Materials Research* 31 (18): 2715–2727.

27 Pervez, S.A., Cambaz, M.A., Thangadurai, V. et al. (2019). Interface in solid-state lithium battery: challenges, progress, and outlook. *ACS Applied Materials & Interfaces* 11 (25): 22029–22050.

28 Krauskopf, T., Hartmann, H., Zeier, W.G. et al. (2019). Toward a fundamental understanding of the lithium metal anode in solid-state batteries – an electrochemo-mechanical study on the garnet-type solid electrolyte $Li_{6.25}Al_{0.25}La_3Zr_2O_{12}$. *ACS Applied Materials & Interfaces* 11 (15): 14463–14477.

29 Shen, Z., Zhang, W., Zhu, G. et al. Design principles of the anode–electrolyte interface for all solid-state lithium metal batteries. *Small Methods* 4: 1900592.

30 Lewis, J.A., Tippens, J., Cortes, F.J. et al. (2019). Chemo-mechanical challenges in solid-state batteries. *Trends in Chemistry* 1 (9): 845–857.

31 Shen, Z., Zhang, W., Zhu, G. et al. (2020). Design principles of the anode–electrolyte interface for all solid-state lithium metal batteries. *Small Methods* 4 (1): 1900592.

32 Sharafi, A., Yu, S., Naguib, M. et al. (2017). Impact of air exposure and surface chemistry on Li–$Li_7La_3Zr_2O_{12}$ interfacial resistance. *Journal of Materials Chemistry A* 5 (26): 13475–13487.

33 Wu, B., Wang, S., Lochala, J. et al. (2018). The role of the solid electrolyte interphase layer in preventing Li dendrite growth in solid-state batteries. *Energy & Environmental Science* 11 (7): 1803–1810.

34 Zhou, C., Samson, A.J., Hofstetter, K. et al. (2018). A surfactant-assisted strategy to tailor Li-ion charge transfer interfacial resistance for scalable all-solid-state Li batteries. *Sustainable Energy & Fuels* 2 (10): 2165–2170.

35 Tsai, C.-L., Roddatis, V., Chandran, C.V. et al. (2016). $Li_7La_3Zr_2O_{12}$ Interface modification for Li dendrite prevention. *ACS Applied Materials & Interfaces* 8 (16): 10617–10626.

36 Han, X., Gong, Y., Fu, K. et al. (2017). Negating interfacial impedance in garnet-based solid-state Li metal batteries. *Nature Materials* 16 (5): 572–579.

37 Sharafi, A., Kazyak, E., Davis, A.L. et al. (2017). Surface chemistry mechanism of ultra-low interfacial resistance in the solid-state electrolyte $Li_7La_3Zr_2O_{12}$. *Chemistry of Materials* 29 (18): 7961–7968.

38 Porz, L., Swamy, T., Sheldon, B.W. et al. (2017). Mechanism of lithium metal penetration through inorganic solid electrolytes. *Advanced Energy Materials* 7 (20): 1701003.

39 Bucci, G., Swamy, T., Chiang, Y.M. et al. (2017). Modeling of internal mechanical failure of all-solid-state batteries during electrochemical cycling, and implications for battery design. *Journal of Materials Chemistry A* 5 (36): 19422–19430.

40 Swamy, T., Park, R., Sheldon, B.W. et al. (2018). Lithium metal penetration induced by electrodeposition through solid electrolytes: example in single-crystal $Li_6La_3ZrTaO_{12}$ garnet. *Journal of the Electrochemical Society* 165 (16): A3648–A3655.

41 Li, G. and Monroe, C.W. (2019). Dendrite nucleation in lithium-conductive ceramics. *Physical Chemistry Chemical Physics* 21 (36): 20354–20359.

42 Yan, G., Nonemacher, J.F., Zheng, H. et al. (2019). An investigation on strength distribution, subcritical crack growth and lifetime of the lithium-ion conductor $Li_7La_3Zr_2O_{12}$. *Journal of Materials Science* 54 (7): 5671–5681.

43 Klinsmann, M., Hildebrand, F.E., Ganser, M. et al. (2019). Dendritic cracking in solid electrolytes driven by lithium insertion. *Journal of Power Sources* 442: 227226.

44 Kim, K.H., Iriyama, Y., Yamamoto, K. et al. (2011). Characterization of the interface between $LiCoO_2$ and $Li_7La_3Zr_2O_{12}$ in an all-solid-state rechargeable lithium battery. *Journal of Power Sources* 196 (2): 764–767.

45 Hänsel, C., Afyon, S., and Rupp, J.L.M. (2016). Investigating the all-solid-state batteries based on lithium garnets and a high potential cathode – $LiMn_{1.5}Ni_{0.5}O_4$. *Nanoscale* 8 (43): 18412–18420.

46 Panahian Jand, S. and Kaghazchi, P. (2018). Theoretical study of cubic-$Li_7La_3Zr_2O_{12}$(001)/$LiCoO_2$(10–14) interface. *MRS Communications* 8 (2): 591–596.

47 Koerver, R., Aygun, I., Leichtweib, T. et al. (2017). Capacity fade in solid-state batteries: interphase formation and chemomechanical processes in nickel-rich layered oxide cathodes and lithium thiophosphate solid electrolytes. *Chemistry of Materials* 29 (13): 5574–5582.

48 Lewis, J.A., Cortes, F.J., Boebinger, M.G. et al. (2019). Interphase morphology between a solid-state electrolyte and lithium controls cell failure. *ACS Energy Letters* 4 (2): 591–599.

49 Zhang, W., Schroder, D., Arlt, T. et al. (2017). (Electro)chemical expansion during cycling: monitoring the pressure changes in operating solid-state lithium batteries. *Journal of Materials Chemistry A* 5 (20): 9929–9936.

50 Luo, W., Gong, Y., Zhu, Y. et al. (2017). Reducing interfacial resistance between garnet-structured solid-state electrolyte and Li-metal anode by a germanium layer. *Advanced Materials* 29 (22): 1606042.

51 Luo, W., Gong, Y., Zhu, Y. et al. (2016). Transition from superlithiophobicity to superlithiophilicity of garnet solid-state electrolyte. *Journal of the American Chemical Society* 138 (37): 12258–12262.

52 Wang, C., Gong, Y., Liu, B. et al. (2017). Conformal, nanoscale ZnO surface modification of garnet-based solid-state electrolyte for lithium metal anodes. *Nano Letters* 17 (1): 565–571.

53 Kim, Y., Jo, H., Allen, J.L. et al. (2016). The effect of relative density on the mechanical properties of hot-pressed cubic $Li_7La_3Zr_2O_{12}$. *Journal of the American Ceramic Society* 99 (4): 1367–1374.

54 Wolfenstine, J., Jo, H., Cho, Y.-H. et al. (2013). A preliminary investigation of fracture toughness of $Li_7La_3Zr_2O_{12}$ and its comparison to other solid Li-ion conductors. *Materials Letters* 96: 117–120.

55 Sharafi, A., Haslam, C.G., Kerns, R.D. et al. (2017). Controlling and correlating the effect of grain size with the mechanical and electrochemical properties of $Li_7La_3Zr_2O_{12}$ solid-state electrolyte. *Journal of Materials Chemistry A* 5 (40): 21491–21504.

56 Wang, A.-N., Nonemacher, J.F., Finsterbusch, M. et al. (2018). Mechanical properties of the solid electrolyte Al-substituted $Li_7La_3Zr_2O_{12}$ (LLZO) by utilizing micro-pillar indentation splitting test. *Journal of the European Ceramic Society* 38 (9): 3201–3209.

57 Pandian, A.S., Chen, X.C., Chen, J. et al. (2018). Facile and scalable fabrication of polymer-ceramic composite electrolyte with high ceramic loadings. *Journal of Power Sources* 390: 153–164.

58 Zhang, J., Zhao, N., Zhang, M. et al. (2016). Flexible and ion-conducting membrane electrolytes for solid-state lithium batteries: dispersion of garnet nanoparticles in insulating polyethylene oxide. *Nano Energy* 28: 447–454.

59 Zhang, W., Nie, J., Li, F. et al. (2018). A durable and safe solid-state lithium battery with a hybrid electrolyte membrane. *Nano Energy* 45: 413–419.

60 Athanasiou, C.E., Jin, M.Y., Ramirez, C. et al. (2020). High toughness inorganic solid electrolytes via the use of reduced graphene-oxide. *Matter* 3 (1): 14–15.

61 Cheng, E.J., Sharafi, A., and Sakamoto, J. (2017). Intergranular Li metal propagation through polycrystalline $Li_{6.25}Al_{0.25}La_3Zr_2O_{12}$ ceramic electrolyte. *Electrochimica Acta* 223: 85–91.

62 Yu, S. and Siegel, D.J. (2017). Grain boundary contributions to Li-ion transport in the solid electrolyte $Li_7La_3Zr_2O_{12}$ (LLZO). *Chemistry of Materials* 29 (22): 9639–9647.

63 Cheng, L., Chen, W., Kunz, M. et al. (2015). Effect of surface microstructure on electrochemical performance of garnet solid electrolytes. *ACS Applied Materials & Interfaces* 7 (3): 2073–2081.

64 Wu, J.-F. and Guo, X. (2017). Origin of the low grain boundary conductivity in lithium ion conducting perovskites: $Li_{3x}La0.6_{7-x}TiO_3$. *Physical Chemistry Chemical Physics* 19 (8): 5880–5887.

65 Barai, P., Higa, K., Ngo, A.T. et al. (2019). Mechanical stress induced current focusing and fracture in grain boundaries. *Journal of the Electrochemical Society* 166 (10): A1752–A1762.

66 Han, F., Westover, A.S., Fan, X. et al. (2019). High electronic conductivity as the origin of lithium dendrite formation within solid electrolytes. *Nature Energy* 4 (3): 187–196.

67 Lawn, B. (1993). *Fracture of Brittle Solids*, Cambridge Solid State Science Series, 2e. Cambridge: Cambridge University Press.

68 Xu, C., Ahmad, Z., Aryanfar, A. et al. (2017). Enhanced strength and temperature dependence of mechanical properties of Li at small scales and its implications for Li metal anodes. *Proceedings of the National Academy of Sciences* 114 (1): 57.

69 Manalastas, W., Rikarte, J., Chater, R.J. et al. (2019). Mechanical failure of garnet electrolytes during Li electrodeposition observed by in-operando microscopy. *Journal of Power Sources* 412: 287–293.

12

Solid-State NMR and EPR Characterization of Transition-Metal Oxides for Electrochemical Energy Storage

Xiang Li[1], Michael Deck[1], and Yan-Yan Hu[1,2]

[1]*Florida State University, Department of Chemistry and Biochemistry, Tallahassee, FL 32306, USA*
[2]*National High Magnetic Field Laboratory, 1800 East Paul Dirac Drive, Tallahassee, FL 32310, USA*

12.1 Introduction

As the demand for energy continues to rise, advanced energy-storage technologies are needed to address the challenges of using renewable energy as an alternative to fossil fuels. To accomplish this, various types of alkali-ion rechargeable batteries have attracted notable attention [1–5]. Among them, lithium-ion batteries have the desirable combination of high-energy density and power density, making them the most popular energy-storage method in applications such as portable electronics and electric cars [6–9]. However, the specific capacity of these batteries is mainly limited by the cathode materials, resulting in extensive studies to search for high-energy-density cathode materials [9].

Layered manganese-based materials with large specific capacity and high potential vs. Li/Li^+ represent state-of-the-art cathodes. Layered Li-rich materials such as Li_2MnO_3 (LMO), $Li_{1.2}Ni_{0.2}Mn_{0.6}O_2$ (LNMO), and $Li_{1.2}Ni_{0.13}Mn_{0.54}Co_{0.13}O_2$ (LNCMO) have attracted tremendous research effort due to their high gravimetric capacity at high voltage [6]. However, after the first cycle, these layered materials suffer from severe voltage decay and capacity fade. These issues are mainly related to structural changes involving transition-metal (TM) migration and O_2 evolution [6, 10]. Transition-metal migration occurs when Li is extracted upon charge [11]. To achieve charge balance, TM ions are oxidized to higher charge states; e.g. Ni^{2+} to Ni^{3+}, then to Ni^{4+}. As a result, the TM ionic radius is reduced, which promotes TM ion migration [12]. Structural studies of these layered Li-rich cathodes have been carried out. They exhibit an O3-type structure in the monoclinic system with a $C2/m$ space group containing layers filled with either transition metal or Li [13]. Fell et al. utilized *in situ* X-ray diffraction to study the dynamic changes in peak positions and lattice parameters of $Li[Ni_{0.2}Li_{0.2}Mn_{0.6}]O_2$ during the first cycle [11]. Meng and coworkers applied *operando* neutron diffraction (ND) to follow the dynamics for both Li and O for Co- and Ni- containing LMO cathodes [14]. The mechanism of O evolution is more complex; therefore, the O redox reactions in Li-rich cathodes

Transition Metal Oxides for Electrochemical Energy Storage, First Edition.
Edited by Jagjit Nanda and Veronica Augustyn.

are not fully understood. Bruce and coworkers proposed that instead of forming a peroxide group, localized electron holes are formed at the O coordinating with Mn^{4+} and Li^+ [15]. Ceder and coworkers predicted with computational studies that a specific Li–O–Li configuration is necessary to create a lone state in O p orbitals and raise the O bonding energy, making the oxidation compete between unpaired O and TM [16]. Because of the small atomic mass of Li and O, direct observations of their local environments using traditional diffraction and electron microscopy techniques remain challenging [17]. Thus, more suitable characterization tools are needed.

Nuclear magnetic resonance (NMR) is a powerful tool to determine the local structural environments of nuclei such as Li, Na, and O [8, 18–24]. The shift in the NMR spectrum allows for distinguishing between different Li sites in TM and Li layers. Due to paramagnetic interactions between Li and unpaired electrons of TM ions, Li in TM layers surrounded by more TM ions are susceptible to stronger interactions, which cause Li NMR shifts to move downfield. In contrast, Li in Li layers is coordinated with fewer TM ions and weaker paramagnetic interactions; thus, its shift occurs at higher fields [23]. The relaxation-corrected peak intensity correlates with the absolute amount of Li. In principle, high-resolution Li NMR can provide insight into Li environments in transition-metal oxides, leading to deeper understanding of Li dynamics during electrochemical cycling. However, strong nucleus-electron spin dipolar interactions result in extremely large shift anisotropy of a few MHz, giving poor spectral resolution, which prevents reliable peak assignments and quantification. Fast magic-angle-spinning (MAS) NMR is rendered ineffective for studying these materials. To address this issue, ^{6}Li NMR is often used for reduced paramagnetic effects due to its smaller gyromagnetic ratio compared to the ^{7}Li isotope [22]. However, NMR sensitivity is compromised due to the low natural abundance of ^{6}Li (7.59%), leading to much longer data acquisition time [22, 25]. This is especially true when only a small amount of electrode materials are obtained from electrochemically cycled battery cells. ^{6}Li isotope enrichment can help to enhance the sensitivity but at a high cost. For instance, Dogan and coworkers performed ^{6}Li NMR to directly study the Li local environments in $0.5^6Li_2MnO_3{\cdot}0.5^6LiMn_{0.5}Ni_{0.5}O_2$ at different states of charge [26]. This study provided insight into Li structural environment changes after cycling; however, *operando* ^{6}Li NMR characterizations are not viable due to poor temporal resolution.

^{7}Li NMR provides much higher sensitivity than ^{6}Li, because of its high natural abundance and larger gyromagnetic ratio. To improve spectral resolution, NMR experiments that combine magic-angle turning (MAT) and phase-adjusted sidebands separation (PASS) are employed [27]. The MATPASS technique enables acquisition of quantitative spectra of paramagnetic electrode materials at relatively low spinning rates without any inference of spinning sidebands (SSBs) [27, 28].

Electron paramagnetic resonance (EPR) is capable of studying the local configurations of different TMs and spin interactions. Furthermore, EPR can accurately detect peroxo-like species, which may be formed in layered cathodes during electrochemical cycling. Recently, EPR spectroscopy and imaging have been employed to probe the mossy structure of Li and radical formation of working batteries [17].

In addition, *operando* NMR and EPR are also discussed in this chapter to provide quantitative analysis during battery cycling. *Operando* NMR can be useful to examine Li/Na evolution and offer insight that cannot be obtained by *ex situ* NMR [28]. *Operando* EPR can simultaneously follow the evolution of cationic TM and anionic O redox reactions during battery operation to probe cation–anion redox correlation in real time [17].

12.2 Brief Introduction of NMR Basics

12.2.1 Nuclear Spins

Nuclear spin is an intrinsic property of a nucleus and a form of angular momentum, often denoted as $\vec{I}$. The nuclear magnetic moment $\vec{\mu}$ can be expressed as:

$$\vec{\mu} = \gamma \vec{I}$$

where γ is the gyromagnetic ratio. Gyromagnetic ratio can be either positive or negative. If the sign is positive, the spin angular momentum is parallel to the magnetic moment. Likewise, if it is negative, the spin angular momentum is antiparallel to the magnetic moment [25, 29]. Table 12.1 lists NMR parameters of commonly studied nuclei in transition-metal-oxide cathodes.

12.2.2 NMR Spin Interactions

There are five important NMR spin interactions: chemical shift, J-coupling, dipolar coupling, quadrupole coupling, and paramagnetic interaction [25, 29].

- *Chemical shift*: Once a molecule is placed in an external magnetic field, the circulating electron motion can generate a current, and this current further generates a relatively small magnetic field compared with the external field, yielding a shift from the Larmor frequency called the chemical shift. Generally, the chemical

Table 12.1 NMR parameters of commonly studied nuclei in transition-metal-oxide cathodes.

Isotope	Natural abundance (%)	Nuclear spin (I)	Gyromagnetic ratio (10^7 rad/T × s)	Quadruple moment (Q/millibarn)	Relative sensitivity (^{1}H = 1.00)
^{1}H	99.99	1/2	26.7522	—	1.00
^{6}Li	7.42	1	3.9371	−0.808	6.31×10^{-4}
^{7}Li	92.58	3/2	10.3976	−40.1	0.27
^{17}O	0.038	5/2	−3.6281	−25.58	1.11×10^{-5}
^{23}Na	100	3/2	7.0808	104	9.27×10^{-2}
^{27}Al	100	5/2	6.9762	146.6	0.207
^{55}Mn	100	5/2	6.6452	4.1042	0.179

Source: Adapted from Levitt and Dechant [25, 30].

shift is caused by the electronic environment around the nucleus of interest. For example, protons in CH_4 and CH_3Cl have different electronic environments, and thus have different chemical shifts. Even for the same molecule, due to orientation dependence, the chemical shift would be slightly different in solids and in liquids [25, 29].

- *J-coupling*: Also called indirect dipole–dipole coupling. J-coupling is created through bonding electrons and plays a very important role in liquid NMR. In solid state, J-coupling is often very small (usually a few Hz) except for heavy isotopes [25, 29].
- *Dipolar coupling*: Unlike J-coupling, dipolar coupling interaction is through space. Each nucleus generates a local magnetic field and these magnetic fields can interact with each other through space [25, 29].
- *Quadrupole coupling*: Occurs in nuclei with spin number $I > 1/2$. They have a nonspherical charge distribution and thus nonvanishing quadrupole moment. The interaction between the electric quadrupole moment and the electric field gradient is called quadrupole coupling [25, 29].
- *Paramagnetic interaction*: In paramagnetic materials, nuclear spins couple with unpaired electrons through two mechanisms: Fermi contact and pseudo contact. Fermi contact is through bonds, while pseudo contact is through space [25, 29].

NMR investigations of high-voltage transition-metal-oxide cathodes mainly involve chemical shift, dipolar coupling, quadrupolar coupling, and paramagnetic interaction. Paramagnetic interactions are most informative for studying these materials but also most challenging for data acquisition and analysis. Therefore, more details are discussed in the following.

12.2.3 Paramagnetic Interactions and Experimental Approaches to Achieve High Spectral Resolution

When nuclear spins are within a magnetic field, they generate induced magnetism, which can be written as:

$$\mu_{\text{induced}} = \mu_0^{-1} \text{V} \chi \mathbf{B}$$

where μ_0 is the magnetic constant, also called vacuum permeability ($\mu_0 = 4\pi \times 10^{-7}\ \text{Hm}^{-1}$), V is the volume of the material, **B** is the external field, and χ is the magnetic susceptibility. The sign of χ could be either positive or negative. If χ is negative, the material is called diamagnetic; otherwise, it is paramagnetic [25].

In paramagnetic LMO systems, the nuclear spins are coupled with unpaired electrons from TM 3d orbitals through the hyperfine nuclear–electron interaction. Moreover, the relaxation of electrons is incredibly fast (less than 0.1 ms), and its net effect is to shift the nuclear spin resonances off their original positions. For example, the Li in TM layers resonates around 1500 ppm in the Li_2MnO_3 system, compared with a typical diamagnetic Li shift at 0 ppm in Li_2CO_3 [28].

Acquiring NMR spectra of paramagnetic materials is very challenging. Due to strong paramagnetic interactions, the SSBs from large paramagnetic shift anisotropy often overlap with themselves and the main resonances. Even fast MAS (MAS > 60 kHz) may not be enough to separate isotropic peaks from SSBs [27].

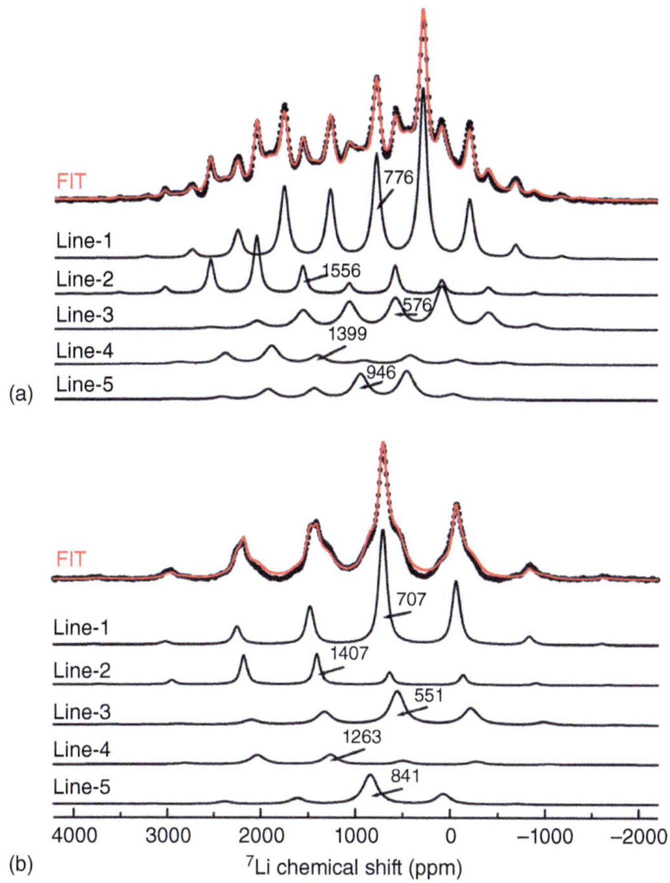

Figure 12.1 ^{7}Li MAS NMR spectra for pristine $Li_{1.2}Mn_{0.61}Ni_{0.18}Mg_{0.01}O_2$ measured at (a) 38, (b) 60 kHz, and the corresponding spectra deconvolution. Source: Buzlukov et al. [31]/with permission of American Chemical Society.

Figure 12.1a shows the ^{7}Li MAS NMR spectra of pristine $Li_{1.2}Mn_{0.61}Ni_{0.18}Mg_{0.01}O_2$ acquired at different spinning rates as well as the corresponding deconvolution analysis. The spectrum collected at 38 kHz MAS shows significantly overlapped peaks due to residual paramagnetic interactions. Peak deconvolution can be performed to extract information from such spectra but with compromised accuracy. This issue can be mitigated by increasing the spinning rate. As seen in Figure 12.1b, by increasing the spin rate from 38 to 60 kHz, the SSBs are separated farther away from each other, resulting in improved spectral resolution for more accurate quantitative analysis [31].

A more effective way to improve spectral resolution is to employ a new NMR technique – projection magic-angle-turning phase-adjusted sideband separation (pjMATPASS). The pulse sequence is shown in Figure 12.2a. Three pulse periods ($t_1/3$) cancel out the chemical shift anisotropy (CSA) and the isotropic NMR shift is refocused at the end of t_1 (or the beginning of t_2 acquisition) [27]. The technique uses the PASS approach to separate SSBs of different orders into different rows in

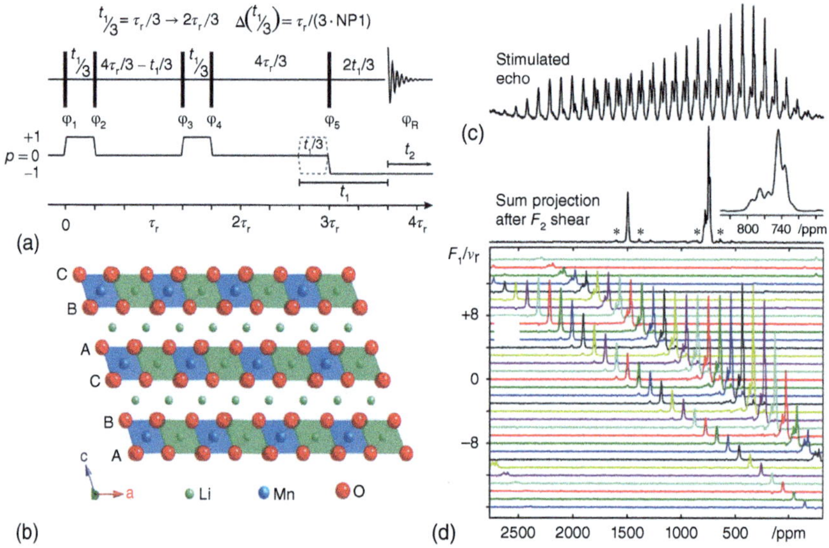

Figure 12.2 (a) Pulse sequence of the pj-MATPASS NMR experiment, where NP1 is the number of t_1 increments. (b) Crystal structure of Li_2MnO_3. (c) Stimulated-echo and (d) pj-MATPASS ^7Li NMR spectrum of Li_2MnO_3. Source: (a, c, d) Hung et al. [27]/with permission of American Chemical Society. (b) Rana et al. [32]/with permission of John Wiley & Sons.

2D pjMATPASS. Spectral shearing during data processing allows the summation of aligned SSBs to deliver a pure isotropic spectrum with enhanced sensitivity.

The successful application of pjMATPASS is demonstrated in Li_2MnO_3. The crystal structure of Li_2MnO_3 is shown in Figure 12.2b; it has a layered monoclinic O3-type structure with a space group of $C2/m$ [32]. Due to the high-temperature synthesis process, stacking faults are generated. In Li_2MO_3, Li ions are surrounded by six M atoms to form a honeycomb pattern. Li ions reside in both TM layers and Li layers, denoted as Li_{TM} and Li_{Li}, respectively. The stimulated-echo NMR spectrum is very broad with numerous spikelets. The spectrum itself is complex and difficult for quantitative analysis of different Li sites in Li_2MnO_3. To overcome this issue, pjMATPASS is performed to separate different orders of SSBs. As seen in Figure 12.2d, the pjMATPASS experiment yields a 2d spectrum with SSBs separated into different rows based on their orders. After spectral shearing, all SSBs can be aligned vertically with NMR peaks at their isotropic positions and the projection of all rows yields a pure isotropic spectrum. The resonance around 1500 ppm is from Li in TM layers [21]. The peak cluster shown around 800 ppm is from Li in Li layers, and the very broad Gaussian line-shape indicates very disordered Li local environments. This is believed to be the result of stacking faults formed during the high-temperature annealing process. Li in TM layers has a paramagnetic interaction with unpaired electrons from Mn^{4+}, which is stronger than that for Li in Li layers. This is because of different degrees of spin density transfers from Mn to Li; therefore, Li in TM layers exhibits a relatively larger paramagnetic shift [27, 28].

12.3 Multinuclear NMR Studies of Transition-metal-oxide Cathodes

12.3.1 Li Extraction and Insertion Dynamics

Figure 12.3 demonstrates the application of pjMATPASS to investigate the lithiation and delithiation mechanisms of Li_2MnO_3. Upon charge, the Li in both Li and TM layers gradually decreases as seen in Figure 12.3c, indicating continuous Li extraction from both layers. Upon discharge, reversible Li insertions are observed in both Li and TM layers. More detailed quantification work is shown in Figure 12.3d. Upon charge, Li extraction preference of Li in TM layer is observed with about 20% Li in TM layers extracted, compared to around 10% Li in Li layers extracted. While upon discharge, only 87% of Li_{TM} is reinserted back to the structure, compared to about 97% Li_{Li} recovery [28].

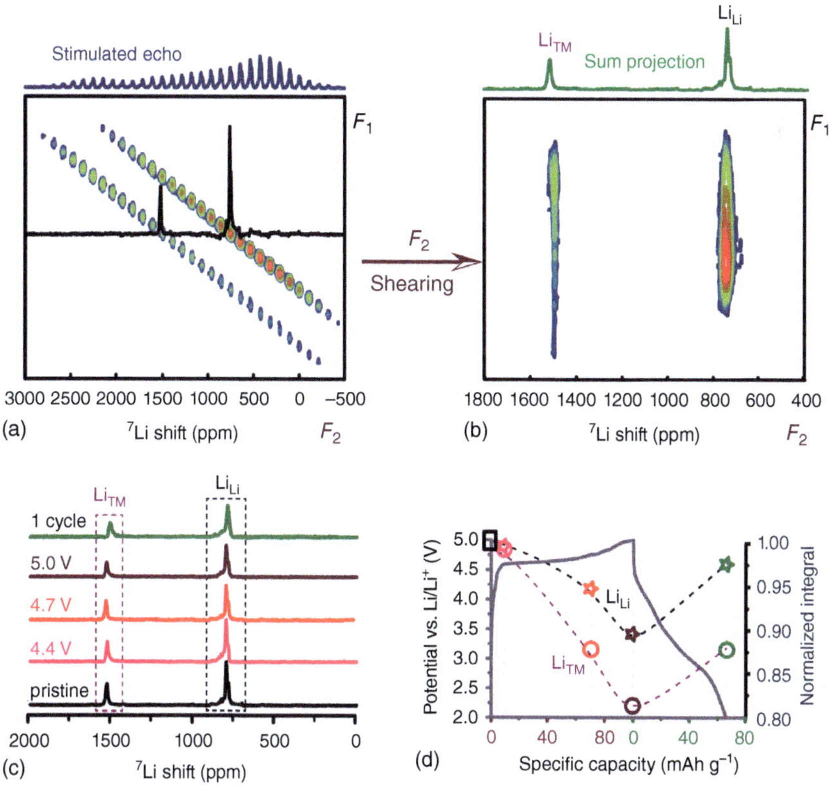

Figure 12.3 pj-MATPASS ^{7}Li NMR spectra of electrochemically cycled Li_2MnO_3. Spectral shearing along the F2 dimension of the spectrum (a) yields the spectrum (b). A stimulated-echo ^{7}Li NMR spectrum is shown on top of (a). (c) ^{7}Li isotropic NMR spectra of Li_2MnO_3 electrodes at different states of charge obtained by pj-MATPASS NMR. (d) Normalized quantification of Li_{Li} and Li_{TM} in Li_2MnO_3 at different states of the first cycle. Source: Li et al. [28]/with permission of American Chemical Society.

Li_2MnO_3 is a relatively simple model compound to study; however, as a cathode material, it suffers from low specific capacity and poor cyclability. $Li_{1.2}Ni_{0.2}Mn_{0.6}O_2$ and $Li_{1.2}Ni_{0.13}Co_{0.13}Mn_{0.56}O_2$ with much higher capacity (over 330 mAh g^{-1}) are two promising cathode candidates derived from Li_2MnO_3. $Li_{1.2}Ni_{0.2}Mn_{0.6}O_2$ and $Li_{1.2}Ni_{0.13}Co_{0.13}Mn_{0.56}O_2$ are solid solutions of Li_2MnO_3 and $LiMO_2$ (M = Mn, Ni, or Co). With other TM ions doped into the Li_2MnO_3 structure, more Li can be extracted due to the positive effects of TM ions in stabilizing the structure and providing more active redox pairs, for instance $Ni^{2+}/Ni^{3+}/Ni^{4+}$ and Co^{3+}/Co^{4+}. However, incorporation of Ni and Co brings more complexity to the Li environments. As illustrated in Figure 12.4, much broader peaks are observed from 250 to 1100 ppm and 1250 to 1700 ppm, indicating more diverse Li local environments in LNMO and LNCMO as compared with LMO [28].

Another example of using lithium NMR to study Li local environments in $LiNi_{1/3}Mn_{1/3}Co_{1/3}O_2$ is shown in Figure 12.5 [33]. ^{6}Li NMR spectra collected from pristine samples with and without isotope enrichment are color-coded by black and blue, respectively. Without ^{6}Li enrichment, the signal from Li in TM layers isn't very clear and therefore higher spectral resolution NMR requires much longer data acquisition or ^{6}Li labeling. On the other hand, small sample quantity from cycled coin cells is enough to perform high-sensitivity ^{7}Li NMR. Note that the overlap between isotropic shifts and their SSBs hinders further spectral analysis, especially for Li in TM layers. With the advantage of MATPASS technique, suppression of SSBs efficiently contributes to identifying different Li local environments. NMR shift calculations of ^{7}Li in TM layers were also performed. By examining the 1st and 2nd coordination shells of Li atoms (as shown in Figure 12.5), each Li is coordinated with six TM ions through Li–O–TM. Different combinations of TMs would lead to varied Li local environments, and therefore different chemical shifts. The contribution of each TM ion to the ^{7}Li NMR shift is listed in Table 12.2. ^{7}Li NMR shifts can be estimated as a linear combination of the effects from each surrounding TM ion. For instance, Li in Li layers resonates around 750 ppm with a

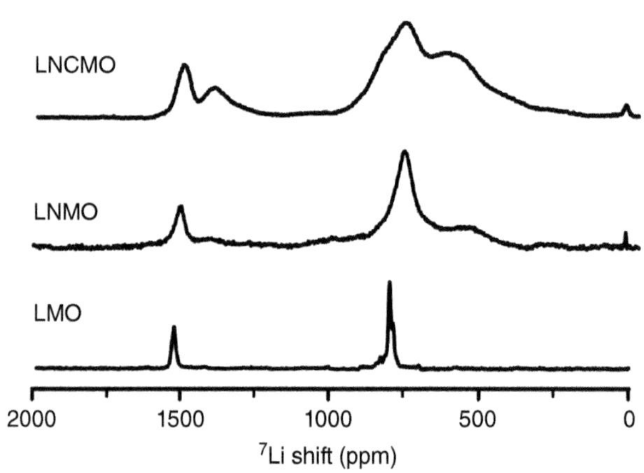

Figure 12.4 ^{7}Li pj-MATPASS NMR spectra of Li_2MnO_3 (LMO), $Li_{1.2}Ni_{0.2}Mn_{0.6}O_2$ (LNMO), and $Li_{1.2}Ni_{0.13}Mn_{0.54}Co_{0.13}O_2$ (LNCMO).

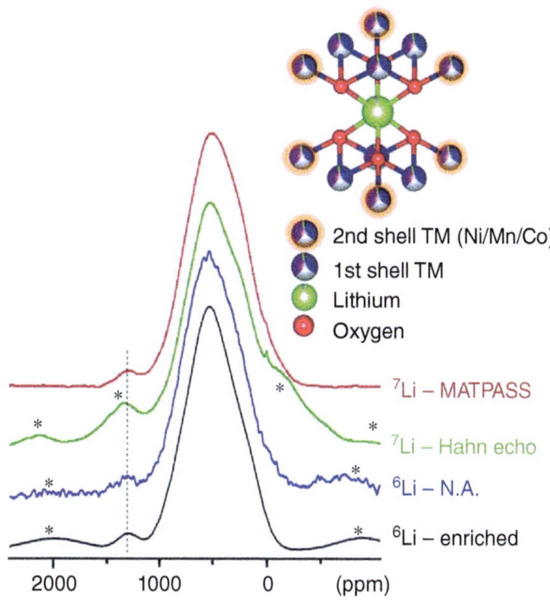

Figure 12.5 ^{6}Li and ^{7}Li NMR spectra on pristine $LiNi_{1/3}Mn_{1/3}Co_{1/3}O_2$. Two ^{6}Li NMR spectra acquired at 46 kHz MAS, and two ^{7}Li NMR spectra acquired at 60 kHz MAS. Inset illustrates the first and second coordination shells of Li ions. Source: Liu et al. [33]/with permission of Royal Society of Chemistry.

Table 12.2 Contribution of each transition-metal ion to ^{7}Li NMR shift [33].

TM ion	First coordination shell (ppm)	Second coordination shell (ppm)
Mn^{4+}	255	−52
Ni^{2+}	−25	120
Co^{3+}	0	0

relatively broad peak, which agrees with Li in this layer being surrounded by three TM ions. While Li in TM layers is surrounded by 5–6 TM ions and the chemical shift is estimated to be around 1300 ppm [33].

Operando NMR can provide insights into real-time monitoring of Li/Na during electrochemical cycling, which cannot be achieved by *ex situ* NMR. Combining temporal resolution helps to follow the Li/Na dynamics. The *operando* experimental set-up is shown in Figure 12.6. The static probe is connected to both the NMR console and a galvanostat, making it possible to measure NMR spectra during battery operation. The *operando* NMR experiments are performed under static conditions, making spectral resolution relatively lower when compared with *ex situ* NMR spectra, but sufficient to monitor Li/Na evolution in real time [28, 34].

As an example, ^{7}Li MAS and *operando* NMR on spinel-type $Li_{1.08}Mn_{1.92}O_4$ (Figure 12.7c) are shown in Figure 12.7a,b, respectively. In Figure 12.7a, several resonances are observed from 500 to 800 ppm. These peaks are assigned to Li in tetrahedral sites. By utilizing fast MAS, the dipole–dipole coupling is partially removed, and the line shape of the spectrum becomes relatively sharp. In contrast, only a broad resonance at 579 ppm is observed under static conditions (Figure 12.7b)

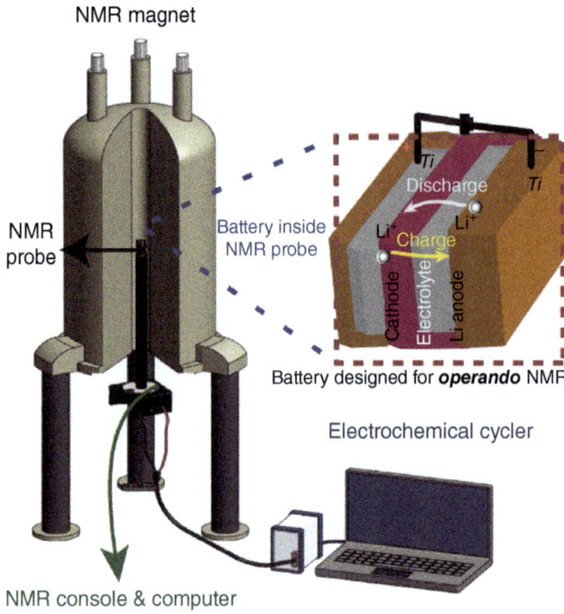

Figure 12.6 Experimental set-up for *operando* NMR.

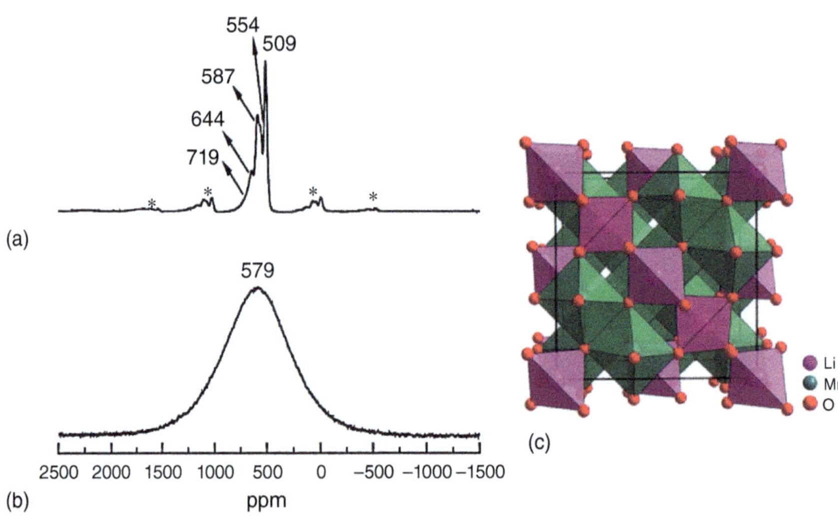

Figure 12.7 The ^{7}Li (a) MAS, (b) static NMR spectra, and (c) crystal structure of pristine $Li_{1.08}Mn_{1.92}O_4$. Source: (a, b) Zhou et al. [34]/with permission of Elsevier (c) Zhang et al. [35]/with permission of Elsevier.

due to strong dipolar interactions. The small shift difference between MAS and static spectra is ascribed to the difference in temperature at which the two experiments are carried out, since the sample is slightly heated during spinning [34].

The *operando* ^{7}Li NMR spectrum of a Li_2MnO_3/Li bag-cell battery is shown in Figure 12.8. The very sharp peak around 250 ppm comes from Li metal, which is

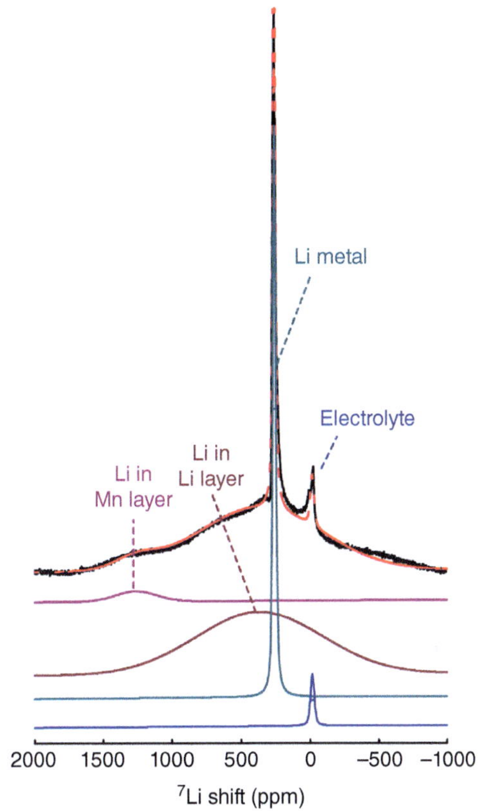

Figure 12.8 *Operando* ^{7}Li NMR spectra acquired on a Li_2MnO_3 vs. Li/Li^+ bag-cell battery. Source: Li et al. [28]/with permission of American Chemical Society.

used as the anode. The broad resonance from −1000 to 2000 ppm is attributed to Li in Li_2MnO_3. Like the *ex situ* ^{7}Li NMR, the main resonance around 500 ppm is Li in Li layers, and the peak at 1400 ppm is from Li in TM layers. The NMR shifts at static conditions are different from the MAS NMR, due to the shift anisotropy and asymmetry parameters. However, they represent the same Li local environments. The peak at 0 ppm is from Li in the electrolyte LP30 (1 M $LiPF_6$ in ethylene carbonate and dimethyl carbonate) [28].

Selective *operando* ^{7}Li NMR spectra of a Li_2MnO_3/Li bag-cell battery are shown in Figure 12.9a where the ^{7}Li NMR intensity of the broad peak gradually decreases during charge, implying Li extraction from Li_2MnO_3. The intensity of Li metal, which resonates at 250 ppm, increases due to Li deposition onto Li anode surface. During discharge, the intensity of the Li_2MnO_3 increases, indicating Li reinsertion into both Li_{Li} and Li_{TM}. It is worth noting that the phase distortion is observed around 250 ppm, as a consequence of the magnetic susceptibility effect of Li microstructure formed on the Li metal surface [28].

The electrochemical profile of the first two cycles for a Li_2MnO_3/Li half-cell battery is shown in Figure 12.9b. Li extraction rates of Li_{Li} and Li_{TM} are shown in Figure 12.9c,d, respectively. At the beginning of charge, Li extraction from the Li layer is relatively slow with a slope of −1.3. After that, Li extraction is slightly faster with a slope of −1.4. However, different Li extraction behavior is observed for Li in

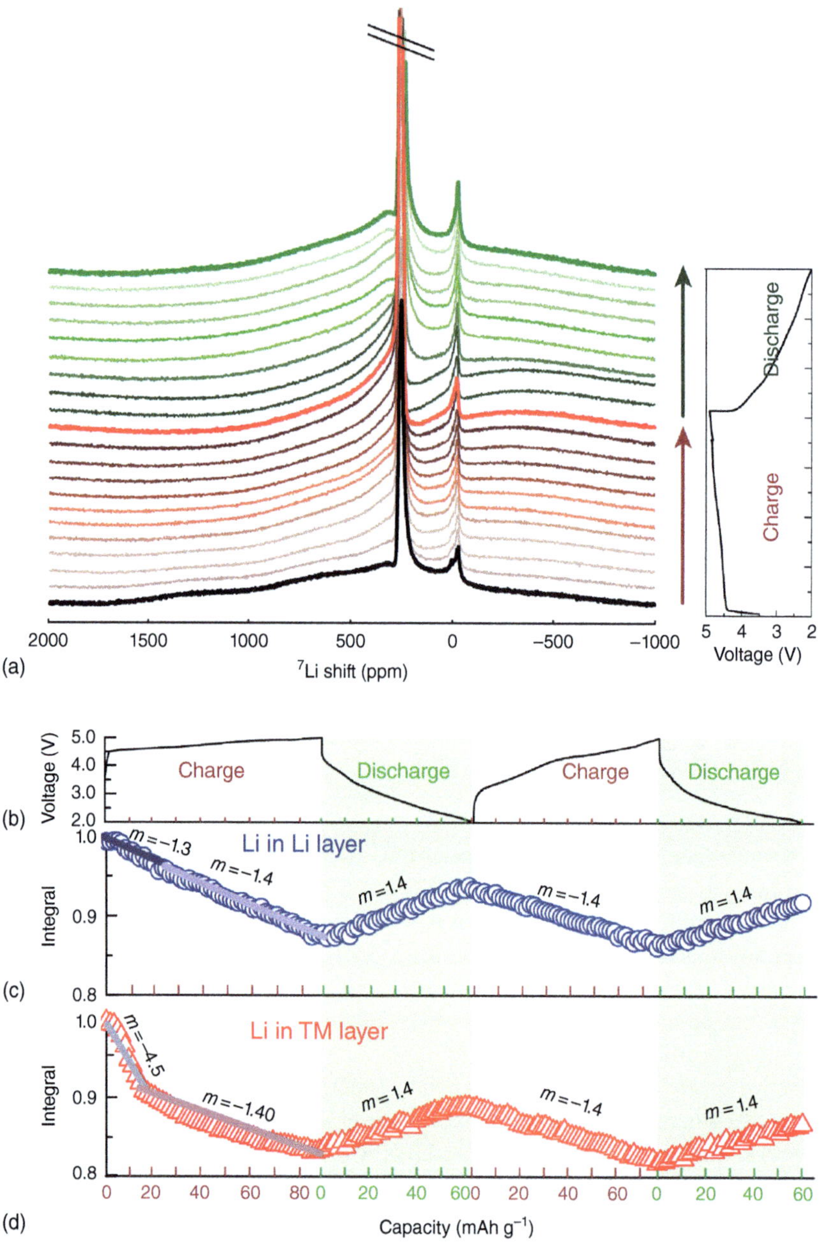

Figure 12.9 (a) *Operando* ^{7}Li NMR spectra during the first cycle with the corresponding electrochemical profile. (b) Electrochemical profile of a Li_2MnO_3/Li half-cell battery with a working voltage range of 2.0–5.0 V. Normalized ^{7}Li integrals of NMR resonances of Li in Li layers (c) and TM layers (d). Source: Li et al. [28]/with permission of American Chemical Society.

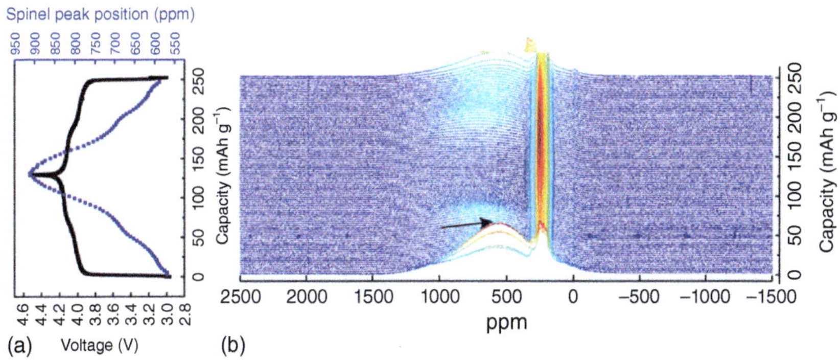

Figure 12.10 (a) *In situ* ^{7}Li NMR spectra of $Li_{1.08}Mn_{1.92}O_4$ with the corresponding electrochemical profile (black line) and spinel peak position (blue dot). (b) ^{7}Li NMR spectra evolution as a function of specific capacity. Source: Zhou et al. [34]/with permission of Elsevier.

the TM layer. Li extraction is fast with a slope of −4.5 at the beginning of charge, and then a slow extraction with a slope of −1.4. As for discharge, both Li_{Li} and Li_{TM} show very similar reinsertion rate with a slope of 1.4 [28].

Figure 12.10 shows another example of *in situ* ^{7}Li NMR, which follows the first cycle of a $Li_{1.08}Mn_{1.92}O_4$ vs. Li/Li^+ bag-cell battery. The electrochemical profile in Figure 12.10a shows two processes: from 3.0 to 4.1 V, 50% of Li is gradually extracted from the tetrahedral sites and Mn^{3+} ions are oxidized to Mn^{4+}; from 4.1 to 4.5 V, only Li is extracted. In Figure 12.10b, the broad resonance around 500 ppm is assigned to Li in the $Li_{1.08}Mn_{1.92}O_4$ electrode. The peak at 280 ppm is from Li metal, and the sharp peak around 0 ppm is from Li in the electrolyte as well as from the solid-electrolyte-interphase (SEI) formed on both the cathode and anode. The broad resonance shifts to downfield as Mn^{3+} is oxidized to Mn^{4+} upon charge and shifts back to high field upon discharge. The Li signal from $Li_{1.08}Mn_{1.92}O_4$ drops significantly at the beginning of charge, followed by a slight increase, and then further decrease till the end of charge. The authors attribute this to the combined effects of changes in transverse relaxation (T_2) and gradual Li extraction [34].

In parallel to the experimental investigations, theoretical work has been performed to predict Li dynamics with first-principle calculations, which show consistent results with those obtained from *in situ* NMR [23, 36]. For example, Kristin Persson and coworkers examined the structural stability of layered Li_2MnO_3 upon charge. The DFT calculation results are shown in Figure 12.11. The red dots represent compounds where the structure has changed from the original layered structure, while the blue dots represent compounds that remain layered during electrochemical cycling. At the beginning of charge ($1 < x < 2$), Li at the 2b site and Li at the 2c site are first extracted, then a fraction of the Li ions at the 4h site is extracted, and lastly all the remaining Li ions at 4h site are extracted, which echoes the results shown in Figure 12.9 [36].

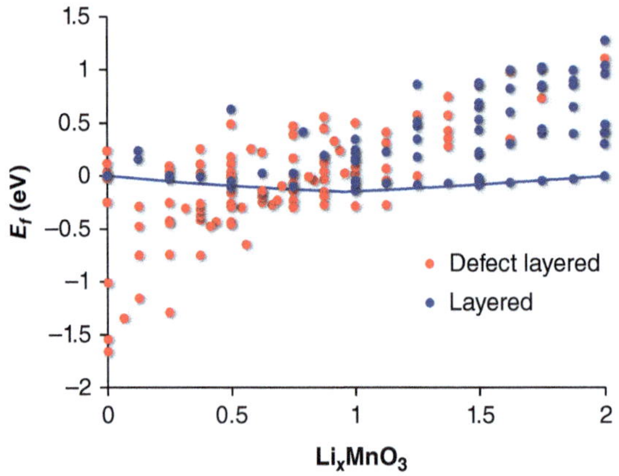

Figure 12.11 The formation energy of Li_xMnO_3 based on DFT calculations where the red and blue dots represent the defect layered and layered structures, respectively. Source: Lee and Persson [36]/with permission of John Wiley & Sons.

12.3.2 O Evolution

A unique feature of Li-rich materials is a voltage plateau observed around 4.5 V during the first charge process, which delivers significant additional capacity [15, 16]. However, a large irreversible capacity loss in the subsequent process and continuous structural changes from the layered structure prevent this material from being used in a wider range of applications. Oxygen plays a key role in producing the extra capacity shown in 3D Li-excess materials. Directly probing oxygen activity during electrochemical processes helps to understand the mechanism that yields extra capacity and cycling stability. Due to strong nuclear–electron dipolar interactions and extremely low natural abundance (0.037%) for NMR-active ^{17}O, the ^{17}O NMR spectrum has a very large shift anisotropy with overlapping SSBs and low sensitivity [37]. To address the low sensitivity issue, ^{17}O isotope enrichment is required, and the experimental set-up is shown in Figure 12.12.

After successful ^{17}O-isotope enrichment, the ^{17}O level inside the material is significantly increased and sufficient for ^{17}O NMR. As an example, ^{17}O-enriched Li_2MnO_3 is studied as a model compound.

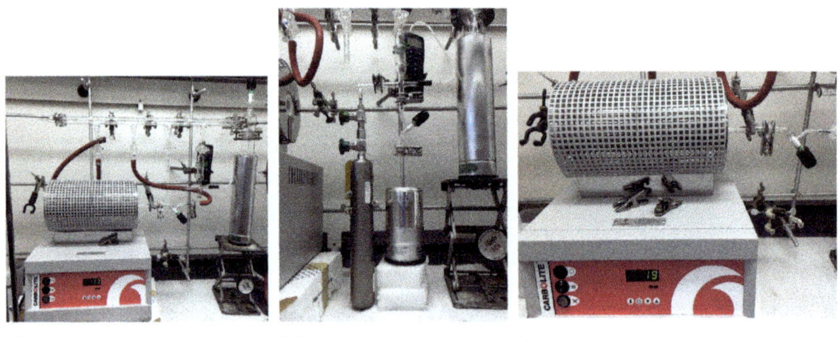

Figure 12.12 ^{17}O isotope enrichment set-up: (a) dehydration, (b) $^{17}O_2$ fill, and (c) ^{16}O–^{17}O exchange at high temperature.

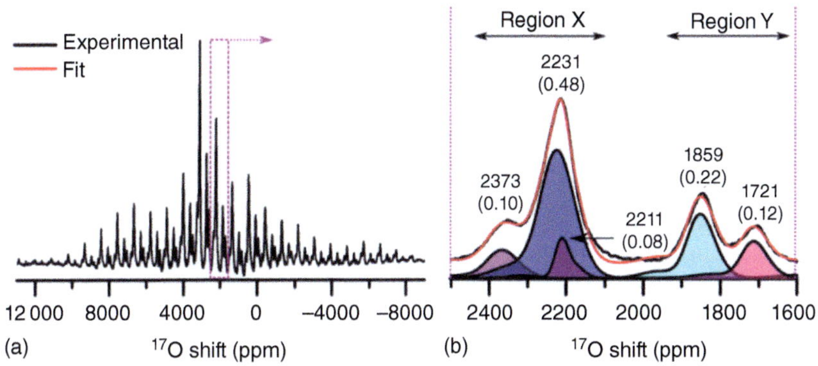

Figure 12.13 (a) ^{17}O NMR spectrum of ^{17}O-enriched Li_2MnO_3 and (b) an expanded view of the isotropic resonances. Source: Seymour et al. [37]/with permission of American Chemical Society.

The ^{17}O NMR variable-offset cumulative spectroscopy (VOCS) spectrum is shown in Figure 12.13a. An expanded view of the isotropic resonances from 1600 to 2400 ppm is shown in Figure 12.13b, which contains two regions: X and Y. Spectral deconvolution helps the identification of the five individual isotropic peaks in these two regions. The most intense resonances at 2231 and 1859 ppm are assigned to O at the 8j and 4i sites in the $C2/m$ structure, respectively. As discussed before, the heating process always generates stacking faults within a $P3_112$ structure, which forms disorder in the $Li_{1/3}Mn_{2/3}$ layer. Therefore, the additional peaks at 1721, 2211, 2373 ppm are assigned to O in those stacking faults [37].

Besides ultra-fast MAS NMR, pj-MATPASS has also been employed to improve the ^{17}O spectral sensitivity. The crystal structure of a promising solid oxide fuel cell material $La_2NiO_{4+\delta}$ is shown in Figure 12.14a. Three different O sites (equatorial, axial, and interstitial) can be clearly observed with the use of the pj-MATPASS technique (Figure 12.14b) [38].

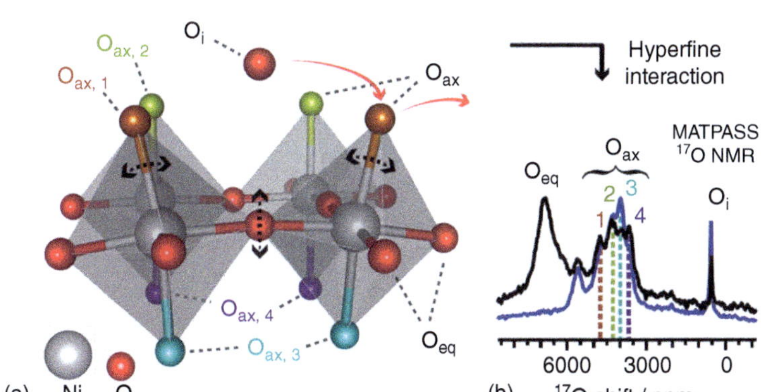

Figure 12.14 (a) Local structural distortion from part of the DFT-optimized $La_{16}Ni_8O_{33}$ supercell. (b) ^{17}O pj-MATPASS NMR spectra with/without quadrupolar filtering. Source: Halat et al. [38]/with permission of American Chemical Society.

12.4 EPR Studies

Operando EPR is a powerful tool to simultaneously follow the evolution of cationic TM and anionic O redox reactions during battery operation with high detection sensitivity. Li_2MnO_3 is typically chosen as the model material to study the evolution of $Mn^{3+/4+}$, together with the O^{2-}/O^{n-} redox couple [17].

The *operando* EPR experimental set-up is shown in Figure 12.15, with the bag-cell battery placed in the center of the EPR cavity (Figure 12.15b). From bottom to top, the EPR spectra of pristine LMO, LMO mixed with active carbon, bare Li metal, and an entire LMO/Li bag-cell battery are presented in Figure 12.15c. The very broad peak is from the Li_2MnO_3 electrode and is caused by interactions between the $Mn^{4+}$and O^{2-} ions, leading to a very symmetric line shape, with a g-factor of ~2.00. Both conductive carbon and Li metal provide very sharp signals, due to their large amount of delocalized electrons, with a g-factor of 2.00. More detailed quantification studies can be performed based on the analysis given above [17].

Selective *operando* EPR spectra of a Li_2MnO_3/Li bag-cell battery are shown in Figure 12.16a. The intensity of Li metal gradually increases upon charge, due to Li deposition onto the Li metal surface. The electrochemical profile is presented in Figure 12.16b. Upon first charge, 0.3 Li is extracted and during discharge 0.21 Li is reinserted. In the second cycle, 0.48 Li is extracted and 0.21 Li is reintercalated upon discharge. The EPR signal integral evolution is presented in Figure 12.16c. At the beginning of the first charge, the Mn^{4+}integration increases from 75% to

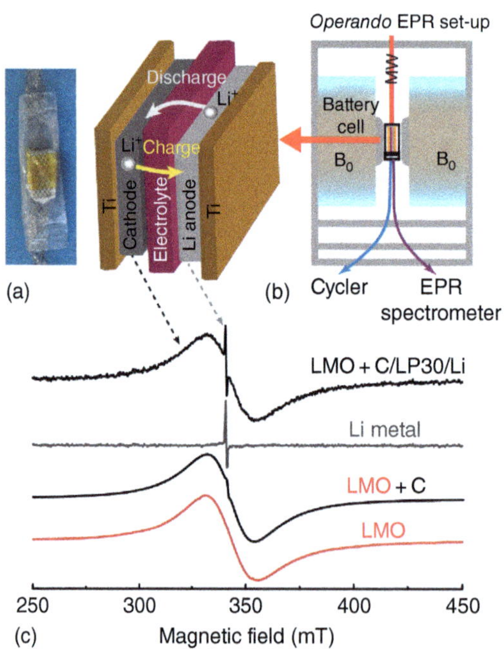

Figure 12.15 *Operando* EPR of high-voltage transition-metal-oxide cathodes. (a) and (b) a bag-cell battery is placed in the center of the resonator cavity, with a connection to a galvanostat. (c) CW-EPR spectra of a bag-cell battery, Li metal, pristine LMO and LMO mixed with conductive carbon. Source: Tang et al. [17]/with permission of American Chemical Society.

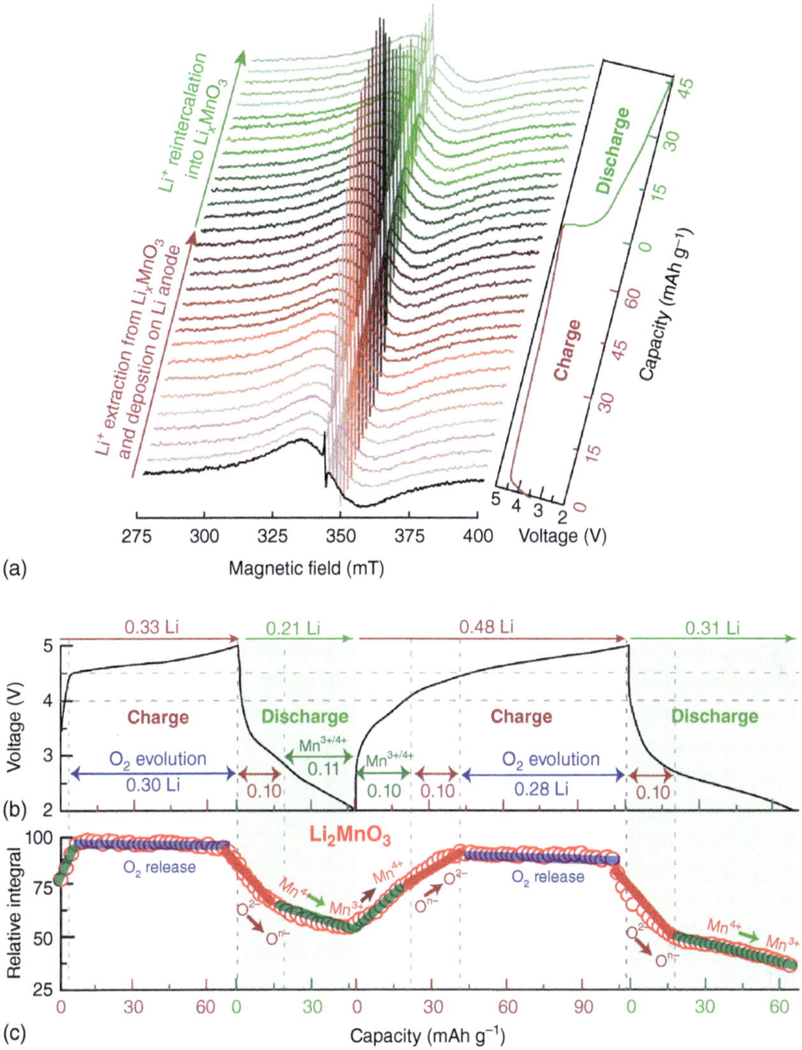

Figure 12.16 (a) Selective *operando* CW-EPR spectra of a Li_2MnO_3/Li bag-cell battery with the corresponding electrochemical profile of the first-cycle. (b) Electrochemical profile of a Li_2MnO_3/Li bag-cell battery as a function of specific capacity. (c) EPR signal integral evolution from Li_2MnO_3 cathode. Source: Tang et al. [17]/with permission of American Chemical Society.

100%, which is due to residual Mn^{3+} oxidized to Mn^{4+}. No further increase is observed at the plateau region, indicating that no Mn redox reaction is associated with this process. Moreover, 0.3 Li is extracted at the same time, which should be accompanied by O redox reaction (from O^{2-} to O_2^{n-}, $n = 1, 2, 3$). Upon discharge, EPR signal gradually drops due to reversible O reduction from paramagnetic O_2^{n-} to diamagnetic O^{2-}. Below 3 V, Mn^{4+} is reduced to Mn^{3+}, leading to a reduction of Mn EPR signal. Similar behavior is observed in the second cycle [17].

12.5 Summary

Ex situ and *operando* nuclear magnetic resonance characterizations are powerful tools to gain fundamental insights into the dynamic charging/discharging mechanisms of high-voltage cathodes. Multinuclear studies lead to a comprehensive understanding of the structural origins of the observed challenges in voltage decay and capacity fade of alkali-ion batteries. Li NMR reveals changes in Li local structural environments upon electrochemical cycling and elucidates structural transformations. ^{17}O NMR permits direct observation of O local structure. In conjunction with EPR measurements of transition metals, the correlations among the elements in the structures and their impact on electrochemical performance can be studied and exploited to create high-energy-density transition-metal cathode materials.

References

1 Armand, M. and Tarascon, J.M. (2008). Building better batteries. *Nature* 451: 652.

2 Tarascon, J.M. and Armand, M. (2021). Issues and challenges facing rechargeable lithium batteries. *Nature* 414: 359–367.

3 Larcher, D. and Tarascon, J.M. (2014). Towards greener and more sustainable batteries for electrical energy storage. *Nature Chemistry* 7: 19.

4 Goodenough, J.B. and Park, K.-S. (2013). The Li-ion rechargeable battery: a perspective. *Journal of the American Chemical Society* 135 (4): 1167–1176.

5 Dunn, B., Kamath, H., and Tarascon, J.-M. (2011). Electrical energy storage for the grid: a battery of choices. *Science* 334 (6058): 928.

6 Whittingham, M.S. (2004). Lithium batteries and cathode materials. *Chemical Reviews* 104 (10): 4271–4302.

7 Etacheri, V., Marom, R., Elazari, R. et al. (2011). Challenges in the development of advanced Li-ion batteries: a review. *Energy & Environmental Science* 4 (9): 3243–3262.

8 Hu, L., Choi, J.W., Yang, Y. et al. (2009). Highly conductive paper for energy-storage devices. *Proceedings of the National Academy of Sciences* 106 (51): 21490.

9 Van der Ven, A. and Ceder, G. (2004). Ordering in $Li_x(Ni_{0.5}Mn_{0.5})O_2$ and its relation to charge capacity and electrochemical behavior in rechargeable lithium batteries. *Electrochemistry Communications* 6 (10): 1045–1050.

10 Rozier, P. and Tarascon, J.M. (2015). Review – Li-rich layered oxide cathodes for next-generation Li-ion batteries: chances and challenges. *Journal of the Electrochemical Society* 162 (14): A2490–A2499.

11 Fell, C.R., Chi, M., Meng, Y.S., and Jones, J.L. (2012). In situ X-ray diffraction study of the lithium excess layered oxide compound $Li[Li_{0.2}Ni_{0.2}Mn_{0.6}]O_2$ during electrochemical cycling. *Solid State Ionics* 207: 44–49.

12 Key, B., Dogan, F., Croy, J.R. et al. (2013). Solid state NMR studies of Li-rich NMC cathodes: investigating structure change and its effect on voltage fade phenomenon. *Meeting Abstracts* MA2013-02 (12): 809.

13 Kim, J.-S., Johnson, C.S., Vaughey, J.T. et al. (2004). Electrochemical and structural properties of $x\mathrm{Li_2M'O_3}\cdot(1-x)\mathrm{LiMn_{0.5}Ni_{0.5}O_2}$ electrodes for lithium batteries (M‘ = Ti, Mn, Zr; $0 \leq x \leq 0.3$). *Chemistry of Materials* 16 (10): 1996–2006.

14 Liu, H., Chen, Y., Hy, S. et al. (2016). Operando lithium dynamics in the Li-rich layered oxide cathode material via neutron diffraction. *Advanced Energy Materials* 6 (7): 1502143.

15 Luo, K., Roberts, M.R., Hao, R. et al. (2016). Charge-compensation in 3d-transition-metal-oxide intercalation cathodes through the generation of localized electron holes on oxygen. *Nature Chemistry* 8 (7): 684–691.

16 Seo, D.-H., Lee, J., Urban, A. et al. (2016). The structural and chemical origin of the oxygen redox activity in layered and cation-disordered Li-excess cathode materials. *Nature Chemistry* 8: 692.

17 Tang, M., Dalzini, A., Li, X. et al. (2017). Operando EPR for simultaneous monitoring of anionic and cationic redox processes in Li-rich metal oxide cathodes. *The Journal of Physical Chemistry Letters* 8 (17): 4009–4016.

18 Feng, X.-Y., Li, X., Tang, M. et al. (2017). Enhanced rate performance of $\mathrm{Li_4Ti_5O_{12}}$ anodes with bridged grain boundaries. *Journal of Power Sources* 354: 172–178.

19 Ma, C., Li, X., Deng, C. et al. (2018). Coaxial carbon nanotube supported $\mathrm{TiO_2}$@$\mathrm{MoO_2}$@carbon core–shell anode for ultrafast and high-capacity sodium ion storage. *ACS Nano* 13 (1): 671–680.

20 Yang, L., Li, X., Ma, X. et al. (2018). Design of high-performance cathode materials with single-phase pathway for sodium ion batteries: a study on P2-$\mathrm{Na}_x(\mathrm{Li}_y\mathrm{Mn}_{1-y})\mathrm{O_2}$ compounds. *Journal of Power Sources* 381: 171–180.

21 Hodges, J.M., Hao, S., Grovogui, J.A. et al. (2018). Chemical insights into PbSe-x%HgSe: high power factor and improved thermoelectric performance by alloying with discordant atoms. *Journal of the American Chemical Society* 140 (51): 18115–18123.

22 Kim, J., Li, W., Philips, B.L., and Grey, C.P. (2011). Phosphate adsorption on the iron oxyhydroxides goethite (α-FeOOH), akaganeite (β-FeOOH), and lepidocrocite (γ-FeOOH): a 31P NMR study. *Energy & Environmental Science* 4 (10): 4298–4305.

23 Lee, Y.J., Wang, F., and Grey, C.P. (1998). ^{6}Li and ^{7}Li MAS NMR studies of lithium manganate cathode materials. *Journal of the American Chemical Society* 120 (48): 12601–12613.

24 Hu, Y.-Y., Liu, Z., Nam, K.-W. et al. (2013). Origin of additional capacities in metal oxide lithium-ion battery electrodes. *Nature Materials* 12: 1130.

25 Levitt, M.H. (2008). *Spin Dynamics: Basics of Nuclear Magnetic Resonance*. Wiley.

26 Dogan, F., Croy, J.R., Balasubramanian, M. et al. (2015). Solid state NMR studies of $\mathrm{Li_2MnO_3}$ and Li-rich cathode materials: proton insertion, local structure, and voltage fade. *Journal of the Electrochemical Society* 162 (1): A235–A243.

27 Hung, I., Zhou, L., Pourpoint, F. et al. (2012). Isotropic high field NMR spectra of Li-ion battery materials with anisotropy >1 MHz. *Journal of the American Chemical Society* 134 (4): 1898–1901.

28 Li, X., Tang, M., Feng, X. et al. (2017). Lithiation and delithiation dynamics of different Li sites in Li-rich battery cathodes studied by operando nuclear magnetic resonance. *Chemistry of Materials* 29 (19): 8282–8291.
29 Keeler, J. (2011). *Understanding NMR Spectroscopy*. Wiley.
30 Dechant, J. (1988). Compiled for the commission on physicochemical symbols, terminology, and units, physical chemistry division, international union of pure and applied chemistry. In: *Quantities, Units and Symbols in Physical Chemistry* (eds. I. Mills, T. Cvitaš, K. Homann, et al.), 1–166. Oxford: Blackwell Scientific Publications. ISBN: 0-632-01773-2.
31 Buzlukov, A., Mouesca, J.-M., Buannic, L. et al. (2016). Li-rich Mn/Ni layered oxide as electrode material for lithium batteries: a ^{7}Li MAS NMR study revealing segregation into (nanoscale) domains with highly different electrochemical behaviors. *The Journal of Physical Chemistry C* 120 (34): 19049–19063.
32 Rana, J., Stan, M., Kloepsch, R. et al. (2014). Structural changes in Li_2MnO_3 cathode material for Li-ion batteries. *Advanced Energy Materials* 4 (5): 1300998.
33 Liu, H., Bugnet, M., Tessaro, M.Z. et al. (2016). Spatially resolved surface valence gradient and structural transformation of lithium transition metal oxides in lithium-ion batteries. *Physical Chemistry Chemical Physics* 18 (42): 29064–29075.
34 Zhou, L., Leskes, M., Ilott, A.J. et al. (2013). Paramagnetic electrodes and bulk magnetic susceptibility effects in the in situ NMR studies of batteries: application to $Li_{1.08}Mn_{1.92}O_4$ spinels. *Journal of Magnetic Resonance* 234: 44–57.
35 Zhang, T., Li, D., Tao, Z., and Chen, J. (2013). Understanding electrode materials of rechargeable lithium batteries via DFT calculations. *Progress in Natural Science: Materials International* 23: 256–272.
36 Lee, E. and Persson, K.A. (2014). Structural and chemical evolution of the layered Li-excess Li_xMnO_3 as a function of Li content from first-principles calculations. *Advanced Energy Materials* 4 (15): 1400498.
37 Seymour, I.D., Middlemiss, D.S., Halat, D.M. et al. (2016). Characterizing oxygen local environments in paramagnetic battery materials via ^{17}O NMR and DFT calculations. *Journal of the American Chemical Society* 138 (30): 9405–9408.
38 Halat, D.M., Dervişoğlu, R., Kim, G. et al. (2016). Probing oxide-ion mobility in the mixed ionic–electronic conductor $La_2NiO_{4+\delta}$ by solid-state ^{17}O MAS NMR spectroscopy. *Journal of the American Chemical Society* 138 (36): 11958–11969.

13

In Situ and *In Operando* Neutron Diffraction of Transition Metal Oxides for Electrochemical Storage

Christophe Didier[1,2], Zaiping Guo[1], Bohang Song[3], Ashfia Huq[3], and Vanessa K. Peterson[1,2]

[1] *University of Wollongong, Institute for Superconducting & Electronic Materials, Squires Way, Wollongong, 2500, NSW, Australia*
[2] *Australia Nuclear Science and Technology Organisation, Lucas Heights, NSW, Australia*
[3] *Oak Ridge National Laboratory, Neutron Scattering Division, 1 Bethel Valley Road, Oak Ridge, TN 37831, USA*

13.1 Introduction

13.1.1 Neutron Diffraction and Transition Metal Oxides

A detailed knowledge of crystal structure is essential to advance the performance of materials needed for applications in the ever-growing field of energy storage. In the context of transition metal oxides, diffraction methods are commonplace for the study of crystal structure, with both single-crystal and powder diffraction methods being routinely applied.

While it is relatively straightforward to determine a crystal structure using single-crystal approaches, many materials cannot be obtained in single-crystal form. The loss of three-dimensional information in powder diffraction data often means that some prior knowledge of the crystal structure is required, leading to structure refinement in place of structure solution. The major breakthrough in realizing the power of powder diffraction for structural analysis was achieved by a method that overcame the peak overlap problem inherent in powder diffraction data [1]. This methodology advancement enabled the determination of the structure of polycrystalline materials, revolutionizing the application of powder diffraction to energy materials studies [2–5].

Both single-crystal and powder diffraction methods yield a space-averaged crystal structure, but the emergence of increasingly complex functional materials means that often deeper structural knowledge is required to understand function, regarding the structure of the material at short-range, over the span of a few unit-cells or less for instance, and therefore requiring approaches beyond traditional methods [6]. To study the short-range structure, the total scattering approach can be used, with this becoming an increasingly invaluable tool in the study of nanocrystalline and disordered materials. Diffraction measurement yields the scattering intensity $I(\boldsymbol{Q})$,

Transition Metal Oxides for Electrochemical Energy Storage, First Edition.
Edited by Jagjit Nanda and Veronica Augustyn.

where $I(\mathbf{Q})$ is proportional to the square of the structure factor $F(\mathbf{Q})$. The structure factor is the sum of intensity at each point of a reciprocal lattice of given periodicity with wave vectors $\mathbf{Q}$ ($=4\pi \sin\theta/\lambda$) having reciprocal space units (typically Å^{-1} and where θ is the scattering angle and λ is the incident radiation wavelength). In this formulation, the $I(\mathbf{Q})$ is often considered as a sum of Bragg peaks and a background function. However, the diffuse scattering information that is modeled as a simple background contains information about the short-range structure that is not taken into account in Bragg diffraction when it is not uniformly repeated over many unit-cells at long-range. The sine Fourier transform of the traditionally measured $F(\mathbf{Q})$ yields the atomic pair distribution function (PDF), which is the density of matter at a given point in space, $\rho(\mathbf{r})$, containing the direct-space information of the vectors $\mathbf{r}$ (typically in Å) [7]. In this formulation, the true instrumental background is subtracted out and the diffuse scattering from the sample is included in the analysis for the structural modeling. Advances in experimental and analysis software have extended the use of the PDF method, making it a relatively commonly applied tool in materials analysis.

Diffraction methods are routinely carried out using X-ray radiation on laboratory-based and relatively cheap instrumentation. Where higher resolution or intensity is required, these measurements can be performed using synchrotron-based radiation at large-scale facilities. However, for many transition metal oxides used in energy-storage systems, the structural information contained in diffraction data using X-ray radiation can be limited by several factors. Transition metals tend to have similar atomic numbers, so they are difficult to distinguish using X-rays that scatter from electrons, especially if they occupy the same crystallographic site. Detailed information may be difficult to obtain from lighter elements, such as oxygen, in the presence of heavier ones, as X-rays are scattered more strongly by heavier elements. Neutrons are thus essential in the study of many transition metal oxides, due to their charge neutrality, magnetic moment, and spin. The charge neutrality of neutrons means that they interact with the atomic nuclei of atoms, yielding sensitivity to isotope and therefore atom type, independently from the electronic configuration. This same property also results in a high penetration of neutrons through many materials, enabling relatively large and more complicated devices to be probed. The magnetic moment of a neutron given by its spin angular momentum can probe the magnetization components of a sample that are perpendicular to the neutron's momentum transfer, and therefore, neutrons can probe the magnetic structure, in addition to the atomic (nuclear) structure. Thus, X-rays and neutrons are complementary tools in the study of transition metal oxides.

The $Ba_{0.5}Sr_{0.5}(Co_{0.7}Fe_{0.3})_{0.6875}W_{0.3125}O_{3-\delta}$ (BSCFW) composite perovskite fuel-cell cathode material exemplifies the need for neutron powder diffraction in the determination of the phase composition and structure of transition metal oxides [8]. While X-ray powder diffraction revealed the coexistence of single-perovskite and double-perovskite phases in the BSCFW material, neutron powder diffraction was required to determine the details of the coexistence of Fe and Co at crystallographic sites, where excellent contrast between Fe, Co, and W atoms is achieved using neutron scattering. Furthermore, the details of oxygen-deficient sites could also be obtained from these data, where neutron scattering data are not dominated by electron-dense atoms as it is in X-ray scattering data. This work took advantage of the higher spatial resolution achieved by synchrotron X-ray compared to neutron instrumentation, and therefore used joint analysis of room-temperature synchrotron X-ray and neutron diffraction data to study the full details of the BSCFW structure. The phase ratios, atomic composition, and oxygen nonstoichiometry of each phase were determined with high accuracy following joint Rietveld refinements. The work revealed disordered *A*-site cations in both single- and double-perovskite cubic phases, with the double-perovskite phase also having magnetic ordering identified by the neutron powder diffraction data. The single-perovskite phase was found to have an unexpectedly high 23.2(7)% oxygen vacancy content, explaining the mixed conductor property.

Another excellent example of the requirement for neutron powder diffraction is the rechargeable battery electrode material $LiNi_{0.5}Mn_{1.5}O_4$ (LNMO), which crystallizes into the $P4_332$ space group when Ni^{2+} and Mn^{4+} cations are ordered and into $Fd\bar{3}m$ space group symmetry when these cations are disordered [9–11]. Given the similar number of electrons between Ni and Mn atoms, neutron diffraction is essential for distinguishing these phases. Further, both calculations and experimental data suggest that understanding the local ordering of these cations is required to optimize the LNMO electrochemical properties, with neutron powder diffraction used to show that the size of domains of ordered cations in the material increased with annealing time, explaining the annealing conditions required for optimal material performance [9].

Modern X-ray and neutron sources, optics, and detectors allow high spatial resolution data, as well as relatively fast data acquisition. Such instrumentation now makes it possible for real-time diffraction measurements to be carried out, accessing fast processes on the microsecond timescale [12, 13], or complex parametric studies [14] of materials under conditions of varying external parameters such as temperature [15, 16]. The BSCFW fuel-cell cathode material was also studied under flowing air at 600 °C to emulate intermediate-temperature solid oxide fuel-cell (SOFC) operating conditions, using joint synchrotron X-ray and neutron powder diffraction data analysis [8]. While the double-perovskite phase oxygen content was noted to remain stoichiometric under these conditions, the single-perovskite phase weight fraction increased from 29.3(2)% to 35.7(2)% under operating conditions, resulting in an increase in W content of this phase from 5.6(3)% to 9.4(4)% to accommodate the disordering of W at the higher temperature.

While this chapter is primarily dedicated to understanding the crystal structure of materials using *in situ* neutron diffraction techniques, it is important to remember that other techniques based on both elastic and inelastic neutron scattering have played, and will continue to play, an important role in the study of transition metal oxide systems. Following is a short description of several of the more prominent techniques relevant to the study of transition metal oxides.

13.1.1.1 Neutron Reflectometry

Neutron reflectometry is a method used to determine interfacial structures buried within samples, with nanoscale resolution in one dimension. Features as thin as

0.5–4 nm, depending on experimental set-up, can be determined with sub-Å precision using neutron reflectometry. This is achieved by determining the depth profile of scattering length density averaged in the plane of the sample, which consists of thin films or their surfaces on highly polished substrates. In the context of electrochemical energy storage systems, these are usually idealized models of surfaces, interfaces, coatings, and intergrain boundaries. The information yielded in these models can be used to understand the structure and transport of ions through surface coatings, and also the distribution of cations on an electrode, in platings, and/or the solid electrolyte interphase. The evolution of the solid electrolyte interphase as it develops was first followed using neutron reflectometry in a working lithium half-cell by Owejan et al. [17]. Given the experimental necessity for atomically flat surfaces in neutron reflectometry, the majority of examples of this research are for relatively simple systems such as tungsten [18], copper [17], and silicon [19]; however, important insights into oxide systems have been made, such as the first by Hirayama et al. [20] for $LiFePO_4$ and Browning et al. [21] who determined *in situ* the thickness and scattering length density profile of the electrode–electrolyte interface for the high-voltage cathode $LiMn_{1.5}Ni_{0.5}O_4$.

13.1.1.2 Small-Angle Neutron Scattering

When the length scale of interest is between 10 and 3000 Å, the diffraction-based technique of choice is small-angle scattering. This technique is ideal for the study of nanoscale structures, particle shape, and interparticle correlations. Small-angle scattering has been used to examine the microstructure of mesoporous materials relevant for energy-storage applications, such as batteries, capacitors, and fuel cells, with transition metal oxides being important materials for these applications. Such experiments have been extended to include time-resolved *in situ* studies, where unlike neutron reflectometry experiments carried out on model thin-film electrode systems, small-angle neutron scattering can be applied to fully operational coin and pouch cells, bridging the gap between fundamental and real-world applied studies. Such studies include carbonaceous and relatively simple electrode materials [22, 23] and on cells containing commercial $LiNi_{0.33}Mn_{0.33}Co_{0.33}O_2$ and graphite electrode materials [24].

13.1.1.3 Quasielastic and Inelastic Neutron Scattering

The dynamics of atoms influence material properties, in a similar concept to the structure–function relations discussed earlier. For many materials pertinent to electrochemical energy storage systems, including transition metal oxides, insight into atom dynamics can be used to understand the origin of important electrochemical properties. Neutron spectroscopy methods usually measure the dynamic structure-factor, $S(\boldsymbol{Q}, \omega)$, which describes scattered neutrons in terms of $\boldsymbol{Q}$, as measured in diffraction, but also the simultaneously measured neutron energy-transfer $\hbar\omega$, where $\hbar = h/2\pi$ and h is Planck's constant [25]. Of particular importance is that the size and geometry of the volume in the sample explored by neutron spectroscopy matches that of dynamic processes pertinent to many materials, including transition metal oxides used for electrochemical energy storage. Unlike diffraction, which takes advantage of coherent neutron scattering processes, it is incoherent neutron scattering that is pertinent to neutron spectroscopy, and so the range of sensitivities and elemental contrast afforded by neutrons has to be assessed in the context of both incoherent and coherent scattering. Inelastic scattering arises from periodic motions of nuclei and quasielastic scattering from diffusive motions, with the latter capturing motions of nuclei that do not return to their origin position within the timescale of the measurement. Inelastic neutron scattering allows vibrational spectroscopy measurements to be made without the selection rules that apply to other vibrational spectroscopy methods, and an inelastic neutron scattering spectrum is very straightforward to calculate, providing a highly useful method for validating a computational model. Because both time and space are probed using neutron spectroscopy, quasielastic neutron scattering provides direct information on the diffusion of ions that is directly relevant to ion transport underpinning device performance in many electrochemical energy storage systems, such as within electrodes and electrolytes. Quasielastic neutron scattering is often used to understand the mechanism of ion diffusion through solid-state ion conductors [26], with Klenk et al. [27] combining incoherent quasielastic neutron scattering and molecular dynamics (MD) simulations to unveil the details of lithium self-diffusion in the model lithium garnet oxide $Li_5La_3Ta_2O_{12}$ and Wind et al. [28] using coherent quasielastic neutron scattering to extract the details of oxygen diffusion in bismuth oxide, which they correlate with molecular dynamics simulation results validated against the generalized density of states as measured in the neutron scattering experiment.

13.1.2 Neutron Diffraction Instrumentation

In comparison with the large number of quanta in the X-ray beams, neutron beams are relatively weak, and when combined with their relatively weak interaction with matter, this necessitates the use of relatively large samples and/or beam areas relative to X-ray diffraction [29]. Intense neutron beams can be produced either by nuclear fission induced in a critical mass of ^{235}U at a reactor or by a spallation process arising from the collision of high-energy protons with a heavy metal target such as mercury, tungsten, and tantalum. A review on neutron sources around the world and some examples of these sources can be found elsewhere [30]. Reactor sources include the High Flux Isotope Reactor (HFIR) at Oak Ridge National Laboratory (ORNL) and the National Institute of Standards and Technology's Center for Neutron Research (NIST NCNR) in the United States, the Institute Laue Langevin (ILL) in France, Forschungsreaktor Munchen II (FRM II) in Germany, and the Open Pool Australian Light-Water (OPAL) reactor facility in Australia. The production of neutrons via spallation can produce a highly energetic neutron beam, and in addition to its reactor source, ORNL in the United States also hosts the world's most intense pulsed source, the Spallation Neutron Source (SNS). Other prominent sources include the Japanese Proton Accelerator Research Complex (J-PARC) in Japan, the China Spallation Neutron Source (CSNS) in China, the ISIS facility at the Rutherford Appleton Laboratory in the United Kingdom, Los Alamos Neutron Science Center (LANSCE) in the United States, and the Swiss Spallation Neutron Source (SINQ) in Switzerland.

Neutron scattering instrumentation is required to measure the number of neutrons scattered (intensity) by a sample as a function of $\boldsymbol{Q}$ and the energy change (E) of the neutron [25]. The intensity as a function of $\boldsymbol{Q}$ and E is proportional to the space and time Fourier transform of the probability of finding two atoms separated by a particular distance at a particular time, and in this way, structural determination is possible. The neutron wavelength and direction of travel have uncertainties that mandate the definition of $\boldsymbol{Q}$ and E with a given precision. Resolution and flux are inversely proportional as a result of the total signal being proportional to the phase space volume within the elliptical resolution volume, and the neutron instrumentation and configuration must be designed to optimize both for any given target experiment.

The distribution of neutron energies produced by both reactor and spallation sources is generally higher than required for scattering experiments, and neutrons are moderated into more useful energy ranges before they are directed to instruments. The different ways in which neutrons are produced impact the distribution of neutron energies available.

A reactor source produces a Maxwellian distribution of neutron energies (and therefore wavelengths). Reactor-based scattering experiments use neutrons in a narrow energy range, and a monochromatic beam is produced using a highly reflective monochromator material in most instruments. This is the constant wavelength approach, and instrumentation is similar to a laboratory-based X-ray diffractometer. The incident wavelength is controlled by changing the scattering angle of the monochromator, making it relatively straightforward to control resolution of both $\boldsymbol{Q}$ and E and to tailor these for experimental requirements. Polarized neutron experiments are easier to implement at reactor-based sources. Reactor sources usually produce a relatively large flux of longer wavelength (slower) neutrons, making them good for studying larger objects and slower dynamics.

At spallation sources neutrons are generated in pulses, and scattering experiments advantageously use these in the time-of-flight (TOF) method, where the neutron velocities are determined from the time taken for each neutron to travel from the moderator to the detector, often using the Laue geometry. These sources produce a constant and relatively small change of neutron wavelengths ($d\lambda/\lambda$), especially at large neutron energies, leading to excellent resolution at high $\boldsymbol{Q}$ and E. The big advantage of the TOF method is the access to very high $\boldsymbol{Q}$ information, which is useful in providing detailed information such as atomic occupancy and displacement parameters and also local (shorter-range) interactions required for PDF measurement. The low flux of neutrons between pulses leads to low background, and good signal to noise. The Laue geometry also enables the full $\boldsymbol{Q}$ range of data collection simultaneously, which is convenient for *in situ* studies.

A neutron diffraction instrument may contain a monochromator (to define the energy of a neutron beam using Bragg's law) or chopper (to define a short pulse or small band of neutron energies), collimators (that define the direction of travel of the neutron), detectors (usually comprising ^{3}He or scintillators), and shielding (to minimize background and radiation exposure to users). This instrumentation is optimized for different kinds of diffraction experiments, with single crystal, powder, and total scattering instrumentation having slightly different requirements for an optimized signal. Where the detector is curved with respect to scattering angle around the sample, cylindrical cells are very much preferred over a flat-plate geometry, providing for better ease of data reduction and correction. This is especially true for instruments using the Laue geometry.

The scattering technique and the nature of the source of neutrons determine how the measurements are carried out and hence also often the choice of material for sample containment. Vanadium is often the primary choice for sample holders for neutron diffraction, as a result of the extremely small coherent cross section of vanadium, which means that it has very low intensity Bragg peaks. By alloying Ti and Zr, which have opposite neutron scattering lengths, a null matrix material can be created, and this alloy can be used also as a sample container. However, V as well as the alloy of Ti and Zr can be a significant source of background in neutron diffraction data as a result of incoherent neutron scattering. The sample container must not only add very little or nothing to the measured neutron scattering in an experiment, but also be chemically and thermally compatible for the experimental conditions. Glassy quartz is frequently used at higher temperatures as it also does not have Bragg peaks. For high-temperature work such as structure studies of transition metal oxide formation, the possibility of sample reaction at elevated temperatures must be considered. Many transition metal oxides used in battery systems contain lithium, which can react with metals and quartz if the temperature is high. In such cases, it is often possible to line the inside of the container with inert material that the sample does not react with or use relatively inert materials such as platinum or Al_2O_3.

13.1.3 *In Situ* and *In Operando* Neutron Diffraction

The structure–function relation is central to advancing materials for application, but to do this we must have knowledge concerning not only the material structure, but also how this is influenced and changed by external stimuli or composition during use in a device or working environment. This knowledge is gained through *in situ* and *in operando* studies. *In situ* ("on-site") experiments measure the system in a particular environment, but not necessarily during change [31]. These tend to be experiments that examine the material that is equilibrated at a known condition. *In operando* (or "working") experiments measure the system while it is operating and capture the material changes that occur under nonequilibrium conditions. These also encompass time-resolved work.

New-generation neutron sources and associated faster instrumentation, combined with their high penetration through sample environments, make neutrons an ideal tool for *in situ* and *in operando* studies, particularly for functional materials [32], with powder diffraction the premier tool used in these studies [33]. Sample environments are often specialized to target the type of scientific problem that is being studied (Figure 13.1), and this will be addressed in more detail in Sections 13.2.1, 13.2.2, 13.3.1, and 13.4 along with examples.

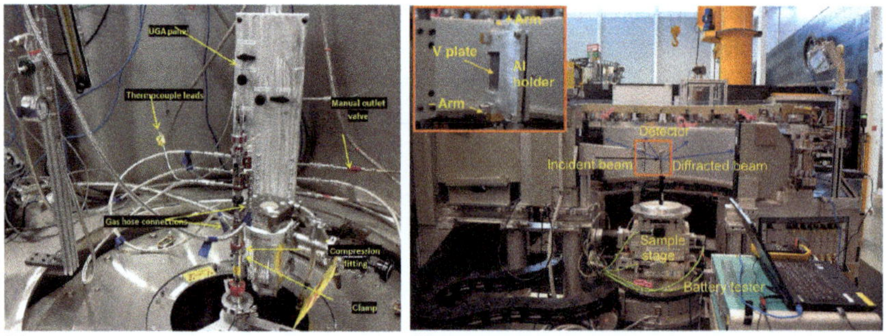

(a) (b)

Figure 13.1 Specialized sample environments for neutron powder diffraction experiments of (a) *in situ* studies of solid oxide fuel-cell materials on the POWGEN diffractometer at the SNS and (b) *in operando* battery research at the Wombat diffractometer at OPAL. Source: (b) Reproduced from Peterson et al. [33] with permission of the International Union of Crystallography.

13.2 Device Operation

13.2.1 Experimental Design and Approach to the Real-Time Analysis of Battery Materials

In the context of transition metal oxides, real-time *in operando* neutron diffraction has only been achieved for electrochemical energy devices such as alkali-ion batteries and SOFCs. In these devices, transition metal oxides are often employed as electrodes and/or electrolyte. The output and performance characteristics, such as voltage, rate capability, capacity, and stability of the entire device, are directly related to the structural phase transitions that the electrode and electrolyte materials undergo during device use. The ability to directly track these modifications in a functioning device under "real-life" conditions provides unparalleled insight into the working mechanism of that device. *In operando* neutron diffraction measurements of electrochemical cells are rather recent as the technique necessitated diffractometers with sufficient temporal resolution [34], with the first of these experiments being performed on the Wombat instrument at the OPAL facility in 2010 [35]. Consequently, the number of *in operando* neutron diffraction experiments has dramatically increased in recent years [36]. A major issue regarding the quality of data gained in such experiments arises because of hydrogen in the electrolyte and separator materials, which results in large background in the diffraction data because of the large incoherent neutron scattering from hydrogen. Substitution of the conventional constituents of these components has been made to reduce hydrogen [34, 37, 38], and several sample environments have been designed to enhance the diffraction signal from the transition metal oxide material of interest within the whole device [36, 39, 40]. The potential to extract crystal structure information under nonequilibrium conditions from *in operando* neutron diffraction experiments is shown in Sections 13.2.2–13.4 through selected examples.

13.2.2 Advancements in Understanding Electrode Structure During Battery Operation

Thanks to the high penetration of neutrons, it is possible to directly investigate transition metal oxides within commercial Li-ion batteries, and several authors have performed *in operando* neutron diffraction measurements using 18650 cylindrical format batteries [41–47]. Despite the presence of hydrogen in the electrolyte and separators, the resulting data are sufficient for the extraction of detailed crystallographic information owing to the large amount of material under observation as a result of the high compaction of the electrode assembly roll in the cell [36]. A study made on the special environment powder diffractometer (SPICA) diffractometer (J-PARC, Japan) [44] demonstrates the very fast acquisition time of one minute, permitting the measurement of nonequilibrium phases in faster operating conditions up to 2 C current rate. The batteries contained $LiNi_{1/3}Mn_{1/3}Co_{1/3}O_2$ and graphite electrodes, and the authors successfully performed a semiautomatic sequential Rietveld analysis and reliably extracted lattice parameters and phase fractions during cycling [44]. They compared *in operando* nonequilibrium structures with the ones measured at constant voltage and after relaxation, and show that the insertion mechanisms are affected by kinetics at high current rate, which lowers the reversible capacity of the battery under fast load. *In operando* neutron diffraction was also measured with 10-second temporal resolution using the Wombat diffractometer (OPAL, Australia) on commercial 18650 cylindrical cells [36]. These observations are relevant to commercial application of batteries that need to sustain fast current rates.

The use of commercial batteries restricts studies to commercially available chemistries and, as several authors mention [34, 41], Rietveld refinements are usually possible but necessitate large restrictions of refined parameters. To resolve this issue, several custom designs inspired from the cylindrical geometry have been proposed [38, 48–50]. Although the replacement of hydrogenated electrolyte with a deuterium-containing equivalent reduces incoherent neutron scattering arising from the ^1H nucleus, counterintuitively, hydrogen-rich polypropylene and polyethylene-based separators (Celgard) may provide a lower background in custom-designed cells due to their lower thickness and higher tensile strength compared with hydrogen-poor equivalents such as polyvinylidene difluoride [38]. *In operando* neutron diffraction data from such custom cells containing $Li(Ni_{0.6}Mn_{0.2}Co_{0.2})O_2$ (NMC622) and Li electrodes under C/20 current using the high resolution powder diffractometer (HRPT) diffractometer (PSI, Switzerland) supported the independent refinement of cell parameters, oxygen and lithium positions, and lithium occupancy in the layered oxide structure at all states of charge [38]. Unfortunately, lengthy acquisition times of one hour were necessary using the HRPT instrument in this experiment, restricting the study to the quasi-equilibrium state, and the cell also exhibited poor electrochemical performance, with the authors suggesting imperfect wrapping as a result of the handmade cell assembly as the underlying cause [38].

Among the proposed *in operando* electrochemical cell designs, those used by Godbole et al. [51], Bianchini et al. [37, 52], and Song et al. [53] attempt to remove all constituents other than the electrode under study from contributing to the diffraction data. In the Bianchini et al. design, the electrode material is placed at the edge of the battery in a container made of the null matrix $Ti_{0.68}Zr_{0.32}$ with the separators and counterelectrode out of the beam, and the electrolyte is replaced with a deuterated equivalent [37]. This cell was used on the D20 diffractometer (ILL, France) with spinel $LiNi_{0.4}Mn_{1.6}O_4$ and Li metal electrodes [52]. Each dataset was collected in 30 minutes and supported the independent refinement of atomic positions, isotropic displacement parameters, and lithium occupancy in the spinel structure at all states of charge after minor restraints. An increase in the atomic displacement parameter for the oxygen atom was observed and attributed to local distortions of the NiO_6 octahedra induced by Jahn–Teller active Ni^{3+}. Although the cell modifications enabled high-quality *in operando* data, it compromised performance, where the high loading (22 mg/cm^2) and thickness (1 mm) of electrodes resulted in strong polarization, reduced capacity, and reversibility in this cell design compared to a conventional coin cell, with the cell operated at the low current of C/20 [52].

Detailed crystallographic information for transition metal oxide electrodes has been obtained from *in operando* neutron diffraction performed using prismatic cells containing the "lithium-rich" spinel-type $Li_{1+y}Mn_2O_4$ and graphite electrodes [54]. Data were collected in five-minute intervals from batteries cycled at C/10 using the Wombat diffractometer at the OPAL facility (ANSTO, Australia). Compared to previous custom-made cells, this battery construction performs better under intermediate current rates and the acquisition time is fast on this instrument. Sequential Rietveld refinement using these data with a few constraints enabled the extraction of phase fractions as well as cell parameters, oxygen position, and the occupancy of lithium at the 8*a* and 16*c* crystallographic sites of the spinel structure, which then could be followed during cycling. During charge, lithium was found to first occupy 16*c* sites, and then progressively relocate to 8*a* sites at higher lithium content, with this mechanism differing during discharge, explaining the relative ease of discharge compared to charge for the material. The authors also show that the composition limit of the material varies under applied current compared to relaxation conditions, illustrating again the importance of *in operando* measurements as opposed to equilibrium or low-current condition experiments. A similar level of crystallographic information was extracted for a spinel $Li_4Ti_5O_{12}$ electrode against a $LiFePO_4$ counterelectrode within a prismatic battery [55]. In this work, a 1-cm-thick pouch cell was prepared using polyvinylidene difluoride separators and deuterated electrolyte [34, 55]. The batteries were cycled at C/10 and data collected in five-minute intervals on the Wombat diffractometer. The authors reliably perform sequential Rietveld refinements with appropriate constraints and extract phase fractions, cell parameters, as well as the occupancy of lithium at 8*a* and 16*c* crystallographic sites. Taking advantage of the negative neutron scattering length of lithium, Fourier difference maps using the *in operando* data revealed the lithium migration pathway and preferential position in the spinel structure of $Li_{4+x}Ti_5O_{12}$ at different states of charge (Figure 13.2). The

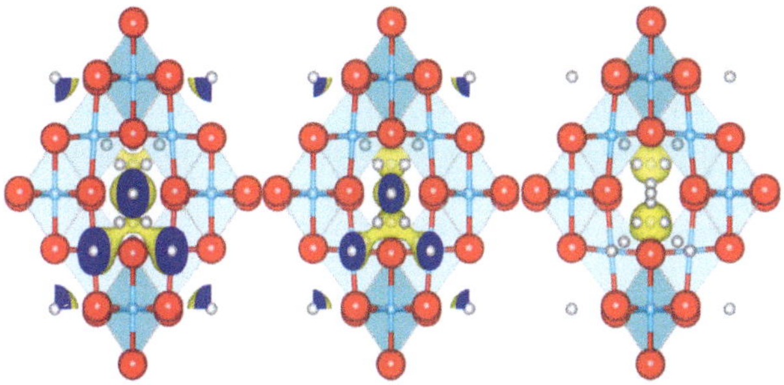

Figure 13.2 Lithium in "small grain size" $Li_{4+x}Ti_5O_{12}$ at different states of charge obtained using *in operando* neutron diffraction. Ti is shown as blue and oxygen in red, with the 8*a* and 16*c* sites available for lithium shown as gray spheres with the nuclear density arising from the lithium shown in yellow. Source: Pang et al. [55]/with permission of American Chemical Society.

authors show that grain size influences the concentration of lithium at 8*a* and 16*c* sites, which results in a larger reversible capacity when grain size is reduced.

The complex transition metal doping in layered oxides commonly used as materials for commercial lithium ion battery electrodes is exemplified by the "lithium-rich NMC" material. This material, commonly described as $Li_{1+x}MO_2$ (M = Co, Ni, Mn, etc.), has a highly complicated structure. Depending on the composition and preparation conditions, cation mixing and ordering within and between the crystal structure layers occurs, influencing the electrochemical performance of the material. Copious literature regarding the quantification of cation mixing exists, with both monoclinic single-phase [56] and two-phase [57] descriptions proposed. This includes atomic-resolution annular bright-field and high-angle annular dark-field scanning transmission electron microscopy that found a layered phase with rhombohedral symmetry $LiMO_2$ (M = Ni, Co, Mn, etc.) to coexist with a second monoclinic Li_2MnO_3-like phase [41]. Crystallographic studies on this material using X-ray diffraction are complicated as Mn, Co, and Ni are virtually indistinguishable. Neutron diffraction yields excellent elemental contrast between these elements and Li, offering the possibility to quantify cation ratios within a phase model. *In operando* neutron diffraction data from a neutron-friendly pouch cell comprised $Li_{1+x}MO_2$ (M = Co, Ni, Mn) and $Li_4Ti_5O_{12}$ electrodes using the Wombat diffractometer was able to be described using a two-phase $LiMO_2{\cdot}Li_2MO_3$ (M = Li, Ni, Co, Mn) model [58]. Although the atomic ratios could not be followed during cycling, the lattice parameters were independently extracted for each phase and, together with *in operando* transmission X-ray microscopy data, refinements show that lattice changes in the $LiMO_2$ phase lead to phase separation and the formation of cracks that are ultimately responsible for capacity fade. The related cobalt-free $Li_{1+x}MO_2$ (M = Li, Ni, Mn, Fe) material was also investigated [59], where the ratio between $LiMO_2$ and Li_2MO_3 phases and the cation mixing was first determined from *ex situ* neutron diffraction. *In operando* neutron diffraction was then performed using a customized cell containing this material against a mesocarbon microbead counter electrode. The cell parameter variations during cycling of the $LiMO_2$ phase were found to be about 80% smaller than in $LiCoO_2$, resulting in improved structural stability and explaining the capacity retention of this material. Notably, the authors observed solvent decomposition at high voltage in this work, as shown by the decrease of the liquid structure factor contribution from the organic electrolyte, which would be too weak to observe using X-ray diffraction.

13.3 Gas and Temperature Studies

13.3.1 Experimental Design and Approach to the *In Situ* Study of Solid Oxide Fuel-Cell (*SOFC*) Electrodes

Several families of compounds such as perovskites, brownmillerites, Ruddlesden–Popper phases, and pyrochlores have been extensively investigated for application as electrode materials for SOFCs [60], with transition metal oxides featuring heavily in this group. Most of these compounds are nonstoichiometric with respect to oxygen, resulting in good ionic and electronic conductivity. The oxygen anion conductivity is highly dependent on both the concentration of oxygen vacancies and the local atomic environment, with the latter determining the mechanism of oxygen mobility. To gain insight into the oxygen conduction mechanism, neutron diffraction is frequently applied due to its sensitivity to oxygen in the presence of heavier elements. In particular, *in situ* neutron diffraction with controlled temperature and partial pressure of O_2 (pO_2) yields information concerning oxygen, including phase stability, crystallographic site ordering, vacancy concentration, and isotropic/anisotropic atomic displacement parameters that are useful in understanding the transportation pathway of oxygen.

The simultaneous control of pO_2 and temperature is required in such a study, with the independent measurement of gas flow and pO_2 also desirable. The requirement for a range of pO_2 from pure O_2 to pure H_2 necessitates safety controls that need to be met in any sample environment design. The standard V and Ti/Zr null scattering containers cannot be used in the presence of hydrogen at high temperature as these react with hydrogen, leading to embrittlement. Thus, quartz is often the material of choice for the sample container due to its stability over a wide temperature range and its amorphous nature, which contributes no Bragg peaks to the data. A good example of such a sample environment is the automated gas environment system (AGES) developed for the POWGEN high resolution neutron powder diffractometer (beamline 11A) at the SNS of ORNL (Figure 13.1a) [61]. The AGES enables the controlled flow and mixing of gases, including nitrogen, oxygen, helium, argon, carbon dioxide, carbon monoxide, hydrogen/deuterium, methane, as well as premade mixtures of 4% hydrogen in helium and air. The input and exhaust gases can be monitored in this environment using a universal gas analyzer and an oxygen sensor. Along with the gas, temperature control between room temperature and 850 °C is available. At the nanoscale ordered materials diffractometer (NOMAD, beamline 1b) of the SNS, another high-temperature gas flow environment was also implemented enabling the

PDF to be measured under these conditions [62]. Similar sample environments for *in situ* gas flow and high-temperature studies are also available at other neutron facilities around the world [63–66].

13.3.2 Advancements in Understanding Solid Oxide Fuel-Cell Electrode Function

The family of $LnBaCo_2O_{5+\delta}$ (Ln = rare earth element) materials with a layered double-perovskite structure are recognized as promising cathode materials for SOFCs because of their high conductivity and high surface exchange rates. Debate concerning the mechanism for rapid mobility of oxygen in these materials was settled using *in situ* neutron powder diffraction experiments in which both pO_2 (10^{-4} to 10^{-1} atm) and temperature (573–852 °C) were controlled [67]. It was shown that in the $PrBaCo_2O_{5+\delta}$ (PBCO) system, fast ion hopping occurs between the oxygen site in the Pr layer and the nearest-neighbor oxygen site in the Co layer. This hopping mechanism is illustrated in Figure 13.3a [67]. The mechanism for excellent oxygen conductivity exhibited by the $Ba_3MoNbO_{8.5}$ material with a hexagonal perovskite derivative structure was also investigated. This material maintains good oxygen conductivity, 2.2×10^{-3} S/cm at 600 °C, over a wide range of pO_2 (10^{-20} to 1 atm) at high temperature [69], with the conductivity increasing linearly with temperature following two distinct regimes over the 300–500 and 500–600 °C discrete temperature ranges. *In situ* neutron powder diffraction experiments showed

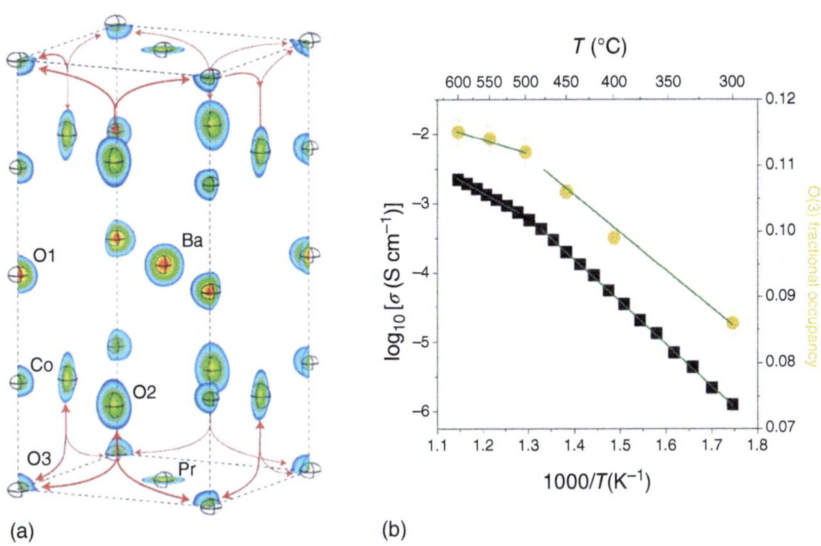

Figure 13.3 (a) Nuclear density and the corresponding determined atomic displacement parameters (20% probability) for $PrBaCo_2O_{5+\delta}$ (with space group *P4/mmm*) under 10^{-1} atm oxygen and at 573 °C. Arrows indicate the oxygen transport pathway. Source: Cox-Galhotra et al. [67]/Elsevier. (b) Thermal variation of the fractional occupancy of O(3) and the Arrhenius plot of the bulk conductivity in $Ba_3MoNbO_{8.5}$. Source: Fop et al. [68]/with permission of The Royal Society of Chemistry.

that rearrangements among the Mo and Nb transition metals (M) and the oxygen in the crystal lattice trigger an increase of the ratio of MO_4 tetrahedra to MO_6 octahedra between 300 and 600 °C, with a modification of the rate of increase at 500 °C reflected in the slope of the conductivity. Such direct structure–property relationship is well represented in Figure 13.3b [68].

13.4 Materials Formation and Synthesis

Synthesizing novel materials with enhanced performance in a wide variety of applications may enable technology leaps. Transition metal oxides feature in many discoveries, including lithium ion battery electrode materials such as $LiFePO_4$ [70] and SOFC cathodes such as $Ba_{0.5}Sr_{0.5}Co_{0.8}Fe_{0.2}O_{3-\delta}$ [71]. While there is a widely accepted basic understanding of design principles established for the synthesis of organic molecular solids, the synthesis of many inorganic materials, including oxides, often relies on serendipity or trial-and-error methods. More recently, theoretical calculations have made great strides in predicting new structures with exceptional properties, but have fallen short in directing synthesis routes. As a result, there is currently substantial activity in using *in situ* diffraction as a tool to study phase stability, intermediate phase formation, and reaction kinetics, with the information gained accelerating the prediction-synthesis-measurement feedback loop toward obtaining materials with optimal performance. While the advantages of using neutrons for this purpose are obvious, one significant disadvantage is the typically lower flux offered by neutron sources compared to synchrotron X-ray sources, often limiting the time resolution to minutes rather than seconds and making extremely fast kinetics difficult to follow.

A rare example of where this limitation may be overcome is in the neutron powder diffraction study using stroboscopic approaches, such as achieved for piezoelectric materials. Piezoelectric transducers in fuel injectors are shown to reduce fuel consumption and CO_2 emissions for internal combustion engines, and these materials exhibit a structural response to electric field. Stroboscopic data acquisition of the commercial lead zirconate titanate on the microsecond timescale under 100 Hz oscillating field was possible using the Wombat detector, and correlated to the electric field, enabling reconstruction of data with sufficient counting statistics in each field state to establish structure [72].

Another potential complication of the application of neutron diffraction in this area is the need for deuteration, particularly in the case of hydrothermal or solvothermal reactions, which is not always possible. On the other hand, large sample volumes can easily be probed using neutrons, which is more representative of scaled-up synthesis that is crucial for further development of applied methods.

The development of suitable sample environment and holders for this research is also challenging, as materials tend to be reactive under these conditions, particularly for solid-state methods often requiring high temperature. Nevertheless, many sample environments for *in situ* neutron diffraction studies of materials synthesis have been developed and reported across various neutron facilities, including targeting solid-state [73–77], hydrothermal/solvothermal [78–83], and ion-exchange [84, 85] reactions. Rao et al. [74] investigated the phase formation mechanism of the $Li_7La_3Zr_3O_{12}$ (LLZO) fast ionic conductor useful as an electrolyte in all-solid-state lithium ion batteries. Using *in situ* neutron diffraction, they found that temperatures over 1000 °C negatively affect the ionic conductivity of the material, which they associated with its partial decomposition at this temperature. The data allowed the lithium occupancy within the phase of various doped LLZO samples to be determined as a function of temperature, which was linked to the ionic conductivity during cooling. In another work, Walton et al. [79] reported the hydrothermal crystallization of perovskite $BaTiO_3$ using neutron diffraction, noting the dominance of a dissolution–precipitation mechanism as the isotropic crystal grows in the solution. Besides the above chemical reactions, ion-exchange processes can also be studied using *in situ* neutron diffraction, such as by Liu et al. [85] who tracked the Li^+/Na^+ exchange process that produces the $Li_2Mg_2P_3O_9N$ lithium ion battery electrode material from $Na_2Mg_2P_3O_9N$. Using Rietveld refinement, a complex phase transition mechanism involving thermodynamically distinct solid solutions, Li^+/Na^+ displacement, and Li^+ migration was revealed to underpin this process.

13.5 Short-Range Structure

Deviations from the long-range ordered structure cause diffuse scattering, which carries a wealth of structural information. The PDF described in the introduction can be obtained from total scattering experiments that simultaneously measure both Bragg and diffuse scattering data. This technique is now extensively used for the structural study of transition metal oxides and other compounds. The high-voltage $LiNi_{0.5}Mn_{1.5}O_4$ electrode material (LNMO) for lithium ion batteries described in Section 13.1.1 is a good example of where short-range structure information is required to understand material function. It was generally accepted that in the disordered phase Ni and Mn are randomly distributed at 16*d* crystallographic sites of the $Fd\bar{3}m$ space group symmetry structure, as opposed to the $P4_332$ space group symmetry structure when Ni and Mn are ordered onto separate sites [9, 86]. Shin et al. observed diffuse scattering from the ordered phase and attributed this to domains of Ni and Mn ordering in the bulk disordered sample [10]. These broad features grow into Bragg peaks as a function of annealing time as these ordered domains grow. Similar features were also reported by Casas-Cabanas et al. [87], who revealed the existence of antiphase boundaries. A later PDF study by Liu et al. [88] addressed questions concerning the local Ni and Mn ordering in disordered and ordered $LiNi_{0.5}Mn_{1.5}O_4$ samples. It was revealed that even in the disordered phase the short-range cation ordering between Ni and Mn occurs within the length scale of a few unit cells, while long-range disorder persists. Short-range transition metal ordering that is disordered at long-range was also observed in $LiMn_xCo_yNi_zO_2$ layered oxide materials using neutron PDF, noting the useful association with ^6Li magic angle spinning nuclear magnetic resonance spectroscopy to determine the local environments around lithium to restrict the simulated model in reverse Monte

Carlo (RMC) calculations [89]. Neutron PDF has also been applied to $LiFePO_4$ phosphates, with authors noting the desirability for *in situ* or *in operando* PDF techniques, which remain technically challenging [90].

Combined with other techniques, neutron-based PDF methods can solve complex structures across the full scale of local-medium–long-range ordering. For instance, Diaz-Lopez et al. [91] demonstrated that RMC approaches can provide a medium-range model with full atomic ordering for $Li_4Mn_2O_5$ using combined PDF and near-edge X-ray absorption spectroscopy data, noting that the $Li_4Mn_2O_5$ material delivers record reversible capacity as a lithium ion battery electrode. Based on the RMC-refined structure, the derived bond valence site energy (BVSE) map allowed visualization of a three-dimension Li-diffusion pathway. Electrode materials for SOFCs have also been investigated using neutron PDF methods. For example, Norberg et al. [92] studied the average and local structure of Bi_2O_3 using both Bragg and diffuse scattering and RMC methods, with this work highlighting the nature of anion disorder within the highly conducting δ phase and its symmetry relationship to the ordered α and β phases. In the solid solution system, $Zr_{0.8}Sc_{0.2-x}Y_xO_{1.9}$ ($0.0 \leq x \leq 0.2$) neutron PDF was combined with molecular dynamics (MD) simulations and RMC analysis to assess short-range ordering of anion vacancies, which unveiled the ionic diffusion process [92].

Although *in operando* neutron powder diffraction has been successfully applied to studies of lithium ion battery and SOFC materials, the application of *in situ/in operando* neutron PDF analysis in these areas is still in its infancy, with PDF examples in this area being performed using total scattering in synchrotron sources [93–98]. Thus far, *in situ* neutron PDF has been used to study transition metal oxide materials of interest for electrochemical energy storage to build a picture of the important structure–property relations, through approaches that include, for example, studying a battery electrode material that is equilibrated at a known state of charge. PDF studies usually require a background subtraction, which is challenging in the *in operando* case, particularly for neutron PDF studies where corrections for inelastic and incoherent neutron scattering are likely to be of importance. These challenges have slowed the development of the neutron PDF technique for the analysis of transition metal oxides in electrochemical applications.

13.6 Outlook

It is anticipated that neutron scattering and especially neutron diffraction, sensitive to both long- and short-range ordering, will continue to play a significant role regarding *in situ* and *in operando* studies of electrochemical storage materials containing transition metals. The historical limitations of neutron powder diffraction for such analyses, including the need for large samples and lengthy data collection times, have been overcome largely by the availability of high-power spallation sources and reactors, and improvement in detector technology leading to the ability to use stroboscopic methods. Extensive work is being carried out at the neutron scattering facilities to develop both sample environment and data analysis tools to tackle some of the challenges arising from this new era of experimental science. In the past, neutron diffraction has made a huge contribution to the study of strongly correlated materials owing to the ability to characterize materials under ultra-low temperatures and/or under magnetic fields. It is expected that, in the future, similar strides will be achieved in studying chemistry, allowing advancements in understanding and leading to vastly improved energy storage materials.

Acknowledgments

This research has been carried out with the support of the Australian Research Council through Discovery Projects DP170102406, DP200101862, and Linkage Project LP160101629. Part of this manuscript has been authored by UT-Battelle, LLC, under contract DE-AC05-00OR22725 with the US Department of Energy (DOE). The publisher, by accepting the article for publication, acknowledges that the US government retains a nonexclusive, paid-up, irrevocable, worldwide license to publish or reproduce the published form of this manuscript, or allow others to do so, for US government purposes. DOE will provide public access to these results of federally sponsored research in accordance with the DOE Public Access Plan (http://energy.gov/downloads/doe-public-access-plan).

References

1 Rietveld, H.M. (1969). A profile refinement method for nuclear and magnetic structures. *Journal of Applied Crystallography* 2: 65.

2 David, W.I.F., Shankland, K., Mc Cusker, L.L., and Baerlocher, C. (2002). *Structure Determination from Powder Diffraction Data*. Oxford: Oxford University Press.

3 Clearfield, A., Reibenspies, J.H., and Bhuvanesh, N. (2008). *Principles and Applications of Powder Diffraction*. Oxford: Blackwell.

4 Dinnebier, R.E. and Billinge, S.J.L. (2008). *Powder Diffraction – Theory and Practice*. Cambridge: Royal Society of Chemistry.

5 Oszlányi, G. and Sütö, A. (2004). Ab initio structure solution by charge flipping. *Acta Crystallographica Section A: Foundations* 60: 134.

6 Billinge, S.J.L. and Levin, I. (2007). The problem with determining atomic structure at the nanoscale. *Science* 316 (5824): 561–565.

7 Egami, T. and Billinge, S.J.L. (2003). *Underneath the Bragg Peaks: Structural Analysis of Complex Materials*. Oxford: Pergamon Press Elsevier.

8 Shin, J.F., Xu, W., Zanella, M. et al. (2017). Self-assembled dynamic perovskite composite cathodes for intermediate temperature solid oxide fuel cells. *Nature Energy* 2: 16214.

9 Kim, J.-H., Huq, A., Chi, M. et al. (2014). Integrated nano-domains of disordered and ordered spinel phases in $LiNi_{0.5}Mn_{1.5}O_4$ for Li-ion batteries. *Chemistry of Materials* 26: 4377–4386.

10 Shin, D.W., Bridges, C.A., Huq, A. et al. (2012). Role of cation ordering and surface segregation in high-voltage spinel $LiMn_{1.5}Ni_{0.5-x}M_xO_4$ (M = Cr, Fe, and Ga) cathodes for lithium-ion batteries. *Chemistry of Materials* 24 (19): 3720–3731.

11 Cabana, J., Casas-Cabanas, M., Omenya, F.O. et al. (2012). Composition-structure relationships in the Li-ion battery electrode material $LiNi_{0.5}Mn_{1.5}O_4$. *Chemistry of Materials* 24: 2952–2964.

12 Eckold, G., Gibhardt, H., Caspary, D. et al. (2003). Stroboscopic neutron diffraction from spatially modulated systems. *Zeitschrift fur Kristallographie* 218: 144.

13 Jones, J.L., Hoffman, M., Daniels, J.E., and Studer, A.J. (2006). Direct measurement of the domain switching contribution to the dynamic piezoelectric response in ferroelectric ceramics. *Applied Physics Letters* 89: 092901.

14 Stinton, G.W. and Evans, J.S.O. (2007). Parametric Rietveld refinement. *Journal of Applied Crystallography* 40: 87.

15 Ali, R. and Yashima, M. (2005). Space group and crystal structure of the perovskite $CaTiO_3$ from 296 to 1720 K. *Journal of Solid State Chemistry* 178 (9): 2867–2872.

16 David, W.I.F., Evans, J.S.O., and Sleight, A.W. (1999). Direct evidence for a low-frequency phonon mode mechanism in the negative thermal expansion compound ZrW_2O_8. *Europhysics Letters* 46 (5): 661.

17 Owejan, J.E., Owejan, J.P., DeCaluwe, S.C., and Dura, J.A. (2012). Solid electrolyte interphase in Li-ion batteries: evolving structures measured in situ by neutron reflectometry. *Chemistry of Materials* 24 (11): 2133–2140.

18 Lee, C.H., Dura, J.A., LeBar, A., and DeCaluwe, S.C. (2018). Direct, operando observation of the bilayer solid electrolyte interphase structure: electrolyte reduction on a non-intercalating electrode. *Journal of Power Sources* 412: 725–735.

19 Jerliu, B., Huger, E., Horisberger, M. et al. (2017). Irreversible lithium storage during lithiation of amorphous silicon thin film electrodes studied by in-situ neutron reflectometry. *Journal of Power Sources* 359: 415–421.

20 Hirayama, M., Yonemura, M., Suzuki, K. et al. (2010). Surface characterization of $LiFePO_4$ epitaxial thin films by X-ray/neutron reflectometry. *Electrochemistry* 78 (5): 413–415.

21 Browning, J.F., Baggetto, L., Jungjohann, K.L. et al. (2014). In situ determination of the liquid/solid interface thickness and composition for the Li ion cathode $LiMn_{1.5}Ni_{0.5}O_4$. *ACS Applied Materials & Interfaces* 6: 18569–18576.

22 Jafta, C.J., Sun, X.-G., Veith, G.M. et al. (2019). Probing microstructure and electrolyte concentration dependent cell chemistry via operando small angle neutron scattering. *Energy & Environmental Science* 12: 1866–1877.

23 Hattendorff, J., Seidlmayer, S., Gasteiger, H.A., and Gilles, R. (2020). Li-ion half-cells studied operando during cycling by small-angle neutron scattering. *Journal of Applied Crystallography* 53: 210–221.

24 Seidlmayer, S., Hattendorff, J., Buchberger, I. et al. (2015). In operando small-angle neutron scattering (SANS) on Li-ion batteries. *Journal of the Electrochemical Society* 162 (2): A3116–A3125.

25 Peterson, V.K. and Kearley, G.J. (2015). *Neutron Applications in Materials for Energy: An Overview*. Springer International Publishing.

26 Liu, X., Chen, Y., Hood, Z.D. et al. (2019). Elucidating the mobility of H^+ and Li^+ ions in $(Li_{6.25-x}H_xAl_{0.25})La_3Zr_2O_{12}$ via correlative neutron and electron spectroscopy. *Energy & Environmental Science* https://doi.org/10.1039/C1038EE02981D.

27 Klenk, M.J., Boeberitz, S.E., Dai, J. et al. (2017). Lithium self-diffusion in a model lithium garnet oxide $Li_5La_3Ta_2O_{12}$: a combined quasi-elastic neutron scattering and molecular dynamics study. *Solid State Ionics* 312: 1–7.

28 Wind, J., Mole, R.A., Yu, D., and Ling, C.D. (2017). Liquid-like ionic diffusion in solid bismuth oxide revealed by coherent quasielastic neutron scattering. *Chemistry of Materials* 29 (17): 7408–7415.

29 Bacon, G.E. (1975). *Neutron Diffraction*, 3e. Oxford: Clarendon Press.

30 Hansen, T.C. and Kohlmann, H. (2014). Chemical reactions followed by in situ neutron powder diffraction. *Zeitschrift fur Anorganische und Allgemeine Chemie* 641 (15): 3044–3063.

31 Bañares, M.A. (2005). Operando methodology: combination of in situ spectroscopy and simultaneous activity measurements under catalytic reaction conditions. *Catalysis Today* 100: 71–77.

32 Peterson, V.K. and Papadakis, C.M. (2015). Functional materials analysis using in situ and in operando X-ray and neutron scattering. *IUCrJ* 2: 292–304.

33 Peterson, V.K., Auckett, J.E., and Pang, W.K. (2017). Real-time powder diffraction studies of energy materials under non-equilibrium conditions. *IUCrJ* 4: 540–554.

34 Sharma, N., Pang, W.K., Guo, Z., and Peterson, V.K. (2015). In situ powder diffraction studies of electrode materials in rechargeable batteries. *ChemSusChem* 8 (17): 2626–2853.

35 Sharma, N., Peterson, V.K., Elcombe, M.M. et al. (2010). Structural changes in a commercial lithium-ion battery during electrochemical cycling: an in situ neutron diffraction study. *Journal of Power Sources* 195 (24): 8528–8266.

36 Liang, G., Didier, C., Guo, Z. et al. (2019). Understanding rechargeable battery function using in operando neutron powder diffraction. *Advanced Materials* 32 (18), 1904528.

37 Bianchini, M., Leriche, J.B., Laborier, J.-L. et al. (2013). A new null matrix electrochemical cell for Rietveld refinements of in-situ or operando neutron powder diffraction data. *Journal of the Electrochemical Society* 160 (11): A2176–A2183.

38 Vitoux, L., Reichardt, M., Sallard, S. et al. (2018). A cylindrical cell for operando neutron diffraction of Li-ion battery electrode materials. *Frontiers in Energy Research* 6: 76.

39 Baran, V., Mulhbauer, M.J., Schulz, M. et al. (2019). In operando studies of rotating prismatic Li-ion batteries using monochromatic wide-angle neutron diffraction. *Journal of Energy Storage* 24: 100772.

40 Goonetilleke, D., Sharma, N., Pang, W.-K. et al. (2019). Structural evolution and high-voltage structural stability of $Li(Ni_xMn_yCo_z)O_2$ electrodes. *Chemistry of Materials* 31: 376–386.

41 Dolotko, O., Senyshyn, A., Mülhbauer, M.J. et al. (2014). Understanding structural changes in NMC Li-ion cells by in situ neutron diffraction. *Journal of Power Sources* 255: 197–203.

42 Senyshyn, A., Dolotko, O., Muhlbauer, M.J. et al. (2013). Lithium intercalation into graphitic carbons revisited: experimental evidence for twisted bilayer behaviour. *Journal of the Electrochemical Society* 160 (5): A3198–A3205.

43 Zinth, V., von Lüders, C., Hofmann, M. et al. (2014). Lithium plating in lithium-ion batteries at sub-ambient temperatures investigated by in situ neutron diffraction. *Journal of Power Sources* 271: 152–159.

44 Taminato, S., Yonemura, M., Shiotani, S. et al. (2016). Real-time observations of lithium battery reactions – operando neutron diffraction analysis during practical operation. *Scientific Reports* 6: 28843.

45 Lee, P.-H., Wu, S.-H., Pang, W.K., and Peterson, V.K. (2018). The storage degradation of an 18650 commercial cell studied using neutron powder diffraction. *Journal of Power Sources* 374: 31–39.

46 Alam, M., Hanley, T., Pang, W.K. et al. (2014). Comparison of the so-called CGR and NCR cathodes in commercial lithium-ion batteries using in situ neutron powder diffraction. *Powder Diffraction* 29: S35–S39.

47 Rodriguez, M.A., Van Banthen, M.H., Ingersoll, D. et al. (2010). In situ analysis of $LiFePO_4$ batteries: signal extraction by multivariate analysis. *Advances in X-Ray Analysis* 53: 57–65.

48 Boulet-Roblin, L., Borel, P., Sheptyakov, D. et al. (2016). Operando neutron powder diffraction using cylindrical cell design: the case of $LiNi_{0.5}Mn_{1.5}O_4$ vs. graphite. *Journal of Physical Chemistry C* 120 (31): 17268–17273.

49 Roberts, M., Biendicho, J.J., Hull, S. et al. (2013). Design of a new lithium ion battery test cell for in-situ neutron diffraction measurements. *Journal of Power Sources* 226: 249–255.

50 Sharma, N., Du, G., Studer, A.J. et al. (2011). In-situ neutron diffraction study of the MoS_2 anode using a custom-built Li-ion battery. *Solid State Ionics* 199–200: 37–43.

51 Godbole, V.A., Hess, M., Villevieille, C. et al. (2012). Circular in situ neutron powder diffraction cell for study of reaction mechanism in electrode materials for Li-ion batteries. *RSC Advances* 3: 757.

52 Bianchini, M., Fauth, F., Suard, E. et al. (2015). Spinel materials for Li-ion batteries: new insights obtained by operando neutron and synchrotron X-ray diffraction. *Acta Crystallographica Section B: Structural Science* B71: 688–701.

53 Song, B., Veith, G.M., Park, J. et al. (2019). Metastable $Li_{1+\delta}Mn_2O_4$ $(0 \leq \delta \leq 1)$ spinel phases revealed by in operando neutron diffraction and first-principles calculations. *Chemistry of Materials* 31: 124–134.

54 Sharma, N., Yu, D., Zhu, Y. et al. (2013). Non-equilibrium structural evolution of the lithium-rich $Li_{1+y}Mn_2O_4$ cathode within a battery. *Chemistry of Materials* 25: 754–760.

55 Pang, W.K., Peterson, V.K., Sharma, N. et al. (2014). Lithium migration in $Li_4Ti_5O_{12}$ studied using in situ neutron powder diffraction. *Chemistry of Materials* 26: 2318–2326.

56 Khushalchand Shukla, A., Ramasse, Q.M., Ophus, C. et al. (2015). Unravelling structural ambiguities in lithium- and manganese-rich transition metal oxides. *Nature Communications* 6: 8711.

57 Yu, H., Ishikawa, R., So, Y.-G. et al. (2013). Direct atomic-resolution observation of two phases in the $Li_{1.2}Mn_{0.567}Ni_{0.166}Co_{0.067}O_2$ cathode material for lithium-ion batteries. *Angewandte Chemie International Edition* 52 (23): 5969–5973.

58 Chen, C.-J., Pang, W.K., Mori, T. et al. (2016). The origin of capacity fade in the $Li_2MnO_3{\cdot}LiMO_2$ (M = Li, Ni, Co, Mn) microsphere positive electrode: an operando neutron diffraction and transmission X-ray microscopy study. *Journal of the American Chemical Society* 138: 8824–8833.

59 Pang, W.K., Kalluri, S., Peterson, V.K. et al. (2014). Electrochemistry and structure of the cobalt-free $Li_{1+x}MO_2$ (M = Li, Ni, Mn, Fe) composite cathode. *Physical Chemistry Chemical Physics* 16: 25377.

60 Tamimi, M.A. and McIntosh, S. (2014). High temperature in situ neutron powder diffraction of oxides. *Journal of Materials Chemistry A* 2 (17): 6015–6026.

61 Kirkham, M., Heroux, L., Ruiz-Rodriguez, M., and Huq, A. (2018). AGES: automated gas environment system for in situ neutron powder diffraction. *Review of Scientific Instruments* 89 (9): 092904.

62 Olds, D., Mills, R.A., McDonnell, M.T. et al. (2018). A high temperature gas flow environment for neutron total scattering studies of complex materials. *Review of Scientific Instruments* 89 (9): 092906.

63 Wang, H.W., Fanelli, V.R., Reiche, H.M. et al. (2014). Pressure/temperature fluid cell apparatus for the neutron powder diffractometer instrument: probing atomic structure in situ. *Review of Scientific Instruments* 85 (12): 125116.

64 Makowska, M.G., Kuhn, L.T., Cleemann, L.N. et al. (2015). Flexible sample environment for high resolution neutron imaging at high temperatures in controlled atmosphere. *Review of Scientific Instruments* 86 (12): 125109.

65 Kandemir, T., Wallacher, D., Hansen, T. et al. (2012). In situ neutron diffraction under high pressure-providing an insight into working catalysts. *Nuclear Instruments and Methods in Physics Research Section A: Accelerators, Spectrometers, Detectors and Associated Equipment* 673: 51–55.

66 al-Wahish, A., Armitage, D., al-Binni, U. et al. (2015). A new apparatus design for high temperature (up to 950 °C) quasi-elastic neutron scattering in a controlled gaseous environment. *Review of Scientific Instruments* 86 (9), 095102.

67 Cox-Galhotra, R.A., Huq, A., Hodges, J.P. et al. (2013). An in-situ neutron diffraction study of the crystal structure of $PrBaCo_2O_{5+\delta}$ at high temperature and controlled oxygen partial pressure. *Solid State Ionics* 249: 34–40.

68 Fop, S., Wildman, E.J., Irvine, J.T.S. et al. (2017). Investigation of the relationship between the structure and conductivity of the novel oxide ionic conductor $Ba_3MoNbO_{8.5}$. *Chemistry of Materials* 29 (9): 4146–4152.

69 Fop, S., Skakle, J.M.S., McLaughlin, A.C. et al. (2016). Oxide ion conductivity in the hexagonal perovskite derivative $Ba_3MoNbO_{8.5}$. *Journal of the American Chemical Society* 138 (51): 16764–16769.

70 Padhi, A.K., Nanjundaswamy, K.S., and Goodenough, J.B. (1997). Phospho-olivines as positive-electrode materials for rechargeable lithium batteries. *Journal of the Electrochemical Society* 144 (4): 1188–1194.

71 Shao, Z.P. and Haile, S.M. (2004). A high-performance cathode for the next generation of solid-oxide fuel cells. *Nature* 431 (7005): 170–173.

72 Pramanick, A., Prewitt, A.D., Cottrell, M.A. et al. (2010). In situ neutron diffraction studies of a commercial, soft lead zirconate titanate ceramic: response to electric fields and mechanical stress. *Applied Physics A* 99 (3): 557–564.

73 Kan, W.H., Huq, A., and Manthiram, A. (2015). Low-temperature synthesis, structural characterization, and electrochemistry of Ni-rich spinel-like $LiNi_{2-y}Mn_yO_4$ ($0.4 \leq y \leq 1$). *Chemistry of Materials* 27 (22): 7729–7733.

74 Rao, R.P., Gu, W.Y., Sharma, N. et al. (2015). In situ neutron diffraction monitoring of $Li_7La_3Zr_2O_{12}$ formation: toward a rational synthesis of garnet solid electrolytes. *Chemistry of Materials* 27 (8): 2903–2910.

75 Kan, W.H., Huq, A., and Manthiram, A. (2016). Exploration of a metastable normal spinel phase diagram for the quaternary Li–Ni–Mn–Co–O system. *Chemistry of Materials* 28 (6): 1832–1837.

76 Kan, W.H., Kuppan, S., Cheng, L. et al. (2017). Crystal chemistry and electrochemistry of $Li_xMn_{1.5}Ni_{0.5}O_4$ solid solution cathode materials. *Chemistry of Materials* 29 (16): 6818–6828.

77 Abeysinghe, D., Huq, A., Yeon, J. et al. (2018). In situ neutron diffraction studies of the flux crystal growth of the reduced molybdates $La_4Mo_2O_{11}$ and $Ce_4Mo_2O_{11}$: revealing unexpected mixed-valent transient intermediates and determining the sequence of events during crystal growth. *Chemistry of Materials* 30 (3): 1187–1197.

78 Walton, R.I., Francis, R.J., Halasyamani, P.S. et al. (1999). Novel apparatus for the in situ study of hydrothermal crystallizations using time-resolved neutron diffraction. *Review of Scientific Instruments* 70 (8): 3391–3396.

79 Walton, R.I., Millange, F., Smith, R.I. et al. (2001). Real time observation of the hydrothermal crystallization of barium titanate using in situ neutron powder diffraction. *Journal of the American Chemical Society* 123 (50): 12547–12555.

80 Walton, R.I., Smith, R.I., and O'Hare, D. (2001). Following the hydrothermal crystallisation of zeolites using time-resolved in situ powder neutron diffraction. *Microporous and Mesoporous Materials* 48 (1–3): 79–88.

81 Ok, K.M., O'Hare, D., Smith, R.I. et al. (2010). New large volume hydrothermal reaction cell for studying chemical processes under supercritical hydrothermal conditions using time-resolved in situ neutron diffraction. *Review of Scientific Instruments* 81 (12): 125107.

82 Ok, K.M., Lee, D.W., Smith, R.I., and O'Hare, D. (2012). Time-resolved in situ neutron diffraction under supercritical hydrothermal conditions: a study of the synthesis of $KTiOPO_4$. *Journal of the American Chemical Society* 134 (43): 17889–17891.

83 Xia, F., O'Neill, B., Ngothai, Y. et al. (2010). A thermosyphon-driven hydrothermal flow-through cell for in situ and time-resolved neutron diffraction studies. *Journal of Applied Crystallography* 43: 511–519.

84 Celestian, A.J., Parise, J.B., Smith, R.I. et al. (2007). Role of the hydroxyl-water hydrogen-bond network in structural transitions and selectivity toward cesium in $Cs_{0.38}(D_{1.08}H_{0.54})SiTi_2O_7\cdot(D_{0.86}H_{0.14})_2O$ crystalline silicotitanate. *Inorganic Chemistry* 46 (4): 1081–1089.

85 Liu, J., Whitfield, P.S., Saccomanno, M.R. et al. (2017). In situ neutron diffraction studies of the ion exchange synthesis mechanism of $Li_2Mg_2P_3O_9N$: evidence for a hidden phase transition. *Journal of the American Chemical Society* 139 (27): 9192–9202.

86 Gryffroy, D., Vandenberghe, R.E., and Legrand, E. (1991). A neutron-diffraction study of some spinel compounds containing octahedral Ni and Mn at a 1:3 ratio. *Materials Science Forum* 79: 785–790.

87 Casas-Cabanas, M., Kim, C., Rodriguez-Carvajal, J., and Cabana, J. (2016). Atomic defects during ordering transitions in $LiNi_{0.5}Mn_{1.5}O_4$ and their relationship with electrochemical properties. *Journal of Materials Chemistry A* 4 (21): 8255–8262.

88 Liu, J., Huq, A., Moorhead-Rosenberg, Z. et al. (2016). Nanoscale Ni/Mn ordering in the high voltage spinel cathode $LiNi_{0.5}Mn_{1.5}O_4$. *Chemistry of Materials* 28 (19): 6817–6821.

89 Zeng, D., Cabana, J., Breger, J. et al. (2007). Cation ordering in $Li[Ni_xMn_xCo_{(1-2x)}]O_2$-layered cathode materials: a nuclear magnetic resonance (NMR), pair distribution function, X-ray absorption spectroscopy, and electrochemical study. *Chemistry of Materials* 19: 6277–6289.

90 Slavinski, W.A., Playford, H.Y., Hull, S. et al. (2019). Neutron pair distribution function study of $FePO_4$ and $LiFePO_4$. *Chemistry of Materials* 31 (14): 5024–5034.

91 Diaz-Lopez, M., Freire, M., Joly, Y. et al. (2018). Local structure and lithium diffusion pathways in $Li_4Mn_2O_5$ high capacity cathode probed by total scattering and XANES. *Chemistry of Materials* 30 (9): 3060–3070.

92 Norberg, S.T., Hull, S., Ahmed, I. et al. (2011). Structural disorder in doped zirconias, part I: the $Zr_{0.8}Sc_{0.2-x}Y_xO_{1.9}$ $(0.0 \leq x \leq 0.2)$ system. *Chemistry of Materials* 23 (6): 1356–1364.

93 Talaie, E., Bonnick, P., Sun, X.Q. et al. (2017). Methods and protocols for electrochemical energy storage materials research. *Chemistry of Materials* 29 (1): 90–105.

94 Stratford, J.M., Mayo, M., Allan, P.K. et al. (2017). Investigating sodium storage mechanisms in tin anodes: a combined pair distribution function analysis, density functional theory, and solid-state NMR approach. *Journal of the American Chemical Society* 139 (21): 7273–7286.

95 Wiaderek, K.M., Borkiewicz, O.J., Pereira, N. et al. (2014). Mesoscale effects in electrochemical conversion: coupling of chemistry to atomic- and nanoscale structure in iron-based electrodes. *Journal of the American Chemical Society* 136 (17): 6211–6214.

96 Doan-Nguyen, V.V.T., Subrahmanyam, K.S., Butala, M.M. et al. (2016). Molybdenum polysulfide chalcogels as high-capacity, anion-redox-driven electrode materials for Li-ion batteries. *Chemistry of Materials* 28 (22): 8357–8365.

97 Borkiewicz, O.J., Shyam, B., Wiaderek, K.M. et al. (2012). The AMPIX electrochemical cell: a versatile apparatus for in situ X-ray scattering and spectroscopic measurements. *Journal of Applied Crystallography* 45: 1261–1269.

98 Chapman, K.W. (2016). Emerging *operando* and X-ray pair distribution function methods for energy materials development. *MRS Bulletin* 41 (3): 231–238.

14

Synchrotron X-ray Spectroscopy and Imaging for Metal Oxide Intercalation Cathode Chemistry

Chixia Tian[1] *and Feng Lin*[2]

[1] *Virginia Tech, Academy of Integrated Science, 800 West Campus Drive, Blacksburg, VA 24061, United States*
[2] *Virginia Tech, Department of Chemistry, 1040 Drillfield Drive, Blacksburg, VA 24061, United States*

14.1 Introduction

Electrochemical energy storage relies on ion storage and redox reactions in electrode materials. The intercalation reaction represents one of the most prolific battery chemistries that are used in practical rechargeable batteries [1]. The intercalation battery chemistry will continue to dominate the market for years to come, especially in electric vehicles. The ion intercalation can take place in many metal oxide materials that provide open lattice sites and redox centers. Ion intercalation disrupts the local geometry by modifying the bond strength, angle, and coordination. The charge-compensating electrons interact with the orbitals of the redox-active ions. Such interaction alters the local electronic structure of the metal oxides. Experimentally, these local changes can be monitored by spectroscopic and scattering methods, especially with synchrotron X-ray capabilities [2, 3]. The interaction between X-rays and matter induces numerous structural and electronic responses in the probed materials, including photoelectrons, fluorescence, Auger electrons, and scattered X-rays, all of which can be monitored for investigating the salient features of battery materials. In general, the intercalation and deintercalation of charged ions and the associated charge-compensating electrons periodically interrupt the local bonding and crystal geometry. Such interruption and its reversibility can determine the major performance metrics of a battery material. The investigation of the intercalation chemistry, and that of nearly all battery chemistries, needs to consider the modulations of electronic structure and crystal symmetry as well as the coupling between the two. The Jahn–Teller effect in Mn-containing electrode materials is a refined example of how the coupling between electronic structure and crystal symmetry may influence battery performance, such as cycle life [4, 5]. Topochemical reactions in battery materials take place at multiple length and time scales, and they are influenced by external parameters such as temperature and electrochemical protocols. The redox reactions at the surface of a material are expected to be distinct from those in the bulk, resulting in potentially different degradation pathways as a function of

Transition Metal Oxides for Electrochemical Energy Storage, First Edition.
Edited by Jagjit Nanda and Veronica Augustyn.

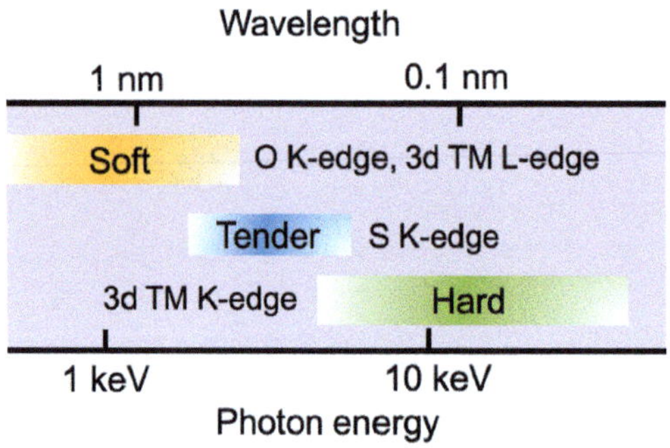

Figure 14.1 Categories of X-rays in terms of wavelength and photon energy, labeled with the accessible edges for spectroscopic measurements. The tender X-rays are located in between soft and hard X-rays, with an energy range of approximately 2–5 keV. There are overlapping regions since the "hardness" of X-rays remains broadly defined in the literature.

depth [6, 7]. Furthermore, most chemical reactions in battery particles show heterogeneous characteristics [8, 9]. For example, electrochemical reactions can initiate at a specific region of a particle that is populated with defects [10–14]. Therefore, *in situ* and *operando* monitoring of intercalation chemistry, with good spatial and temporal resolution, become critical and sometimes indispensable to understand and improve cell chemistry. It is the hope of battery scientists that materials design can go beyond the conventional trial-and-error approach, especially beyond morphological control such as engineering nanostructures. The ideal materials design can be informed by deep understanding of materials electrochemistry through advanced analytical studies [15–18]. Often times, the process of designing materials can be accelerated by computational methods [19, 20]. It becomes clear that one of the most powerful design methods for battery materials is through combining computation/prediction, synthesis, advanced characterization, and performance evaluation.

In today's synchrotron facilities, synchrotron X-rays are generated by bending fast-moving electrons by magnetic fields. The electrons travel with an energy in the GeV scale and a speed close to that of light. The advantages of synchrotron radiation include the continuously tunable photon energy, ranging from soft X-rays to hard X-rays, depending on their energy or capability of penetrating matter (Figure 14.1). Hard X-rays, with an energy of a few kiloelectron volt up to tens of kiloelectron volt, can penetrate relatively thick and high-Z matter (Z is the atomic mass), which allows for elaborate *in situ* and *operando* spectroscopic, scattering, and imaging studies of battery cells under operating conditions. Hard X-rays allow for bulk structural determination through X-ray diffraction (XRD) (crystal structure) and hard X-ray absorption spectroscopy (XAS) (electronic structure, global transition metal [TM] oxidation state, and coordination environment). Furthermore, spatially resolved X-ray imaging allows for mapping of the state-of-charge distribution, compositional distribution, and crystal defects in individual battery particles or particle ensembles [18, 21–23]. Soft X-rays range from tens of eVs up to about 2 keV and are suitable for probing the surface chemistry of battery materials [24, 25]. According to our previous studies, such depth sensitivity is ideal for studying the electrode surface chemistry (surface TM oxidation state, TM3d–O2p orbital hybridization) [6, 25–27]. However, the *in situ* and *operando* capabilities of soft X-ray-based techniques are rather limited in battery research. The intermediate X-rays between soft and hard X-rays are usually termed tender X-rays. Compared to soft and hard X-rays, tender X-rays are relatively less explored for battery research. However, tender X-rays can be excellent probes for sulfur chemistries, such as lithium sulfur batteries. For example, researchers applied S K-edge XAS to track the speciation of polysulfides as the battery underwent electrochemical cycling [28–30]. In recent years, tender X-ray photoelectron spectroscopy (XPS) has also been applied to study electrochemical interfaces under ambient pressure conditions [31].

The widespread application of synchrotron X-rays for battery chemistries did not nucleate until the pioneering work by McBreen and coworkers [32–36]. In the past 20 years, the techniques have expanded from simple scattering and absorption to a rich family of sophisticated and powerful analytical techniques that were highlighted in a review in 2017 [2]. Importantly, the experimental design and data analysis methods have advanced drastically over the last few years. For example, machine-learning and data-mining methods have been reported by several groups to obtain previously unnoticed but functionally important minor phases in battery cells [37]. A complete summary of the synchrotron X-ray/battery field is beyond the scope of this chapter and probably deserves a book if not more. Therefore, in this chapter, we focus on several spectroscopy and spectroscopic imaging techniques that are applied to the intercalation chemistry of metal oxide cathode materials. We specifically highlight soft and hard XAS, full-field X-ray imaging, and tomographic imaging. With regard to the intercalation chemistry of metal oxides, we limit the discussion to only layered and spinel oxides. Readers are encouraged to refer to reviews on synchrotron X-ray techniques for battery science [2, 24, 38, 39].

14.2 X-ray Absorption Spectroscopy

14.2.1 Soft X-ray Absorption Spectroscopy

For metal oxide battery materials, soft XAS is a perfect tool for studying surface and interfacial chemistry with excellent surface sensitivity and depth profiling capability. Most metal oxide-based battery materials are within the probing range of soft XAS. Specifically, probing the occupancies of TM 3d and oxygen (O) 2p orbitals can provide information on the charge compensation mechanism at the particle surface. The interaction between the incident X-rays and battery materials can generate different signals that enable different detection modes. The most popular detection modes include Auger electron yield (AEY), total electron yield (TEY), and fluorescence yield (FY), as shown in Figure 14.2. Due to the limited escape length of Auger electrons, the AEY mode probes the top 2 nm at the surface. Upon X-ray excitation, electrons escape the surface leaving behind a positively charged material. The electrons from an external circuit flow into the sample to neutralize the positively charged material (i.e. TEY mode), which can be measured using the sample current. The TEY mode typically probes metal oxides from the top 10 nm at the surface. The FY mode measures the fluorescence with an energy equivalent to the energy difference between the ground and excited states. The FY mode typically probes around 50–100 nm of the surface. Given that most commercially used

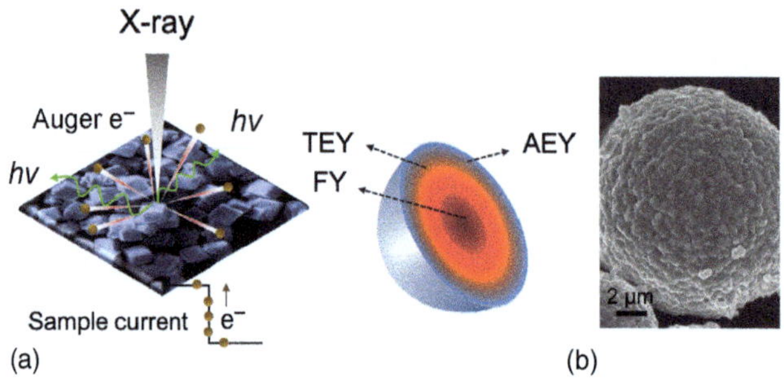

Figure 14.2 (a) Schematic representation of the soft XAS signal collection and the corresponding probing depth using three popular detection modes: Auger electron yield (AEY, 1–2 nm), total electron yield (TEY, 2–10 nm), and fluorescence yield (FY, 50–100 nm) Source: Lin, F. et al., [25]/Royal Society of Chemistry. (b) A typical oxide cathode secondary particle (tens of micrometers) that consists of nanosized primary particles (a few hundred nanometers). The soft XAS detects the surface and subsurface regions of oxide cathode particles Source: Lin et. al [25]/with permission of Royal Society of Chemistry.

intercalating metal oxide electrode materials have particle sizes above 100 nm, the FY mode is effectively a subsurface-sensitive technique compared to hard XAS. The soft XAS directly measures the dipole-allowed 2p to 3d transition for TM L-edges and 1s to 2p transition for O K-edge, which makes soft XAS a perfect method to probe the charging mechanism of intercalating metal oxide electrodes. In addition, interfacial reactions between the electrode surface and the electrolyte involve electron and oxygen transfer as well as redox phase transformation [6, 26, 40]. These changes are detectable by soft XAS. Since the electronic structure and crystal symmetry are coupled in this case, one may use the electronic information probed by soft XAS to precisely infer the crystal symmetry at the electrode surface. In addition, soft XAS is an ensemble-averaged technique that can overcome the statistic challenge of spatially resolved imaging techniques such as transmission electron microscopy (TEM). Lin and coworkers investigated the surface degradation of $LiNi_xMn_yCo_zO_2$ (NMC) materials under high-voltage cycling (Figure 14.3) [6]. The buildup of the surface reconstruction layer, primarily in the form of the rocksalt phase (Figure 14.3a–c), resulted in a reduced surface that was measured by soft XAS TM L-edge (Figure 14.3d,e). It was found that the surface-reduced layer continued to grow in the first 20 cycles, as monitored by the gradually increasing low-energy shoulders at the Mn and Co L-edges (Figure 14.3d,e). Such a surface reconstruction layer contributed to the impedance growth of the cell.

Since the TM ion mobility changes as the TM undergoes reduction, surface reconstruction can potentially lead to different degrees of TM dissolution into the electrolyte. For example, Mn^{2+} is more soluble than Mn^{4+} in the conventional carbonate-based electrolyte. Recently, using soft XAS and XPS, Lin and coworkers reported Mn dissolution and deposition on the anode surface for sodium ion cathode materials (Figure 14.4a–c) [27]. There was a positive correlation between the Mn deposition on the anode surface and the Mn reduction on the cathode surface. The surface reconstruction can be potentially more significant when the particle size gets smaller or when chemomechanical failure (i.e. cracks) leads to

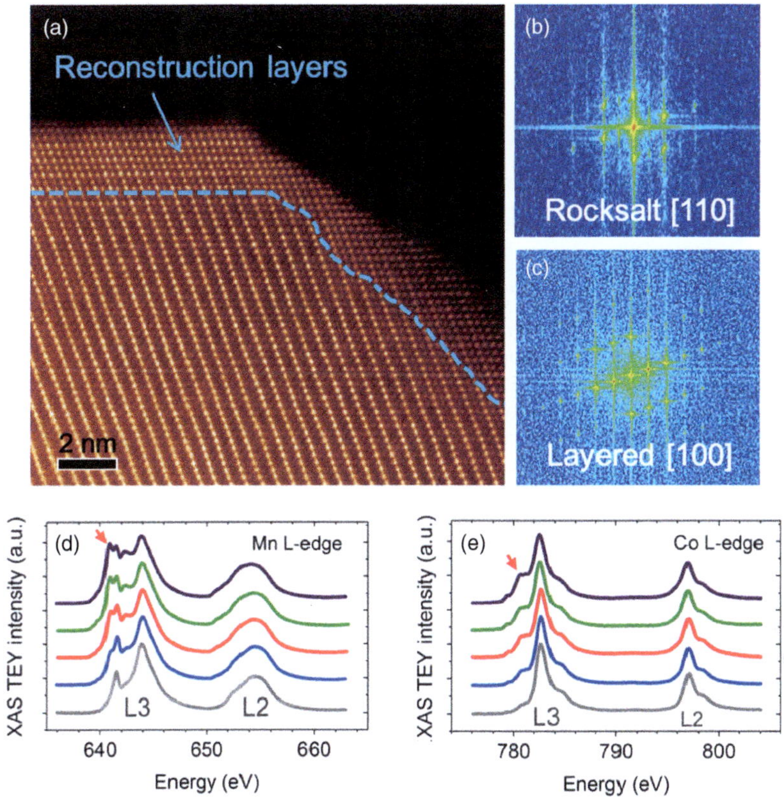

Figure 14.3 Surface reconstruction in NMC-layered oxide materials studied by coupling scanning transmission electron microscopy and soft XAS. (a) STEM image of an NMC primary particle after one cycle (2.0–4.7 V vs. Li/Li^+); the blue arrow indicates the surface reconstruction layer. (b, c) Fast Fourier transform (FFT) results showing the surface reconstruction rocksalt layer ([110] zone axis) and the NMC-layered structure ([100] zone axis), respectively in (b). (d) Soft XAS Mn L-edge of the NMC electrode as a function of cycle number. (e) Soft XAS Co. L-edge of the NMC electrode as a function of cycle number. The spectra in (d, e) were collected from pristine, and after 1, 2, 5, 20 cycles, from bottom to top. The growing low-energy shoulders, indicated by the red arrows, represent the reduction of Mn and Co. Source: Lin et al. [6]/with permission of Springer Nature.

more exposed surfaces (Figure 14.4d,e). Furthermore, the surface metal reduction and reconstruction can take place simply by soaking the cathode particles in the electrolyte or electrolyte solvent [8, 25, 41]. Mu et al. stirred $NaNi_{1/3}Fe_{1/3}Mn_{1/3}O_2$ particles in the electrolyte solvent to generate the surface reconstruction layer prior to fabricating the particles into an electrode (Figure 14.5a,b) [41]. The authors found that the pre-passivated surface indeed lowered the initial capacity because of the impedance buildup. However, the pre-passivated surface also allowed for better cycle life (Figure 14.5c,d). It was likely that the surface reconstruction layer produced by the pre-passivation not only created impedance for sodium intercalation/deintercalation but also slowed down the interfacial side reactions. The authors also found that the effectiveness of Ti doping in extending the cycle life of

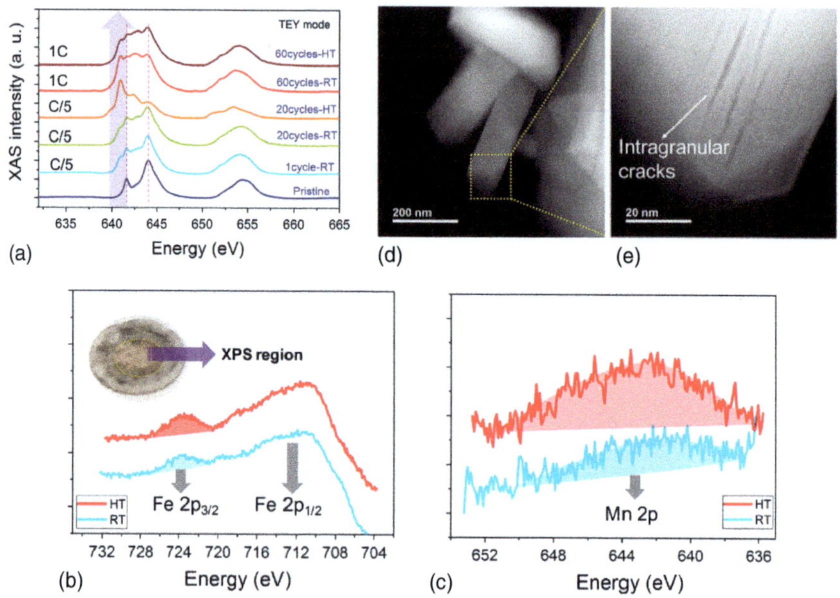

Figure 14.4 Evolution of the electronic structure and morphology of O3-$NaNi_{1/3}Fe_{1/3}Mn_{1/3}O_2$ after electrochemical cycling. (a) Soft XAS Mn L-edge after various numbers of cycles; the cells cycled for 1 and 20 cycles were at C/5 in 2.0–4.0 V, whereas the cells cycled for 60 cycles were cycled at 1C in 2.0–4.0 V (HT = 45 °C, RT = 25 °C). (b) Fe2p and (c) Mn2p XPS spectra from the light-yellow region of the separator after 20 cycles at 25 °C and 45 °C; the inset shows the trace transition metal deposit with a typical light-yellow color after 20 cycles at 45 °C. The intensities of Fe 2p and Mn 2p peaks were nominalized by the C1s peaks. (d, e) STEM images of cathode particles after 60 cycles at 1C in 2.0–4.0 V. Source: Mu et al. [27]/with permission of John Wiley & Sons.

this material was due to the inhibition of Mn dissolution [27]. In the cases presented above, ensemble-averaged soft XAS provided a critical method to overcome the limitations of scanning transmission electron microscopy–electron energy loss spectroscopy (STEM–EELS) that can only probe small regions in a limited number of particles, even though STEM provides visual evidence for the structural change. Therefore, coupling soft XAS with STEM and EELS is a powerful strategy to understand the chemical and structural properties of cathode materials, especially at the surface region (Figure 14.6).

Soft XAS is also a high-resolution method to probe the state of charge (SOC) in an electrode. *In situ* and *operando* soft XAS, a method that allows direct probing of TM oxidation state, can be coupled with coulometry to study relaxation effects. To date, however, *in situ* and *operando* soft XAS have not been widely used in the battery field primarily due to the limited penetration depth of soft X-rays and short electron mean free paths. One of the successful examples was demonstrated by W. Yang's group at the Advanced Light Source, Lawrence Berkeley National Laboratory [42]. As shown in Figure 14.6a, the authors used a polyethylene oxide–based polymer electrolyte and probed the soft XAS signal from the current collector end. High-quality soft XAS

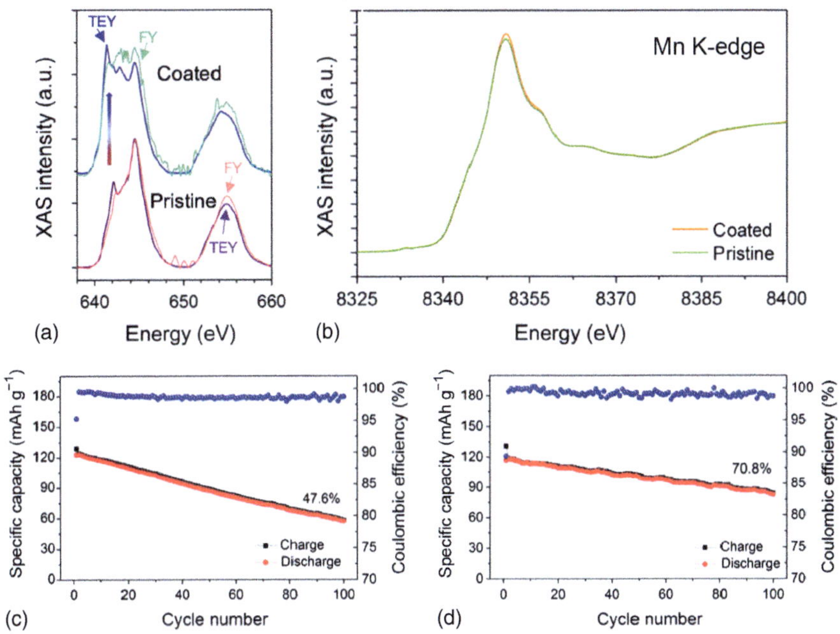

Figure 14.5 (a) Soft XAS Mn L-edge spectra collected in the FY and TEY modes for pristine and coated $NaNi_{1/3}Fe_{1/3}Mn_{1/3}O_2$ powders. (b) Hard XAS Mn K-edge spectra for pristine and coated $NaNi_{1/3}Fe_{1/3}Mn_{1/3}O_2$ powders. There was no change in the hard XAS, but there are changes in the XAS, indicating that the "cocktail" treatment only modified the surface. Long-term cycling performance for cells containing (c) pristine $NaNi_{1/3}Fe_{1/3}Mn_{1/3}O_2$ and (d) coated $NaNi_{1/3}Fe_{1/3}Mn_{1/3}O_2$ materials at 1C between 2.0 and 4.0 V. The coated $NaNi_{1/3}Fe_{1/3}Mn_{1/3}O_2$ sample was prepared by stirring the pristine $NaNi_{1/3}Fe_{1/3}Mn_{1/3}O_2$ powder in a "cocktail" solvent (weight ratio: propylene carbonate [PC]/ethylene carbonate [EC]/dimethyl carbonate [DMC]/*N*-methyl-2-pyrrolidone [NMP] = 1/1/1/1) with continuous stirring for 24 hours. Subsequently, the coated $NaNi_{1/3}Fe_{1/3}Mn_{1/3}O_2$ powder was centrifuged and dried in vacuum oven overnight at 100 °C. Source: Mu et al. [41]/with permission of Royal Society of Chemistry.

spectra in the total fluorescence yield (TFY) mode were obtained. Nickel in the NMC electrode was oxidized and reduced during charging and discharging, respectively (Figure 14.6b–d). Such an *in situ* observation was consistent with those reported using *ex situ* soft XAS [25]. The *in situ* soft XAS offers dynamic measurements that can be used to study the relaxation effect. In the same study, W. Yang and coworkers found that the NMC electrode reached the SOC uniformity across the electrode much faster than the $LiFePO_4$ electrode. For the latter, it took tens of hours to reach the uniform SOC (Figure 14.6e).

Utilizing anion redox activity has been at the frontier of battery research in the past few years, especially with the development of lithium/manganese-rich layered oxides and disordered rocksalt oxides [20, 43–45]. According to W. Yang's research, resonant inelastic X-ray scattering (RIXS) remains one of the most direct techniques to probe oxygen redox activity [46]. Nevertheless, soft XAS O K-edge can provide additional information to complete the picture, especially in the strongly correlated

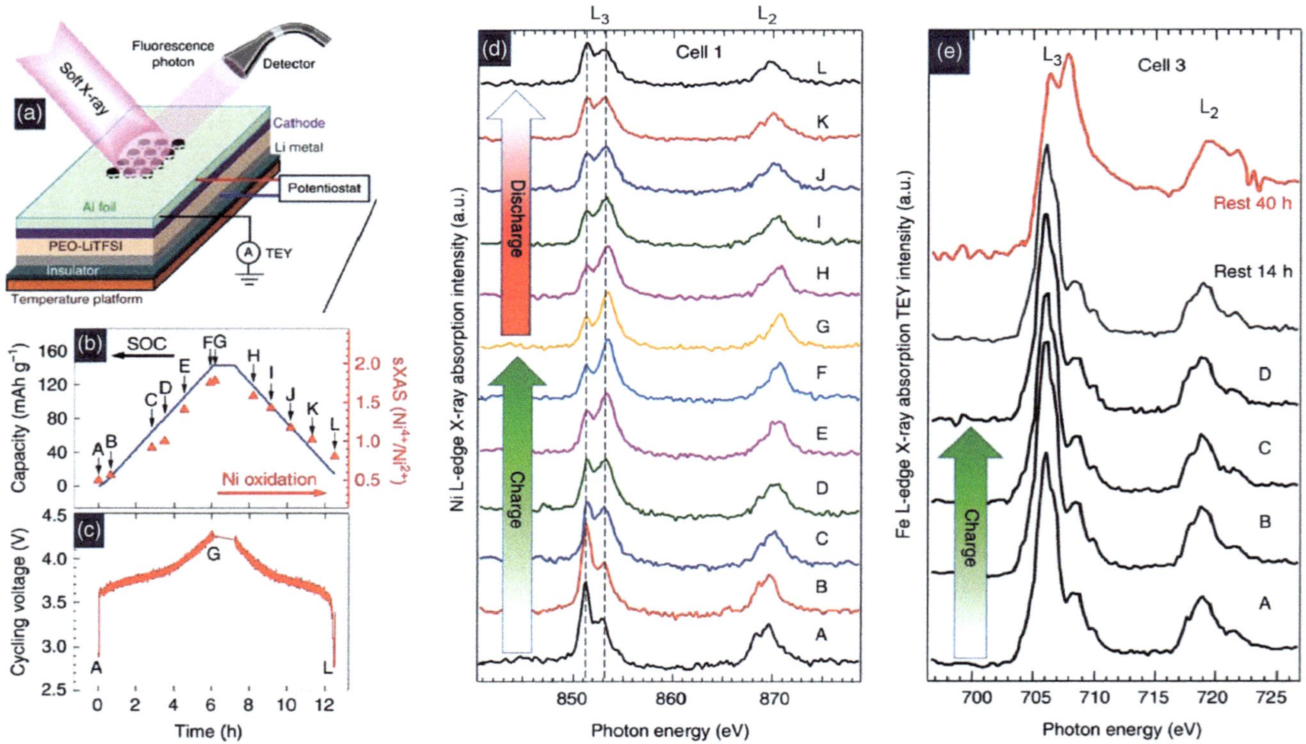

Figure 14.6 (a) Schematic representation of the *in situ* soft XAS experimental setup developed by W. Yang and coworkers. (b) The nickel oxidation state as a function of the obtained capacity calculated using the charge–discharge curves in (c) for the NMC333 material. (d) The soft XAS Ni L-edge spectra collected for the charging and discharging specified in (b, c), these spectra were used to calculate the Ni oxidation states in (b). (e) The soft XAS Fe L-edge spectra collected upon charging the $LiFePO_4$ material. The cell was rested for 14 and 40 hours to collect two additional Fe L-edge spectra. The comparison between NMC and $LiFePO_4$ showed that the latter had a significant relaxation effect. Source: Liu et al. [42]/with permission of Springer Nature.

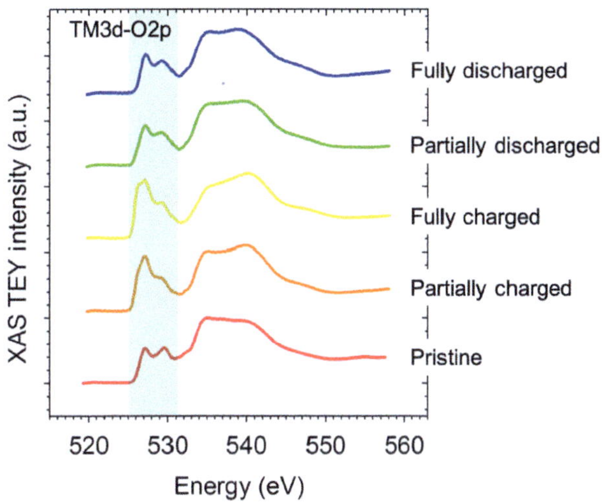

Figure 14.7 Soft XAS O K-edge spectra collected from $Li_xNi_{0.4}Mn_{0.4}Co_{0.18}Ti_{0.02}O_2$ at different states of charge: From the bottom to the top, $x \approx 1$, 0.6, 0.2, 0.7, and 0.9. The pre-edge feature associated with the TM3d-O2p hybridization is highlighted in the figure. Source: Based on Lin et al. [25]/with permission of Royal Society of Chemistry.

systems with TM3d and O2p hybridization [25, 45]. In the conventional layered oxides, such as NMC materials, the pre-edge features of the O K-edge soft XAS provide information about the covalency between TM and O, and the crystal field splitting. Figure 14.7 shows the changes of the O K-edge during charging of an NMC electrode [25]. The TM3d–O2p hybridization (covalency) got significantly enhanced upon charging and decreased upon discharging. Doping is an effective method to reduce the hybridization (covalency) between TM and O. For example, Ti and Al can form strong ionic bonding with oxygen anions and thus have been found to stabilize the battery performance of many layered oxide cathode materials. However, the direct experimental spectroscopic evidence for the reduced hybridization (covalency) has not been reported.

Although not the focus of this chapter, we would like to highlight that soft XAS has been widely used for studying the surface chemistry of oxide-based solid ionic conductors, especially for $Li_{7-x}La_3Zr_2O_{12}$ (LLZO). The surface of LLZO is highly basic and can easily react with CO_2 and H_2O from the ambient air. Cheng et al. carefully probed the LLZO surface using soft XAS and found that the surface was more populated with Li_2CO_3 when the sample was processed in air (Figure 14.8a) [47]. There was not much Li_2CO_3 in the bulk, as shown by the weaker π^*(C=O) intensity in the FY mode (Figure 14.8b). The authors then compared Li_2CO_3 formation on the large- and small-grain LLZO and found that more Li_2CO_3 was formed on the large-grain LLZO (Figure 14.8c). Such determination was made possible by comparing the C and O K-edge spectra with those of the standard Li_2CO_3 sample. Li_2CO_3 creates a large interfacial impedance between LLZO and electrodes. Subsequently, Cheng et al. applied various methods to eliminate the surface Li_2CO_3. One of the most effective and economical methods involves low-temperature thermal

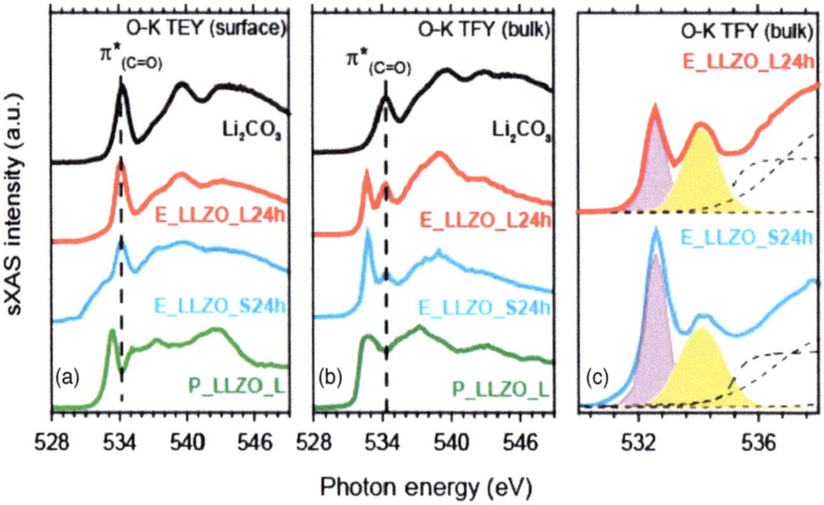

Figure 14.8 Soft XAS O K-edge spectra collected from Al-substituted LLZO with different grain sizes in the pristine state (P_LLZO_L) and exposed to air for 24 hours (E_LLZO_L24h, E_LLZO_S24h). O K-edge of Li_2CO_3 is provided as a reference spectrum. (a) TEY mode, (b) FY mode, and (c) direct comparison between large and small grains in the FY mode. The π^* transition associated with the C=O bond in CO_3^{2-} is labeled. Source: Cheng et al. [47]/with permission of American Chemical Society.

annealing [48]. Li_2CO_3 was readily decomposed at a temperature as low as 250 °C. The surface chemistry and structure of LLZO were completely restored.

Electrode–electrolyte interphases are functionally important in alkali metal ion batteries. The most studied case seems to be the solid–electrolyte interphase (SEI) on the graphite anode. Nevertheless, the nature of the SEI is still under debate. The challenge of fully understanding the SEI is associated with the complex nature of such an interphase and the lack of techniques that can resolve the multiscale heterogeneity of SEIs [49]. The cathode–electrolyte interphases (CEIs) can potentially be more complicated and dynamic. This is likely due to the poor oxidative stability of CEIs when the cathode is highly charged at high voltages. The most popular X-ray spectroscopic method for studying these interphases has been XPS due to its wide availability. However, most lab source XPS is limited to a single energy, such as Al Kα and Mg Kα. Synchrotron XPS, with tunable incident X-ray energy, can potentially provide depth-profiling of the interphases. Certainly, soft XAS, with AEY, TEY, and FY modes, can offer additional depth profiling. In addition, soft XAS is usually more sensitive to the local structure and coordination compared to XPS. Thus, we expect that the application of soft XAS for probing interphases will further increase in the coming years.

14.2.2 Hard X-ray Absorption Spectroscopy

As shown above, soft XAS offers a powerful toolkit for probing the surface and interfacial chemistry in batteries [50]. Battery electrodes are typically tens to hundreds of micrometers thick, and the surface only represents a small fraction of the electrode. To probe the chemical and structural transformations of the entire electrode, hard X-ray-based techniques, such as X-ray diffraction and hard XAS, become indispensable. Hard XAS can be generally divided into two regions, that is X-ray absorption near edge structure (XANES) and extended X-ray absorption fine structure (EXAFS) [51]. The former is typically used for analyzing the oxidation state of the probed element. The edge position and the shape of the edge can be sensitive to the local geometry. EXAFS represents the region that is typically 20–50 eV above the absorption edge. EXAFS is generated as the excited core electrons are scattered by the neighboring atoms. The constructive and destructive interferences of these electron waves contribute to the formation of EXAFS. EXAFS reveals structural and geometric information of the probed element with local specificity. The elements in amorphous samples can also be studied, and the result is typically an averaged effect. With the proper fitting, EXAFS data can be used to obtain accurate local environments, including coordination number, interatomic distance, and geometry surrounding the probed atom. Therefore, the combination of XANES and EXAFS can give the most salient information regarding the charge compensation mechanism, local environment, and structural stability. Hard XAS typically probes TM K-edges for most intercalating electrode materials. The K-edge XAS involves the excitation of 1s electrons. The first ever *in situ* hard XAS measurement was performed at Brookhaven National Lab by McBreen and coworkers in the late 1980s. Over the past 30 years, hard XAS has gradually become the go-to technique for studying battery redox chemistry. Compared to other synchrotron techniques, hard XAS measurements are relatively easier to perform and the spectra are easily interpreted with readily available software package. However, special caution should be taken when building EXAFS lattice models to investigate local structural ordering in a highly disordered material.

One may couple soft XAS with hard XAS to derive the depth-dependent charging mechanism and degradation processes. For example, the conclusion that surface degradation dominated the performance fading of NMC materials during the initial cycles was based on the combination of hard and soft XAS [6, 25, 26]. Specifically, soft XAS identified that the surface reconstruction was caused by the TM reduction and surface oxygen loss. However, there was no bulk structural or chemical changes, as probed by hard XAS. In a recent study, we used a combination of synchrotron X-ray techniques to provide one of the first comprehensive studies for the surface and bulk chemistry of sodium ion-layered cathode materials (Figure 14.9) [7]. $Na_{0.9}Cu_{0.2}Fe_{0.28}Mn_{0.52}O_2$ materials represent a class of promising materials that are stable in air and thus can lower the cost of materials storage and transport [52]. Hard XAS showed that Cu and Fe were responsible for the overall charge compensation, which was consistent with the seminal study done by Hu and coworkers [52]. The surface of the cathode material, however, underwent completely different redox processes. The surface redox process was dominated by the Mn redox and the Mn3d-O2p hybridization. Upon charging, the surface Mn, probed by soft XAS, initially underwent significant reduction (Figure 14.9b) that was opposite to the inactive nature of Mn in the bulk identified by hard XAS (Figure 14.9a). This was because the surface underwent reductive degradation. In subsequent studies,

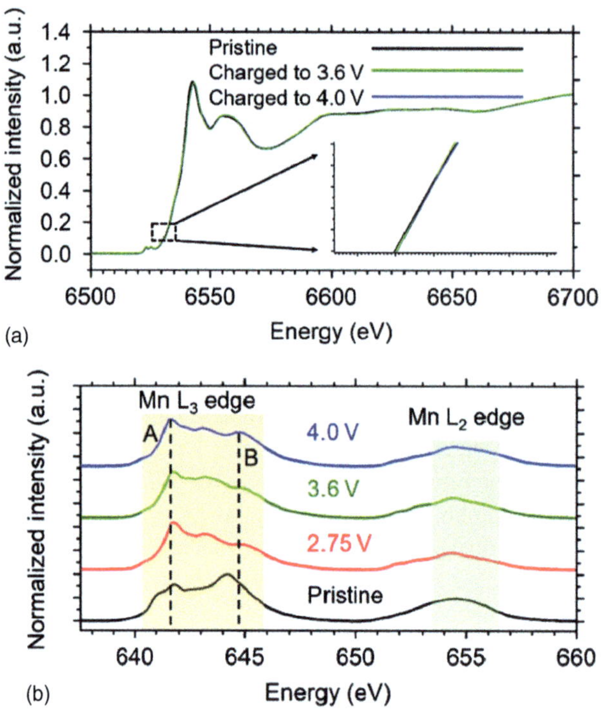

Figure 14.9 (a) Hard XAS Mn K-edge of $Na_{0.9}Cu_{0.2}Fe_{0.28}Mn_{0.52}O_2$ at different voltages during charging from open circuit voltage (OCV) to 4.0 V vs. Na/Na^+. (b) Soft XAS Mn L-edge at different voltages during charging from OCV to 4.0 V vs. Na/Na^+. The A and B features are added to show the relative intensity change between the low- and high-energy regions of the Mn L_3-edge. Source: Rahman et al. [7]/with permission of Royal Society of Chemistry.

we found that doping and surface pre-passivation enhanced the surface stability without compromising the bulk redox properties (identified by hard XAS) [27, 41]. The above examples made use of hard XAS as the auxiliary technique to resolve the role of the surface chemistry compared to that of the bulk chemistry.

Next, we discuss the key roles that hard XAS has played in understanding the bulk redox chemistry and local environments. Battery safety is a key issue for practical applications. There are numerous reasons causing battery hazards. The use of flammable liquid electrolytes is one of the major causes. However, the electrode materials, especially under heating conditions, also contribute significantly to the combustion of liquid electrolytes. Layered oxide materials can offer high energy density. At highly charged states, the TM3d–O2p hybridized states are activated due to extensive TM oxidation, which makes layered oxides unstable [53]. Such instability leads to surface reconstruction at normal battery-operating temperatures or complete bulk transformation to spinel/rocksalt phases under thermal heating conditions. Both processes result in oxygen loss. There have been extensive efforts to stabilize the oxygen anions through compositional modifications, which has led to different families of lithium-layered oxides. $LiNi_{1-x-y}Co_xAl_yO_2$ (NCA), $LiNi_{1-x-y}Mn_xCo_yO_2$ (NMC), and Co-free $LiNi_{1-x}M_xO_2$ (M = metal, but not Co)

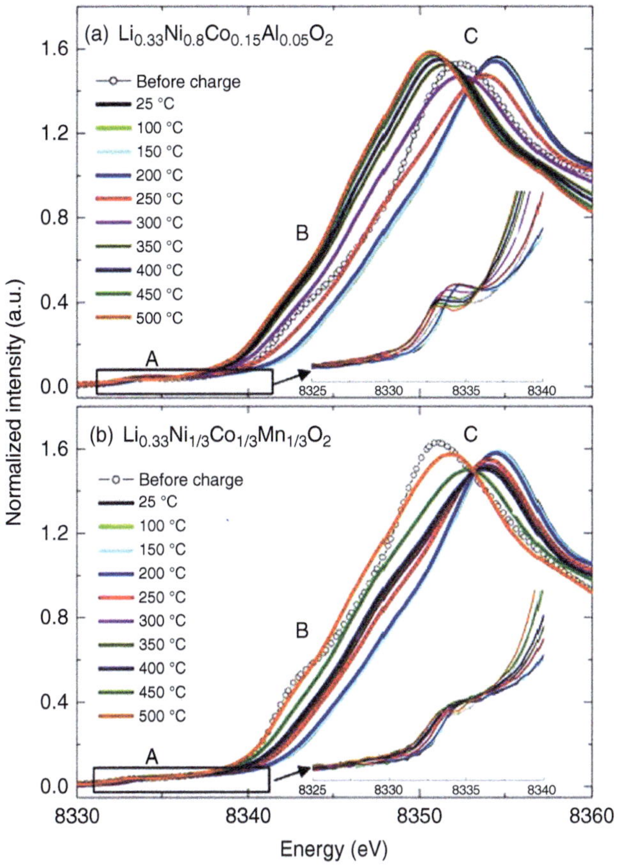

Figure 14.10 Hard XAS Ni K-edge of (a) $Li_{0.33}Ni_{0.8}Co_{0.15}Al_{0.05}O_2$ and (b) $Li_{0.33}Ni_{1/3}Co_{1/3}Mn_{1/3}O_2$ upon heating from room temperature to 500 °C. The pre-edge regions are amplified and shown as insets in the respective panel. Source: Nam et al. [54]/ with permission of John Wiley & Sons.

are among the most studied materials. Understanding how each metal cation migrates under heating conditions provides important insights into the origin of thermal instability and enables stabilization methods. X. Yang and coworkers showed the different migration pathways of different TMs in NMC and NCA materials under identical charged states and thermal conditions [54]. Figure 14.10 shows the Ni K-edge XANES of the overcharged (a) $Li_{0.33}Ni_{0.8}Co_{0.15}Al_{0.05}O_2$ and (b) $Li_{0.33}Ni_{1/3}Co_{1/3}Mn_{1/3}O_2$ electrodes during heating, a process used to mimic the thermal runaway condition. Upon heating, Ni in both samples underwent gradual reduction, as indicated by the red shift of the absorption edge. The intensity of the pre-edge feature is associated with the site symmetry of the probed atom. Typically, metals in the tetrahedral site give rise to stronger pre-edge peaks than those in the octahedral site. There was no major change in the Ni K-edge pre-edge feature, indicating that nickel remained in the octahedral site throughout the heating process. This was likely because nickel cations remained in the octahedral

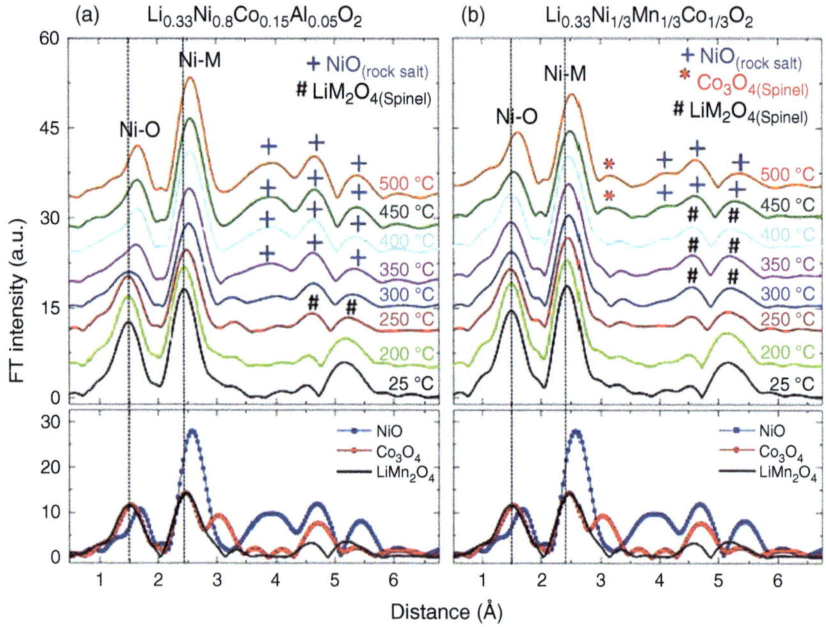

Figure 14.11 Ni K-edge EXAFS analysis of (a) $Li_{0.33}Ni_{0.8}Co_{0.15}Al_{0.05}O_2$ and (b) $Li_{0.33}Ni_{1/3}Co_{1/3}Mn_{1/3}O_2$ upon heating from room temperature to 500 °C. The reference EXAFS patterns are provided to compare with the experimental data. Source: Nam et al. [54]/with permission of John Wiley & Sons.

site of either the $R\bar{3}m$-layered structure or the $Fm\bar{3}m$ rocksalt structure. The rate of nickel reduction remained close before 200 °C. However, nickel reduction was slower in $Li_{0.33}Ni_{1/3}Co_{1/3}Mn_{1/3}O_2$ when the temperature was above 250 °C. In the $Li_{0.33}Ni_{1/3}Co_{1/3}Mn_{1/3}O_2$ material, the formation of LiM_2O_4-type spinel retarded the nickel reduction. The authors then used Ni K-edge EXAFS to compare the local structural changes in these materials, as shown in Figure 14.11. One can calculate the local structural fingerprints, such as bond length, and compare the results with reference samples to infer the phases that are present. The NCA material quickly transformed to the rocksalt phase above 300 °C, whereas rocksalt formation did not take place below 450 °C in the NMC material. This was consistent with the XANES results that the nickel reduction occurred more rapidly in NCA. The Co K-edge XANES and EXAFS underwent different evolution compared with those of Ni K-edge (Figure 14.12). First, the pre-edge features experienced visible changes upon heating. In the NCA material, the pre-edge intensity first increased and then decreased, indicating the migration of Co first to the neighboring tetrahedral site (spinel) and then to the octahedral site in the Li layer, completely transforming to the rocksalt phase. In the NMC material, however, the pre-edge intensity continuously increased upon heating up to 500 °C, which suggested that the continuous transformation from the layered structure to the spinel structure for Co. The authors also determined that Co was in the Co_3O_4 spinel phase, which was stable as high as 500 °C.

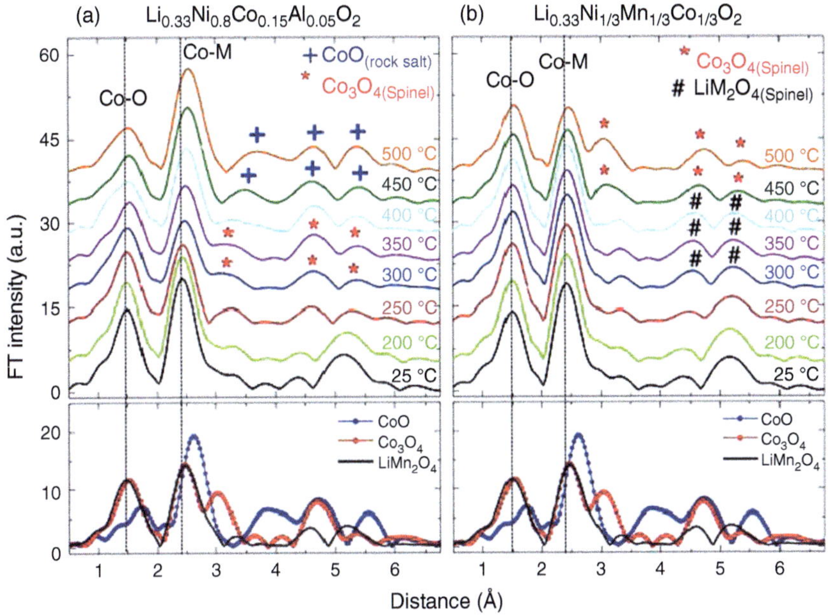

Figure 14.12 Co K-edge EXAFS analysis of (a) $Li_{0.33}Ni_{0.8}Co_{0.15}Al_{0.05}O_2$ and (b) $Li_{0.33}Ni_{1/3}Co_{1/3}Mn_{1/3}O_2$ upon heating from room temperature to 500 °C. The reference EXAFS patterns are provided to compare with the experimental data. Source: Nam et al. [54]/with permission of John Wiley & Sons.

Using hard XAS for investigating charge compensation mechanisms and studying local structural changes upon battery charging and discharging have been widely reported. It has been the prime technique for this purpose. Especially when a new material is proposed, the charge compensation mechanism is usually studied using hard XAS. Ceder and coworkers have recently shown the capability of distinguishing the anionic activity from the cationic activity in Mn^{2+}/Mn^{4+} double redox in disordered rocksalt materials [45]. The data interpretation is straightforward for cases like this. One would usually compare the experimental spectra with the spectra from reference compounds. It is important to note that the metal cation in the reference compound should have the same local coordination environment as that in the studied material. In materials with multiple redox-active TMs, hard XAS can resolve not only the contribution of each redox couple at different voltages, but also the rate-limiting redox couple. For example, one may apply a constant voltage above the potential of all the redox couples in a material, then monitor the rate of oxidation for each TM, the one that undergoes the slowest oxidation is most likely the rate-limiting TM redox couple. X. Yang and coworkers used this approach to study the redox reactions in a Li/Mn-rich NMC material ($Li_{1.2}Ni_{0.15}Co_{0.1}Mn_{0.55}O_2$) [55]. They found that Ni and Co redox couples were much faster than Mn. This study, however, assumed the Li/Mn-rich materials were composite materials. Later studies have found that the oxygen activity also contributed significantly to the capacity in these materials.

Finally, we emphasize that hard XAS techniques are still evolving, and their applications in batteries are rapidly growing. Ultrafast XAS can be extremely powerful to study fast dynamics in battery materials, including intermediate and metastable phases. Furthermore, hard XAS can be potentially operated as the surface-sensitive probing technique. Research in this area has been focused on hard XAS in the reflection mode [56]. Due to the space limitations, the present chapter cannot cover all aspects of hard XAS.

14.3 Real-Space X-ray Spectroscopic Imaging

X-ray imaging was started at the time of X-ray discovery. The first-ever X-ray imaging was generated by Wilhelm Röntgen in 1895. In the early days, the X-ray imaging was primarily based on different attenuation lengths of X-rays in different materials. With the development of synchrotron X-ray spectroscopy, especially those with high spatial resolution, X-ray imaging has evolved into a family of techniques that are sensitive not only to the morphological attributes of the probed materials but also to the compositional distribution and electronic structure. The interaction between synchrotron X-ray imaging and battery science has opened the era of "seeing is believing" at the mesoscale that is typically challenging for electron microscopy. The best spatial resolution of synchrotron X-ray imaging can range from a few nanometers to tens of nanometers. Thus far, synchrotron X-ray spectroscopic imaging has been used to study battery chemistries at the length scales from primary particles, to secondary particles, to electrodes and the cell level [57–60]. Over the last few years, spectroscopic imaging has become multidimensional because of three-dimensional reconstruction methods and the improved speed of data acquisition. In this section, we highlight a few studies that probe the heterogeneity of battery secondary particles using full-field X-ray imaging. Finally, we discuss our studies using tomographic imaging to study phase transformation and chemomechanical properties of layered oxide materials.

14.3.1 2D Full-Field X-ray Imaging

Both soft and hard X-rays can be utilized for full-field X-ray imaging. The typical spatial resolution is tens of nanometers for hard X-rays and <15 nm for soft X-rays [2]. Intercalating metal oxides are typically polycrystalline materials that consist of many primary particles [8, 61]. The polycrystalline secondary particles are susceptible to chemomechanical breakdown, especially under highly charged states. Although the charging and discharging of intercalating materials induce much smaller volume change than conversion materials, such volume changes are still typically large enough to create mechanical stress in secondary particles resulting in the formation of microcracks along grain boundaries [62]. The first example is the chemomechanical response of NMC materials undergoing fast charging (Figure 14.13) [63]. In this measurement, we used a single energy (i.e. 8 keV) X-ray source to probe NMC secondary particles undergoing various cycling conditions.

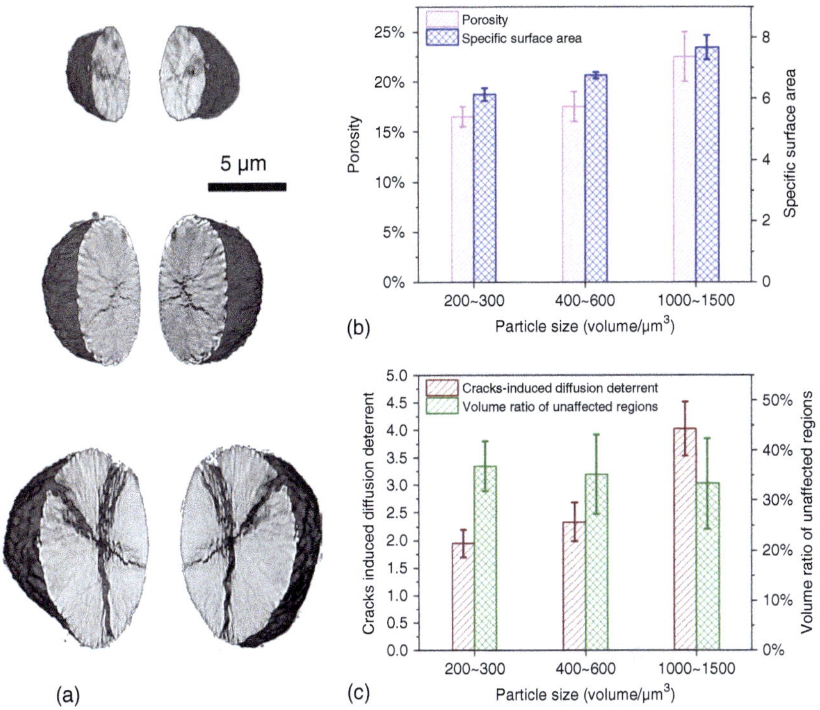

Figure 14.13 (a) Visualization and quantification of the chemomechanical breakdown of NMC particles of different sizes, after being cycled at 10C for 50 times (2.5–4.5 V vs. Li/Li^+). (b) The porosity and the specific crack surface area as functions of particle size. (c) The crack induced diffusion deterrent and the unaffected regions as functions of particle size. Source: Xia et al. [63]/with permission of Elsevier.

To visualize the interior of the particles, we created 3D tomographic reconstruction based on the large number of individual X-ray images. The individual images are taken with a 0.5 second exposure time. The samples were rotated to allow for 360 individual images that are 0.5° apart from each other. We found that the chemomechanical breakdown was largely correlated with the particle size.

As shown in Figure 14.13, larger particles in general underwent more significant microcracking, leading to larger porosity and higher specific surface area (Figure 14.13b). Furthermore, the formation of microcracks was largely dependent on the charging rates. Faster charging rates resulted in more significant buildup of microcracks. The possible explanation is that for fast charging, the SOC distribution became more heterogeneous and resulted in more mechanical stress buildup. In a practical battery, the increased exposed surface, caused by microcracking, can potentially lead to more drastic electrode–electrolyte side reactions. This is typically not desirable in cathode materials, as new surface reconstruction layers can form along the cracked area. Another outcome of microcracking is that the electron diffusion pathways are interrupted, leading to the crack-induced diffusion deterrent (Figure 14.13c).

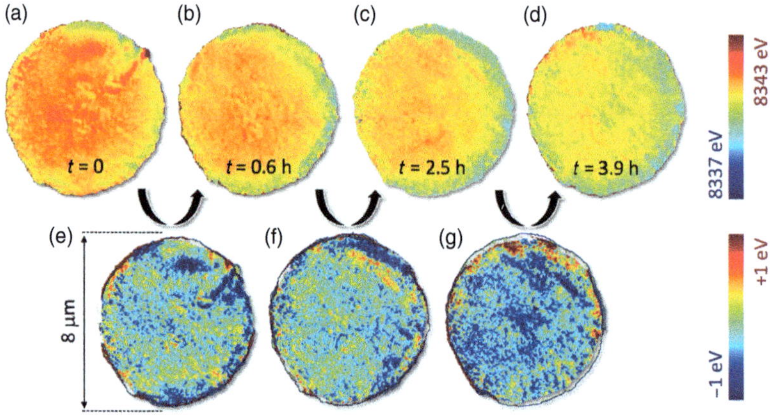

Figure 14.14 Ni K-edge XAS mapping of a $Li_{0.5}Ni_{0.6}Mn_{0.2}Co_{0.2}O_2$ secondary particle heated at 380 °C. (a–d) show the 2D distribution of Ni valence state as a function of time (color coded to the corresponding color map with red and blue indicating more oxidized and more reduced domains, respectively). (e–g) show the differential valence maps between (a) and (b), (b) and (c), (c) and (d), respectively. The diameter of the studied particle is about 8 μm. Source: Figures are used with permission from Wei et al. [64].

A previous study found that the microcrack formation was also related to oxygen loss and surface reconstruction [53]. Upon thermal heating, oxygen loss took place. The oxygen loss generated spinel and rocksalt phases and created local lattice mismatch, thus leading to the formation of microcracks. In a recent study using full-field synchrotron Ni K-edge XANES imaging, Wei et al. investigated the distribution of oxygen loss (Ni reduction) in charged NMC particles upon thermal heating (Figure 14.14) [64]. Overall, Ni experienced continuous reduction upon thermal heating. However, the reduction was highly heterogeneous. Based on these recent studies, one may conclude that the heterogeneous chemomechanical breakdown of polycrystalline NMC particles is associated with the stress buildup from the lattice volume change as well as surface reconstruction upon cycling or heating [53] (Figure 14.15).

In a recent study, Tian et al. investigated the SOC distribution in electrochemically charged and chemically delithiated NMC secondary particles [8]. The authors used valence-sensitive transmission X-ray microscopy (TXM) to visualize the distribution of nickel oxidation state. The study found that the distribution of nickel oxidation state was heterogeneous in both electrochemically charged and chemically delithiated samples. Since nickel is the primary redox-active TM in the NMC material, the heterogenous distribution of nickel oxidation state represents the heterogeneous SOC distribution in NMC secondary particles. The TXM measurements were performed after the particles had been equilibrated for a few days, but the heterogenous SOC distribution was still present. Therefore, the presence of grain boundaries and other structural defects did not allow for homogenization of the lithium ion distribution. The authors further compared the surface chemistry of the electrochemically charged and chemically delithiated samples. There was more significant

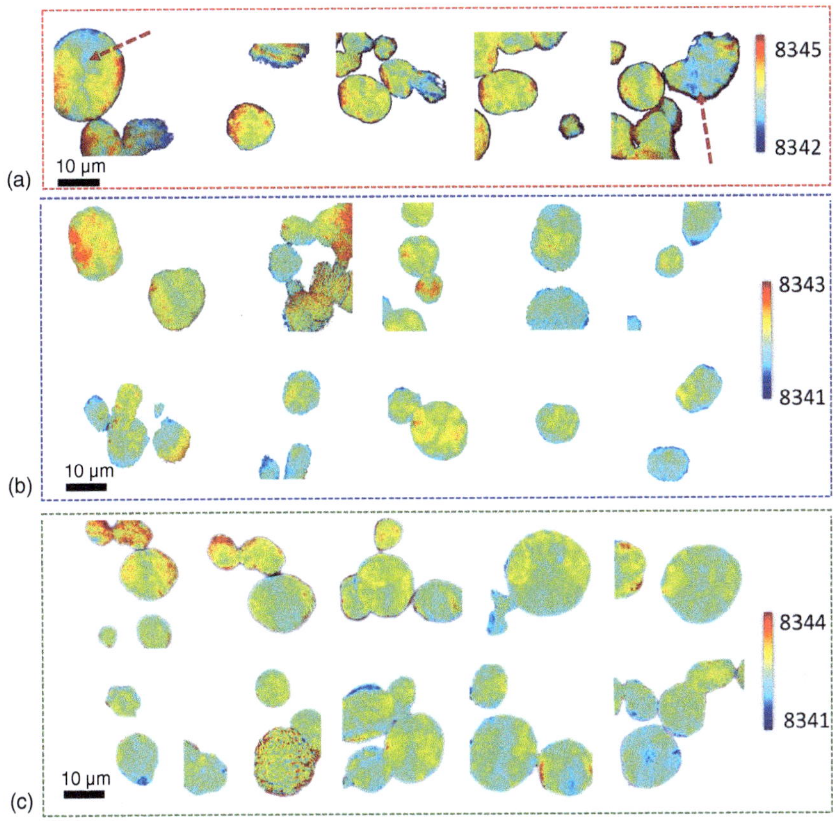

Figure 14.15 Ni K-edge XAS mapping of (a) chemically delithiated NMC-622; (b) electrochemically charged NMC-622 electrode; and (c) electrochemically discharged NMC-622 electrode. Arrows in (a) point to regions of microcracks, which are less oxidized than surrounding areas. Source: Tian et al. [8]/with permission of Elsevier.

TM reduction in the electrochemically charged sample due to the presence of electrolyte and the cathode-electrolyte side reactions in the charged state.

The direct visualization of SOC and morphological defects (e.g. cracks) is complementary to bulk measurements, such as hard XAS. However, there are a few challenges. First, the speed of data acquisition is currently insufficient for *operando* imaging measurements, especially under fast charging conditions. Second, the limited field of view makes the statistical analysis challenging.

Scanning X-ray imaging is another rapidly growing synchrotron X-ray imaging technique that has been widely used to study battery electrochemistry. In the scanning method, the X-ray beam is focused onto a small spot, and the probed particles are scanned in the focal plane. The interaction between X-ray and the sample can be monitored using different detectors, including transmission, fluorescence, and even diffraction. Chueh and coworkers have investigated extensively the charging mechanism of $LiFePO_4$ materials using the FY X-ray microscopy [22, 65]. The recently developed ptychography technique can potentially provide even better spatial resolution by taking advantage of the coherence property of the incident X-rays. Readers are referred to a few examples of this technique used in batteries [66, 67].

14.3.2 X-ray Tomographic Imaging

Battery chemistry has multidimensional characteristics. It is important to understand how different regions of battery particles respond to electrochemical cycling. Our earlier sections have made it clear that the surface redox behaviors are typically distinct from those in the bulk particles. Furthermore, due to the heterogeneous nature of polycrystalline secondary particles, the redox behaviors are different depending on the physical location in the particle. The 2D TXM studies discussed in Section 14.3.1 clearly showed the nearly random distribution of redox-active sites in polycrystalline NMC particles. The distribution of the redox-active sites and the propagation of redox reactions in these materials govern the key battery performance metrics, including energy density, stability, and power density. It is difficult, if not impossible, to reveal the bulk redox behaviors of battery particles in a nondestructive way using conventional characterization methods.

In this section, we discuss the application of 3D TXM in two different categories: 3D compositional mapping and 3D valence state mapping [68].

In recent years, increasing research efforts have been directed toward controlling the chemical distribution and microstructure of battery secondary particles. Most intercalating oxide materials are multicomponent compounds that consist of three or more metal cations. Different metal cations show different reactivity with the electrolyte due to the different metal–oxygen bonding characteristics. In lithium ion NMC materials, Ni plays the major role in the capacity, whereas Mn stabilizes the layered structure and promotes thermal stability. The well-known fading mechanism in these materials involves the surface metal reduction and structural reconstruction to the rocksalt structure. Sun and coworkers have done numerous studies to tune the distribution of Ni, Mn, and Co in NMC secondary particles and consistently found that the combination of Mn-rich surface and Ni-rich bulk could improve the cycling and thermal stability of these materials [69–71]. The group has also expanded the design method to sodium-layered oxides and observed a similar stabilization mechanism. In a recent study, we used a spray pyrolysis method to synthesize an NMC material with highly heterogeneous distribution of all three TMs, as shown in Figure 14.16. Because of the different reactivity between each TM and the carrier gas during the synthesis, the resulting material was rich in Mn at the surface, thus leading to improved cycling stability at high voltages.

Another typical application of 3D TXM compositional mapping is to quantify the change of elemental distribution in battery particles after long-term cycling. Yang et al. were among the first group of researchers to use this capability (Figure 14.17) [73]. The authors investigated Mn segregation after long-term cycling of Li/Mn-rich layered oxides. It was found that Mn segregation to the surface of the secondary particles completely modified the particle surfaces and resulted in decreased battery performance, including voltage fade. Rahman et al. designed a highly heterogeneous $Na_{0.9}Cu_{0.2}Fe_{0.28}Mn_{0.52}O_2$ sodium cathode material, which was able

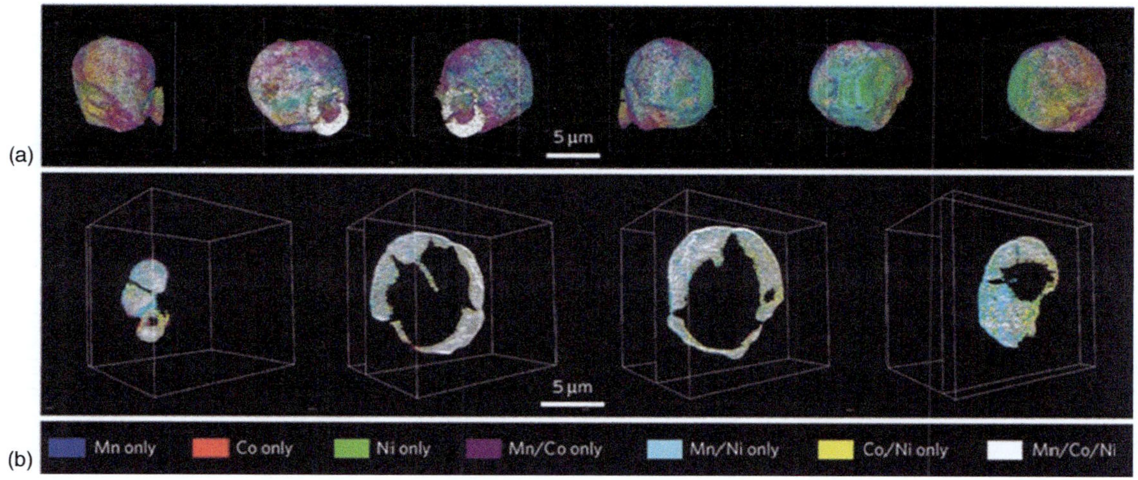

Figure 14.16 Elemental association maps of NMC442 synthesized by spray pyrolysis. (a) 3D rendering of the elemental associations viewing the particles at different angles. (b) 2D slices of the elemental associations cut through at different depths of the imaged particles. The colors representing the elemental associations are shown at the bottom. Source: Lin et al. [72]/with permission of Springer Nature.

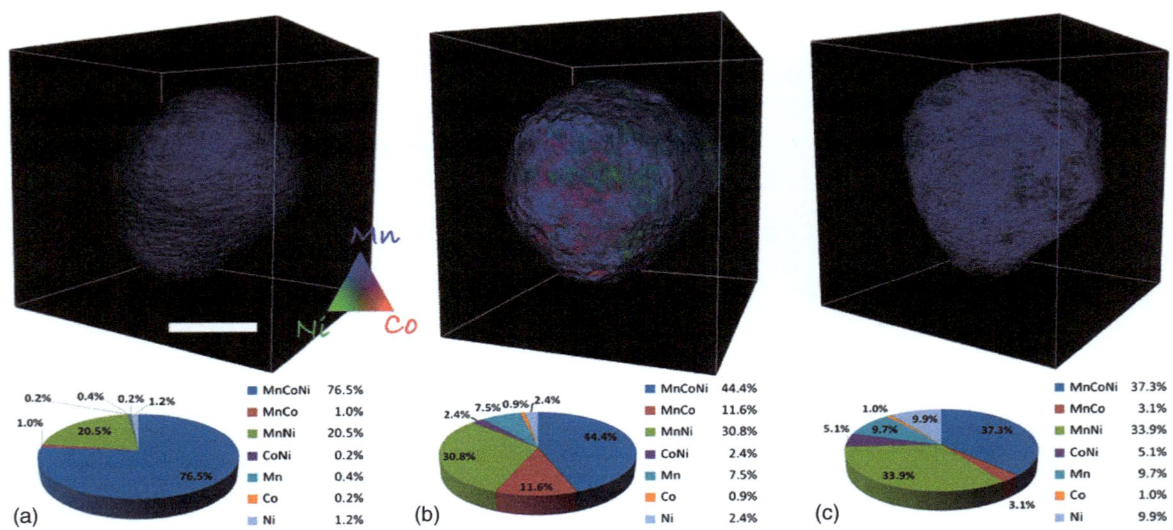

Figure 14.17 3D elemental maps of $Li_{1.2}Mn_{0.525}Ni_{0.175}Co_{0.1}O_2$ (a) in the pristine state, (b) after 1 cycle, and (c) after 200 cycles. The respective elemental associations are shown on the bottom. The scale bar is 5 μm. Source: Yang et al. [73]/with permission of American Chemical Society.

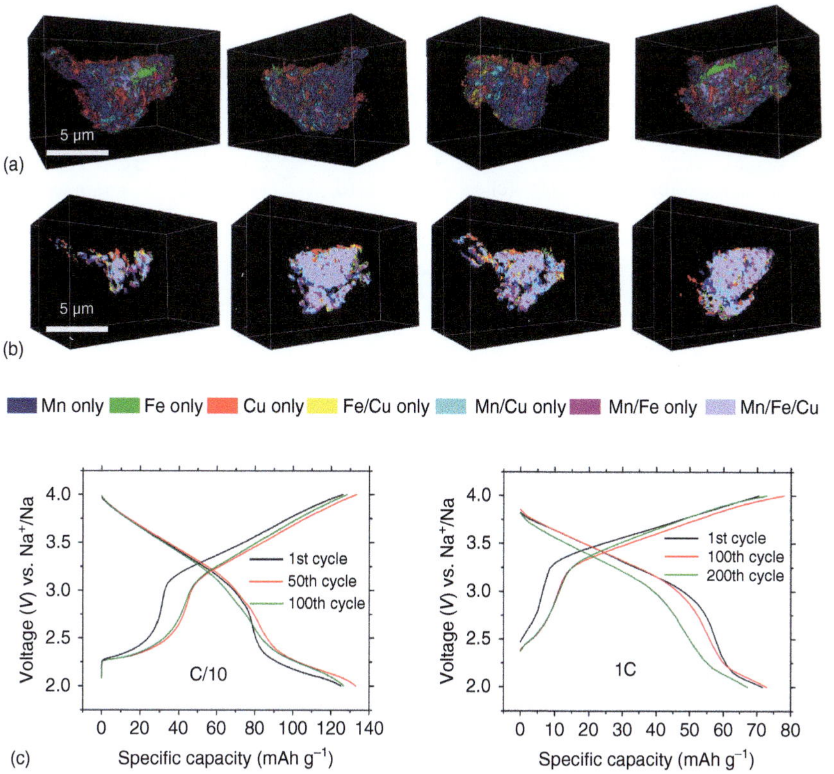

Figure 14.18 3D elemental association maps of $Na_{0.9}Cu_{0.2}Fe_{0.28}Mn_{0.52}O_2$: (a) from different angles showing that the surface is dominated by single metal and binary metal associations, and (b) cutting through the particle at different depths. (c) Charge–discharge curves at different cycles for C/10 and 1C rates between 2.0 and 4.0 V vs. Na/Na^+. Source: Rahman et al. [7]/with permission of Royal Society of Chemistry.

to deliver surprisingly stable performance (Figure 14.18) [7]. They combined soft XAS, hard XAS, and 3D TXM to quantify the Mn segregation in the highly heterogeneous sodium-layered cathode material. The surface TM3d-O2p hybridization was dominated by the Mn3d–O2p coordination. Upon cycling, the surface Mn underwent significant reduction. After long-term cycling, the authors observed significant buildup of reduced Mn at the surface. Such Mn surface segregation after long-term cycling could be attributed to the migration of the bulk Mn to the surface, as well as the dissolution and redeposition of reduced Mn onto the surface.

As shown above, 3D TXM is an excellent technique to quantify 3D compositional distribution with a good spatial resolution (30 nm × 30 nm × 30 nm). To date, the technique has been only used for particles above a few microns. Therefore, coupling TXM with TEM will then resolve the composition distribution from both primary and secondary particles. On the other hand, the data acquisition speed is currently too slow to measure many particles during a given beamtime. Therefore, the statistical analysis of the results needs to be further improved. Fortunately, we have other synchrotron techniques that can be used to corroborate the TXM results.

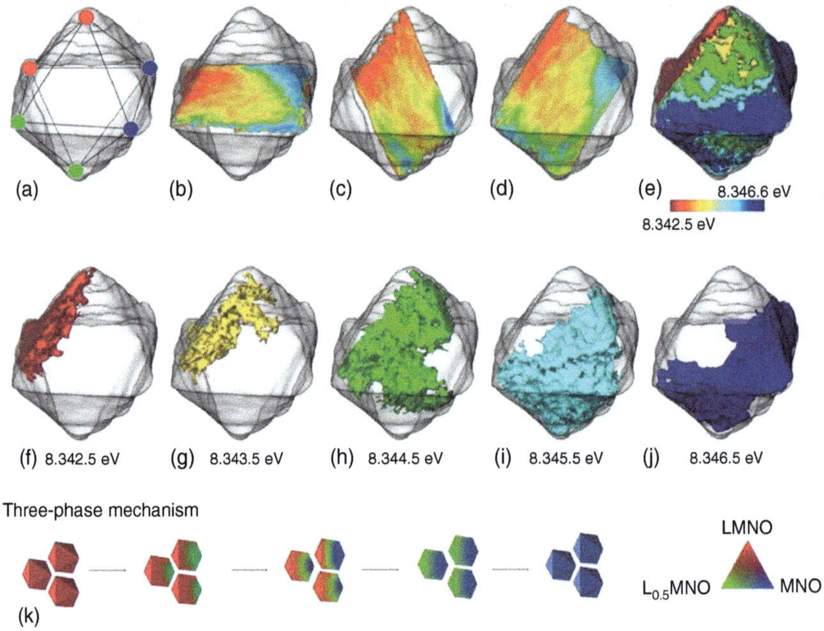

Figure 14.19 Ni oxidation state mapping in a $Li_{0.51}Ni_{0.5}Mn_{1.5}O_4$ particle: (a–d) the internal oxidation state heterogeneity illustrated using the diagonal slices, (e–j) surfaces of the 3D Ni oxidation state map, and (k) the proposed three-phase reaction mechanism. Source: Kuppan et al. [11]/Springer Nature/CC BY 4.0.

The TXM operated using the valence-sensitive mode can be regarded as adding energy resolution to the TXM in the composition-sensitive mode. The resulting XAS spectrum from each voxel (30 nm × 30 nm × 30 nm) can be analyzed in a way that is close to a typical hard XAS spectra. However, due to the poorer energy resolution compared to conventional bulk XAS, the best analysis of TXM-XAS has been limited to the absorption edge or the whiteline and thus only the oxidation state. In a typical experiment, one can determine the oxidation state of a probed element based on the position of the absorption edge. Kuppan et al. applied TXM to investigate the phase transformation behavior of $LiNi_{0.5}Mn_{1.5}O_4$ single crystals upon electrochemical charging (Figure 14.19) [11]. According to the study, $LiNi_{0.5}Mn_{1.5}O_4$ is a phase-change material with three different phases, $LiNi_{0.5}Mn_{1.5}O_4$, $Li_{0.5}Ni_{0.5}Mn_{1.5}O_4$, and $Ni_{0.5}Mn_{1.5}O_4$, where the nickel oxidation states are Ni^{2+}, Ni^{3+}, and Ni^{4+}, respectively. The authors used the Ni oxidation state to identify the presence of these phases. As shown in Figure 14.19, the study led to two major conclusions: (i) The nucleation of new phases in $LiNi_{0.5}Mn_{1.5}O_4$ single crystals was faster for the (100) facets. (ii) Upon charging, there were regions where all three phases were present. The authors also supported their TXM phase results with XRD.

The layered to rocksalt phase transformation can also be studied using TXM, because there is nickel oxidation state change during such transformation. If we can study the phase transformation in 3D, then we will be able to establish the phase

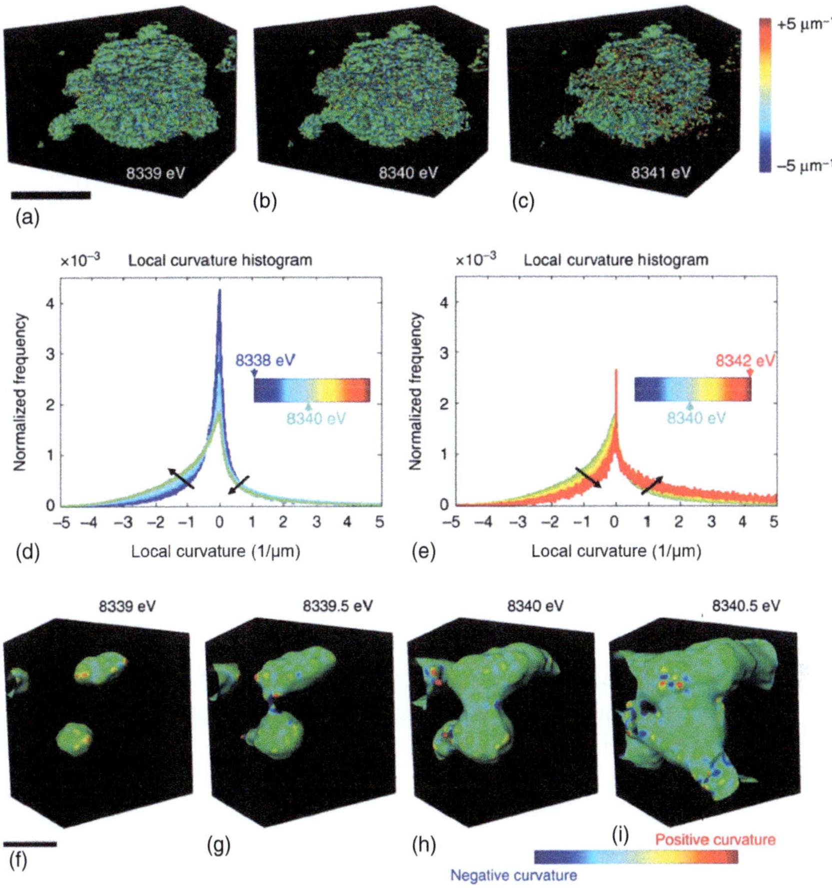

Figure 14.20 Evolution of local valence curvatures upon phase transformation in a charged NMC cathode. Isosurfaces of the 3D Ni valence state maps at (a) 8339 eV, (b) 8340 eV, and (c) 8341 eV. The scale bar is 10 μm. (d, e) The histograms for the changes of local valence curvatures as a function of the selected energy value (d) from 8338 to 8340 eV and (e) from 8340 to 8342 eV are shown. Visualization of the phase front propagation at a selected localized region. The isosurface develops as the thresholding Ni K-edge energy value is changed, as labeled on the top of each image (f–i). Source: Mu et al. [74]/Springer Nature/CC BY 4.0.

propagation behaviors. Recently, we performed an inaugural study to map the phase propagation topography in charged NMC materials under thermal heating conditions. We connected the voxels that showed an identical nickel oxidation state (absorption-edge position) and formed the isosurfaces of the 3D Ni valence state map (Figures 14.20a–c). The phase propagation proceeded from the edge energy 8339 to 8340 eV and then to 8341 eV. There were changes in the local curvature of each topography. We summarized the results in Figure 14.20d,e. During the early stage of the phase propagation, the negative curvature increased. During the later stage, the negative curvature decreased, and the positive curvature increased. Such a dynamic change of local valence state curvature was directly visualized with a selected region (Figure 14.20f–i). Basically, multiple nucleation sites formed in the early stage and then merged. Upon the initial merging, the negative curvature formed. With the continuous phase transformation, negative curvatures disappeared, and positive curvatures appeared. This was the first time that the valence state curvature was introduced to describe the solid-state redox phase transformation in battery materials. We expect that the method can be applied to study battery phase transformations under electrochemical cycling conditions. Since the phase transformation in most oxide cathode materials is composition dependent, future studies should couple local chemical and structural heterogeneity with the phase propagation behaviors. The valence state curvature method is still at its infant stage, and more studies are needed to expand its application.

14.4 Conclusion

In this chapter, we discussed the application of synchrotron X-ray techniques in studying battery electrochemistry, with a focus on intercalating oxide materials that are studied in the Lin Lab. Synchrotron X-ray techniques have been developed into powerful and indispensable characterization methods for battery science. The scope of the present chapter is necessarily narrow due to the limited space. We, therefore, focus our discussion on only a few popular techniques: soft XAS, hard XAS, and X-ray spectroscopic imaging. The capability of *in situ* and *operando* experiments using hard XAS has helped unravel many dynamic behaviors in battery materials regarding local electronic structure and symmetry. However, *in situ* and *operando* experiments using soft XAS and full-field X-ray spectroscopic imagining have been limited thus far, primarily due to the limited penetration depths of soft XAS and the inferior data acquisition speed of TXM. Although not extensively discussed in the chapter, the potential beam damage associated with these measurements must be considered when interpreting the experimental results. Furthermore, it remains a grand challenge to *in situ* characterize electrochemical interphases in batteries. With the recent development of ambient pressure XPS [31], this challenge could be partially resolved. Finally, each technique has its own limitations; therefore, we emphasize that combining multiple characterization techniques, especially the techniques at different length scales, is extremely important to get the full picture of a probed system.

In situ and *operando* measurements can be more advanced than *ex situ* measurements due to their capabilities in capturing dynamic, transient phenomena. However, the proper design of *in situ* and *operando* electrochemical cells is sometimes nontrivial. For hard XAS discussed here, typical coin cells and pouch cells can be used directly as *in situ* cells. For example, it is very common to drill holes on the stainless steel cell components to allow X-rays to go through coin cells. For pouch cells, as long as the stacked layers still allow X-rays to go through with a reasonable transmitted X-ray intensity, *in situ* measurements can be conducted. *In situ* soft XAS measurements have remained a challenge for liquid electrolyte batteries, due to the limited penetration depth of soft X-rays and the requirement of ultrahigh vacuum during the measurements. In general, TXM techniques using hard X-rays have similar sample requirements as hard XAS. However, for the 3D reconstruction, the sample needs to be rotated at a wide-angle range and X-rays should have ample penetration at each of the angles. With some advanced data analysis methods, the angle range can be reduced.

In summary, the critical role that synchrotron X-ray techniques play in battery science has been well documented by thousands of publications in the past 25 years, and it is expected to grow even faster in the coming years.

References

1 Whittingham, M.S. (2004). Lithium batteries and cathode materials. *Chemical Reviews* 104 (607): 4271–4301.

2 Lin, F., Liu, Y., Yu, X. et al. (2017). Synchrotron X-ray analytical techniques for studying materials electrochemistry in rechargeable batteries. *Chemical Reviews* 117 (21): 13123–13186.

3 Yang, Z., Mu, L., Hou, D. et al. (2021). Probing dopant redistribution, phase propagation, and local chemical changes in the synthesis of layered oxide battery cathodes. *Advanced Energy Materials* 11 (1): 2002719.

4 Li, X., Wang, Y., Wu, D. et al. (2016). Jahn–Teller assisted Na diffusion for high performance Na ion batteries. *Chemistry of Materials* 28 (18): 6575–6583.

5 Li, X., Ma, X., Su, D. et al. (2014). Direct visualization of the Jahn–Teller effect coupled to Na ordering in $Na_{5/8}MnO_2$. *Nature Materials* 13 (6): 586–592.

6 Lin, F., Markus, I.M., Nordlund, D. et al. (2014). Surface reconstruction and chemical evolution of stoichiometric layered cathode materials for lithium-ion batteries. *Nature Communications* 5: 3529.

7 Rahman, M.M., Xu, Y., Cheng, H. et al. (2018). Empowering multicomponent cathode materials for sodium ion batteries by exploring three-dimensional compositional heterogeneities. *Energy & Environmental Science* 11: 2496–2508.

8 Tian, C., Xu, Y., Nordlund, D. et al. (2018). Charge heterogeneity and surface chemistry in polycrystalline cathode materials. *Joule* 2 (3): 464–477.

9 Tian, C., Nordlund, D., Xin, H.L. et al. (2018). Depth-dependent redox behavior of $LiNi_{0.6}Mn_{0.2}Co_{0.2}O_2$. *Journal of the Electrochemical Society* 165 (3): A696–A704.

10 Lin, F., Nordlund, D., Weng, T.-C. et al. (2014). Phase evolution for conversion reaction electrodes in lithium-ion batteries. *Nature Communications* 5 (1): 3358.

11 Kuppan, S., Xu, Y., Liu, Y., and Chen, G. (2017). Phase transformation mechanism in lithium manganese nickel oxide revealed by single-crystal hard X-ray microscopy. *Nature Communications* 8: 14309.

12 Xu, Z., Hou, D., Kautz, D.J. et al. (2020). Charging reactions promoted by geometrically necessary dislocations in battery materials revealed by in situ single-particle synchrotron measurements. *Advanced Materials* 32 (37): 2003417.

13 Lin, F. and Yang, Z. (2021). Heterogeneous, defect-rich battery particles and electrodes: why do they matter, and how can one leverage them? *Journal of Physical Chemistry C* 125 (18): 9618–9629.

14 Sun, Y., Gorobstov, O., Mu, L. et al. (2021). X-ray nanoimaging of crystal defects in single grains of solid-state electrolyte $Li_{7-3x}Al_xLa_3Zr_2O_{12}$. *Nano Letters* 21 (11): 4570–4576.

15 Tian, C., Lin, F., and Doeff, M.M. (2018). Electrochemical characteristics of layered transition metal oxide cathode materials for lithium ion batteries: surface, bulk behavior, and thermal properties. *Accounts of Chemical Research* 51 (1): 89–96.

16 Cabana, J., Kwon, B.J., and Hu, L. (2018). Mechanisms of degradation and strategies for the stabilization of cathode–electrolyte interfaces in Li-ion batteries. *Accounts of Chemical Research* 51 (2): 299–308.

17 Lu, J., Wu, T., and Amine, K. (2017). State-of-the-art characterization techniques for advanced lithium-ion batteries. *Nature Energy* 2 (3): 17011.

18 Wei, C., Xia, S., Huang, H. et al. (2018). Mesoscale battery science: the behavior of electrode particles caught on a multispectral X-ray camera. *Accounts of Chemical Research* 51 (10): 2484–2492.

19 Hwang, B.J., Tsai, Y.W., Carlier, D., and Ceder, G. (2003). A combined computational/experimental study on $LiNi_{1/3}Co_{1/3}Mn_{1/3}O_2$. *Chemistry of Materials* 15 (19): 3676–3682.

20 Lee, J., Urban, A., Li, X. et al. (2014). Unlocking the potential of cation-disordered oxides for rechargeable lithium batteries. *Science* 343 (6170): 519–522.

21 Wang, J., Chen-Wiegart, Y.K., and Wang, J. (2014). In operando tracking phase transformation evolution of lithium iron phosphate with hard X-ray microscopy. *Nature Communications* 5: 4570.

22 Lim, J., Li, Y., Alsem, D.H. et al. (2016). Origin and hysteresis of lithium compositional spatiodynamics within battery primary particles. *Science* 353 (6299): 566–571.

23 Gent, W.E., Li, Y., Ahn, S. et al. (2016). Persistent state-of-charge heterogeneity in relaxed, partially charged $Li_{1-x}Ni_{1/3}Co_{1/3}Mn_{1/3}O_2$ secondary particles. *Advanced Materials* 28 (31): 6631–6638.

24 Liu, X., Yang, W., and Liu, Z. (2014). Recent progress on synchrotron-based in-situ soft X-ray spectroscopy for energy materials. *Advanced Materials* 26 (46): 7710–7729.

25 Lin, F., Nordlund, D., Markus, I.M. et al. (2014). Profiling the nanoscale gradient in stoichiometric layered cathode particles for lithium-ion batteries. *Energy & Environmental Science* 7: 3077–3085.

26 Steiner, J.D., Mu, L., Walsh, J. et al. (2018). Accelerated evolution of surface chemistry determined by temperature and cycling history in nickel-rich layered cathode materials. *ACS Applied Materials & Interfaces* 10 (28): 23842–23850.

27 Mu, L., Feng, X., Kou, R. et al. (2018). Deciphering the cathode-electrolyte interfacial chemistry in sodium layered cathode materials. *Advances Energy Materials* 1801975.

28 Wujcik, K.H., Wang, D.R., Pascal, T.A. et al. (2017). In situ X-ray absorption spectroscopy studies of discharge reactions in a thick cathode of a lithium sulfur battery. *Journal of the Electrochemical Society* 164 (2): A18–A27.

29 Zhang, L., Sun, D., Feng, J. et al. (2017). Revealing the electrochemical charging mechanism of nanosized Li_2S by in situ and operando X-ray absorption spectroscopy. *Nano Letters* 17 (8): 5084–5091.

30 Zhao, E., Nie, K., Yu, X. et al. (2018). Advanced characterization techniques in promoting mechanism understanding for lithium-sulfur batteries. *Advanced Functional Materials* 28 (38): 1707543.

31 Favaro, M., Jeong, B., Ross, P.N. et al. (2016). Unravelling the electrochemical double layer by direct probing of the solid/liquid interface. *Nature Communications* 7: 12695.

32 McBreen, J., O'Grady, W.E., and Pandya, K.I. (1988). EXAFS: a new tool for the study of battery and fuel cell materials. *Journal of Power Sources* 22 (3–4): 323–340.

33 Skotheim, T.A., Yang, X.Q., Xue, K.H. et al. (1992). X-ray absorption studies on organo-disulfide redox cathodes. *Electrochimica Acta* 37 (9): 1635–1637.

34 Yang, X.Q., Xue, K.H., Lee, H.S. et al. (1992). X-ray-absorption studies of organodisulfide redox polymeric electrodes. *Physical Review B* 45 (10): 5733–5736.

35 Mcbreena, J., Yang, X.Q., Lee, H.S., and Okamoto, Y. (1994). XAS studies of PEO based polymer electrolytes. *MRS Proceedings* 369: 559.

36 McBreen, J., Mukerjee, S., and Yang, X.Q. (1998). In situ synchrotron X-ray studies of battery and fuel cell materials. *Synchrotron Radiation News* 11 (3): 18–22.

37 Xu, Y., Hu, E., Zhang, K. et al. (2017). In situ visualization of state-of-charge heterogeneity within a $LiCoO_2$ particle that evolves upon cycling at different rates. *ACS Energy Letters* 2 (5): 1240–1245.

38 Gong, Z. and Yang, Y. (2018). The application of synchrotron X-ray techniques to the study of rechargeable batteries. *Journal of Energy Chemistry* 27 (6): 1566–1583.

39 Wang, L., Wang, J., and Zuo, P. (2018). Probing battery electrochemistry with in operando synchrotron X-ray imaging techniques. *Small Methods* 2 (8): 1700293.

40 Lin, F., Nordlund, D., Pan, T. et al. (2014). Influence of synthesis conditions on the surface passivation and electrochemical behavior of layered cathode materials. *Journal of Materials Chemistry A* 2 (46): 19833–19840.

41 Mu, L., Rahman, M.M., Zhang, Y. et al. (2018). Surface transformation by a "cocktail" solvent enables stable cathode materials for sodium ion batteries. *Journal of Materials Chemistry A* 6 (6): 2758–2766.

42 Liu, X., Wang, D., Liu, G. et al. (2013). Distinct charge dynamics in battery electrodes revealed by in situ and operando soft X-ray spectroscopy. *Nature Communications* 4: 2568.

43 Seo, D.-H., Lee, J., Urban, A. et al. (2016). The structural and chemical origin of the oxygen redox activity in layered and cation-disordered Li-excess cathode materials. *Nature Chemistry* 8 (7): 692–697.

44 Yabuuchi, N., Takeuchi, M., Nakayama, M. et al. (2015). High-capacity electrode materials for rechargeable lithium batteries: Li_3NbO_4-based system with cation-disordered rocksalt structure. *Proceedings of the National Academy of Sciences of the United States of America* 112 (25): 7650–7655.

45 Lee, J., Kitchaev, D.A., Kwon, D.-H. et al. (2018). Reversible Mn^{2+}/Mn^{4+} double redox in lithium-excess cathode materials. *Nature* 556 (7700): 185–190.

46 Yang, W. and Devereaux, T.P. (2018). Anionic and cationic redox and interfaces in batteries: advances from soft X-ray absorption spectroscopy to resonant inelastic scattering. *Journal of Power Sources* 389: 188–197.

47 Cheng, L., Wu, C.H., Jarry, A. et al. (2015). Interrelationships among grain size, surface composition, air stability, and interfacial resistance of Al-substituted $Li_7La_3Zr_2O_{12}$ solid electrolytes. *ACS Applied Materials & Interfaces* 7 (32): 17649–17655.

48 Cheng, L., Liu, M., Mehta, A. et al. (2018). Garnet electrolyte surface degradation and recovery. *ACS Applied Energy Materials* 1 (12): 7244–7252.

49 Nanda, J., Yang, G., Hou, T. et al. (2019). Unraveling the nanoscale heterogeneity of solid electrolyte interphase using tip-enhanced Raman spectroscopy. *Joule* 3 (8): 2001–2019.

50 Rahman, M.M., McGuigan, S., Li, S. et al. (2021). Chemical modulation of local transition metal environment enables reversible oxygen redox in Mn-based layered cathodes. *ACS Energy Letters* 6: 2882–2890.

51 Zheng, X., Xu, Z., Li, S. et al. (2021). Reversible Mn/Cr dual redox in cation-disordered Li-excess cathode materials for stable lithium ion batteries. *Acta Materialia* 212: 116935.

52 Mu, L., Xu, S., Li, Y. et al. (2015). Prototype sodium-ion batteries using an air-stable and Co/Ni-free O3-layered metal oxide cathode. *Advanced Materials* 27 (43): 6928–6933.

53 Mu, L., Lin, R., Xu, R. et al. (2018). Oxygen release induced chemomechanical breakdown of layered cathode materials. *Nano Letters* 18 (5): 3241–3249.

54 Nam, K.-W., Bak, S.-M., Hu, E. et al. (2013). Combining in situ synchrotron X-ray diffraction and absorption techniques with transmission electron microscopy to study the origin of thermal instability in overcharged cathode materials for lithium-ion batteries. *Advanced Functional Materials* 23 (8): 1047–1063.

55 Yu, X., Lyu, Y., Gu, L. et al. (2014). Understanding the rate capability of high-energy-density Li-rich layered $Li_{1.2}Ni_{0.15}Co_{0.1}Mn_{0.55}O_2$ cathode materials. *Advanced Energy Materials* 4: 1300950.

56 Takamatsu, D., Koyama, Y., Orikasa, Y. et al. (2012). First in situ observation of the $LiCoO_2$ electrode/electrolyte interface by total-reflection X-ray absorption spectroscopy. *Angewandte Chemie, International Edition* 51 (46): 11597–11601.

57 Ebner, M., Marone, F., Stampanoni, M., and Wood, V. (2013). Visualization and quantification of electrochemical and mechanical degradation in Li ion batteries. *Science (80-)* 342 (6159): 716–720.

58 Finegan, D.P., Scheel, M., Robinson, J.B. et al. (2015). In-operando high-speed tomography of lithium-ion batteries during thermal runaway. *Nature Communications* 6 (1): 6924.

59 Ruther, R.E., Samuthira Pandian, A., Yan, P. et al. (2017). Structural transformations in high-capacity $Li_2Cu_{0.5}Ni_{0.5}O_2$ cathodes. *Chemistry of Materials* 29 (7): 2997–3005.

60 Spence, S., Lee, W.K., Lin, F., and Xiao, X. (2021). Transmission X-ray microscopy and its applications in battery material research-a short review. *Nanotechnology* 32 (44): 442003.

61 Yan, P., Zheng, J., Gu, M. et al. (2017). Intragranular cracking as a critical barrier for high-voltage usage of layer-structured cathode for lithium-ion batteries. *Nature Communications* 8: 14101.

62 Jiang, Z., Li, J., Yang, Y. et al. (2020). Machine-learning-revealed statistics of the particle-carbon/binder detachment in lithium-ion battery cathodes. *Nature Communications* 11: 2310.

63 Xia, S., Mu, L., Xu, Z. et al. (2018). Chemomechanical interplay of layered cathode materials undergoing fast charging in lithium batteries. *Nano Energy* 53: 753–762.

64 Wei, C., Zhang, Y., Lee, S.-J. et al. (2018). Thermally driven mesoscale chemomechanical interplay in $Li_{0.5}Ni_{0.6}Mn_{0.2}Co_{0.2}O_2$ cathode materials. *Journal of Materials Chemistry A* 6 (45): 23055–23061.

65 Li, Y., El Gabaly, F., Ferguson, T.R. et al. (2014). Current-induced transition from particle-by-particle to concurrent intercalation in phase-separating battery electrodes. *Nature Materials* 13 (12): 1149–1156.

66 Yu, Y.-S., Farmand, M., Kim, C. et al. (2018). Three-dimensional localization of nanoscale battery reactions using soft X-ray tomography. *Nature Communications* 9 (1): 921.

67 Wolf, M., May, B.M., and Cabana, J. (2017). Visualization of electrochemical reactions in battery materials with X-ray microscopy and mapping. *Chemistry of Materials* 29 (8): 3347–3362.

68 Xu, Z., Jiang, Z., Kuai, C. et al. (2020). Charge distribution guided by grain crystallographic orientations in polycrystalline battery materials. *Nature Communications* 11: 83.

69 Sun, Y.-K., Chen, Z., Noh, H.-J. et al. (2012). Nanostructured high-energy cathode materials for advanced lithium batteries. *Nature Materials* 11 (11): 942–947.

70 Kim, U.-H., Lee, E.-J., Yoon, C.S. et al. (2016). compositionally graded cathode material with long-term cycling stability for electric vehicles application. *Advanced Energy Materials* 6 (22): 1601417.

71 Sun, Y.-K., Myung, S.-T., Park, B.-C. et al. (2009). High-energy cathode material for long-life and safe lithium batteries. *Nature Materials* 8 (4): 320–324.

72 Lin, F., Nordlund, D., Li, Y. et al. (2016). Metal segregation in hierarchically structured cathode materials for high-energy lithium batteries. *Nature Energy* 1 (1): 15004.

73 Yang, F., Liu, Y., Martha, S.K. et al. (2014). Nanoscale morphological and chemical changes of high voltage lithium-manganese rich NMC composite cathodes with cycling. *Nano Letters* 14 (8): 4334–4341.

74 Mu, L., Yuan, Q., Tian, C., Zhang, K., Liu, J., Pianetta, P., Doeff, M.M., Liu, Y., and Lin, F. (2018) Propagation topography of redox phase transformations in heterogeneous layered oxide cathode materials. *Nature Communications*, DOI: https://doi.org/10.1038/s41467-018-05172-x.

15

Atomic-Scale Simulations of the Solid Electrolyte $Li_7La_3Zr_2O_{12}$

Seungho Yu[1,2] *and Donald J. Siegel*[1,3,4,5]

[1] *University of Michigan, Mechanical Engineering Departments, 2350 Hayward St, Ann Arbor, MI 48109, United States*
[2] *Korea Institute of Science and Technology (KIST), Clean Energy Research Division, Energy Storage Research Center, 5, Hwarang-ro 14-gil, Seongbuk-gu, Seoul 02792, Republic of Korea*
[3] *University of Michigan, Materials Science & Engineering, 2300 Hayward St, Ann Arbor, MI 48109, United States*
[4] *University of Michigan, Applied Physics Program, 450 Church St., 1425 Randall Laboratory, Ann Arbor, MI 48109, United States*
[5] *University of Michigan, University of Michigan Energy Institute, 2301 BONISTEEL BLVD, Ann Arbor, MI 48109, United States*

15.1 Introduction

15.1.1 Motivation

Lithium ion batteries are now widely used in portable electronics, and are transitioning to new applications in electric vehicles and stationary energy storage systems [1]. To accelerate this transition, it is desirable to develop batteries with higher energy densities. Optimization of Li-ion systems has largely focused on improving electrode properties, for example, by raising the voltage of cathodes and increasing the capacity of anodes [2, 3]. A promising, yet less examined alternative strategy is to replace the battery's liquid electrolyte with a fast ion-conducting solid [4]. A viable solid-state electrolyte could unlock several high-capacity battery concepts. For example, employing Li metal anodes will enable emerging cell chemistries such as Li–S and Li–air [5, 6].

The substitution of flammable liquid electrolytes with stable solid-state compounds also offers safety advantages [7, 8]. Inorganic solid electrolytes (SE) are intrinsically non-flammable, in contrast to the behavior of carbonate liquid electrolytes. Additionally, Li dendrite formation in liquid electrolytes can be suppressed (in principle) by the use of a stiff SE [9]. Finally, the wide electrochemical window of some solids could enable the use of high-voltage cathodes, resulting in additional gains in energy density [4]. Carbonate-based liquid electrolytes typically have electrochemical windows that result in oxidative decomposition at voltages greater than 4.3 V [4]. Solid electrolytes with a window beyond 5 V would enable high energy density cathodes such as spinel $LiMn_{1.5}Ni_{0.5}O_4$ which operates at 4.7 V [10].

Transition Metal Oxides for Electrochemical Energy Storage, First Edition.
Edited by Jagjit Nanda and Veronica Augustyn.

15.1.2 Solid Electrolytes

A viable SE should satisfy several performance requirements simultaneously [11]: it should possess high Li-ion conductivity, ~1 mS cm^{-1} or higher, with low electronic conductivity; chemical and electrochemical stability against the Li metal anode and state-of-the-art cathodes (5 V or higher vs. Li/Li$^+$) is also essential; finally, it should exhibit sufficient mechanical properties.

Sulfide materials presently exhibit higher ionic conductivities than oxides, which has been suggested to arise from the higher polarizability of the sulfide ion [12]. For example, $Li_{10}GeP_2S_{12}$ (LGPS) exhibits an ionic conductivity of 12 mS cm^{-1} at room temperature, which is comparable to conventional liquid electrolytes [13]. However, sulfides are hygroscopic and form toxic H_2S when exposed to moisture [14]. On the other hand, oxides generally exhibit high stability against air, allowing for more convenient handling.

Among Li-ion-conducting oxides, NaSICON (Na Super Ionic CONductor)-type SE such as $Li_{1.3}Al_{0.3}Ti_{1.7}(PO_4)_3$ have been reported to exhibit high conductivity, ~1 mS cm^{-1} [15]; however, this compound is unstable in contact with Li [16]. Perovskites with the nominal formulation $Li_{3x}La_{2/3-x}TiO_3$ show high conductivity in the 0.1–1 mS cm^{-1} range [17, 18]. Nevertheless, these materials exhibit high grain boundary resistance and are unstable in contact with Li [11, 18]. At present, perhaps the most promising Li-ion-conducting oxides are garnets with nominal composition $Li_7La_3Zr_2O_{12}$ [19, 20]. Commonly referred to as LLZO, this oxide exhibits a favorable combination of high conductivity (~1 mS cm^{-1}), with early reports indicating chemical stability against Li metal [21–23].

15.1.3 $Li_7La_3Zr_2O_{12}$ (LLZO)

At room temperature, stoichiometric LLZO adopts a tetragonal crystal structure (space group $Ia\bar{4}d$) [24]. This polymorph is of less interest for solid electrolyte applications than the cubic polymorph, as the former is limited by low ionic conductivity (~10^{-6} S cm^{-1}) due to its fully ordered Li distribution. However, supervalent doping stabilizes the (typically high-temperature) cubic crystal phase at room temperature (space group $Ia\bar{3}d$), which has a much higher conductivity, ~10^{-4}–10^{-3} S cm^{-1} [21, 23, 25]. Dopants are also expected to increase the degree of vacancy disorder within the Li sub-lattice, leading to enhanced hopping paths for Li$^+$ and an increase in conductivity. For example, substitution of 0.2–0.24 moles of aluminum (nominal oxidation state of 3+) for lithium stabilizes the cubic phase and creates 0.4–0.48 moles of lithium vacancies per LLZO formula unit [21]. In this phase the ionic conductivity is increased to 0.4 mS cm^{-1} at 298 K. Density functional and molecular dynamics (MD) calculations demonstrated that the transition from the tetragonal to cubic phase occurs for Li vacancy concentrations greater than 0.4–0.5 per LLZO formula unit [26]. An alternative to Al doping is substitution of Ta^{5+} on Zr^{4+} sites. Substitution of Ta^{5+} is preferred to Al doping, since Ta, unlike Al, does not reside on the Li sublattice (potentially impeding Li migration), and thus yields higher conductivity. In fact, $Li_{6.5}La_3Zr_{1.5}Ta_{0.5}O_{12}$ (0.5 Li vacancy per formula unit) has achieved a conductivity close to 1 mS cm^{-1} at room temperature [22].

15.1.4 Challenges

There are several challenges that limit the use of LLZO in practical batteries.

Interfacial resistance: The interfacial resistance between a Li metal anode and LLZO is a critical factor for developing efficient solid-state batteries. While several studies have reported significantly higher interfacial resistance than in conventional (liquid-electrolyte) Li-ion batteries [27, 28], coating of the LLZO surface has dramatically lowered this resistance [29, 30]. The reduction of interfacial resistance has been correlated with enhanced Li wettability of the LLZO surface, but the underlying mechanisms are not well understood [29, 30]. It remains to be seen if interfacial coatings are required to achieve low interfacial resistance.

Li dendrite formation: Li metal cells based on liquid electrolytes exhibit the formation of Li dendrites during Li plating (during charge), which leads to an internal short circuit between the electrodes [31, 32]. Use of a stiff solid electrolyte in place of a liquid electrolyte can in principle suppress Li dendrite formation. According to the model of Monroe and Newman [9, 33], a solid electrolyte having a shear modulus that is a factor of approximately two times greater than the shear modulus of metallic Li should suppress dendrite initiation. Nevertheless, Li dendrites could still form even in the presence of an electrolyte that satisfies Monroe's criterion if microstructural features in the solid electrolyte result in inhomogeneous Li plating. For example, recent studies showed that Li dendrites can penetrate LLZO membranes along its grain boundary network, resulting in cell failure [34]. This surprising result indicates that microstructural features, and not elastic properties alone [9], should be considered in the design of practical SE.

Grain boundary resistance: The typical procedure for the synthesis of SE results in a polycrystalline solid rather than a single crystal monolith. Grain boundaries (GBs) between individual crystallites can influence of the transport properties due to their different structure and potentially distinct chemical composition compared to the bulk. Regarding ion transport at GBs, earlier studies have shown that ion diffusivity can be enhanced or suppressed at GBs [35, 36]. In LLZO, several studies showed that the presence of GB can decrease the Li ion conductivity [37, 38], although the magnitude of this effect can be small and depends on the processing conditions. At present, the atomic-scale processes occurring at GB in LLZO, including transport mechanisms and local elastic properties, remain poorly understood.

As a step toward overcoming those challenges, the work presented in this study characterizes several properties of LLZO using atomic-scale simulations based on density functional theory (DFT) and (classical) MD [39–43]. The knowledge gained from these studies will hopefully aid in the development of strategies for improving solid-state batteries employing LLZO, as well as for other classes of SE.

15.2 Elastic Properties of $Li_7La_3Zr_2O_{12}$

A viable SE should exhibit mechanical properties that allow for its manufacture and use in battery applications. In particular, the elastic moduli of a SE are of key importance. First, it has been hypothesized that a SE's shear modulus impacts the tendency for dendrites to form on the anode surface during plating [9]. Linear elasticity analyses performed by Monroe and Newman suggest that the shear modulus of the SE should be at least twice that of Li metal to prevent dendrite nucleation [9]. Secondly, Young's modulus impacts the fracture strength of a SE. Based on Griffith theory, the fracture strength can be estimated using Young's modulus and the surface energy [44].

Unfortunately, few investigations of the mechanical properties of LLZO have been reported. Ni et al. [45] measured the room-temperature elastic moduli (Young's modulus, E, shear modulus, G, bulk modulus, B, and Poisson's ratio, ν) of Al-doped cubic LLZO ($Li_{6.24}Al_{0.24}La_3Zr_2O_{11.98}$, volume fraction porosity, P = 0.03) using resonant ultrasound spectroscopy (RUS) [45]. The values reported were E = 149.8 GPa, G = 59.6 GPa, and B = 102.8 GPa. On the other hand, the elastic properties of Ta-doped LLZO are not available in the literature. Similarly, a full evaluation of the elastic constants of LLZO also appears to be absent.

The present analyses aim to close these gaps by determining the elastic constants and moduli for Al- and Ta-doped LLZO using first-principles computation. The calculations have been validated by comparing the computed values with those reported by experimental measurements. In addition, the elastic properties of metallic Li were re-visited computationally. The data were used with the elasticity analyses of Monroe and Newman [9], and in combination with the measured LLZO elastic properties, to assess the effectiveness of LLZO in suppressing dendrite formation at the Li–LLZO interface.

Figure 15.1a shows the calculated total energy of Al-doped LLZO as a function of cell volume, and the associated fit of this data to the Murnaghan equation of state. Similarly, Figure 15.1b–d shows the total energy as a function of uniaxial, monoclinic, and orthorhombic strain, respectively; solid lines represent fits to a quadratic polynomial. A $2 \times 2 \times 2$ k-point mesh (four irreducible k-points) was determined to be sufficient to achieve converged values for the bulk modulus and elastic constants; this k-point mesh was also adopted for calculations of the elastic constants, C_{11}, C_{12}, and C_{44}.

The calculated elastic constants and moduli of Al-doped LLZO are summarized in Table 15.1. The calculated B, E, and G at 0 K are 112.4, 162.6, and 64.6 GPa, respectively. The elastic moduli of oxides typically decrease with increasing temperature. Representative decreases for common oxides between 0 and 298 K are: ~4% for MgO, ~1% for Al_2O_3 [46], and ~9% for Li_2O [47]. Based on the trends for other oxides, the calculated modulus of LLZO at 0 K was reduced by ~5% to estimate its behavior at room temperature. Adopting this scaling factor, we arrive at room-temperature values of 154.5 and 61.4 GPa for E and G, respectively. DFT calculations were also performed to evaluate the elastic constants of Ta-doped LLZO. The elastic constants derived from the three methods have very similar values. As shown in Table 15.1, the calculated B, E, and G are 99.2, 154.9, and 62.5 GPa, respectively. Assuming a 5% softening of moduli between 0 and 298 K, the predicted values of E and G at room temperature for Ta-doped LLZO are 147.2 and 59.4 GPa, respectively.

From nanoindentation experiments, Young's modulus (E) of the Al-doped LLZO is 150.3 ± 2.2 GPa; for Ta-doped LLZO a value of 153.8 ± 2.7 GPa was measured [39].

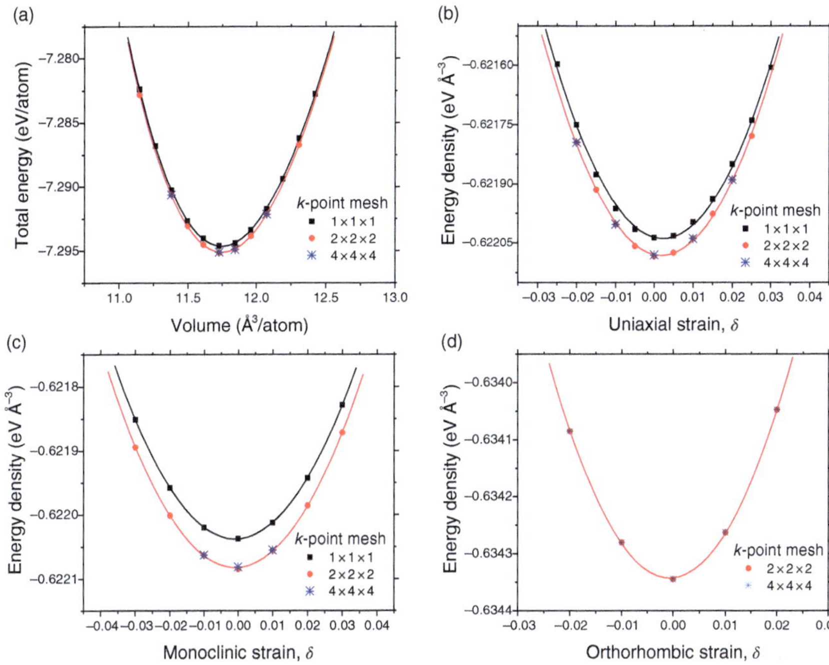

Figure 15.1 Calculated total energy (or energy density) of Al-doped LLZO as a function of the (a) cell volume, (b) uniaxial strain, (c) monoclinic strain, and (d) orthorhombic strain. Source: Yu et al. [39]/with permission of American Chemical Society.

Because the size of each nanoindentation is approximately 500 nm – which is 2–100 times smaller than the diameter of each grain – the measured elastic moduli values are approximately equivalent to an average over several single grain/quasi-single crystal measurements performed on random orientations. Thus, when comparing the calculated (154.5 GPa) and measured average quasi-single crystal values (150.3 GPa), excellent agreement is achieved (Table 15.1). At 298 K the Al-doped and Ta-doped calculated and measured quasi-single crystal elastic moduli are within 3% and 5%, respectively.

Good agreement is also obtained with impulse-excitation experiments, where E for Al-doped LLZO was found to be 146.1 ± 0.8 GPa and 139.9 ± 2.7 GPa for Ta-doped LLZO (Table 15.1). Because the impulse excitation technique interrogates a relatively large volume of the specimen beneath the transducer (8 mm diameter) and the LLZO grains were in this case less than 50 μm in diameter, the measured elastic moduli represent isotropic bulk polycrystalline values that include <2% porosity.

Experimentally, G measured by nanoindentation is 59.8 ± 0.9 GPa for Al-doped LLZO and 61.2 ± 1.1 GPa for the Ta-doped sample (Table 15.1). Similar values were reported by impulse excitation: 58.1 ± 0.3 GPa for Al-doped LLZO and 55.7 ± 0.8 GPa for Ta-doped LLZO. Importantly, the data in Table 15.1 demonstrate that the calculated and measured (by both impulse excitation and nanoindentation) E and G moduli are within 5%. In addition, from Table 15.1 it can be observed that experimental

Table 15.1 Elastic properties of Al-doped and Ta-doped LLZO.[a]

	C_{11}	C_{12}	C_{44}	B	E	G	v
Al-doped LLZO							
DFT (0 K)	187.0	75.1	71.0	112.4	162.6	64.6	0.26
DFT extrapolated (298 K)					154.5	61.4	
Impulse excitation (298 K)				100.2 ± 0.6	146.1 ± 0.8	58.1 ± 0.3	
Nanoindentation (298 K)					150.3 ± 2.2	59.8 ± 0.9	
RUS (298 K) Ref. [45]				102.8 ± 0.3	149.8 ± 0.4	59.6 ± 0.1	0.257 ± 0.002
Ta-doped LLZO							
DFT (0 K)	169.8	63.9	69.8	99.2	154.9	62.5	0.24
DFT extrapolated (298 K)					147.2	59.4	
Impulse excitation (298 K)				96.0 ± 1.4	139.9 ± 2.1	55.7 ± 0.8	
Nanoindentation (298 K)					153.8 ± 2.7	61.2 ± 1.1	

a) The elastic constants and moduli are expressed in GPa.

and calculated E and G values for Al-doped LLZO are consistent with literature values reported from RUS measurements of similar Al-doped LLZO [45].

The average of the calculated, nanoindentation, and impulse excitation measurements for E (Table 15.1 at 298 K) for Al and Ta-doped LLZO is 150 and 147 GPa, respectively. Using $v = 0.26$, these elastic moduli averages give average shear moduli of 60 and 58 GPa for the Al- and Ta-doped LLZO, respectively. The significance of these values will be discussed below after revisiting the shear moduli data for Li.

As previously discussed, in the model of Monroe et al., a solid electrolyte having a shear modulus that is a factor of 2 greater than the shear modulus of metallic Li should suppress dendrite initiation [9]. To facilitate this comparison, the shear modulus for Li was also calculated, and a value of 4.25 GPa was predicted. Thus, based on Monroe's model, a solid electrolyte having a room-temperature shear modulus greater than approximately 8.5 GPa is required to suppress dendrite formation. The averaged shear moduli of Al- and Ta-doped LLZO, 58–60 GPa, are more than 10 times greater than the shear modulus of Li and, therefore, should suppress Li dendrite formation based on Monroe's model. Nevertheless, scenarios may exist in which dendrites could form even in the presence of a solid electrolyte that satisfies Monroe's criterion. This could occur, for example, in cases where significant porosity exists at the electrolyte/anode interface or due to microstructural inhomogeneities in the SE. In other words, satisfying Monroe's criterion should be viewed as a necessary, but not sufficient condition for a viable solid electrolyte.

15.3 Potential Failure Modes Arising from LLZO Microstructure

Li metal penetration has been recently reported in several SEs, such as β-Li_3PS_4 [48], $Li_2(OH)_{0.9}F_{0.1}Cl$ [49], and $Li_7La_3Zr_2O_{12}$ (LLZO) [34, 50]. This behavior occurred despite the fact that the measured densities and moduli for these materials were very high: in the case of LLZO, $G \sim 60$ GPa, which is more than an order of magnitude larger than that of Li metal, while the measured density was 98% of the theoretical value, indicating minimal contributions from porosity. Importantly, the metal penetrants were observed to follow pathways through the SE that coincided with the grain boundary network [34], or with the location of other microstructural features such as pores or surface cracks [48, 51].

The penetration of a stiff ceramic SEs by soft Li is a surprising observation, and raises the question, "how is this possible?" Experimental data indicating contributions from the SE's grain boundaries, pores, and surface flaws suggest that *microstructural features* play a critical role in the formation and propagation of dendrites in SE. Existing models for dendrite nucleation in the presence of a SE do not account for microstructural inhomogeneities – the SE is typically assumed to be homogeneous [52, 53].

Figure 15.2 illustrates three possible scenarios by which microstructural features can promote inhomogeneous Li plating, resulting in metal penetration through a solid electrolyte. The first scenario, shown on the left of Figure 15.2, arises from poor wetting at the Li/SE interface [29, 41]. Li metal will wet a "clean" LLZO surface; however, non-wetting behavior is observed when surface contaminant phases such as LiOH and Li_2CO_3 are present. The Li wettability of LLZO has been examined as a means to explain the origin of high interfacial resistance in the presence of a contamination layer. DFT calculations in Figure 15.3 indicate that Li strongly wets LLZO, but not Li_2CO_3 and LiOH, which is consistent with experimental measurements of contact angle [41]. The removal of surface contamination dramatically reduces the interfacial resistance to $2\,\Omega\,cm^2$, comparable to solid–liquid interfaces in Li-ion cells [41].

Surface non-wetting is also hypothesized to also contribute to inhomogeneous plating. The presence of surface contaminants results in the focusing of Li-ions toward the limited contact points where the SE directly interfaces with the Li anode. This current focusing results in inhomogeneous electrodeposition of Li, and ultimately to the penetration of Li metal into the SE. A similar mechanism has been proposed for the penetration of sodium metal through β-Al_2O_3 [54]. In the case of LLZO, cleaning of the SE surface was found to delay the onset of Li penetration

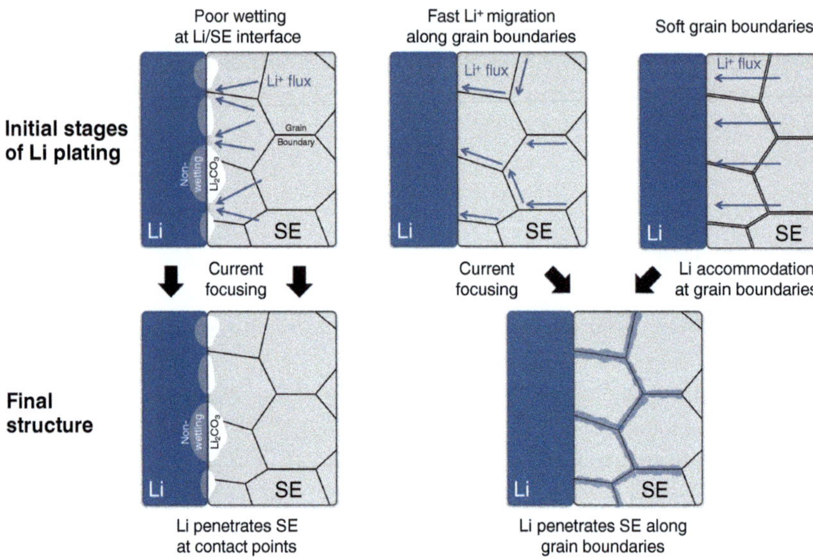

Figure 15.2 Potential microstructural contributions to inhomogeneous Li plating from a solid electrolyte. Left and middle: current focusing mechanisms resulting from poor interfacial wetting (left) or (middle) fast Li migration along GBs. Right: Li penetration along elastically-soft GBs. Source: Yu et al. [43]/with permission of American Chemical Society.

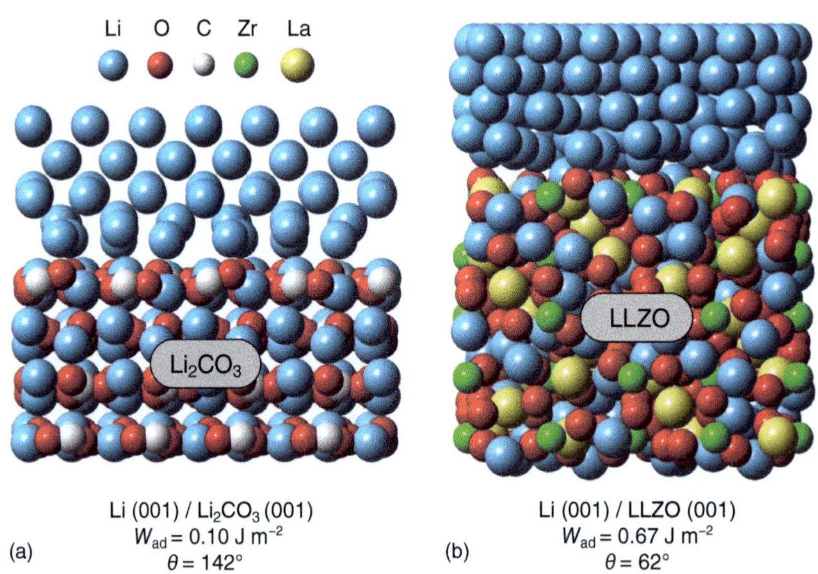

Figure 15.3 Calculated work of adhesion (W_{ad}), contact angle (θ), and atomic structure for the (a) Li–Li_2CO_3 and (b) Li–LLZO interfaces. Source: Sharafi et al. [41]/with permission of American Chemical Society.

to higher current densities [41]. Nevertheless, dendrites were still observed at current densities (0.6 mA cm^{-2}) far below those needed for a practical battery (3–10 mA cm^{-2}) [55]. Thus, although increasing interfacial wetting is helpful, it appears to be insufficient on its own to eliminate the dendrite problem: other mechanisms appear to be contributing.

A second scenario that potentially results in dendrite formation is shown in the middle portion of Figure 15.2. This mechanism also involves current focusing, and is caused by fast Li-ion migration along GBs [42]. In the presence of fast GB transport, Li plating would occur preferentially in regions where GBs intersect the electrode surface. Assuming the arrival rate of Li at these intersections is faster than its lateral migration away (i.e. parallel to the interface), then the resulting "pile-up" of Li could nucleate dendrites.

This "fast GB diffusion" hypothesis was examined by calculating the rate of Li-ion migration along three low-energy tilt GBs($\Sigma3$ & $\Sigma5$) of LLZO using MD. These calculations, summarized in Figure 15.4, reveal that Li transport is generally slower in the GB region compared to the bulk. The magnitude of the GB's effect on Li-ion transport is sensitive to the GB structure and (as expected) temperature. For example, Li-ion diffusion is comparable in all three GBs at high temperatures (>900 K), and only slightly slower than bulk diffusion. In contrast, room-temperature diffusion differs significantly between the $\Sigma3$ and $\Sigma5$ systems: diffusion in the more compact $\Sigma3$ boundary remains relatively fast (~50% of the bulk rate), while transport in the $\Sigma5$ boundaries is roughly two orders of magnitude slower than the bulk (Table 15.2). These trends are reflected in the activation energies for diffusion, which in the $\Sigma5$ boundaries are up to 35% larger than in bulk LLZO, and are equal to the bulk for the $\Sigma3$ boundary. Diffusion within the $\Sigma5$ boundaries is isotropic, while anisotropic transport is observed in the $\Sigma3$ system at lower temperatures (<700 K). In the latter system, intra-plane diffusion at 300 K is up to four times faster than the bulk, while transboundary diffusion is ~200 times slower.

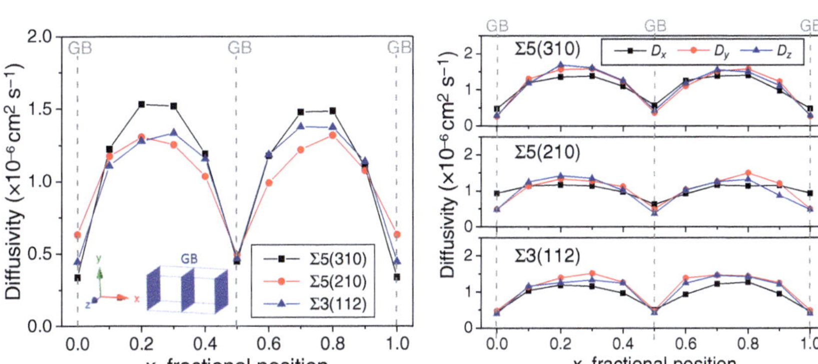

Figure 15.4 Calculated Li ion diffusivity for the three computational cells containing different GBs. (a) Total diffusivity, and (b) diffusivity decomposed along each Cartesian direction at 1000 K as a function of position normal to the GB plane. Source: Yu et al. [42]/with permission of American Chemical Society.

Table 15.2 Calculated transport properties (activation energy for diffusion, E_a, and diffusivity, D) in bulk LLZO and in three GBs.

		$T = 1000$ K				$T = 700$ K				$T = 300$ K			
System	E_a (eV)	D	D_x	D_y	D_z	D	D_x	D_y	D_z	D	D_x	D_y	D_z
		(10^{-7} cm^2 s^{-1})				(10^{-7} cm^2 s^{-1})				(10^{-13} cm^2 s^{-1})			
Bulk	0.52	13.5	—	—	—	1.01	—	—	—	9.16	—	—	—
Σ5(310)	0.64	4.30	5.57	3.61	3.72	0.22	0.24	0.17	0.24	0.15	0.11	0.10	0.28
Σ5(210)	0.71	5.50	7.35	4.90	4.23	0.17	0.21	0.16	0.14	0.03	0.05	0.02	0.02
Σ3(112)	0.52	5.10	5.40	5.35	4.55	0.50	0.18	0.81	0.52	4.71	0.04	40.0	8.12

Li-ion diffusivities are reported for calculations at 700 and 1000 K, and extrapolated to 300 K. The total diffusivity (D) and the diffusivity projected along the three Cartesian coordinates are reported: D_y and D_z correspond to diffusion within the GB plane; D_x corresponds to diffusion normal to the GB plane.

Our observation of mixed GB transport contributions – some boundaries support fast diffusion (comparable to bulk diffusion), while others are slow – is consistent with the moderate GB resistance observed in polycrystalline LLZO samples processed at high temperatures [38]. These data also suggest that higher-energy GBs with less-compact structures should penalize Li-ion conductivity to a greater degree. Additional experimental study is needed to quantify the presence, and characterize the structure, of high-energy boundaries as a function of processing temperature. Such studies would enable computational characterization of those systems, similar to the analyses performed here on low-energy GB.

The present MD calculations reveal that Li transport is generally reduced in the GB region of LLZO. Thus, current focusing along GBs due to fast GB transport does not appear to be a mechanism that can result in Li dendrite penetration. Hence, this hypothesis has been ruled out.

Finally, it is possible that "softening" of the SE in the vicinity of GBs provides a mechanism for dendrite initiation/penetration. This hypothesis represents a third microstructure-related scenario, and is shown on the right in Figure 15.2. Here, "softening" is defined as a reduction in the elastic moduli close to the GBs. These effects could arise from deviations in density and atomic structure (i.e. distinct bonding relative to the bulk) near the GB plane. This scenario is inspired by the elasticity model of Monroe and Newman [9], yet accounts for inhomogeneities in elastic properties. Unlike the first two scenarios from Figure 15.2, in which current focusing drives inhomogeneous Li deposition, in the present case Li deposition is initially homogeneous (i.e. the Li-ion current is *not* focused). However, as plating progresses, Li will accumulate preferentially in softer regions near electrode/GB junctions via interfacial diffusion away from the stiffer, GB-free regions. The resulting Li protrusions will generate locally stronger electric fields that will focus subsequent Li deposition. Indeed, prior studies have indicated the possibility of softening at GBs [56, 57]. If these effects are also present in SEs, then they could contribute to dendrite penetration of SEs.

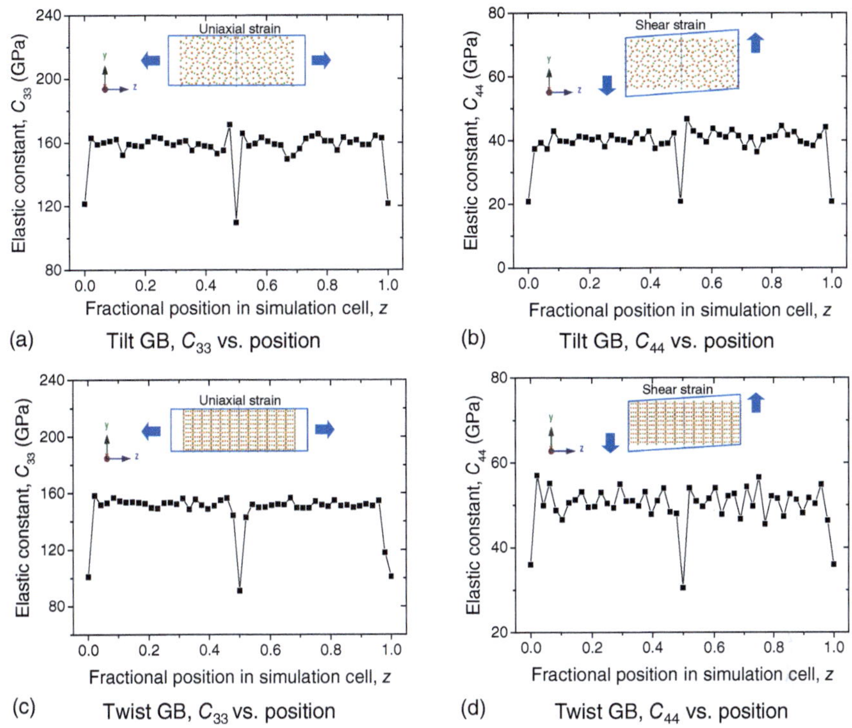

Figure 15.5 Calculated elastic constants C_{33} and C_{44} at 300 K as a function of position normal to the GB planes for the (a, b) the $\Sigma 5$ symmetric tilt GB cell and (c, d) the $\Sigma 5$ twist GB cell.

The calculation results shown in Figure 15.5 demonstrate that significant softening can occur at GBs in SEs [43]. Adopting LLZO as a model, MD simulations on two, low-energy $\Sigma 5$ tilt and twist GBs were conducted. The elastic constants associated with uniaxial strain perpendicular to the GB plane and with shear parallel to the GB were calculated at 300 K. These calculations indicate a severe softening in the immediate vicinity of the GB: elastic constants are observed to be up to 50% smaller at GBs than in the bulk (Table 15.3). Thus, nanoscale softening attributed to microstructural features such as GBs may explain *why* these features are susceptible to metal penetration during electrodeposition. Additional study is needed to characterize the elastic properties of other types of grain boundaries, account for potential variations in grain boundary composition arising from impurities or segregation, and to explore effects in solids beyond LLZO.

In total, this work highlights scenarios by which a solid electrolyte's microstructure can contribute to inhomogeneous electrodeposition and dendrite penetration. At present these scenarios have received limited attention in the battery literature. A better understanding of the role of microstructure will aid in the design of long-lived solid-state batteries.

Table 15.3 Calculated elastic constants, C_{33} and C_{44} (in GPa), in the bulk and GB regions for the Σ5 tilt and twist GB simulation cells.

Elastic constant	Σ5 tilt GB			Σ5 twist GB		
	Bulk	GB	Δ (%)	Bulk	GB	Δ (%)
C_{33}, uniaxial	159	115	−28	152	96	−37
C_{44}, shear	40	21	−48	51	33	−35

15.4 Conclusions

LLZO is one of the most promising Li SE due to its favorable combination of high conductivity and chemical stability against Li metal. However, there are several challenges that limit the use of LLZO in practical batteries. These include adhesion/wettability at electrode interfaces, interfacial resistance, inhomogeneous transport and mechanical properties arising from microstructural imperfections, and dendrite suppression. As a step toward overcoming those challenges, the work presented in this chapter characterizes several properties of LLZO using DFT and MD calculations. The knowledge gained will enable the development of strategies for improving solid-state batteries employing LLZO, as well as for other classes of SE.

The elastic properties of LLZO and Li metal were examined to assess the effectiveness of LLZO in suppressing dendrite formation at the Li–LLZO interface. Based on the model of Monroe and Newman, LLZO is predicted to be stiff enough to suppress lithium dendrite formation. For example, linear elasticity models suggest that a solid electrolyte having a shear modulus greater than approximately 8.5 GPa can suppress dendrite formation on a Li anode. The present calculations yield averaged shear moduli for Al- and Ta-doped LLZO of 58–60 GPa, far exceeding the targeted value. Nevertheless, it is important to recognize that dendrites could still form even in the presence of a solid electrolyte that satisfies Monroe's criterion due to contribution from microstructural features such as porosity at the interface, poor interfacial wetting, or microstructural features in the bulk of the SE. Thus, the Monroe–Newman criterion should be considered as a necessary but insufficient condition for achieving dendrite resistance: contributions from microstructure should also be considered.

Current focusing caused by fast Li-ion migration along GBs is explored as one possible scenario resulting in dendrite formation. In the presence of fast GB transport, Li plating would occur preferentially in regions where GBs intersect the electrode surface. Assuming the arrival rate of Li at these intersections is faster than its lateral migration away, then the resulting "pile-up" of Li could nucleate dendrites. This "fast GB diffusion" hypothesis was studied by calculating the rate of Li-ion migration along three low-energy tilt GBs (Σ3 & Σ5) of LLZO. Our MD results reveal that Li-ion diffusion is comparable in all three tilt GBs at high temperatures (>900 K), and only slightly slower than bulk diffusion. In contrast, room-temperature diffusion differs significantly between the Σ3 and Σ5 systems: diffusion in the more compact

$\Sigma3$ boundary remains relatively fast (~50% of the bulk rate), while transport in the $\Sigma5$ boundaries is roughly two orders of magnitude slower than the bulk. Diffusion within the $\Sigma5$ boundaries is isotropic, while anisotropic transport is observed in the $\Sigma3$ system at lower temperatures (<700 K). In the latter system, intra-plane diffusion at 300 K is up to four times faster than the bulk, while transboundary diffusion is ~200 times slower. Generally, Li transport is reduced in the GB region. Thus, current focusing along GBs due to faster GB Li-ion transport does not appear to explain Li dendrite penetration.

As a second mechanism we explore interfacial wettability at the Li/LLZO interface. The Li wettability of LLZO is examined as a means to explain the origin of high interfacial resistance in the presence of a contamination layer. DFT calculations indicate that Li strongly wets LLZO, but not Li_2CO_3 and LiOH, which is consistent with experimental measurements of contact angle. The presence of surface contaminants results in the focusing of Li-ions toward the limited contact points where the SE directly interfaces with the Li anode. This current focusing results in inhomogeneous electrodeposition of Li, and ultimately to the penetration of Li metal into the SE.

As a third mechanism we hypothesize that softening of the LLZO in the vicinity of GBs can provide a mechanism for dendrite penetration. These effects could arise from deviations in density and atomic structure near the GB plane. MD calculations are performed to examine the softening using two, low-energy $\Sigma5$ tilt and twist GBs. The elastic constants associated with uniaxial strain perpendicular to the GB plane and with shear parallel to the GB were calculated at 300 K. These calculations indicate a severe softening in the immediate vicinity of the GB: elastic constants are observed to be up to 50% smaller at GBs than in the bulk. We propose that nanoscale softening attributed to microstructural features such as GBs may explain why these features are susceptible to metal penetration during electrodeposition.

Acknowledgements

S.Y. and D.J.S. acknowledge financial support from the US Department of Energy, Advanced Research Projects Agency – Energy, grant no. DE-AR0000653 and by the U.S. Department of Energy (DOE) Office of Energy Efficiency and Renewable Energy (EERE) Vehicle Technologies Office (VTO) and Advanced Battery Material Research (BMR) program under Grant DE- EE-0006821.

References

1 Armand, M. and Tarascon, J.-M. (2008). Building better batteries. *Nature* 451: 652.

2 Larcher, D. and Tarascon, J.M. (2015). Towards greener and more sustainable batteries for electrical energy storage. *Nature Chemistry* 7 (1): 19–29.

3 Dunn, B., Dunn, B., Kamath, H., and Tarascon, J. (2011). Electrical energy storage for the grid: a battery of choices. *Science Magazine* 334 (6058): 928–936.

4 Li, J., Ma, C., Chi, M. et al. (2015). Solid electrolyte: the key for high-voltage lithium batteries. *Advanced Energy Materials* 5 (4): 1–6.

5 Bruce, P.G., Freunberger, S.A., Hardwick, L.J., and Tarascon, J.-M. (2012). Li–O_2 and Li–S batteries with high energy storage. *Nature Materials* 11 (1): 19–29.

6 Liu, Y., He, P., and Zhou, H. (2018). Rechargeable solid-state li–air and li–s batteries: materials, construction, and challenges. *Advanced Energy Materials* 8 (4): 1–22.

7 Roth, E.P. and Orendorff, C.J. (2012). How electrolytes influence battery safety. *Interface Magazine* 21 (2): 45–49.

8 Goodenough, J.B. and Kim, Y. (2010). Challenges for rechargeable Li batteries. *Chemistry of Materials* 22 (3): 587–603.

9 Monroe, C. and Newman, J. (2005). The impact of elastic deformation on deposition kinetics at lithium/polymer interfaces. *Journal of the Electrochemical Society* 152 (2): A396.

10 Manthiram, A., Chemelewski, K., and Lee, E.-S. (2014). A perspective on the high-voltage $LiMn_{1.5}Ni_{0.5}O_4$ spinel cathode for lithium-ion batteries. *Energy & Environmental Science* 7 (4): 1339.

11 Bachman, J.C., Muy, S., Grimaud, A. et al. (2016). Inorganic solid-state electrolytes for lithium batteries: mechanisms and properties governing ion conduction. *Chemical Reviews* 116 (1): 140–162.

12 Kanno, R., Hata, T., Kawamoto, Y., and Irie, M. (2000). Synthesis of a new lithium ionic conductor, thio-LISICON-lithium germanium sulfide system. *Solid State Ionics* 130 (1): 97–104.

13 Kamaya, N., Homma, K., Yamakawa, Y. et al. (2011). A lithium superionic conductor. *Nature Materials* 10 (9): 682–686.

14 Muramatsu, H., Hayashi, A., Ohtomo, T. et al. (2011). Structural change of Li2S–P2S5 sulfide solid electrolytes in the atmosphere. *Solid State Ionics* 182 (1): 116–119.

15 Arbi, K., Mandal, S., Rojo, J.M., and Sanz, J. (2002). Dependence of ionic conductivity on composition of fast ionic conductors $Li_{1+x}Ti_{2-x}Al_x(PO_4)_3$, $0 \le x \le 0.7$. A parallel NMR and electric impedance study. *Chemistry of Materials* 14 (3): 1091–1097.

16 Hartmann, P., Leichtweiss, T., Busche, M.R. et al. (2013). Degradation of NASICON-type materials in contact with lithium metal: formation of mixed conducting interphases (MCI) on solid electrolytes. *Journal of Physical Chemistry C* 117 (41): 21064–21074.

17 Bohnke, O. (2008). The fast lithium-ion conducting oxides $Li_{3x}La_{2/3-x}TiO_3$ from fundamentals to application. *Solid State Ionics* 179 (1–6): 9–15.

18 Yao, X., Huang, B., Yin, J. et al. (2015). All-solid-state lithium batteries with inorganic solid electrolytes: review of fundamental science. *Chinese Physics B* 25 (1).

19 Murugan, R., Thangadurai, V., and Weppner, W. (2007). Fast lithium ion conduction in garnet-type $Li_7La_3Zr_2O_{12}$. *Angewandte Chemie, International Edition* 46 (41): 7778–7781.

20 Geiger, C.A., Alekseev, E., Lazic, B. et al. (2011). Crystal chemistry and stability of "$Li_7La_3Zr_2O_{12}$" garnet: a fast lithium-ion conductor. *Inorganic Chemistry* 50 (3): 1089–1097.

21 Rangasamy, E., Wolfenstine, J., and Sakamoto, J. (2012). The role of Al and Li concentration on the formation of cubic garnet solid electrolyte of nominal composition $Li_7La_3Zr_2O_{12}$. *Solid State Ionics* 206: 28–32.

22 Thompson, T., Sharafi, A., Johannes, M.D. et al. (2015). A tale of two sites: on defining the carrier concentration in garnet-based ionic conductors for advanced Li batteries. *Advanced Energy Materials* 5 (11): 1–9.

23 Thompson, T., Wolfenstine, J., Allen, J.L. et al. (2014). Tetragonal vs. cubic phase stability in Al-free Ta doped $Li_7La_3Zr_2O_{12}$ (LLZO). *Journal of Materials Chemistry A* 2 (33): 13431–13436.

24 Awaka, J., Kijima, N., Hayakawa, H., and Akimoto, J. (2009). Synthesis and structure analysis of tetragonal $Li_7La_3Zr_2O_{12}$ with the garnet-related type structure. *Journal of Solid State Chemistry* 182 (8): 2046–2052.

25 Allen, J.L., Wolfenstine, J., Rangasamy, E., and Sakamoto, J. (2012). Effect of substitution (Ta, Al, Ga) on the conductivity of $Li_7La_3Zr_2O_{12}$. *Journal of Power Sources* 206: 315–319.

26 Bernstein, N., Johannes, M.D., and Hoang, K. (2012). Origin of the structural phase transition in $Li_7La_3Zr_2O_{12}$. *Physical Review Letters* 109 (20): 1–5.

27 Sharafi, A., Meyer, H.M., Nanda, J. et al. (2016). Characterizing the Li-$Li_7La_3Zr_2O_{12}$ interface stability and kinetics as a function of temperature and current density. *Journal of Power Sources* 302: 135–139.

28 Cheng, L., Chen, W., Kunz, M. et al. (2015). Effect of surface microstructure on electrochemical performance of garnet solid electrolytes. *ACS Applied Materials & Interfaces* 7 (3): 2073–2081.

29 Han, X., Gong, Y., Fu, K. et al. (2017). Negating interfacial impedance in garnet-based solid-state Li metal batteries. *Nature Materials* 16 (5): 572–579.

30 Tsai, C.L., Roddatis, V., Chandran, C.V. et al. (2016). $Li_7La_3Zr_2O_{12}$ interface modification for Li dendrite prevention. *ACS Applied Materials & Interfaces* 8 (16): 10617–10626.

31 Tarascon, J.-M. and Armand, M. (2010). Issues and challenges facing rechargeable lithium batteries. In: *Materials for Sustainable Energy*, 171–179. UK: Macmillan Publishers Ltd.

32 Xu, Y.-N. and Ching, W.Y. (1999). Electronic structure of yttrium aluminum garnet $Y_3Al_5O_{12}$. *Physical Review B* 59 (16): 10530–10535.

33 Monroe, C. and Newman, J. (2003). Dendrite growth in lithium/polymer systems. *Journal of the Electrochemical Society* 150 (10): A1377.

34 Cheng, E.J., Sharafi, A., and Sakamoto, J. (2017). Intergranular Li metal propagation through polycrystalline $Li_{6.25}Al_{0.25}La_3Zr_2O_{12}$ ceramic electrolyte. *Electrochimica Acta* 223: 85–91.

35 Williams, N.R., Molinari, M., Parker, S.C., and Storr, M.T. (2015). Atomistic investigation of the structure and transport properties of tilt grain boundaries of UO_2. *Journal of Nuclear Materials* 458: 45–55.

36 Sun, L., Marrocchelli, D., and Yildiz, B. (2015). Edge dislocation slows down oxide ion diffusion in doped CeO_2 by segregation of charged defects. *Nature Communications* 6: 1–10.

37 Tenhaeff, W.E., Rangasamy, E., Wang, Y. et al. (2014). Resolving the grain boundary and lattice impedance of hot-pressed $Li_7La_3Zr_2O_{12}$ garnet electrolytes. *ChemElectroChem* 1 (2): 375–378.

38 David, I.N., Thompson, T., Wolfenstine, J. et al. (2015). Microstructure and Li-Ion conductivity of hot-pressed cubic $Li_7La_3Zr_2O_{12}$. *Journal of the American Ceramic Society* 98 (4): 1209–1214.

39 Yu, S., Schmidt, R.D., Garcia-Mendez, R. et al. (2016). Elastic properties of the solid electrolyte $Li_7La_3Zr_2O_{12}$ (LLZO). *Chemistry of Materials* 28 (1): 197–206.

40 Sharafi, A., Yu, S., Naguib, M. et al. (2017). Impact of air exposure and surface chemistry on Li–$Li_7La_3Zr_2O_{12}$ interfacial resistance. *Journal of Materials Chemistry A* 5 (26): 13475–13487.

41 Sharafi, A., Kazyak, E., Davis, A.L. et al. (2017). Surface chemistry mechanism of ultra-low interfacial resistance in the solid-state electrolyte $Li_7La_3Zr_2O_{12}$. *Chemistry of Materials* 29 (18): 7961–7968.

42 Yu, S. and Siegel, D.J. (2017). Grain boundary contributions to Li-Ion transport in the solid electrolyte $Li_7La_3Zr_2O_{12}$ (LLZO). *Chemistry of Materials* 29 (22): 9639–9647.

43 Yu, S. and Siegel, D.J. (2018). Grain boundary softening: a potential mechanism for lithium metal penetration through stiff solid electrolytes. *ACS Applied Materials & Interfaces* 10 (44): 38151–38158.

44 Sakanoi, R., Shimazaki, T., Xu, J. et al. (2014). Communication: different behavior of Young's modulus and fracture strength of CeO_2: density functional theory calculations. *The Journal of Chemical Physics* 140 (12).

45 Ni, J.E., Case, E.D., Sakamoto, J.S. et al. (2012). Room temperature elastic moduli and vickers hardness of hot-pressed LLZO cubic garnet. *Journal of Materials Science* 47 (23): 7978–7985.

46 Wachtman, J.B., Tefft, W.E., Lam, D.G., and Apstein, C.S. (1961). Exponential temperature dependence of Young's modulus for several oxides. *Physics Review* 122 (6): 1754–1759.

47 Dovesi, R., Roetti, C., Freyria-Fava, C. et al. (1991). On the elastic properties of lithium, sodium and potassium oxide. An ab initio study. *Chemical Physics* 156 (1): 11–19.

48 Porz, L., Swamy, T., Sheldon, B.W. et al. (2017). Mechanism of lithium metal penetration through inorganic solid electrolytes. *Advanced Energy Materials* 7 (20): 1–12.

49 Li, Y., Zhou, W., Chen, X. et al. (2016). Mastering the interface for advanced all-solid-state lithium rechargeable batteries. *Proceedings of the National Academy of Sciences* 113 (47): 13313–13317.

50 Ren, Y., Shen, Y., Lin, Y., and Nan, C.W. (2015). Direct observation of lithium dendrites inside garnet-type lithium-ion solid electrolyte. *Electrochemistry Communications* 57: 27–30.

51 Wu, B., Wang, S., Lochala, J. et al. (2018). The role of the solid electrolyte interphase layer in preventing Li dendrite growth in solid-state batteries. *Energy & Environmental Science* 11 (7): 1803–1810.

52 Barai, P., Higa, K., and Srinivasan, V. (2017). Effect of initial state of lithium on the propensity for dendrite formation: a theoretical study. *Journal of the Electrochemical Society* 164 (2): A180–A189.

53 Tikekar, M.D., Archer, L.A., and Koch, D.L. (2016). Stabilizing electrodeposition in elastic solid electrolytes containing immobilized anions. *Science Advances* 2 (7).

54 Virkar, A.V., Viswanathan, L., and Biswas, D.R. (1980). On the deterioration of β''-alumina ceramics under electrolytic conditions. *Journal of Materials Science* 15 (2): 302–308.

55 Albertus, P., Babinec, S., Litzelman, S., and Newman, A. (2018). Status and challenges in enabling the lithium metal electrode for high-energy and low-cost rechargeable batteries. *Nature Energy* 3 (1): 16–21.

56 Adams, J.B., Wolfer, W.G., and Foiles, S.M. (1989). Elastic properties of grain boundaries in copper and their relationship to bulk elastic constants. *Physical Review B* 40 (14): 9479–9484.

57 Nomura, M. and Adams, J.B. (1994). Mechanical properties of twist grain boundaries in Cu. *Interface Science* 2 (2): 137–146.

16

Machine-Learning and Data-Intensive Methods for Accelerating the Development of Rechargeable Battery Chemistries: A Review

Austin D. Sendek[1,2], *Ekin D. Cubuk*[3], *Brandi Ransom*[2], *Jagjit Nanda*[4], and *Evan J. Reed*[2]

[1]*Aionics, Inc., 221 10th Street, Evanston, WY 82930, San Mateo, CA, USA*
[2]*Stanford University, Department of Materials Science and Engineering, 496 Lomita Mall, Suite 102, Stanford, CA 94305, USA*
[3]*Google Research, Brain Team, 1600 Amphitheatre Pkwy, Mountain View, CA 94043, USA*
[4]*Chemical Sciences Division, Oak Ridge National Laboratory, 1 Bethel Valley Road, Oak Ridge TN 37831, USA*

16.1 Introduction

Today's commercially available lithium-ion battery has given us a cheap and efficient means of energy storage, but it has not been an easy road. Nearly 20 years passed from invention to first commercialization, and another 30 years has passed to the current day where electric vehicles are beginning to approach cost parity with internal combustion vehicles. Taking decades to bring a new materials-based invention to market and to scale, however, is common across most materials commercialization efforts and is not unique to batteries per se; it is generally a consequence of trial-and-error materials design and time-intensive experimental validation. Rechargeable batteries and related electrochemical energy storage devices such as electric double-layer capacitors (EDLCs) present an especially difficult design challenge, however, as the structure–property relationships for many important properties are not well understood, and batteries necessarily contain several different materials interacting with each other in perhaps unpredictable ways. Furthermore, experimental validation of candidate materials or designs often requires developing new synthesis routes and/or running weeks-long cycling tests.

As climate change and environmental instability hasten efforts to design cheap and efficient energy storage technologies, the pressure has risen to find ever-better battery materials in less time. This has led to a surge in new materials discovery methods, which complement the fundamental characterization that has previously driven discovery.

At the core of this change is the advent of density functional theory (DFT) based on the work of Hohenberg, Kohn, and Sham beginning in the 1960s [1, 2]. DFT employs approximations to Schrodinger-like equations to compute the electronic

Transition Metal Oxides for Electrochemical Energy Storage, First Edition.
Edited by Jagjit Nanda and Veronica Augustyn.

ground state of materials with generally useful fidelity to experimental measurements on bulk properties such as formation energy [3, 4]. Recent improvements in the accuracy, speed, and accessibility of DFT calculations allow us to predict fundamental materials properties in a semiautomated manner, and more effectively explore new materials spaces. At the same time, parallel advancements in robotic technology have enabled the deployment of high-throughput semi- or fully autonomous experimental setups that are capable of synthesizing and testing many materials over time. One such setup has been commercialized by Wildcat Discovery Technologies in San Diego, USA. Another proposal is ChemOS, which aims to automate chemistry and materials experiments [5]. Both of these approaches have enabled the generation of a massive amount of data on materials properties. With the large volume of battery materials data available today, the potential has never been greater for data-driven methods to accelerate battery materials design.

The field of high-throughput experimentation on battery materials is still relatively young, and over the last several years most high-throughput battery materials discovery efforts have relied on DFT calculation. Many implementations of high-throughput DFT calculations for battery properties have emerged in recent years. These include calculations of voltage [6, 7], ionic conductivity [8, 9], band gap [10], mechanical properties [11], electrochemical stability window in solids [12], the highest occupied molecular orbital/lowest unoccupied molecular orbital (HOMO/LUMO) gap in liquids [13], and structural dimensionality in solids [14, 15]. These studies generally aim to identify promising materials for further experimental study by downselecting from large sets of candidate materials. This approach is powerful by virtue of the sheer amount of candidates that can be considered in time. For example, in 2013, Wolverton and coworkers computed voltage, expansion, and capacity in 515 candidate anodes [7]; in 2014, Cheng et al. computed redox potentials and solvation energies in 1400 organic molecules [16]; in 2015, Ceder and coworkers computed voltage and thermodynamic O_2 release temperatures in 1400 candidate cathodes [6]; in 2016, Ong and coworkers computed the full elastic tensor in 23 well-known electrolytes [11]; in 2018, Sendek et al. computed ionic conductivity in 39 solid electrolytes [8]. To understand how much of an acceleration this represents, consider that before high-throughput DFT methods were widely used (e.g. before the late 2000s), it was common practice to publish one ionic conductivity measurement per paper. For ion conductor discovery, high-throughput DFT may enable the study of >10× more materials per time.

The high-throughput DFT approach requires that the property of interest be computed on a short timescale. For example, thermodynamic properties such as voltage or formation energy can be computed directly from the electronic and ionic ground state of a material and do not require integration of quantities in time. Transport properties that involve kinetic information like ionic conductivity are comparatively expensive to compute using molecular dynamics approaches. This means DFT computation of voltage in thousands of candidates is tractable on the weeks-to-months timescale, while DFT-based molecular dynamics calculations of the ionic conductivity may be computed in only tens of candidates in the same amount of time – hence Sendek et al.'s 39 ionic conductivities [8] vs. Cheng et al.'s 1400 redox potentials [16].

One strategy for dealing with this is to identify new parameters that are correlated with the property of interest but can be computed with DFT in a fraction of the time; for example, in 2019, Shao–Horn and coworkers screened over 14 000 solids for ionic conductivity and electrochemical stability by computing reduced-order lattice dynamics descriptors [17].

It remains an open question the extent to which DFT is a useful predictive tool for materials screening or design. Although DFT generally shows good alignment with bulk single-crystal thermodynamic and transport properties, the materials synthesized and used in real devices may deviate from perfect crystals in substantial ways. For example, two different realizations of ostensibly identical synthesis procedures can result in different experimental measurements of ionic conductivity, sometimes by over an order of magnitude. This is likely due to defects, impurities, and microstructure in the experimentally measured crystals that are not present in the DFT calculations.

As our ability to generate materials data has increased dramatically over the last decade from both high-throughput DFT and experiments, so too has the opportunity to train accurate machine-learning models to predict properties in new materials without direct DFT or experimental investigation of those materials. This requires judicious data aggregation, management, and dissemination. Databases of experimental materials data have existed as long as the field itself; today, researchers can access broad amounts of materials data from such databases as National Institute of Standards and Technology (NIST) [18] and the Inorganic Crystal Structure Database (ICSD) [19]. In recent years, several databases that aggregate data from computational sources have emerged as well, including the Materials Project [20], the Open Quantum Materials Database [21], and A-FLOW [22]. The Materials Project, led by researchers at Lawrence Berkeley National Laboratory, has several tools geared specifically toward battery research, including modules for computing intercalation voltage in candidate cathodes from formation energy calculations (https://materialsproject.org/docs/battery) and electrochemical stability windows in candidate solid electrolytes from grand potential phase diagrams (https://materialsproject.org/docs/phasediagram#Grand_Potential_Phase_Diagrams_2). The Electrolyte Genome Project [13] also provides calculations of oxidation and reduction potentials in candidate liquid electrolyte organic molecules.

The aggregation and dissemination of large sets of DFT and experimental data has enabled research efforts that analyze tens of thousands of candidates – in effect, all known materials – in a way that would have been unthinkable only decades ago, before the advent of such databases and the proliferation of machine-learning approaches. For example, two such studies involving the use of data-driven approaches to electrochemical stability calculations for batteries were published in April 2019: a study published in arXiv by Sendek et al. [25] leveraged the Materials Project's database of DFT grand potential phase diagrams to compute [12, 23, 24] the electrochemical stability windows of over 6600 candidate solid Li-ion electrolytes [25]. That same month, Fitzhugh et al. published work in *Advanced Energy Materials* leveraging the same data set and approaches to compute electrochemical stability in over 67 000 candidate Li-ion electrode materials, to find promising electrode candidates to pair with a $Li_{10}SiP_2S_{12}$ solid electrolyte [26]. In both studies, the data and tools provided by the Materials Project enabled screening of vast numbers of candidates, while statistical analysis of the screening results led to the identification of new correlations among materials properties.

16.2 Machine-Learning Methods and Algorithms

With faster data generation, significant efforts are arising to build the data-driven approaches necessary to improve the rates of materials discovery. Data-driven analyses that are trained on existing data are generally much faster to evaluate than physics-based models or experimental approaches, and therefore can allow us to refine, understand, and discover new materials at a pace previously unthinkable. Through our ability to collect, disseminate, and screen data, machine-learning approaches offer a route to accelerate the materials discovery process.

At its core, machine learning is a broad set of approaches for performing statistical regression on known data to make predictions of new data. For battery applications, this typically involves training models (or fitting functions) to predict materials properties, and then using the models to predict properties in new, unknown materials.

Any machine-learning approach begins with a training set. In the case of supervised learning, the training set consists of a set of data with features (inputs) and labels (outputs), and an ML model is trained to map inputs to outputs. The features describe, e.g. the chemical and structural characteristics of a material that are likely to play some role in determining the output property of interest. They can include bond lengths, number of nearest-neighbor atoms for specific atom types, electronegativites of atoms, and a large spectrum of other physical characteristics. Most of the physics and chemistry of the problem is encapsulated in the choice of features.

In the case of unsupervised learning, the training set consists only of features with no labels, and the job of the ML model is to identify structure among the features. A third option, semisupervised learning, embraces both labeled and unlabeled data. Although rarely used for materials applications, a recent study by Cheon et al. reported excellent performance in predicting the likelihood of a material exhibiting a layered phase given only the chemical formula using a semisupervised learning approach [27].

The number of supervised learning methods is vast, as this term incorporates any mathematical function that maps an input vector to a continuous scalar ("regression") or discrete variable ("classification"). The most common regression methods include ordinary least squares (OLS) linear regression, kernel-based and regularized regression models like ridge and lasso regression, locally weighted regression, support vector machine regression, and random forest regression. Common classification methods include logistic regression, support vector machine classification, random forest classification, and softmax classification.

The amount of structure a model imposes on the data is referred to as its level of bias or variance; low flexibility models have high bias and low variance, while highly flexible models have low bias and high variance. High-bias models can loosely be thought of as having the fewest free-fitting parameters, or having the smallest number of input features that describe physical and chemical characteristics of the material. Linear models like OLS and logistic regression are high-bias models, as linearity is assumed of the data. Support vector machines, however, do not necessarily assume the data to be linear and are therefore higher variance models. In a battery research context, supervised learning models are generally used to predict a material- or cell-level property from a number of easily measurable inputs.

Common unsupervised learning methods include principal component analysis (PCA), t-distributed stochastic neighbor embedding (t-SNE), and k-means clustering. Unsupervised learning is typically used to understand correlations among features or to understand data distributions in high-dimensionality data spaces that cannot be visualized. Unsupervised learning can also be used to reduce the dimensionality of a feature space before a supervised learning approach is applied, but this is potentially risky as dimensions with high correlation to the label may be inadvertently removed if the label is not considered. Typically, unsupervised learning is used either to facilitate the application of a supervised learning approach, or to analyze the results of one. Since the goal of ML approaches in the battery field is often to make a performance prediction, supervised learning approaches tend to be a larger focus; the remainder of this review will focus on supervised learning approaches.

In general, ML models perform better when training sets are larger. It is essential that the training set encompasses enough of the materials space in question to learn to interpolate within the space and extrapolate beyond the space. Models are typically trained by optimizing over the cross-validation error, i.e. the predictive error computed by cyclically removing a fraction of the training set, training on the remainder, and validating on the removed set. Once a model is chosen, it is common practice to validate on a test or hold-out set, which is a set of known data that have been held out from the entire training process. However, the data set size is often quite small in many materials problems, and it may not be beneficial or possible to hold out a fraction of the training set. If a test set is not viable, taking the cross-validation error as a proxy for the test set error is often done. In any case, the training error (e.g. predictive error on the training set) should not be taken as an indicator of model error, as the model has "seen" this data and the error will generally be much lower than in new, "unseen" data.

Model training often involves searching for the optimal model flexibility that neither overfits the data nor underfits the data. Underfitting occurs when the model does not capture enough information, and increasing the flexibility or complexity of the model will increase the model's performance. Overfitting is the opposite phenomenon: the model is more complex than the underlying structure of the training data, and reducing its complexity will improve model performance (Figure 16.1).

Because model flexibility increases with the number of input features and the physical and chemical descriptions of the problem are largely encoded in the features, the need to avoid overfitting places limitations on the number of features or amount of chemical and physical description that can be incorporated into the model. Avoiding overfitting with small data sets can be such a restrictive constraint that the relevant physical features must be known a priori. Larger data sets may

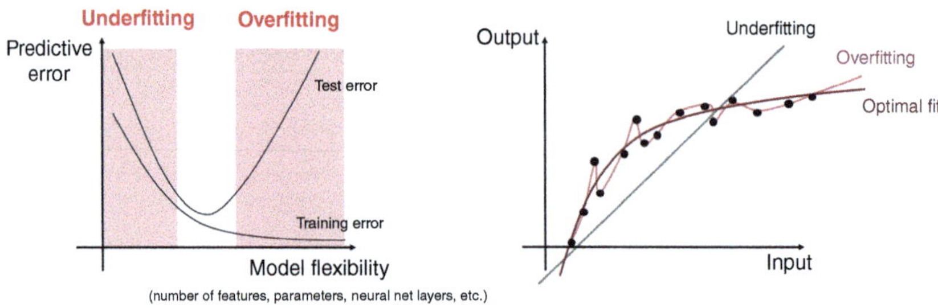

Figure 16.1 Schematic illustration of underfitting vs. overfitting. Left pane: The training error will continually decrease as the flexibility of a model increases, but the test error will tend to be high at low flexibility and high flexibility, with a minimum in between. High error at low flexibility corresponds to underfitting, while high error at high flexibility corresponds to overfitting. Right pane: Flexibility is related to the number of free parameters to fit in a model, which can include number of features, polynomial degree, layers in a neural network, etc. Underfitting results when the model cannot capture enough information about the data (gray line on right plot), while overfitting results when the model captures too much information about the data (pink line on right plot).

enable the postulation of some physical features may or may not be related to the property being predicted.

Overfitting is the phenomenon whereby the observed cross-validation or hold-out error is lower than the actual test error when the model is applied to new data. Overfitting can be difficult to identify because typical model performance metrics may indicate high accuracy even though the model cannot generalize well. The computed error on a hold-out set is more likely to identify overfitting than the cross-validation error, but sometimes hold-out errors can be unreliable, especially if approaching the boundaries of probably approximately-correct (PAC) learning theory [28]. Overfitting is a significant danger when training ML models on small training sets with fewer than 1000 data points, but the use of high-bias models with small numbers of features can reduce the likelihood of overfitting. Because of this, most applications of ML to battery materials studies employ relatively high-bias regression models rather than more complex deep-learning approaches. Deep-learning methods, including the use of neural networks, typically need training sets of thousands to millions of data points before overfitting is unlikely. Since data sets of this size are rarely available in materials applications, deep-learning approaches may not always be suited to a materials ML problem.

The application of machine-learning methods in the battery field has grown significantly over the past few years. Unlike high-throughput DFT or experiments, trained machine-learning models can evaluate a property of interest in tens of thousands of materials within a single day. The fidelity of these predictions to "ground-truth" DFT or experimental observations depends on the goodness of fit of the ML model to the ground-truth data, but in any case ML models are typically much faster to apply than any alternative approach to property prediction. This makes ML approaches attractive in cases where the number of candidate materials is so large that evaluation with DFT or experimental methods is infeasible (Figure 16.2).

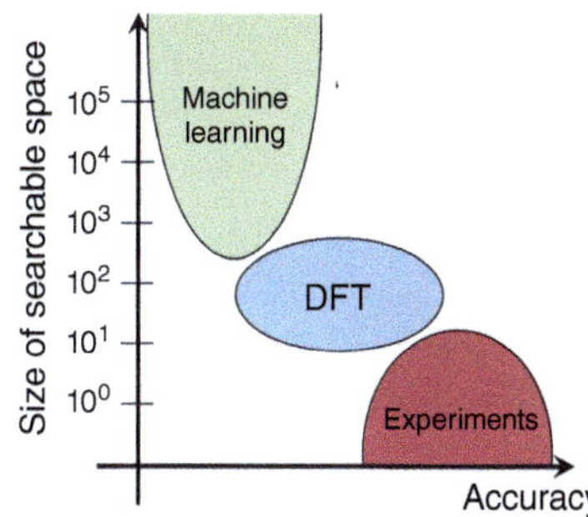

Figure 16.2 Speed vs. accuracy. On the speed vs. accuracy axes of materials discovery, machine learning represents the fastest approach. While experiments enable the study of tens of materials at a time, and DFT enables the study of hundreds to thousands of materials at a time, ML approaches enable the study of millions to billions of materials. The tradeoff is a potential loss in accuracy in ML models vs. experimental or DFT approaches.

Below we discuss a few cases where ML and artificial intelligence (AI) methods have been successfully applied for screening rechargeable battery materials.

16.3 Lithium-Ion-Conducting Solid Electrolytes

Solid lithium ion conductors are a promising application for machine-learning-enhanced materials design. The first work utilizing statistical approaches to aid in the study of solid ion conductors was published by Jalem et al. in 2012 [29]. The authors utilized DFT nudged elastic band calculations to compute the activation barriers for lithium diffusion in a number of olivine-type materials. To predict how atomic substitutions affect the diffusion barrier without re-running DFT, the authors incorporated a partial least squares (PLS) model to learn to predict the energy barrier from material structure. The features were generated based on scientific intuition of ion conductivity in olivine structures. Follow-on work utilized neural networks instead of PLS [30]. In a different approach, in 2013 Fujimura et al. combined high-throughput DFT molecular dynamics calculations with machine-learning-based modeling to optimize the composition of LISICON materials for maximal ionic conductivity [31].

In 2017, a subset of the authors (ADS, EDC, EJR, hereafter referred to as Sendek et al.) published work in which an ML model was trained to predict ionic conductivity from structure using the majority of published experimental conductivity measurements available at the time [32]. Only 40 data points could be aggregated, covering a broad materials space. The authors trained a logistic regression model on these 40 experimental data points to predict a binary indicator of whether a material would experimentally exhibit an ionic conductivity of 10^{-4} S cm^{-1} or higher at room temperature. The model utilized 20 structure-based features, designed based on scientific intuition and general structure–property relationships proposed in the literature for ionic conductivity.

To train the model, the authors minimized the leave-one-out cross-validation error over the space of models trained on all possible feature combinations, resulting in a five-feature model. The five-feature model then only requires structural information as inputs to make a conductivity prediction, and thus is several orders of magnitude faster to calculate than a DFT calculation or experimental assessment of ionic

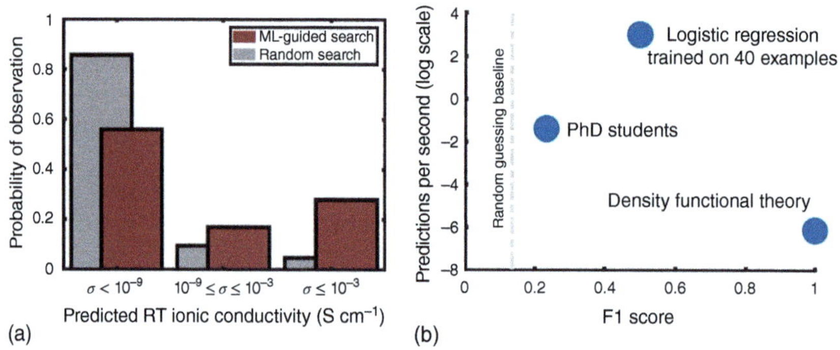

Figure 16.3 Acceleration from ML-guided search. The histograms on the left demonstrate the improved ionic conductivities observed in new materials when guided by an ML model trained on 40 experimental data points. On the right, the ML model shows significant improvement in both accuracy and speed over human intuition. Reprinted (adapted) with permissions from Sendek, A. D. et al. Copyright 2019, American Chemical Society.

conductivity in a new material. Additionally, the model is broadly applicable to any material where Li is the diffusive species and the structure is known. This enabled screening of more than 12 000 materials to identify the most promising candidate ion conductors.

To validate the model, in follow-on work, Sendek et al. used density functional theory-molecular dynamics (DFT-MD) to compute ionic conductivities in 19 of the most promising ML-predicted candidates as well as 21 randomly chosen candidates [8]. The ML-guided search identified materials with significantly higher ionic conductivities than the randomly chosen materials: 45% of the ML-guided materials vs. 14% of the randomly chosen materials. The F1 score of the model was reported to be 0.5, representing a 3× improvement over completely random guesswork and a 2× improvement over the average F1 score of human intuition (Figure 16.3, from Ref. [8]).

In follow-on work, this ionic conductivity model was paired with DFT-based predictions of electrochemical oxidation and reduction potentials in the Materials Project database to screen for promising solid electrolytes with both high ionic conductivity and wide electrochemical stability windows (Figure 16.4) [25]. The results of this analysis suggested that chloride-based materials were the most likely of all anion families to simultaneously optimize for conductivity and stability.

One limitation of any structural model is that the structure of candidate materials must be known, frequently precluding predictions on materials that have yet to be synthesized and structurally characterized. In addition to constraining searches to only known (characterized) materials, structural information takes time to extract and makes ML predictions on large numbers of candidates slower. In contrast, models built only on the chemical composition of materials (e.g. the chemical formula) can be evaluated many times faster, and can be applied to new materials spaces where materials are hypothesized to exist, but their structures are unknown.

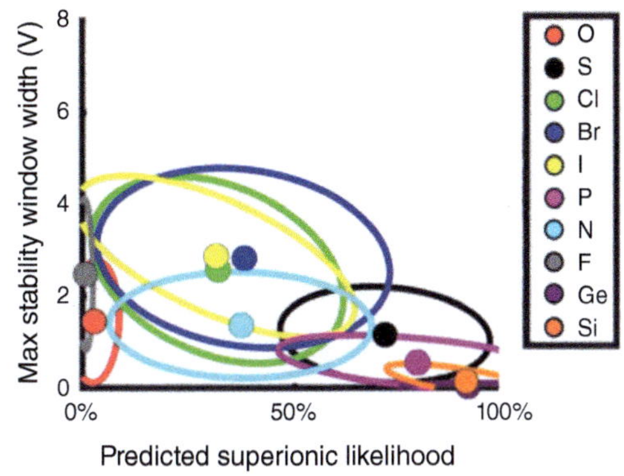

Figure 16.4 Ionic conductivity vs. electrochemical stability. Combining Sendek et al.'s ML model for predicting ionic conductivity with DFT-based calculations of electrochemical stability illuminates new quantitative trends between anion identity and solid electrolyte performance. The Cl-, Br-, and I-based electrolytes perform especially well in this analysis. Source: Reprinted with permissions from Sendek, A. D. et al. [25]. Copyright 2020, American Chemical Society.

Such a model was reported by Cubuk et al. in 2019 [33]. Cubuk et al. trained a model requiring only compositional information (using the feature set defined by Ward et al. [34]) on the same training set of 40 known ion conductors, but reported this to perform significantly worse than structural features. This is perhaps not surprising, given that compositional features capture far less chemical information than structural features. To train a structure-free model that still retained comparable predictive power to the structural model, the authors developed a transfer-learning approach for training a structure-free model on the 12 000+ predictions of the structural model. This approach showed only minimal loss in accuracy from the original structural model, while being capable of generalizing to any material where only the chemical formula is known. This enabled screening of billions of compositions to identify promising new materials that have yet to be synthesized or characterized.

Elemental features are derived from the periodic table, but are not tailored to the specific task of interest. To train a battery-property prediction model end to end, Cubuk et al. replaced the elemental properties derived from the periodic table with a new set of elemental properties that are learned from data (Figure 16.5). This work is reminiscent of word vectors that have been influential in natural language processing, where a vector that describes each word is learned from data [35–37]. The advantage of this approach is that elemental properties are learned directly from conductivity data, which allow a more efficient representation. A similar idea has also been explored by Zhou et al. [38] where the authors used machine learning to construct the periodic table from data.

Currently, many different approaches to incorporating ML into solid electrolyte screening are under development by the community. Zhang et al. applied unsupervised learning approaches to first cluster known superionic materials into discrete groups in multidimensional data space [39]. Candidate materials were then predicted to be superionic by their proximity to clusters of known superionics. Jalem et al. have expanded on their previous work to perform Bayesian optimization-guided DFT calculations of ionic conductivity over hundreds of candidates [40, 41]. Beyond ionic conductivity, recent work by Ahmad et al. applied machine-learning models to screening solid electrolyte materials for mechanical properties [42].

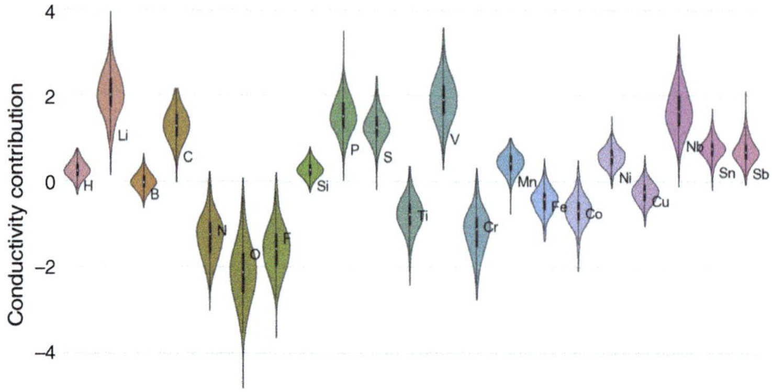

Figure 16.5 A new periodic table for Li-ion conductivity. In Cubuk et al.'s work using atom2vec representations of materials, the authors compute correlations between the presence of certain atoms in a material and the material's ionic conductivity. Positive values indicate atoms that are expected to increase Li-ion conductivity when included in a crystal, while negative values indicate those atoms expected to decrease Li-ion conductivity. Source: Reprinted with permissions from Cubuk et al. [33] Copyright 2019, AIP Publishing.

16.4 Liquid Electrolytes

The problems facing liquid electrolyte design are largely similar to those facing solid electrolytes: unclear composition–property correlations and many candidates to screen. Work in rational design of liquid electrolytes for rechargeable batteries and EDLCs was pioneered in the mid-2010s by Martin Korth and coworkers [43–45] in which candidate solutions are screened based on their known or predicted properties. The property prediction element of this work was done with the COSMOtherm software package, which makes property predictions based largely on quantum chemical models rather than structure-based ML models. Regardless, this work underscores the large volume of data on liquid electrolytes and their properties, which may be utilized in future ML-based efforts.

16.5 Cathode Design

Unlike solid electrolytes, electrode discovery has seen far less incorporation of ML-assisted discovery approaches and remains fertile ground for the development of new data-driven models. Cathodes represent a particularly thorny optimization problem, as many different and potentially conflicting performance metrics are required and it is difficult to optimize for them all simultaneously: voltage, capacity, volume change, ionic conductivity, electronic conductivity, cost, and more.

Notable recent advancements in cathode design include Cunha et al.'s predictions of Li-ion nickel manganse cobalt oxide (NMC) cathode performance from input materials properties and manufacturing protocols [46], Joshi et al.'s predictions of cathode voltage from crystal structure (for both Li- and non-Li cathodes) [47], and Baliyan and Imai's interpretation of the spectral signatures for Li-ion electrodes [48].

The large composition and structure space of candidate cathodes and many performance metrics of interest make this a promising area for future work. The Materials Project's Battery Explorer toolkit may be a valuable tool in this work with its large datasets of materials properties as well as option to search for either intercalation or conversion reactions. Taking a CoO_2 cathode, for example, the Battery Explorer currently offers DFT-predicted voltage, capacity, stability metrics and more for Li^+, Ca^{2+}, K^+, Na^+, Mg^{2+}, Al^{3+}, Zn^{2+}, and Y^{3+} intercalation.

16.5.1 Anodes

Anode materials generally undergo a greater variety of energy storage reaction mechanisms, from intercalation to conversion and alloying. Anode design for Li-ion batteries presents a fundamentally different design problem than cathode design: while the ideal cathode has yet to be identified, it is generally agreed that from an energy standpoint, elemental Li is an ideal anode due to its low voltage and high capacity. However, achieving reversible cycling of elemental Li safely, at high current densities, and over many hundreds to thousands of cycles remains a major challenge. With the fundamental chemistry relatively fixed for Li metal anodes, the main approaches to addressing these stability and performance issues with Li metal anodes tend to go beyond anode materials design, e.g. anode coatings [49], smart separators [50] with flame retardant properties [51], Li nanostructures [52], and rigid host matrices for embedding Li [53].

However, ML approaches have played a notable role in the study of lithiation in amorphous silicon anodes. DFT calculation is not well suited to this area of study because the nonperiodic structure of amorphous silicon necessitates the use of large computational cells to properly simulate molecular dynamics. Although DFT studies of smaller computational cells can provide significant insights [54], recent efforts have leveraged deep-learning models trained to predict interatomic potentials from large datasets of crystal energies; these models offer a way forward for simulating computational cells consisting of thousands to millions of atoms. Notable work in this field has been published by Onat et al. [55] and Artrith et al. [56–59].

16.6 Beyond Lithium

A similarly underdeveloped and promising area for ML approaches is in the discovery of new nonlithium battery materials, including Na- and K-intercalation electrodes and electrolytes. Significant knowledge of Li-ion chemistries currently exists, and ML may help in extrapolating that knowledge to alternative cations. A common approach that does not utilize ML is to begin with known high-performance Li-based materials, substitute in alternative cations experimentally or *in silico*, and then measure their performance. This approach has yielded successes, e.g. the identification and synthesis of the fast Na-ion conductor $Na_{10}GeP_2S_{12}$, inspired from $Li_{10}GeP_2S_{12}$ [60]. However, this approach may miss promising non-Li materials with no existing Li analogs from which to take inspiration. In line with the work of Joshi et al. [47], ML approaches represent a robust data-driven method for exploring new non-Li-based materials without being limited to structures that are directly analogous with known Li-based structures.

Building a model that can generalize over new mobile ions and new sublattices would require featurizing the mobile ion, and including structures with different mobile ions in the training set. Without either of these, the model may not learn to generalize the effect of the mobile ion on the material's properties.

16.7 Electrochemical Capacitors

EDLCs are relatively simple energy storage devices that consist of high surface area carbon electrodes separated by a nonaqueous electrolyte. New liquid electrolytes with high ionic conductivities and electrochemical and thermal stability windows would significantly increase the energy storage of these devices, since the energy of a capacitor varies with the square of the potential difference between the symmetric carbon electrodes ($E = \frac{1}{2} CV^2$, where C is the capacitance and V is the potential difference). The search for new liquid electrolyte compositions for EDLCs is well suited for ML approaches, particularly with growing interest in water-in-salt and deep eutectic solvent electrolytes. MXenes are emerging 2D materials that exhibit pseudocapacitive behavior, that is, highly reversible redox reactions that lead to charge/discharge timescales similar to those of EDLCs but with higher capacitance. The search for new MXenes also lends itself to ML approaches due to the large composition space of the parent MAX phases [61].

16.8 Application of ML in Life Cycle Degradation

Cell- and system-level properties of batteries represent significantly different time- and length-scales than atom-based models, but recent work has reported promising results from incorporating ML approaches to these properties. With different scales come different degrees to which we understand the physics underlying the property or process in question, and therefore the feature sets used in ML must be different.

The field of using machine learning to predict cell-level metrics, including capacity degradation, state-of-health, and state-of-charge in batteries from cycling data, has a rich history, and we point the reader to the several existing reviews of the field [62–64].

Severson and Attia et al. report using early cycling data to predict long-term capacity degradation from early cycling data with high accuracy [65]. In this work, information-rich features were constructed from the time-series cycling data, based on the authors' scientific understanding of battery degradation signatures. This approach enabled predictions of cycle life (on the order of 1000 cycles in their dataset) from the first 100 cycles, representing an acceleration of approximately 10× in the ability to predict capacity fade. This work is particularly notable due to the high prediction accuracy, the deliberately large and diverse training set used, and the fact that ML model was able to make predictions so early in the cycling test, well before the human eye could discern predictive patterns in the capacity vs. cycle number data.

16.9 Conclusion and Future Outlook

The last decade has seen the development and application of many new ML approaches for accelerating rechargeable battery and electrochemical capacitor research and development. Although the field is still in its infancy, recent work shows the exciting potential for data-driven modeling to accelerate design in cathodes, anodes, liquid electrolytes, solid electrolytes, fast charging, and more. Efforts in this area are likely to transition from identification of the best materials to identification of the best places to look for undiscovered materials over the coming decade.

As databases grow and become more accessible, the opportunities for building new, rapid ML models will grow correspondingly. Ongoing efforts in data aggregation and dissemination will be critical to the continued growth of ML-based approaches. In reviewing the current field, several "best practices" themes emerge: models perform best when they capture as much pre-existing domain knowledge as possible; the flexibility of a model should be considered carefully so as not to train a model that is too flexible or too rigid for the under training set (thus resulting in overfitting and underfitting, respectively); and simulation and/or experimental validation of ML predictions will be essential to validating the model and broadly demonstrating its value.

Acknowledgments

Austin D. Sendek, Brandi Ransom, and Evan J. Reed acknowledge funding support from the Accelerated Materials Design and Discovery program at the Toyota Research Institute. Brandi Ransom acknowledges funding support as a Knight–Hennessy Scholar, and from NSF Graduate Research Fellowship 2019257807. Jagjit Nanda acknowledges funding support from Assistant Secretary for Energy Efficiency and Renewable Energy, Office of Vehicle Technologies of the U.S. Department of Energy under Batteries Materials Program (BMR) Program.

References

1 Kohn, W. and Sham, L.J. (1965). Self-consistent equations including exchange and correlation effects. *Physics Review* 140: A1133–A1138.

2 Hohenberg, P. and Kohn, W. (1964). Inhomogeneous electron gas. *Physics Review* 136: B864–B871.

3 Perdew, J.P., Burke, K., and Ernzerhof, M. (1996). Generalized gradient approximation made simple. *Physical Review Letters* 77: 3865–3868.

4 Wu, Z. and Cohen, R.E. (2006). A more accurate generalized gradient approximation for solids. *Physical Review B* 73: 235116.
5 Roch, L.M., Häse, F., Kreisbeck, C. et al. (2018). ChemOS: orchestrating autonomous experimentation. *Science Robotics* 3: eaat5559.
6 Jain, A., Hautier, G., Ong, S.P. et al. (2015). Relating voltage and thermal safety in Li-ion battery cathodes: a high-throughput computational study. *Physical Chemistry Chemical Physics* 17: 5942–5953.
7 Kirklin, S., Meredig, B., and Wolverton, C. (2013). High-throughput computational screening of new Li-ion battery anode materials. *Advanced Energy Materials* 3: 252–262.
8 Sendek, A.D., Cubuk, E.D., Antoniuk, E.R. et al. (2018). Machine learning-assisted discovery of solid Li-ion conducting materials. *Chemistry of Materials* 31 (2): 342–352. https://doi.org/10.1021/acs.chemmater.8b03272.
9 Zhu, Z., Chu, I.-H., and Ong, S.P. (2017). $Li_3Y(PS_4)_2$ and $Li_5PS_4Cl_2$: new lithium superionic conductors predicted from silver thiophosphates using efficiently tiered ab initio molecular dynamics simulations. *Chemistry of Materials* 29: 2474–2484.
10 Chan, M.K.Y. and Ceder, G. (2010). Efficient band gap prediction for solids. *Physical Review Letters* 105: 196403.
11 Deng, Z., Wang, Z., Chu, I.-H. et al. (2016). Elastic properties of alkali superionic conductor electrolytes from first principles calculations. *Journal of the Electrochemical Society* 163: A67–A74.
12 Richards, W.D., Miara, L.J., Wang, Y. et al. (2016). Interface stability in solid-state batteries. *Chemistry of Materials* 28: 266–273.
13 Qu, X., Jain, A., Rajput, N.N. et al. (2015). The electrolyte genome project: a big data approach in battery materials discovery. *Computational Materials Science* 103: 56–67.
14 Cheon, G., Duerloo, K.-A.N., Sendek, A.D. et al. (2017). Data mining for new two- and one-dimensional weakly bonded solids and lattice-commensurate heterostructures. *Nano Letters* 17: 1915–1923.
15 Mounet, N., Gibertini, M., Schwaller, P. et al. (2018). Two-dimensional materials from high-throughput computational exfoliation of experimentally known compounds. *Nature Nanotechnology* 13: 246–252.
16 Cheng, L., Assary, R.S., Qu, X. et al. (2015). Accelerating electrolyte discovery for energy storage with high-throughput screening. *Journal of Physical Chemistry Letters* 6: 283–291.
17 Muy, S., Voss, J., Schlem, R. et al. (2019). High-throughput screening of solid-state Li-ion conductors using lattice-dynamics descriptors. *iScience* 16: 270–282.
18 Hernandez, P. (2017). Materials genome initiative. *NIST* https://www.nist.gov/mgi. Accessed 1 February 2020.
19 https://icsd.fiz-karlsruhe.de/index.xhtml;jsessionid=E771D99B39845E3B90AE31E5A4913127 ICSD. https://icsd.fiz-karlsruhe.de/. Accessed 1 February 2020.

20 Jain, A., Ong, S.P., Hautier, G. et al. (2013). Commentary: The materials project: a materials genome approach to accelerating materials innovation. *APL Materials* 1: 11002.

21 Saal, J.E., Kirklin, S., Aykol, M. et al. (2013). Materials design and discovery with high-throughput density functional theory: the open quantum materials database (OQMD). *JOM* 65: 1501–1509.

22 Curtarolo, S., Setyawan, W., Hart, G.L.W. et al. (2012). AFLOW: an automatic framework for high-throughput materials discovery. *Computational Materials Science* 58: 218–226.

23 Ong, S.P., Jain, A., Hautier, G. et al. (2010). Thermal stabilities of delithiated olivine MPO_4 (M = Fe, Mn) cathodes investigated using first principles calculations. *Electrochemistry Communications* 12: 427–430.

24 Ong, S.P., Wang, L., Kang, B., and Ceder, G. (2008). Li–Fe–P–O_2 phase diagram from first principles calculations. *Chemistry of Materials* 20: 1798–1807.

25 Sendek, A.D., Cheon, G., Pasta, M., and Reed, E.J. (2020). Quantifying the search for solid Li-ion electrolyte materials by anion: a data-driven perspective. *The Journal of Physical Chemistry C* 124 (15): 8067–8079.

26 Fitzhugh, W., Wu, F., Ye, L. et al. (2019). A high-throughput search for functionally stable interfaces in sulfide solid-state lithium ion conductors. *Advanced Energy Materials* 9: 1900807.

27 Cheon, G., Cubuk, E.D., Antoniuk, E.R. et al. (2018). Revealing the spectrum of unknown layered materials with superhuman predictive abilities. *Journal of Physical Chemistry Letters* 9: 6967–6972.

28 Valiant, L. G. (1984). A theory of the learnable. *Proceedings of the Sixteenth Annual ACM Symposium on Theory of Computing*. https://dl.acm.org/doi/abs/10.1145/800057.808710.

29 Jalem, R., Aoyama, T., Nakayama, M., and Nogami, M. (2012). Multivariate method-assisted *ab initio* study of olivine-type $LiMXO_4$ (main group M^{2+}–X^{5+} and M^{3+}–X^{4+}) compositions as potential solid electrolytes. *Chemistry of Materials* 24: 1357–1364.

30 Jalem, R., Nakayama, M., and Kasuga, T. (2013). An efficient rule-based screening approach for discovering fast lithium ion conductors using density functional theory and artificial neural networks. *Journal of Materials Chemistry A* 2: 720–734.

31 Fujimura, K., Seko, A., Koyama, Y. et al. (2013). Accelerated materials design of lithium superionic conductors based on first-principles calculations and machine learning algorithms. *Advanced Energy Materials* 3: 980–985.

32 Sendek, A.D., Yang, Q., Cubuk, E.D. et al. (2017). Holistic computational structure screening of more than 12000 candidates for solid lithium-ion conductor materials. *Energy & Environmental Science* 10: 306–320.

33 Cubuk, E.D., Sendek, A.D., and Reed, E.J. (2019). Screening billions of candidates for solid lithium-ion conductors: a transfer learning approach for small data. *The Journal of Chemical Physics* 150: 214701.

34 Ward, L., Agrawal, A., Choudhary, A., and Wolverton, C. (2016). A general-purpose machine learning framework for predicting properties of inorganic materials. *npj Computational Materials* 2: 1–7.

35 Mikolov, T., Chen, K., Corrado, G. & Dean, J. *Efficient Estimation of Word Representations in Vector Space*. arXiv:1301.3781v3 (2013).

36 Pennington, J., Socher, R., and Manning, C.D. (2014). Glove: global vectors for word representation. *Computational Linguistics* 1532–1543. https://doi.org/10.3115/v1/D14-1162.

37 Bottou, L. (2014). From machine learning to machine reasoning. *Machine Learning* 94: 133–149.

38 Zhou, Q., Tang, P., Liu, S. et al. (2018). Learning atoms for materials discovery. *Proceedings of the National Academy of Sciences of the United States of America* 115: E6411–E6417.

39 Zhang, Y., He, X., Chen, Z. et al. (2019). Unsupervised discovery of solid-state lithium ion conductors. *Nature Communications* 10: 1–7.

40 Jalem, R., Kanamori, K., Takeuchi, I. et al. (2018). Bayesian-driven first-principles calculations for accelerating exploration of fast ion conductors for rechargeable battery application. *Scientific Reports* 8: 1–10.

41 Nakayama, M., Kanamori, K., Nakano, K. et al. (2019). Data-driven materials exploration for Li-ion conductive ceramics by exhaustive and informatics-aided computations. *Chemical Record* 19: 771–778.

42 Ahmad, Z., Xie, T., Maheshwari, C. et al. (2018). Machine learning enabled computational screening of inorganic solid electrolytes for suppression of dendrite formation in lithium metal anodes. *ACS Central Science* 4: 996–1006.

43 Husch, T. and Korth, M. (2015). Charting the known chemical space for non-aqueous lithium–air battery electrolyte solvents. *Physical Chemistry Chemical Physics* 17: 22596–22603.

44 Schütter, C., Husch, T., Viswanathan, V. et al. (2016). Rational design of new electrolyte materials for electrochemical double layer capacitors. *Journal of Power Sources* 326: 541–548.

45 Schütter, C., Husch, T., Korth, M., and Balducci, A. (2015). Toward new solvents for EDLCs: from computational screening to electrochemical validation. *Journal of Physical Chemistry C* 119: 13413–13424.

46 Cunha, R.P., Lombardo, T., Primo, E.N., and Franco, A.A. (2020). Artificial intelligence investigation of NMC cathode manufacturing parameters interdependencies. *Batteries & Supercaps* 3: 60–67.

47 Joshi, R.P., Eickholt, J., Li, L. et al. (2019). Machine learning the voltage of electrode materials in metal-ion batteries. *ACS Applied Materials & Interfaces* 11: 18494–18503.

48 Baliyan, A. and Imai, H. (2019). Machine learning based analytical framework for automatic hyperspectral Raman analysis of lithium-ion battery electrodes. *Scientific Reports* 9: 1–14.

49 Zheng, G., Wang, C., Pei, A. et al. (2016). High-performance lithium metal negative electrode with a soft and flowable polymer coating. *ACS Energy Letters* 1: 1247–1255.

50 Wu, H., Zhuo, D., Kong, D., and Cui, Y. (2014). Improving battery safety by early detection of internal shorting with a bifunctional separator. *Nature Communications* 5: 1–6.

51 Liu, K., Liu, W., Qiu, Y. et al. (2017). Electrospun core–shell microfiber separator with thermal-triggered flame-retardant properties for lithium-ion batteries. *Science Advances* 3: e1601978.

52 Liu, N., Lu, Z., Zhao, J. et al. (2014). A pomegranate-inspired nanoscale design for large-volume-change lithium battery anodes. *Nature Nanotechnology* 9: 187–192.

53 Lin, D., Liu, Y., Liang, Z. et al. (2016). Layered reduced graphene oxide with nanoscale interlayer gaps as a stable host for lithium metal anodes. *Nature Nanotechnology* 11: 626–632.

54 Cubuk, E.D. and Kaxiras, E. (2014). Theory of structural transformation in lithiated amorphous silicon. *Nano Letters* 14: 4065–4070.

55 Onat, B., Cubuk, E.D., Malone, B.D., and Kaxiras, E. (2018). Implanted neural network potentials: application to Li–Si alloys. *Physical Review B* 97: 94106.

56 Artrith, N. (2019). Machine learning for the modeling of interfaces in energy storage and conversion materials. *Journal of Physics: Energy* 1: 32002.

57 Artrith, N., Urban, A., and Ceder, G. (2018). Constructing first-principles phase diagrams of amorphous Li_xSi using machine-learning-assisted sampling with an evolutionary algorithm. *The Journal of Chemical Physics* 148: 241711.

58 Artrith, N., Urban, A., and Ceder, G. (2017). Efficient and accurate machine-learning interpolation of atomic energies in compositions with many species. *Physical Review B* 96: 14112.

59 Artrith, N., Urban, A., Wang, Y. & Ceder, G. Atomic-scale factors that control the rate capability of nanostructured amorphous Si for high-energy-density batteries. arXiv:1901.09272 (2019).

60 Richards, W.D., Tsujimura, T., Miara, L.J. et al. (2016). Design and synthesis of the superionic conductor $Na_{10}SnP_2S_{12}$. *Nature Communications* 7: 1–8.

61 Zhan, C., Sun, W., Xie, Y. et al. (2019). Computational discovery and design of MXenes for energy applications: status, successes, and opportunities. *ACS Applied Materials & Interfaces* 11: 24885–24905.

62 Wu, L., Fu, X., and Guan, Y. (2016). Review of the remaining useful life prognostics of vehicle lithium-ion batteries using data-driven methodologies. *Applied Sciences* 6: 166.

63 Si, X.-S., Wang, W., Hu, C.-H., and Zhou, D.-H. (2011). Remaining useful life estimation – a review on the statistical data driven approaches. *European Journal of Operational Research* 213: 1–14.

64 Waag, W., Fleischer, C., and Sauer, D.U. (2014). Critical review of the methods for monitoring of lithium-ion batteries in electric and hybrid vehicles. *Journal of Power Sources* 258: 321–339.

65 Severson, K.A., Attia, P.M., Jin, N. et al. (2019). Data-driven prediction of battery cycle life before capacity degradation. *Nature Energy* 4: 383–391.

Index

Transition Metal Oxides for Electrochemical Energy Storage, First Edition.
Edited by Jagjit Nanda and Veronica Augustyn.

d

e

u

v

w

x

y